Springer Undergraduate Mathematics Series

Editor-in-Chief
Endre Süli, Oxford, UK

Series Editors
Mark A. J. Chaplain, St. Andrews, UK
Angus Macintyre, Edinburgh, UK
Shahn Majid, London, UK
Nicole Snashall, Leicester, UK
Michael R. Tehranchi, Cambridge, UK

The Springer Undergraduate Mathematics Series (SUMS) is a series designed for undergraduates in mathematics and the sciences worldwide. From core foundational material to final year topics, SUMS books take a fresh and modern approach. Textual explanations are supported by a wealth of examples, problems and fully-worked solutions, with particular attention paid to universal areas of difficulty. These practical and concise texts are designed for a one- or two-semester course but the self-study approach makes them ideal for independent use.

Tom Carroll

Geometric Function Theory

A Second Course in Complex Analysis

Tom Carroll
School of Mathematical Sciences
University College Cork
Cork, Ireland

ISSN 1615-2085 ISSN 2197-4144 (electronic)
Springer Undergraduate Mathematics Series
ISBN 978-3-031-73726-8 ISBN 978-3-031-73727-5 (eBook)
https://doi.org/10.1007/978-3-031-73727-5

Mathematics Subject Classification: 52A55, 53A35, 30-01

© The Editor(s) (if applicable) and The Author(s), under exclusive license to Springer Nature Switzerland AG 2024

This work is subject to copyright. All rights are solely and exclusively licensed by the Publisher, whether the whole or part of the material is concerned, specifically the rights of translation, reprinting, reuse of illustrations, recitation, broadcasting, reproduction on microfilms or in any other physical way, and transmission or information storage and retrieval, electronic adaptation, computer software, or by similar or dissimilar methodology now known or hereafter developed.
The use of general descriptive names, registered names, trademarks, service marks, etc. in this publication does not imply, even in the absence of a specific statement, that such names are exempt from the relevant protective laws and regulations and therefore free for general use.
The publisher, the authors and the editors are safe to assume that the advice and information in this book are believed to be true and accurate at the date of publication. Neither the publisher nor the authors or the editors give a warranty, expressed or implied, with respect to the material contained herein or for any errors or omissions that may have been made. The publisher remains neutral with regard to jurisdictional claims in published maps and institutional affiliations.

This Springer imprint is published by the registered company Springer Nature Switzerland AG
The registered company address is: Gewerbestrasse 11, 6330 Cham, Switzerland

If disposing of this product, please recycle the paper.

For Marian

Preface

Analysis in the tradition of Hardy and Littlewood was the main item on the undergraduate mathematics menu at University College Cork in the 1980s. It is little surprise then, particularly having moved to England to study with Walter Hayman and Phil Rippon, that complex analysis became the branch of mathematics where I feel most at home. While there are many sides to complex analysis, or to function theory as it is known to some, it is the geometric perspective on the subject that interests me particularly. The geometric branch of complex analysis is known as *Geometric Function Theory*, hence the title of this book.

There is general agreement on the list of main topics that comprise a first course in complex analysis. These include the basic properties of analytic functions, contour integrals, Cauchy's Theorem for a disk, Cauchy's Integral Formula, and Taylor's Theorem (that analytic functions are locally given by power series) followed by some of the most elegant theorems in mathematics: Liouville's Theorem (every bounded entire function is constant), the Fundamental Theorem of Algebra, the Maximum Modulus Theorem, the Residue Theorem, the Argument Principle, and Rouché's Theorem. While different authors take different approaches, emphasising some aspects over others, the general syllabus of a first course in complex analysis is well-established.

This consensus evaporates when it comes to a second course in complex analysis as there are numerous legitimate routes that one may follow. The present textbook is one of many possible 'second courses' in complex analysis and, therefore, inevitably omits much more than it contains. My overarching intention has been to set out as clearly and carefully as I can *some* key concepts that arise naturally as one continues to engage with complex analysis beyond that first course. I have focused on the concrete rather than the general, and emphasised calculations believing that abstract theory makes little sense without first meeting the examples from which it emerged. In this, complex analysis has a distinct advantage: the subject showcases general techniques and theories in a concrete setting. I therefore hope that, as well as covering in detail certain key concepts in geometric function theory, this textbook can provide some context to students' subsequent study in analysis and geometry.

Acknowledgements I would like to thank most sincerely: Finbarr Holland, Donal Hurley and Brian Twomey (University College Cork), Phil Rippon (The Open University), as well as Paddy Barry and Walter Hayman should they be looking down on us from above, to whom I principally owe my education in analysis; my friend and fellow student Pat McCarthy who has answered so many of my questions over the years; Joaquim Ortega-Cerdà for helping out when I got stuck; Eleanor Lingham for her unwavering support and for keeping me on track; the reviewers of an early draft whose incisive comments and suggestions have been most valuable; my fellow complex analysts generally for their comradeship over many years; and Rémi Lodh at Springer for his infinite patience and forbearance—without Rémi this book would never have seen the light of day.

Cork, Ireland
September 2024

Tom Carroll

Contents

1	**Introduction**		1
	1.1 Why Study Complex Analysis?		1
	1.2 Prerequisites		2
	1.3 Overall Summary of the Contents		3
	1.4 What Is Not in This Book		5
	1.5 Notation		5
2	**The Complex Plane: Preparatory Topics**		7
	2.1 Circles and Straight Lines in the Complex Plane		7
		2.1.1 A Parameterised Family of Circles	9
	2.2 Möbius Transformations		11
	2.3 The Schwarz Reflection Principle		15
		2.3.1 The Schwarz Reflection Principle for Lines	16
		2.3.2 Reflection in a Circle	19
		2.3.3 The Schwarz Reflection Principle for Circles	22
	2.4 Singularities of Analytic Functions		24
3	**The Riemann Sphere**		27
	3.1 Stereographic Projection		29
		3.1.1 Circles Under Stereographic Projection	34
	3.2 The Topology of the Extended Complex Plane		36
	3.3 The Riemann Sphere		38
		3.3.1 The Riemann Sphere as a Riemann Surface	38
		3.3.2 Analytic Functions on the Riemann Sphere	39
		3.3.3 Meromorphic Functions on the Riemann Sphere	41
	3.4 The Spherical Metric		46
		3.4.1 General Construction of a Riemannian Metric on a Planar Domain	46
		3.4.2 The Infinitesimal Spherical Metric	51
		3.4.3 The Spherical Metric on $\mathbb{C}_\infty$	55
		3.4.4 Isometries of the Spherical Metric	59

		3.4.5	Geodesics for the Spherical Metric	65
	3.5	Some Spherical Geometry		72
4	**The Hyperbolic Disk**			**77**
	4.1	The Basic Schwarz Lemma		78
	4.2	Automorphisms of the Disk		81
	4.3	Automorphisms of a Half-Plane		85
	4.4	The Schwarz-Pick Lemma: First Version		89
	4.5	The Hyperbolic Metric in the Disk		91
		4.5.1	Construction of the Hyperbolic Metric	91
		4.5.2	Geodesics, and a Formula for Hyperbolic Distance	92
	4.6	The Schwarz-Pick Lemma: Second Version		97
	4.7	The Hyperbolic Metric in a Half-Plane		98
5	**Normal Families and Value Distribution**			**103**
	5.1	The Arzelà-Ascoli Theorem		104
	5.2	The Space of Continuous Functions on a Planar Domain		113
	5.3	Convergence of Sequences of Analytic and Meromorphic Functions		121
	5.4	The Theorems of Montel, Marty and Zalcman		128
	5.5	Montel's Second Theorem and the Great Picard Theorem		135
6	**Simply Connected Domains, the Riemann Mapping Theorem and Conformal Mapping**			**141**
	6.1	The Homotopic Form of Cauchy's Theorem		142
	6.2	Simply Connected Domains		148
	6.3	The Riemann Mapping Theorem		152
	6.4	The Hyperbolic Metric in a Simply Connected Domain		160
		6.4.1	Construction of the Hyperbolic Metric	161
		6.4.2	The Schwarz-Pick Lemma: Third Version	164
		6.4.3	Monotonicity of the Hyperbolic Metric	168
	6.5	Some Special Domains and Their Hyperbolic Metrics		169
		6.5.1	The Half-Plane	169
		6.5.2	The Strip	170
		6.5.3	Sectors, Including the Slit-Plane	171
		6.5.4	The Half-Disk and the Slit-Disk	172
	6.6	Boundary Correspondence Under Conformal Mappings		174
		6.6.1	Proof of Theorem 6.12	177
	6.7	Conformal Mapping of Annuli		182
7	**Runge's Theorem and Further Characterisations of Simply Connected Domains**			**189**
	7.1	Runge's Theorem		189
	7.2	A General Integral Formula		195
	7.3	Winding Numbers and Further Characterisations of Simply Connected Domains		197

8	**Univalent Functions: The Basics**		203
	8.1	The Classes **S** and Σ of Univalent Functions	203
	8.2	Bieberbach's Coefficient Estimate and the Koebe 1/4-Theorem	211
		8.2.1 The Area Theorem and Bieberbach's Coefficient Estimate	211
		8.2.2 The Koebe 1/4-Theorem	216
	8.3	Growth and Distortion Theorems for Functions in the Class **S**	220
9	**Carathéodory Convergence of Domains and Hyperbolic Geodesics**		227
	9.1	Carathéodory Convergence of Domains	228
	9.2	Comparison Results for the Hyperbolic Metric	237
		9.2.1 The Laplacian and the Hyperbolic Metric	237
		9.2.2 Estimates for Solutions of $\Delta u = e^{2u}$	241
		9.2.3 Half-Planes and Disks Are Hyperbolically Convex	249
10	**Uniformisation of Planar Domains**		257
	10.1	Covering Spaces and the Monodromy Theorem	257
	10.2	The Modular Function	271
		10.2.1 A Series of Reflections	272
		10.2.2 Proof of Theorem 10.5: The Reflected Mappings	277
		10.2.3 The Automorphism Group of the Modular Function	280
	10.3	The Picard Theorems: Reprise	287
	10.4	Uniformisation of Planar Domains	288

Glossary	293
Solutions to the Exercises	295
References	349
Index	351

Chapter 1
Introduction

The subject of this textbook is *geometric function theory*, the study of complex analysis with an emphasis on the geometric viewpoint and on geometric results.

1.1 Why Study Complex Analysis?

That Complex Analysis is a beautiful branch of mathematics will be apparent to anyone who has taken a first course in the subject. This first course stands out in the undergraduate curriculum for its wealth of attractive results. Examples include Cauchy's Theorem and the Residue Theorem. Liouville's Theorem, to the effect that every bounded analytic function is constant, is elegant in its simplicity. It is nevertheless a powerful result since, for example, it leads with little effort to a proof of the Fundamental Theorem of Algebra which states that every polynomial has a full set of complex roots, counting multiplicity. Contour integrals together with the Residue Theorem can be used to evaluate otherwise challenging real integrals such as $\int_0^\infty dx/(1+x^4)$. The material in this textbook adds significantly, it is hoped, to this cursory list.

Complex Analysis has been described as the calculus of the plane. Some aspects of calculus only make sense when viewed from the wider perspective of the complex plane rather than the narrower view of the real line (real axis) alone. Why does the power series $\sum_{n=0}^\infty (-1)^n x^{2n}$ for $f(x) = 1/(1+x^2)$ only have radius of convergence 1 even though f is a perfectly good function on the whole real line? The reason is that it is the restriction to the real line of the complex power series $\sum_{n=0}^\infty (-1)^n z^{2n}$, which is the power series representation for the complex function $f(z) = 1/(1+z^2)$ and this function has singularities of modulus 1 at i and $-$i. The method of integration by partial fractions, somewhat mysterious in the context of a calculus class, is really the Laurent series expansion in disguise.

© The Author(s), under exclusive license to Springer Nature Switzerland AG 2024
T. Carroll, *Geometric Function Theory*, Springer Undergraduate Mathematics Series, https://doi.org/10.1007/978-3-031-73727-5_1

Complex Analysis has a multitude of applications, especially in a two-dimensional setting: these include applications in fluid dynamics, signal processing, in control theory and in physics, for example in quantum mechanics. Conformal mapping is used to solve boundary value problems in mathematical physics. Two-dimensional problems often become more tractable if expressed using complex numbers and functions. One such example is the use of complex coordinates to prove results in Euclidean Geometry.

Complex analysis is interconnected with many other fields of mathematics and is therefore essential for further study in several areas including, for example, harmonic analysis, differential geometry, algebraic geometry, and analytic number theory. An analyst or geometer who doesn't have a good working knowledge of complex analysis is at a disadvantage. Within the roughest division of mathematics into the broad areas of analysis/geometry/algebra, complex analysis sits primarily within the sphere of analysis yet has strong overlap with geometry. The link with geometry derives most of all from conformal mapping: analytic, one-to-one maps from one domain in the plane onto another. As a consequence and as will be explained in detail in this textbook, every simply connected domain has a natural 'hyperbolic' geometry that is invariant under all conformal maps of the domain onto itself. The branch of complex analysis that focuses on geometric aspects is called *Geometric Function Theory* and is the theme of this second course in complex analysis.

A second course in complex analysis, as well as being interesting and stimulating in its own right, is an excellent transition course to graduate level mathematics. We will meet several advanced topics in this textbook but always in the concrete setting of complex functions of a complex variable. Moreover, though it traces its origins back to the seventeenth century, complex analysis is still very much an active area of research. One such branch of research is complex dynamics, the study of the iteration of complex functions which gives rise to beautiful fractal pictures such as that of the Mandelbrot set or many and varied Julia sets. Incidentally, Fatou and Julia sets are fundamental to complex dynamics and their study makes essential use of the material in Chap. 5 on normal families. The Fatou set of, say, an entire function is the set of points where the iterates of the function locally form a normal family, with the Julia set being the complement of the Fatou set.

In short, a healthy knowledge of complex analysis is an advantage to any mathematician.

1.2 Prerequisites

It is assumed that the reader has taken a first course in complex analysis. John Howie's book *Complex Analysis* in this series covers everything the reader needs to know in this regard and more. When a result from a first course in complex analysis is needed, we refer the reader to [16]. The committed reader would do well to also keep Stein and Shakarchi's wonderful textbook [25] close at hand; though it begins at the beginning, it takes the reader far into some of the most elegant areas of complex analysis. Nehari [20] is a classic textbook on conformal mapping.

As mentioned earlier, complex analysis is interconnected with many other fields of mathematics and, in addition to contributing to other areas of mathematics, it draws on these areas as well. As a consequence, it is impractical to study complex analysis in isolation. General analysis techniques feature heavily, as do topology and functional analysis. Potential Theory and partial differential equations are never far away. I have attempted to incorporate as much as possible of this general subject matter when it is needed, with the constraint of not wishing to overly digress from the main topic of complex function theory. Nevertheless, some analysis and topology is unavoidable and the reader should be aware of this. If viewed as an opportunity rather than an inconvenience, it may be that readers' broad mathematical knowledge and confidence will benefit from this extra work.

1.3 Overall Summary of the Contents

As well as the Euclidean geometry of the complex plane with its familiar straight lines, triangles and circles, two non-flat geometries are fundamental to geometric function theory, namely the constant positive curvature geometry of the sphere and the constant negative curvature geometry of the disk. These geometries are the subjects of the first two main chapters. We construct the geometry of the sphere, which is the spherical geometry of the extended complex plane, in Chap. 3 by means of stereographic projection. The hyperbolic geometry of the disk is constructed in Chap. 4 based on Schwarz's Lemma. Each is an example of a *Riemannian geometry* in which one first defines the length of curves and then defines the distance between two points to be the length of the shortest curve joining them (more precisely, this distance is defined as the infimum of the length of all curves that join the points in the hope of subsequently proving that the infimum is achieved). This approach is familiar to us from Euclidean geometry in which the distance between two points is the length of the straight line segment joining them. One should then ask what the 'straight lines' in our new geometries on the sphere and on the disk look like. We should also ask what the isometries might be: the isometries of Euclidean geometry are the *rigid motions* of translation and rotation. These and similar questions are discussed in detail in these two chapters.

From the viewpoint of Riemannian geometry, Chaps. 3 and 4 cover similar ground. In Chap. 3, therefore, we take a more general approach that saves some work in Chap. 4. One basic difference is that in the case of the sphere we first construct the metric and then ask what the isometries might be, whereas in the case of the disk we decide in advance, guided by Schwarz's Lemma, what the isometries should be and then construct the metric. A key result in Chap. 4 is the Schwarz-Pick Lemma— analytic maps of the disk into itself are contractions in the hyperbolic metric.

When the space of analytic functions on a domain is equipped with the topology of uniform convergence on each compact subset of the domain, subsets of the space of analytic functions with compact closure are (modulo one technicality) called *normal families* among the complex analysis community. We treat normal families in detail in Chap. 5 and prove the Arzelà-Ascoli Theorem. This leads to a beautifully simple criterion due to Montel for a family of analytic functions to be normal—in the presence of analyticity, equicontinuity comes for free. Normal families of meromorphic functions are also treated, which then allows us to obtain the classic results of value distribution up to and including the Great Picard Theorem: the latter is a far-reaching extension of the Casorati-Weierstrass Theorem from a first course in complex analysis.

In Chap. 6, we begin with the general form of Cauchy's Theorem that states that the integrals of an analytic function around two closed curves have the same value if the curves are homotopic. A simply-connected domain is, in layman's terms, one with 'no holes' so that every closed curve in the domain can be shrunk to a point without tearing the curve or leaving the domain. Thus, the integral of an analytic function around a closed curve in a simply connected domain always evaluates to zero. We prove the Riemann Mapping Theorem which states that any simply connected domain can be mapped onto the disk by a conformal mapping. This mapping can then be used to transfer the hyperbolic metric in the disk onto a general simply connected domain, leading to a more general version of the Schwarz-Pick Lemma. We compute the hyperbolic metric for a number of explicit example domains. The situation for multiply connected domains is not as straightforward. We illustrate this by studying conformal mapping of annuli and showing how, in contrast with the situation for simply connected domains, not all annuli are conformally equivalent. To do this, we make use of the Schwarz Reflection Principle and a result on boundary correspondence under a conformal map.

A number of properties of a planar domain that are equivalent to being simply connected are listed in Chap. 6. For example, the property that every non-zero analytic function in the domain has an analytic logarithm is equivalent to the domain being simply connected. In Chap. 7, we work through Runge's Theorem on approximation of analytic functions by rational functions and by polynomials and use these ideas to prove another equivalent condition for a domain to be simply connected, namely that its complement relative to the Riemann sphere is connected.

The remaining three chapters can be read more or less independently of each other. Chapter 8 is an introduction to the classical theory of univalent functions. In particular, the Koebe 1/4-Theorem and Bieberbach's coefficient estimate are obtained. In Chap. 9 we return to the central theme of conformal mapping and characterise the convergence of a sequence of conformal mappings in terms of the geometric 'convergence' of the sequence of image domains. The relevant notion of convergence of a sequence of domains is known as 'Carathéodory convergence'. We also take the opportunity to describe to some extent what geodesics/'straight lines' look like in the hyperbolic geometry of a simply connected domain and reproduce some results of Jørgensen in this vein.

Our main goal in the final Chap. 10 is to prove the Uniformisation Theorem for planar domains, namely that the disk is the universal covering space for any planar domain whose complement contains at least two points. The proof, which is based on the paper [11] and the account in [30], makes essential use of the modular function, the covering of the twice punctured plane by the upper half-plane. Consequently, we need to spend some time on general covering spaces before then constructing the modular function. Though not essential, we describe in detail the deck transformations for the modular group.

1.4 What Is Not in This Book

As pointed out earlier, writing a second course in complex analysis necessitates considerable choice in the selection of topics. Even if every topic covered here has earned its place, that is not to say that many other topics could lay equal claim to being included, even under the more restrictive heading of geometric function theory. Notably absent is any in-depth discussion of conformal invariants, other than the hyperbolic metric, in the sense of Ahlfors' classic *Conformal Invariants* [1]. It has been mentioned that the spherical metric on the extended complex plane has constant positive curvature and that the hyperbolic metric in the disk has constant negative curvature. However, we do not treat curvature here. Curvature, and what the Schwarz Lemma has to say about curvature, is given expert treatment in [1] and in Steven Krantz's *Complex analysis: the geometric viewpoint* [19]. A proper treatment of Riemann surfaces, abstract or otherwise, is omitted except in the special case of the Riemann sphere. The circle of ideas around the Denjoy–Wolff theorem on iteration of self-maps of the disk and the Julia-Wolff-Carathéodory theorem would fit naturally in this book. And there are further omissions. Nevertheless, it is hoped that the choice of material set out here will encourage readers to continue their study of geometric function theory and complex analysis in general and may lighten the load should they choose to do so.

1.5 Notation

We denote the real number system by $\mathbb{R}$ and the complex numbers by $\mathbb{C}$. The real and imaginary parts of a complex number $z = x + iy$ will be denoted by $\Re z$ and $\Im z$, respectively, so that $x = \Re z$ and $y = \Im z$. The modulus of z is $|z| = \sqrt{x^2 + y^2}$ and its complex conjugate is $\bar{z} = x - iy$. Then $|z|^2 = z\bar{z}$.

A disk in the complex plane with centre a and radius r will be denoted by $D(a, r)$, and the corresponding circle by $C(a, r)$. Thus,

$$D(a, r) = \{z : |z - a| < r\} \text{ and } C(a, r) = \{z : |z - a| = r\}.$$

When the circle $C(a, r)$ is viewed as a curve it is described counterclockwise as in, for example, the parameterisation $\gamma(t) = a + re^{it}, 0 \leq t \leq 2\pi$. A punctured disk is denoted by $D'(a,r)$ so that $D'(a,r) = D(a,r) \setminus \{a\}$. The *unit disk* $D(0,1)$ will be denoted by $\mathbb{D}$ and sometimes the unit circle $C(0,1)$ will be denoted by $\mathbb{T}$.

A *domain* is an open, connected set in the complex plane $\mathbb{C}$. For example, both the unit disk $\mathbb{D}$ and the upper half-plane $\mathbb{H} = \{z : \Im z > 0\}$ are domains.

A curve γ is the continuous image of a closed interval. If the curve is parameterised by $\gamma(t)$, $t \in [a,b]$, whereby its speed $\gamma'(t)$ is continuous and non-zero on $[a,b]$, the curve is said to be *smooth*. We say that γ is *piecewise-smooth* if $\gamma'(t)$ exists, is non-zero and is continuous for all $t \in [a,b]$ with at most finitely many exceptions. The curve is *closed* if $\gamma(a) = \gamma(b)$. Practically without exception, all curves are assumed to be piecewise smooth. As is commonly the case, we blur the distinction between the curve, which is properly a function from a closed interval on the real line into the complex plane, and its *trace* which is a subset of the complex plane. For example, we may talk of a 'curve γ in a domain D' meaning that the trace or image of the curve lies inside the domain D.

We write $B(x,r) = \{y \in \mathbb{R}^3 : |y - x| < r\}$ for the ball in three dimensional Euclidean space with centre x and radius r. The sphere $\{y \in \mathbb{R}^3 : |y - x| = r\}$ with that centre and radius is denoted by $S(x,r)$.

We write $\text{int}(E)$ for the interior of a set E in a metric space (X,d). We write $\text{cl}(E)$ for the closure of E in (X,d), reserving the notation $\overline{E} = \{\overline{z} : z \in E\}$ for the reflection of a subset E of the complex plane in the real axis. The only exception will be the notation $\overline{D(a,r)}$ for the closure of a disk, so that $\overline{D(a,r)} = \{z : |z| \leq r\}$. A *neighbourhood* of a point x in X is a set that is open and connected in (X,d) and contains x.

A function that is analytic in a domain and is also one-to-one (or injective) there is said to be *univalent* in the domain. The term *conformal mapping* is understood to be a synonym for 'univalent function'.

Chapter 2
The Complex Plane: Preparatory Topics

In this preparatory chapter, we gather together some relatively elementary material that will be used later. This material is also a worthwhile refresher of the very basics of working with complex numbers. Circles play a central role in geometric function theory, as do Möbius transformations which map circles to circles. These are the first two topics that we treat in detail. Next we work through the Schwarz Reflection Principle. This sets out circumstances in which a function analytic on one side of a straight line or circle can be extended by reflection across the line or circle in such a way that the resulting function remains analytic. The final section is a short review of the three types of singularities an analytic function may have.

2.1 Circles and Straight Lines in the Complex Plane

The circle in the complex plane with centre a and radius r has equation $|z - a| = r$. Now $|z - a| = r$ if and only if

$$r^2 = |z - a|^2 = (z - a)(\bar{z} - \bar{a}) = |z|^2 + |a|^2 - \bar{a}z - a\bar{z},$$

that is if and only if $|z|^2 - \bar{a}z - a\bar{z} + |a|^2 - r^2 = 0$. This is of the form $A|z|^2 + \overline{B}z + B\bar{z} + C = 0$ with

$$A = 1, \quad B = -a, \quad C = |a|^2 - r^2.$$

Notice that A and C are real and that $|B|^2 - AC = |a|^2 - \left(|a|^2 - r^2\right) = r^2 > 0$.

Conversely, suppose that $A\,|z|^2 + \overline{B}\,z + B\,\overline{z} + C = 0$ with A and C real, and with $|B|^2 - AC > 0$. Then, if $A \neq 0$,

$$A\,|z|^2 + \overline{B}\,z + B\,\overline{z} + C = A\left(|z|^2 + \frac{\overline{B}}{A}z + \frac{B}{A}\overline{z} + \frac{C}{A}\right) = A\left(\left|z + \frac{B}{A}\right|^2 - \frac{|B|^2}{A^2} + \frac{C}{A}\right).$$

Here we used that A is real. The equation $A\,|z|^2 + \overline{B}\,z + B\,\overline{z} + C = 0$ therefore becomes

$$\left|z + \frac{B}{A}\right|^2 = \frac{|B|^2 - AC}{A^2}.$$

This is a circle with centre $a = -B/A$ and radius r where $r^2 = (|B|^2 - AC)/A^2$ (note that $|B|^2 - AC > 0$ by assumption).

If $A = 0$, then the locus is that of $\overline{B}\,z + B\,\overline{z} + C = 0$ with C real and $B \neq 0$ (which is what the condition $|B|^2 - AC > 0$ reduces to in this case). Writing $z = x + iy$, $\overline{z} = x - iy$ and setting $B = a + ib$, the equation $\overline{B}\,z + B\,\overline{z} + C = 0$ is equivalent to

$$0 = (a - ib)(x + iy) + (a + ib)(x - iy) + C = 2ax + 2by + C,$$

which is the equation of a straight line since not both a and b are zero. Similarly, the equation of a straight line $ax + by + c = 0$ has the form $\overline{B}z + B\overline{z} + C = 0$ with $C = c$ and $B = (a + ib)/2$, with $B \neq 0$.

We gather the results of these computations in the following proposition.

Proposition 2.1 *Any circle or straight line in the complex plane can be described by an equation*

$$A\,|z|^2 + \overline{B}\,z + B\,\overline{z} + C = 0, \tag{2.1}$$

where A and C are real, and where $|B|^2 - AC > 0$. In the case of a straight line we have $A = 0$ and in the case of a circle we have $A \neq 0$.

Conversely, if A and C are real, and $|B|^2 - AC > 0$, then the locus of (2.1) is a circle when $A \neq 0$ and a straight line when $A = 0$. When $A \neq 0$, the circle has centre $-B/A$ and radius $\sqrt{(|B|^2 - AC)/A^2}$.

The locus of Eq. (2.1), under the conditions given in Proposition 2.1, will be referred to as a *generalised circle*, so that generalised circles encompass both circles and straight lines. This is reasonable as a straight line can be thought of as the limiting case of a circle; for example, the imaginary axis can be viewed as the limit of the circles $C(x, x)$ with centre x on the positive real axis and radius x as $x \to \infty$.

2.1 Circles and Straight Lines in the Complex Plane

2.1.1 A Parameterised Family of Circles

We now study a parametrised family of circles that will arise later in the context of hyperbolic geodesics in simply connected domains. We fix two distinct points z_1 and z_2 in the complex plane. For $k > 0$, consider the locus $C(z_1, z_2, k)$ of all complex numbers z that satisfy

$$\left|\frac{z - z_1}{z - z_2}\right| = k. \tag{2.2}$$

Notice that $C(z_1, z_2, k) = C(z_2, z_1, 1/k)$. When $k = 1$, the locus $C(z_1, z_2, 1)$ is described by $|z - z_1| = |z - z_2|$. This is the set of all points that are equidistant from z_1 and z_2, that is the perpendicular bisector of the line segment $[z_1, z_2]$. The locus is a circle when $k \neq 1$ as we will now see.

Proposition 2.2 *When $k \neq 1$, the locus of (2.2) is a circle $C(a, r)$ with centre*

$$a = \frac{z_1 - k^2 z_2}{1 - k^2} \tag{2.3}$$

and radius

$$r = \frac{k}{|1 - k^2|}|z_2 - z_1|. \tag{2.4}$$

Proof Suppose then that $k \neq 1$ and that z lies on $C(z_1, z_2, k)$. We have

$$|z - z_1|^2 = k^2 |z - z_2|^2$$

which when expanded becomes

$$|z|^2 + |z_1|^2 - z_1\bar{z} - \bar{z}_1 z = k^2\left(|z|^2 + |z_2|^2 - z_2\bar{z} - \bar{z}_2 z\right).$$

This has the form

$$(1 - k^2)|z|^2 + (k^2\bar{z}_2 - \bar{z}_1)z + (k^2 z_2 - z_1)\bar{z} + (|z_1|^2 - k^2|z_2|^2) = 0,$$

which is (2.1) with

$$A = 1 - k^2, \quad B = k^2 z_2 - z_1, \quad C = |z_1|^2 - k^2|z_2|^2.$$

Here A and C are both real. Moreover,

$$\begin{aligned}|B|^2 - AC &= |k^2 z_2 - z_1|^2 - (1-k^2)(|z_1|^2 - k^2|z_2|^2) \\ &= \left(k^4|z_2|^2 + |z_1|^2 - k^2 \bar{z}_1 z_2 - k^2 z_1 \bar{z}_2\right) + k^2|z_1|^2 - k^4|z_2|^2 \\ &\quad + k^2|z_2|^2 - |z_1|^2 \\ &= k^2\left(|z_1|^2 + |z_2|^2 - \bar{z}_1 z_2 - z_1 \bar{z}_2\right) \\ &= k^2|z_2 - z_1|^2 > 0.\end{aligned}$$

The locus of (2.2) is therefore a circle when $k \neq 1$. The formulas (2.3) and (2.4) for its centre and radius now follow from Proposition 2.1. □

Exercise 2.1 ▷ Show that, for $k \neq 1$,

$$z_1 - a = \frac{k^2}{1-k^2}(z_2 - z_1) \quad \text{and} \quad z_2 - a = \frac{1}{1-k^2}(z_2 - z_1).$$

Deduce that $|z_1 - a| = kr$ and that $|z_2 - a| = r/k$.
▷ Show that a, z_1 and z_2 are collinear.

We can get a feel for how these circles might look by examining the case $z_1 = 1$ and $z_2 = 0$, that is $C_k = C(1, 0, k)$. In this case, C_k has centre $1/(1-k^2)$ and radius $r = k/|1-k^2|$. Shown below are the circles corresponding to $k = 1/3$, $k = 1/2$, and $k = 2/3$, namely $C(9/8, 3/8)$, $C(4/3, 2/3)$ and $C(9/5, 6/5)$. Notice that the circles in Fig. 2.1 are nested and increase as k increases. This is clear if we introduce the notation $D(z_1, z_2, k)$ for the locus all complex numbers z that satisfy

$$\left|\frac{z - z_1}{z - z_2}\right| < k. \tag{2.5}$$

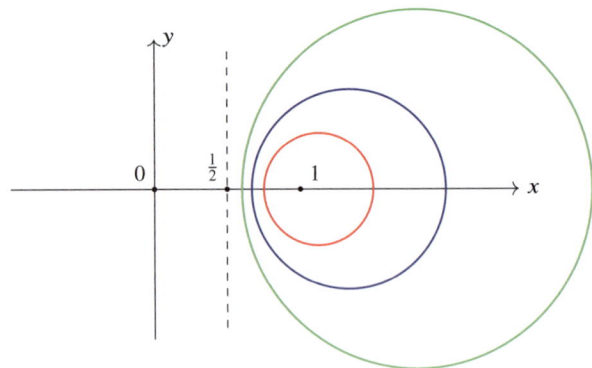

Fig. 2.1 Circles $C(1, 0, k)$ with $k = 1/3$ in red, $k = 1/2$ in blue, and $k = 2/3$ in green. The dashed line is the perpendicular bisector of $[0, 1]$, the line $\Re z = 1/2$

For $0 < k < 1$, this is simply a disk bounded by the circle $C(z_1, z_2, k)$; for $k = 1$ it is a half-plane containing z_1 and bounded by the perpendicular bisector of $[z_1, z_2]$; for $k > 1$, we can write $D(z_1, z_2, k)$ as the locus of $|z - z_2|/|z - z_1| > 1/k$, which is the exterior of the disk $C(z_2, z_1, 1/k)$. The following lemma is immediate from the definition (2.5) of the region $D(z_1, z_2, k)$.

Lemma 2.1 *For z_1 and z_2 fixed, $D(z_1, z_2, k)$ forms an increasing sequence of regions as k increases in the range $(0, \infty)$. That is, $D(z_1, z_2, k_1) \subset D(z_1, z_2, k_2)$ if $0 < k_1 < k_2 < \infty$.*

As k increases from 0 to 1, the regions $D(z_1, z_2, k)$ increase continuously to fill out the half-plane containing z_1 and bounded by the perpendicular bisector of $[z_1, z_2]$.

Exercise 2.2 Describe the regions $D(z_1, z_2, k)$ as k increases from 1 to ∞.

2.2 Möbius Transformations

Möbius transformations are elementary transformations that, nevertheless, play a central role in geometric function theory. The isometries of the spherical and the hyperbolic metrics that we will study in Chaps. 3 and 4 are, in each case, special Möbius transformations. Even if the natural home of Möbius transformations is the Riemann sphere, we first take an elementary view as would often be the case in a first course in complex analysis.

Definition 2.1 A *Möbius transformation M* has the form

$$M(z) = \frac{az+b}{cz+d}, \; z \in \mathbb{C}, \text{ where } ad - bc \neq 0. \tag{2.6}$$

That $ad - bc \neq 0$ is the condition for M not to be constant. For any non-zero complex number ζ, the expression $M(z) = (\zeta az + \zeta b)/(\zeta cz + \zeta d)$ is the same Möbius transformation as (2.6). If we choose $\zeta^2 = 1/(ad - bc)$, we can normalise the coefficients of the Möbius transformation so that $ad - bc = 1$. This is a common, and very useful, normalisation that can simplify calculations. Möbius transformations are discussed in [16, Sect. 11.3].

Making appropriate choices of a, b, c, and d gives rise to special maps:

$M(z) = e^{i\theta} z, \quad (\theta \in \mathbb{R})$ (a rotation);

$M(z) = rz, \quad (r > 0)$ (a dilation);

$M(z) = z + b, \; (b \in \mathbb{C})$ (a translation);

$M(z) = 1/z, \quad$ (inversion).

The map $M(z) = az$, $a \in \mathbb{C}$, is a composition of a rotation and a dilation. This can be seen by writing $a = re^{i\theta}$ in polar coordinates.

Lemma 2.2 *Every Möbius transformation can be decomposed into a composition of rotations, dilations, translations and the inversion $z \to 1/z$.*

The Möbius transformation (2.6) *is invertible and its inverse is the Möbius transformation*

$$M^{-1}(z) = \frac{dz - b}{-cz + a}, \quad z \in \mathbb{C}. \tag{2.7}$$

The composition of two Möbius transformations is also a Möbius transformation, so that Möbius transformations form a group.

Proof If $c = 0$ in (2.6) then $M(z) = (a/d)z + (b/d)$. This is the dilation cum rotation $z \to (a/d)z$, composed with the translation $z \to z + b/d$. If $c \neq 0$, the decomposition is made explicit by the identity

$$\frac{az + b}{cz + d} = \frac{a}{c} - \frac{1}{c}\left(\frac{ad - bc}{cz + d}\right). \tag{2.8}$$

Möbius transformations are one-to-one (that is, injective), as it is easy to check that $z_1 = z_2$ if $M(z_1) = M(z_2)$. By direct computation, the inverse function is the Möbius transformation given by (2.7).

It is immediate that the composition $M_2 \circ M_1$ is a Möbius transformation if both M_1 and M_2 are when M_2 is either a dilation/rotation, a translation or an inversion since, in these cases, $M_2 \circ M_1$ becomes αM_1, $M_1 + \beta$, ($\alpha, \beta \in \mathbb{C}$), and $1/M_1$ respectively. The general case that $M_2 \circ M_1$ is a Möbius transformation whenever both M_1 and M_2 are Möbius transformations follows if we decompose M_2 into a composition of rotations, dilations, translations, and inversions.

That Möbius transformations form a group under composition is now immediate as it is closed under composition, contains the identity function, and each Möbius transformation has an inverse which is again a Möbius transformation. □

Möbius transformations can also be viewed through the lens of matrices. We associate the matrix

$$A_M = \begin{pmatrix} a & b \\ c & d \end{pmatrix} \text{ with the Möbius transformation } M(z) = \frac{az + b}{cz + d}.$$

This association is not one-to-one since the coefficients of M can be multiplied by any non-zero number which doesn't affect M but does affect the matrix A_M. This, however, causes little difficulty. Note that the condition $ad - bc \neq 0$ on M becomes the condition $\det(A_M) \neq 0$ on the matrix A_M.

Exercise 2.3 Verify that composition of Möbius transformations corresponds to multiplication of the corresponding matrices. That is, if $M_1(z) = (a_1z + b_1)/(c_1z +$

2.2 Möbius Transformations

d_1) and $M_2(z) = (a_2z + b_2)/(c_2z + d_2)$ then the matrix corresponding to $M_2 \circ M_1$ is the product $A_{M_2} \times A_{M_1}$ of the matrices corresponding to M_2 and M_1.

We have, however, been skating over a technical difficulty. If $c = 0$ in (2.6) (and we normalise so that $d = 1$) then $M(z) = az + b$, which is a dilation/rotation by a followed by translation by b and all is fine. But if $c \neq 0$ then $-d/c$ is a perfectly good complex number yet $M(-d/c)$ is not defined as we're attempting to divide by 0. Looking more closely, the numerator of $M(z)$, that is $az + b$, is finite and non-zero at $z = -d/c$. In fact, it has the value $(bc - ad)/c$ and recall that $ad - bc \neq 0$ by assumption. It follows that $|M(z)| \to \infty$ as $z \to -d/c$. With this justification, we *define* $M(-d/c) = \infty$ where we may as well take ∞ to stand for an *arbitrarily large complex number*. It will only be in the next chapter on the Riemann sphere that we can put flesh on the bones of this entity ∞, and realise it explicitly.

What value might we then assign to $M(\infty)$? If $M(z) = az + b$ then it we should surely set $M(\infty) = \infty$. If $c \neq 0$ in (2.6), then

$$M(z) = \frac{a + b/z}{c + d/z} \to \frac{a}{c} \text{ as } z \to \infty,$$

and so it makes sense to set $M(\infty) = a/c$.

The *extended complex plane* is the complex plane with the addition of the point ∞, that is $\mathbb{C} \cup \{\infty\}$, and is here denoted by $\mathbb{C}_\infty$. We refer to ∞ in this context as the *point at infinity*. Addition, multiplication and division are generalised to the extended complex plane in the natural way. For example, $a + \infty = \infty$ and $a/\infty = 0$, $a \in \mathbb{C}$. For $a \neq 0$, $a \times \infty = \infty$ and $a/0 = \infty$. The expressions $0 \times \infty$ and $0/0$ are left undefined, however.

Consider the image of the line L given by $\Re z = 1$ under the inversion $M(z) = 1/z$. Writing $z = 1 + iy$, we have $M(1 + iy) = u + iv$ where $u = 1/(1 + y^2)$ and $v = -y/(1 + y^2)$. Then $(u - 1/2)^2 + v^2 = 1/4$ and so the image of the line is contained in the circle $C = C(1/2, 1/2)$. As y changes from 0 to $+\infty$, $M(z)$ traces out the lower semicircle of C from 1 to 0, while $M(z)$ traces out the upper semicircle from 1 to 0 as y goes from 0 to $-\infty$. However, we never have $M(z) = 0$ and so the image of the line L under the inversion $M(z) = 1/z$ is the circle $C(1/2, 1/2)$ minus the point 0. Since $1/\infty = 0$, the image is the full circle, that is $M(L \cup \{\infty\}) = C$, only when we include the extra point at infinity. This suggests that we revise our concept of 'generalised circle' to include the point at infinity in the case of a straight line. To be explicit, a *generalised circle* in the extended complex plane $\mathbb{C}_\infty$ is either a circle in the complex plane, or a line in the complex plane together with the point at infinity. One special such line is the *extended real line* $\mathbb{R}_\infty = \mathbb{R} \cup \{\infty\}$, the real line with the point at infinity added. This revised definition modifies how we should interpret (2.1) in the context of the extended complex plane: if $A \neq 0$ in (2.1), we are dealing with an ordinary circle and the point at infinity does not feature; if $A = 0$ then to all points in the complex plane that satisfy $\overline{B}z + B\overline{z} + C = 0$ we need to add the point at infinity. This is reasonable if we rewrite $\overline{B}z + B\overline{z} + C = 0$ as $\overline{B}/\overline{z} + B/z + C/|z|^2 = 0$.

Exercise 2.4 Show that each Möbius transformation (2.6) is a bijection from $\mathbb{C}_\infty$ to $\mathbb{C}_\infty$ with inverse (2.7). (Hint: consider the cases $c = 0$ and $c \neq 0$ separately.)

Exercise 2.5 Find an explicit formula for the Möbius transformation in each case:

▷ a Möbius transformation that maps 1 to ∞, maps ∞ to -1, and maps 0 to 1;
▷ a Möbius transformation that fixes ∞, maps 0 to -1, and maps 1 to 0;
▷ a Möbius transformation that fixes 0, maps ∞ to 1, and maps 1 to i;
▷ a Möbius transformation that fixes ∞, and swaps z_1 and z_2, $(z_1, z_2 \in \mathbb{C})$.

Exercise 2.6 ▷ Find the image of the generalised line $\Re z = x_0$ ($-\infty < x_0 < \infty$) under the inversion $M(z) = 1/z$. (Hint: when $x_0 \neq 0$, the image is a circle!)

▷ Show that if a Möbius transformation M fixes three points in the extended complex plane (in that there are distinct points α, β, γ in $\mathbb{C}_\infty$ for which $M(\alpha) = \alpha$, $M(\beta) = \beta$, and $M(\gamma) = \gamma$) then it is the identity $M(z) = z$, $z \in \mathbb{C}_\infty$.

▷ Show that a Möbius transformation is determined by any three of its values. In other words, if M_1 and M_2 are Möbius transformations, if α, β, γ are distinct points in $\mathbb{C}_\infty$ and if $M_1(\alpha) = M_2(\alpha)$, $M_1(\beta) = M_2(\beta)$, and $M_1(\gamma) = M_2(\gamma)$ then $M_1 = M_2$. (Hint: consider $M = M_1^{-1} \circ M_2$.)

The next result, while relatively straightforward to verify, has important consequences in the sequel.

Proposition 2.3 *The image of a generalised circle in $\mathbb{C}_\infty$ under a Möbius transformation is again a generalised circle.*

Proof Let T be a generalised circle given as the locus of, say,

$$A|z|^2 + \overline{B}z + B\overline{z} + C = 0, \tag{2.9}$$

where A and C are real, B is complex and $|B|^2 > AC$. Let M be a Möbius transformation. To show that the image $T' = M(T)$ of the generalised circle T under M is also a generalised circle, it is sufficient to show this for the special cases when M is an inversion, or M is a dilation / rotation, or M is a translation. This is because, by (2.8), any Möbius transformation is a composition of these three basic Möbius transformations.

Since $M: \mathbb{C}_\infty \to \mathbb{C}_\infty$ is a bijection, a point $w \in T'$ if and only if $M^{-1}(w) \in T$, that is if and only if

$$A|M^{-1}(w)|^2 + \overline{B}M^{-1}(w) + B\overline{M^{-1}(w)} + C = 0. \tag{2.10}$$

We need to verify that (2.10) represents a generalised circle.

It is straightforward to check that (2.10) is the locus of a generalised circle if $M(z) = z + b$ (so that $M^{-1}(w) = w - b$) or if $M(z) = az$, $a \neq 0$, (so that $M^{-1}(w) = w/a$). In each of these cases, $M(\infty) = \infty$ and so we need only consider

2.3 The Schwarz Reflection Principle

z and w complex. In any case, it is geometrically clear that the image of a generalised circle under a translation or a dilation / rotation is another generalised circle.

The final case is to check that (2.10) is a generalised circle when M is the inversion $M(z) = 1/z$. Here, $M^{-1}(w) = 1/w$ where $1/0 = \infty$ and $1/\infty = 0$. For $w \neq 0$ or ∞, (2.10) becomes

$$\frac{A}{|w|^2} + \frac{\overline{B}}{w} + \frac{B}{\overline{w}} + C = 0$$

or, multiplying across by $|w|^2 = w\,\overline{w}$,

$$C\,|w|^2 + B\,w + \overline{B}\,\overline{w} + A = 0. \tag{2.11}$$

Of course, the condition $|B|^2 > CA$ holds so that (2.11), hence (2.10), describes a generalised circle when M is an inversion.

Digging a little deeper, suppose that $A = 0$ in (2.9) in which case T is a straight line through ∞. If $C \neq 0$ the line T doesn't pass through 0 and (2.11) represents a circle in the complex plane that passes through $0 = 1/\infty$. If $C = 0$ then the line T does pass through 0 and (2.11) represents a line that passes through $0 = 1/\infty$ and through $\infty = 1/0$.

If $A \neq 0$ then T is a circle in the complex plane. If $C \neq 0$ this circle doesn't pass through 0 and (2.11) also represents a circle in the complex plane. If $C = 0$ the circle T does pass through 0 and now (2.11) represents a line through $\infty = 1/0$. □

Exercise 2.7 Find all Möbius transformations that fix the extended real line $\mathbb{R}_\infty$. Show that all such Möbius transformations M satisfy $M(z) = \overline{M(\overline{z})}$.

2.3 The Schwarz Reflection Principle

We aim in this section to show how, in certain circumstances, a function analytic on one side of a generalised circle can be extended across that circle. The main tool we need is Morera's Theorem [16, Theorem 7.8], a converse of Cauchy's Theorem. The majority of the work is carried out in the next subsection which deals with reflection of an analytic function across a line. After that, we discuss reflection in circles. Finally, we use Möbius transformations to generalise the Schwarz Reflection Principle from lines to circles.

2.3.1 The Schwarz Reflection Principle for Lines

Recall that, for a subset E of $\mathbb{C}$, we write $\overline{E}$ for the set $\{z : \overline{z} \in E\}$, so that $\overline{E}$ is the reflection of E in the real axis. The set E is said to be *symmetric about the real axis* if $\overline{E} = E$, that is if $\overline{z} \in E$ whenever $z \in E$.

Recall also that $\mathbb{H}$ denotes the upper half-plane $\mathbb{H} = \{z : \Im z > 0\}$. Let D be a domain that is symmetric about the real axis. Let $J = D \cap \mathbb{R}$, $D^+ = D \cap \mathbb{H}$ and $D^- = D \cap \overline{\mathbb{H}}$. Since D is symmetric, $\overline{D^-} = D^+$. Since D is a domain, J is an open subset of the real line (and so consists of at most countably many disjoint open intervals). Also D is the disjoint union of D^+, J, and D^-. A typical situation is shown in Fig. 2.2.

Theorem 2.1 (Schwarz Reflection Principle for a Line) *Suppose that the domain D in the complex plane is symmetric about the real axis. Set D^+, D^- and J as above. Suppose that f is continuous on $D^+ \cup J$, is analytic on D^+, and takes real values on J. Then f can be extended to an analytic function on all of D.*

If, in addition, f is one-to-one on $D^+ \cup J$ and if $f(D^+)$ lies completely in either the upper half-plane $\mathbb{H}$ or its reflection $\overline{\mathbb{H}}$ then the extension of f to all of D is univalent and, moreover, $f(D) = f(D^+) \cup f(J) \cup \overline{f(D^+)}$.

Proof We set

$$\widetilde{f}(z) = \begin{cases} f(z), & z \in D^+ \cup J; \\ \overline{f(\overline{z})}, & z \in D^-. \end{cases} \quad (2.12)$$

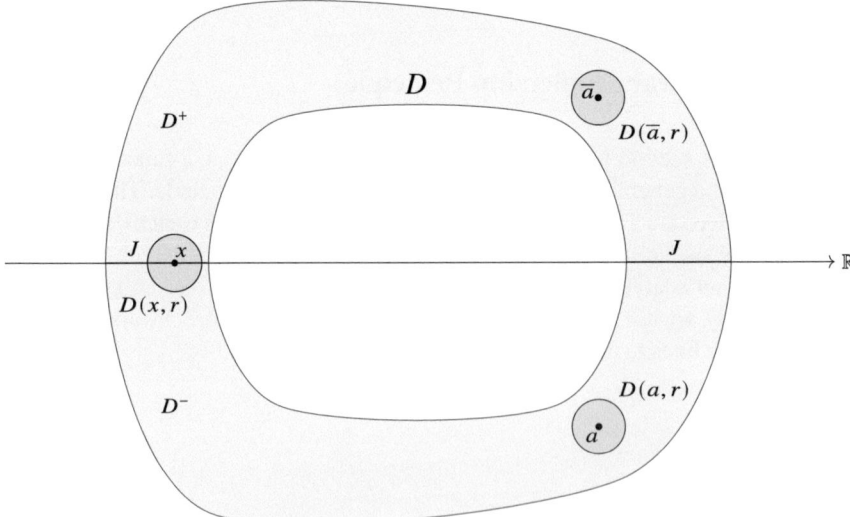

Fig. 2.2 An example of a domain symmetric about the real axis with notation from the proof of the Schwarz Reflection Principle, Theorem 2.1

2.3 The Schwarz Reflection Principle

We wish to show that $\tilde{f}$ is analytic in D. It is certainly analytic in D^+ since it coincides with f there. Next we show that $\tilde{f}$ is analytic at an arbitrary point a in D^-. Choose a disk $D(a, r)$ with $D(a, r) \subset D^-$. Since $D(\bar{a}, r)$ lies in D^+ and f is analytic at $\bar{a}$, f has a Taylor expansion about $\bar{a}$, say

$$f(w) = \sum_{n=0}^{\infty} a_n (w - \bar{a})^n, \quad w \in D(\bar{a}, r).$$

For $z \in D(a, r)$, we have $\bar{z} \in D(\bar{a}, r)$ and so

$$\tilde{f}(z) = \overline{f(\bar{z})} = \overline{\sum_{n=0}^{\infty} a_n (\bar{z} - \bar{a})^n} = \sum_{n=0}^{\infty} \overline{a_n} (z - a)^n,$$

demonstrating that $\tilde{f}$ has a power series expansion in a disk about a and hence that $\tilde{f}$ is analytic at a.

Our final task is to show that $\tilde{f}$ is analytic at an arbitrary point x in J. We first show that $\tilde{f}$ is continuous on J: (this is where we use the assumption that f takes real values on J). By the assumption that f is continuous on $D^+ \cup J$, if $\{w_n\}_{n=1}^{\infty}$ is a sequence in D^+ that converges to x in J then $f(w_n) \to f(x)$ as $n \to \infty$. Let $\{z_n\}_{n=1}^{\infty}$ be a sequence in D^- that converges to x. Then $\{\overline{z_n}\}_{n=1}^{\infty}$ is a sequence in D^+ that converges to x and so (with $w_n = \overline{z_n}$ above)

$$\tilde{f}(z_n) = \overline{f(\overline{z_n})} \to \overline{f(x)} = f(x), \quad \text{as } n \to \infty,$$

where the assumption that f is real on the real axis gave $\overline{f(x)} = f(x)$. Thus, for any sequence $\{z_n\}_{n=1}^{\infty}$ in D that converges to x the sequence $\{\tilde{f}(z_n)\}_{n=1}^{\infty}$ converges to $f(x)$, showing that $\tilde{f}$ is continuous on J.

To establish analyticity of $\tilde{f}$ at x, choose a disk $D(x, r)$ contained in D and let T be any triangle in $D(x, r)$ as shown in Fig. 2.3a. Our goal is to show that the integral of $\tilde{f}$ around T equals 0: this, together with the continuity of $\tilde{f}$ on $D(x, r)$, shows that the assumptions of Morera's Theorem [16, Theorem 7.8] are satisfied and we may conclude that $\tilde{f}$ is analytic in $D(x, r)$ and, in particular, at x.

If T lies either entirely in D^+ (or in D^-) then $\int_T \tilde{f} = 0$ since $\tilde{f}$ is analytic in D^+ (and in D^-) so that Cauchy's Theorem applies. The essential case, therefore, is when T crosses (or even touches) the real axis, as shown in Fig. 2.3a. We add crosscuts just above and below the real axis at a small positive distance, say $\varepsilon/2$, from the real axis. The integral of $\tilde{f}$ over the triangle T is, in this way, written as the sum of (at most) three integrals around closed curves. One of these lies entirely in D^+ where $\tilde{f}$ is analytic and therefore this integral equals 0 by Cauchy's Theorem. A second lies entirely in D^- where we have shown $\tilde{f}$ to be analytic and again the integral around this closed curve equals 0. Thus, the integral of $\tilde{f}$ over T equals the integral of $\tilde{f}$ over a closed curve C_ε formed by the two crosscuts and by two short line segments from sides of the original triangle T as shown in Fig. 2.3b. The

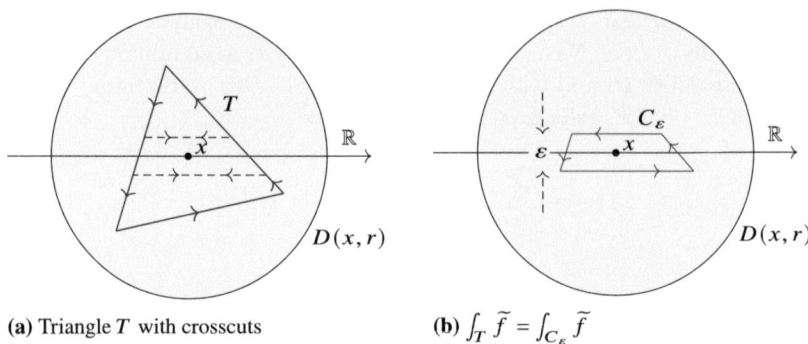

(a) Triangle T with crosscuts   (b) $\int_T \tilde{f} = \int_{C_\varepsilon} \tilde{f}$

Fig. 2.3 Triangle T with crosscuts and contour C_ε from the proof of Theorem 2.1. (a) Triangle T with crosscuts. (b) $\int_T \tilde{f} = \int_{C_\varepsilon} \tilde{f}$

crosscuts are a distance ε apart, as shown. As $\varepsilon \to 0^+$, the integral over the crosscut in the lower half-plane converges to the integral of $\tilde{f}$ over an interval on the real line, this by the continuity of $\tilde{f}$ in $D(x, r)$, while the integral over the crosscut in the upper half-plane converges to the integral of $\tilde{f}$ over the same interval on the real line but in the opposite direction. Moreover, as $\varepsilon \to 0^+$, the integrals over the sides also vanish as their lengths vanish. Hence, $\int_{C_\varepsilon} \tilde{f} \to 0$ as $\varepsilon \to 0^+$. Since $\int_T \tilde{f} = \int_{C_\varepsilon} \tilde{f}$ for each ϵ we can conclude, as we had hoped, that $\int_T \tilde{f} = 0$.[1]

The final step is to verify the statement regarding the univalence of the extension with the additional assumptions that f is one-to-one on $D^+ \cup J$ and that $f(D^+)$ lies entirely within the upper half-plane $\mathbb{H}$ (if $f(D^+)$ lies entirely in the lower half-plane $\overline{\mathbb{H}}$ then we consider $-f$ instead of f). We have shown that $\tilde{f}$, defined on D by (2.12), is analytic on D and now proceed to check univalence. Suppose that z and w lie in D and that $\tilde{f}(z) = \tilde{f}(w)$. We show that $z = w$ by checking each of a number of cases.

- If z and w both lie in $D^+ \cup J$ then $z = w$ since $\tilde{f} = f$ on $D^+ \cup J$ and f is one-to-one there.
- If z and w both lie in D^- then, by (2.12), $f(\bar{z}) = \overline{\tilde{f}(z)}$ and $f(\bar{w}) = \overline{\tilde{f}(w)}$. Since $\tilde{f}(z) = \tilde{f}(w)$ we have $f(\bar{z}) = f(\bar{w})$ with both $\bar{z}$ and $\bar{w}$ in D^+. Since f is one-to-one in D^+, $\bar{z} = \bar{w}$ and so $z = w$.
- The case $z \in D^+ \cup J$ and $w \in D^-$ (and vice versa) doesn't arise in the sense that $\tilde{f}(z) = \tilde{f}(w)$ never holds in this case. To see this, suppose that $w \in D^-$. Then $\tilde{f}(w) = \overline{f(\bar{w})}$ is a point in $\overline{f(D^+)}$ since $\bar{w} \in D^+$. Since, by assumption, $f(D^+)$ lies in the upper half-plane $\mathbb{H}$, the point $\tilde{f}(w)$ lies in the lower half-plane. If $z \in D^+ \cup J$ then $\tilde{f}(z) = f(z)$ lies either in the upper half-plane $\mathbb{H}$ (if $z \in D^+$)

---

[1] As often happens in these arguments, we showed here that $\int_{C_\varepsilon} \tilde{f} \to 0$ as $\varepsilon \to 0^+$, hence that $\int_T \tilde{f} = 0$ which in turn implies that, in fact, $\int_{C_\varepsilon} \tilde{f}$ is actually *equal* to 0 for each ε. This can seem a little strange at first sight.

2.3 The Schwarz Reflection Principle

or on the real axis (if $z \in J$). The imaginary parts of $\tilde{f}(z)$ and $\tilde{f}(w)$ are therefore different and so $\tilde{f}(z)$ and $\tilde{f}(w)$ cannot be equal.

This shows that f is univalent in D and that $f(D^-)$ coincides with $\overline{f(D^+)}$. □

Suppose that D is a domain that is symmetric about the real axis and that f is analytic in D and is real on the real axis. Then, by the Uniqueness Principle, f has to satisfy

$$f(\bar{z}) = \overline{f(z)}, \quad z \in D. \tag{2.13}$$

This is because (2.12) extends f analytically from D^+ to D, and so f has to be given by (2.12) on D^-.

Exercise 2.8 Consider the case of the function $f(z) = z^2$ on $\{z : |z| < 1, \Im z \geq 0\}$. What does this example tell us about the Schwarz Reflection Principle?

What if again $f(z) = z^2$ but this time on $\{z : |z| < 1, 0 \leq \arg z < 3\pi/4\}$?

2.3.2 Reflection in a Circle

We are familiar with reflection in a line. Two points z and w are symmetric in a line L if each is the reflection of the other in L. A line being a special case of a 'generalised circle', what about reflection in other generalised circles, namely Euclidean circles in the plane? Let $C = C(a, r)$ be such a circle with centre a and radius r. The Möbius transformation

$$M(z) = -i\frac{z - (a - r)}{z - (a + r)}, \quad z \in C_\infty, \tag{2.14}$$

maps C onto the extended real line $\mathbb{R}_\infty$ as shown in Fig. 2.4 (this since $M(a-r) = 0$, $M(a-ir) = 1$ and $M(a+r) = \infty$ and so, by Proposition 2.3, $M(C)$ is a generalised

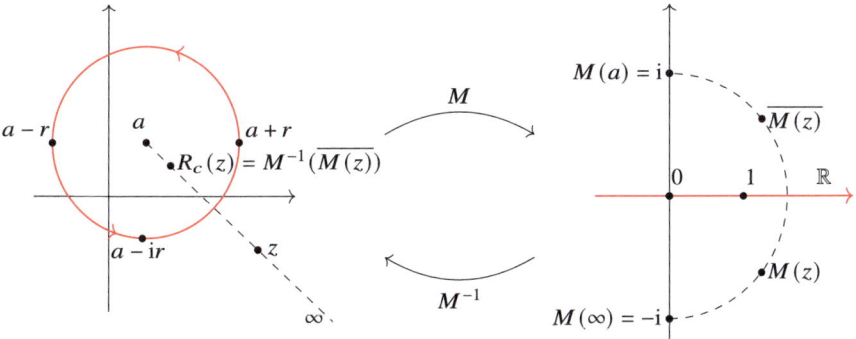

Fig. 2.4 Reflection in a circle realised via reflection in the real line

circle through 0, 1 and ∞, in other words the extended real line). The inverse of this transformation is

$$M^{-1}(w) = \frac{(a+r)w + i(a-r)}{w+i}, \quad w \in \mathbb{C}_\infty.$$

A point z in $\mathbb{C}_\infty$ will be mapped by M to $w = M(z)$. The reflection of this point in $\mathbb{R}_\infty = M(C)$ is $\overline{M(z)}$, with the understanding that ∞ is its own reflection in the extended real line. It is natural to call the point $M^{-1}(\overline{M(z)})$ the *reflection of z in the circle C*. In other words, if R_C is reflection in the circle and R_L is reflection in the real axis, then

$$R_C(z) = (M^{-1} \circ R_L \circ M)(z), \quad z \in \mathbb{C}_\infty.$$

For example, points on the circle C do not move under reflection in C. The centre a is sent by M to i, then to $-$i under reflection in $\mathbb{R}$, then to ∞ under M^{-1}, so that $R_C(a) = \infty$. Also, $(R_C \circ R_C)(z) = z$ for each z in $\mathbb{C}_\infty$.

Lemma 2.3 *The reflection $R_C(z)$ of a point z, other than a or ∞, in the circle $C(a, r)$ is*

$$R_C(z) = a + \frac{r^2}{\overline{z} - \overline{a}} = a + \frac{r^2}{|z-a|^2}(z-a). \tag{2.15}$$

The reflected point $R_C(z)$ lies on the ray from a through z.

Proof For points other than a and ∞,

$$R_C(z) = \frac{(a+r)\overline{M(z)} + i(a-r)}{\overline{M(z)} + i}$$

$$= \frac{(a+r)i\frac{\overline{z}-(\overline{a}-r)}{\overline{z}-(\overline{a}+r)} + i(a-r)}{i\frac{\overline{z}-(\overline{a}-r)}{\overline{z}-(\overline{a}+r)} + i}$$

$$= \frac{a\overline{z} - a\overline{a} + r^2}{\overline{z} - \overline{a}}$$

$$= a + \frac{r^2}{\overline{z}-\overline{a}} = a + \frac{r^2}{|z-a|^2}(z-a).$$

This formula exhibits $R_C(z)$ as $a + \lambda(z-a)$ with $\lambda > 0$, and shows that a, z and $R_C(z)$ are collinear. $\square$

The formula (2.15) also shows that

$$(R_C(z) - a)\,\overline{(z-a)} = r^2,$$

which justifies the following definition.

2.3 The Schwarz Reflection Principle

Definition 2.2 Points z and w (neither equal to a or ∞) are said to be *symmetric in the circle* $C(a, r)$ ($a \in \mathbb{C}, r > 0$) if

$$(w - a)(\bar{z} - \bar{a}) = r^2. \tag{2.16}$$

The centre a and the point at infinity are defined to be symmetric in the circle $C(a, r)$. Reflection R_C in a circle C, defined on $\mathbb{C}_\infty$, sends a point z to the point symmetric to z with respect to C.

Exercise 2.9 Show that the condition for two points z and w to be symmetric in the imaginary axis is that $w = -\bar{z}$.

Exercise 2.10 In the context of generalised circles, a Euclidean straight line can be thought of as a limiting case of Euclidean circles as the radius of the circle tends to infinity. For example, as $\lambda \to \infty$, the circle $C_\lambda = \{z : |z - i\lambda| = \lambda\}$ approximates the real line. To see that this behaviour interacts well with reflection show that, for fixed z in $\mathbb{C}$,

$$R_{C_\lambda}(z) \to \bar{z} \text{ as } \lambda \to \infty.$$

Exercise 2.11 ▷ Denote the circle $C(a, r)$ by C. Show that reflection R_C in this circle can be written as $R_C(z) = \overline{M(z)}$ for a certain Möbius transformation M. Deduce that the reflection of a generalised circle in C is again a generalised circle.
▷ Describe the reflection of the circle $C(a + r/2, r/2)$ in C.
▷ Find the reflection of the half-plane $H = \{z : \Re z > 2\}$ in the unit circle.

Exercise 2.12

▷ Referring back to (2.2) and Exercise 2.1, show that z_1 and z_2 are symmetric in the circle $C(z_1, z_2, k)$.
▷ Conversely, suppose that z_1 and z_2 are symmetric in the circle $C(a, r)$ where z_1 lies inside C but is distinct from a. Show that $C(a, r)$ can viewed as the circle $C(z_1, z_2, k)$ where $k = |z_1 - a|/r$.

The explicit choice of Möbius transformation in (2.14) that maps the circle $C(a, r)$ onto $\mathbb{R}_\infty$ leads to an explicit formula (2.15) for reflection but otherwise doesn't matter. We could equivalently define z and w to be symmetric in the circle $C = C(a, r)$ if, whenever $\widetilde{M}$ is a Möbius transformation mapping C onto $\mathbb{R}_\infty$, the points $\widetilde{M}(z)$ and $\widetilde{M}(w)$ are symmetric in $\mathbb{R}$. To see that this definition doesn't depend on the choice of $\widetilde{M}$, let M_1 and M_2 be two such Möbius transformations. Then $M_2 \circ M_1^{-1}$ is a Möbius transformation that fixes $\mathbb{R}_\infty$. By Exercise 2.7,

$$\overline{(M_2 \circ M_1^{-1})(z)} = (M_2 \circ M_1^{-1})(\bar{z}), \text{ for each } z.$$

Replace z by $M_1(z)$ and suppose that $M_1(z) = \overline{M_1(w)}$ (that is, $M_1(z)$ and $M_1(w)$ are symmetric in $\mathbb{R}$). Then,

$$M_2(z) = (M_2 \circ M_1^{-1})(M_1(z)) = (M_2 \circ M_1^{-1})(\overline{M_1(w)})$$
$$= \overline{(M_2 \circ M_1^{-1})(M_1(w))} = \overline{M_2(w)},$$

so that $M_2(z)$ and $M_2(w)$ are also symmetric in $\mathbb{R}$. This prompts the following formal definition.

Definition 2.3 Let C be a generalised circle and z and w be points in $\mathbb{C}_\infty$. We say that z and w are symmetric in C if, whenever M is a Möbius transformation sending C onto the extended real line $\mathbb{R}_\infty$, then $M(z)$ and $M(w)$ are symmetric in the real line. Here the point at infinity is said to be its own reflection in the real line.

This definition covers both symmetry in Euclidean lines and Euclidean circles and, in the latter case, agrees with the explicit formula (2.15) by the immediately preceding discussion.

Though it shouldn't be surprising considering how we defined reflection in a circle, it is a fundamental fact that the property of points being reflections of each other in a generalised circle is preserved under Möbius transformations.

Proposition 2.4 *Let M be a Möbius transformation and let C be a generalised circle. Then, two complex numbers z and w are symmetric in C if and only if $M(z)$ and $M(w)$ are symmetric in the generalised circle $M(C)$.*

Proof It is enough to prove the 'only if' direction as the 'if' direction then follows with C replaced by $M(C)$, z and w replaced by $M(z)$ and $M(w)$ respectively, and M replaced by M^{-1}.

Let M be a Möbius transformation, let C be a generalised circle, and let z and w be symmetric in C. Choose a Möbius transformation $\widetilde{M}$ that maps C onto the extended real line $\mathbb{R}_\infty$. Then $\widetilde{M}(z)$ and $\widetilde{M}(w)$ are symmetric in the real line. The Möbius transformation $\widetilde{M} \circ M^{-1}$ sends $M(C)$ to the extended real line and sends $M(z)$ and $M(w)$ to $\widetilde{M}(z)$ and $\widetilde{M}(w)$, respectively. By definition, then, $M(z)$ and $M(w)$ are symmetric in $M(C)$. □

2.3.3 The Schwarz Reflection Principle for Circles

Now that we are familiar with reflection in a circle, we can extend the Schwarz Reflection Principle from reflection in the real axis to reflection in a circle. Recall that, for a circle C, the reflection of the point z in C is denoted by $R_C(z)$. In exact analogy with reflection of a set in the real axis we denote by $R_C(E)$, for a subset E of $\mathbb{C}_\infty$, the set $\{z : R_C(z) \in E\}$, so that $R_C(E)$ is the reflection of E in the circle C. The set E is said to be *symmetric in C* if $R_C(E) = E$, that is if $R_C(z) \in E$ whenever $z \in E$.

2.3 The Schwarz Reflection Principle

Putting together the Schwarz Reflection Principle for the real line and the invariance of reflection in circles under Möbius transformations, we can prove a Schwarz Reflection Principle for circles.

Theorem 2.2 (Schwarz Reflection Principle for Generalised Circles) *Let C_1 and C_2 be generalised circles, with C_i^+ and C_i^- denoting the two components of $\mathbb{C} \setminus C_i$, $i = 1, 2$. Let D be a domain that is symmetric in C_1 and set $D^+ = D \cap C_1^+$, $D^- = D \cap C_1^-$, and $J = D \cap C_1$. Suppose that f is continuous on $D^+ \cup J$, is analytic on D^+, and maps J into C_2. Then f can be extended to an analytic function $\tilde{f}$ on all of D.*

Moreover, if f is one-to-one on $D^+ \cup J$ and if $f(D^+)$ is either completely contained in C_2^+ or C_2^- then the extension $\tilde{f}$ is univalent on D with $\tilde{f}(D^-) = R_{C_2}(f(D^+))$.

Proof Choose a Möbius map M_1 that sends the extended real line onto C_1 and maps the upper half-plane $\mathbb{H}$ onto C_1^+. Choose a Möbius map M_2 that sends C_2 onto the extended real line and maps C_2^+ onto the upper half-plane $\mathbb{H}$. Let $\tilde{D} = M_1^{-1}(D)$. Then, by Definition 2.3, $\tilde{D}$ is a domain that is symmetric in the real line. As before, we set $\tilde{D}^+ = \tilde{D} \cap \mathbb{H}$, $\tilde{D}^- = \tilde{D} \cap \overline{\mathbb{H}}$, and $\tilde{J} = \tilde{D} \cap \mathbb{R}$. By construction, $M_1(\tilde{D}^+) = D^+$, $M_1(\tilde{D}^-) = D^-$, and $M_1(\tilde{J}) = J$. On $\tilde{D}^+ \cup \tilde{J}$, set $F = M_2 \circ f \circ M_1$. This sequence of mappings is shown in Fig. 2.5. Then, F is continuous on $\tilde{D}^+ \cup \tilde{J}$ and analytic on $\tilde{D}^+$. Moreover, for $z \in \tilde{J}$, we have $M_1(z) \in C_1$, so $f(M_1(z)) \in C_2$,

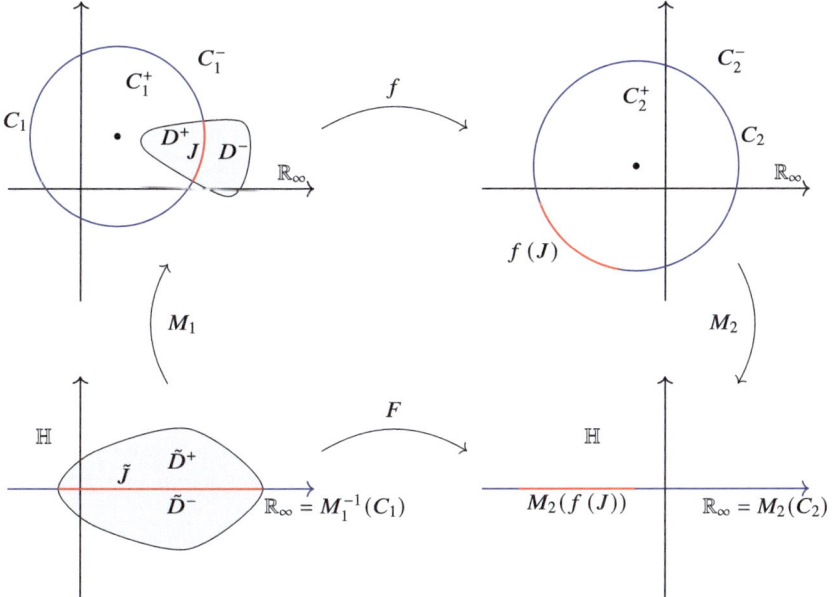

Fig. 2.5 Mappings used in the proof of the Schwarz Reflection Principle for circles

and then $F(z) \in \mathbb{R}$. That is, F satisfies the assumptions of the Schwarz Reflection Principle for the real line (Theorem 2.1). The function F can therefore be extended to a function $\tilde{F}$ analytic on all of $\tilde{D}$. But then $\tilde{f}$, defined by

$$\tilde{f}(z) = (M_2^{-1} \circ \tilde{F} \circ M_1^{-1})(z), \quad z \in D,$$

is analytic in D and agrees with f on $D^+ \cup J$.

Suppose that, in addition, f is one-to-one on $D^+ \cup J$ and $f(D^+)$ is completely contained in either C_2^+ or C_2^-. Since Möbius transformations are univalent, F is one-to-one on $\tilde{D}^+ \cup \tilde{J}$. Moreover, tracing the mappings shows that $F(\tilde{D}^+)$ lies entirely in either $\mathbb{H}$ or $\overline{\mathbb{H}}$. Thus, by Theorem 2.1, the extension $\tilde{F}$ of F is univalent in $\tilde{D}$. It follows that the extension $\tilde{f}$ of the original function f is univalent in D and that $\tilde{f}(D^-) = R_{C_2}(f(D^+))$. □

Exercise 2.13 In the context and notation of Theorem 2.2, and denoting by $R_C(z)$ the reflection of a point z in the generalised circle C, deduce from (2.13) that

$$f(z) = R_{C_2}(f(R_{C_1}(z))), \quad z \in D. \tag{2.17}$$

Exercise 2.14 Does there exist a function f that is continuous on $\mathbb{D} \cup C(0, 1)$, is analytic on the unit disk $\mathbb{D}$, and for which $f(z) = \bar{z}$ on $C(0, 1)$?

2.4 Singularities of Analytic Functions

We recall, in this short section, the notions of removable singularity, pole, and essential singularity.

A complex-valued function f has an *isolated singularity* at $a \in \mathbb{C}$ if f is analytic in the punctured disk $D'(a, r)$ for some positive r. In this case, f can be expanded as a Laurent series about a,

$$f(z) = \sum_{n=-\infty}^{\infty} c_n (z - a)^n.$$

There are three cases to consider depending on whether there are none, finitely many, or infinitely many non-zero coefficients c_n for $n < 0$.

Removable Singularity If $c_n = 0$ for all indices $n < 0$ then f can be written as the convergent power series $f(z) = \sum_{n=0}^{\infty} c_n (z-a)^n$ for z near a, and f becomes analytic at a if we simply set $f(a) = c_0$.

2.4 Singularities of Analytic Functions

Pole If only finitely many c_n, for n negative, are non-zero then there is $N \in \mathbb{N}$ such that

$$f(z) = \sum_{n=-N}^{\infty} c_n (z-a)^n$$

$$= \frac{c_{-N}}{(z-a)^N} + \frac{c_{-N+1}}{(z-a)^{N-1}} + \cdots + \frac{c_{-1}}{z-a} + c_0 + c_1(z-a) + \cdots$$

with $c_{-N} \neq 0$. In this case, f is said to have a pole of order N at a. We can write f near a as

$$f(z) = \frac{1}{(z-a)^N}\left[c_{-N} + c_{-N+1}(z-a) + \cdots c_0(z-a)^N + \cdots\right] = \frac{1}{(z-a)^N} g(z)$$

where g is analytic at a and doesn't vanish at a. In particular, $|f(z)| \to +\infty$ as $z \to a$. For example, the Möbius transformation (2.12) with $c \neq 0$ has a pole of order 1, a simple pole, at $-d/c$.

Essential Singularity In this case, $c_n \neq 0$ for infinitely many negative n. Then, by the Casorati-Weierstrass Theorem [16, Theorem 8.10], $f\left(D'(a,\delta)\right)$ is dense in $\mathbb{C}$ for each positive δ. There is, therefore, no way to define $f(a)$ so that f is continuous at a. For example,

$$f(z) = e^{1/z} = 1 + \frac{1}{z} + \frac{1}{2!z^2} + \frac{1}{3!z^3} + \cdots$$

has an essential singularity at 0. Since, for example, $f(x) \to +\infty$ as x approaches 0 from the right through real values whereas $f(x) \to 0$ as x approaches 0 from the left, we see that $\lim_{z \to 0} f(z)$ does not exist.

In Chap. 5 we will prove a significant extension of the Casorati-Weierstrass Theorem. We will prove the Great Picard Theorem showing that a function that has an essential singularity at a will, in any neighbourhood of a, assume every complex value, with at most one exception. That there has to be an exceptional value is shown by the function $f(z) = e^{1/z}$ which has an essential singularity at 0 but never takes the value 0.

Chapter 3
The Riemann Sphere

When working with Möbius transformations in the introduction, it was natural to introduce the notion of a 'complex number at infinity', the *point at infinity* ∞. In so doing, the domain and the range of a Möbius transformation became the extended complex plane $\mathbb{C}_\infty = \mathbb{C} \cup \{\infty\}$, and every Möbius transformation became a bijection. If we had restricted ourselves to just the complex plane $\mathbb{C}$ then, as we saw in Sect. 2.2, the image of the line $L = \{z : \Re z = 1\}$ under the inversion $M(z) = 1/z$ would be the circle $C(1/2, 1/2)$ but with the origin missing. Setting $1/\infty = 0$, the image under this inversion of the generalised line $L \cup \{\infty\}$ is the whole circle $C(1/2, 1/2)$, a much more satisfactory situation.

In this chapter, the point at infinity is given physical form so that we may have a way of visualising it. The construction itself is a simple one. Picture the complex numbers $\mathbb{C}$ as the complex plane, infinite in all directions. Sit a ball at the origin on top of this plane. Now fold the plane up around this ball—you'll need to shrink the plane as you go out from the origin so that it folds neatly around the ball. This can be done in many ways whereby the surface of the ball except for a single point is covered—there has to be at least one omitted point—that's the point on the surface of the ball that we may visualise as the point at infinity. In this way, the extended complex plane $\mathbb{C} \cup \{\infty\}$ can be visualised as the surface of a ball, that is a sphere, where all points on the sphere, except one point, correspond to ordinary complex numbers and the remaining point corresponds to ∞. Of course, from the point of view of the sphere, all points are geometrically the same.

We implement an explicit version of this folding in which the complex plane passes through the centre of the ball rather than having the ball sit on top of the plane. This is a matter of preference—either scenario is perfectly fine. Now a line from the North Pole of the sphere to a point in the complex plane will pass through a unique point on the sphere—see Fig. 3.1. Mapping the point in the complex plane to this point on the sphere, we succeed in folding the complex plane onto the sphere with the North Pole removed. Conversely, a line from the North Pole through a second point on the sphere will pass through a unique point in the complex plane. Mapping

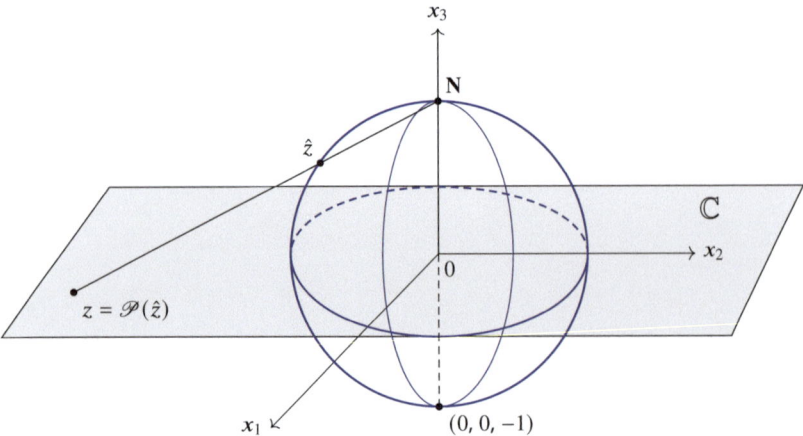

Fig. 3.1 Stereographic projection

the point on the sphere to this point in the complex plane, we effectively flatten the sphere minus the North Pole onto the complex plane. This simple geometric construction, called *stereographic projection*, is a bijection between the sphere with the North Pole removed and the complex plane. It is then natural to identify the North Pole as the point at infinity. Topologically, the complex plane is equivalent to the surface of a ball with one point removed and the extended complex plane to the entire surface of the ball, which is compact. This is a special case of the Alexandroff one-point compactification of a topological space—the non-compact space $\mathbb{C}$ is made compact by the addition of one extra notional point.

These are two complementary viewpoints of $\mathbb{C}_\infty$; first, the complex plane $\mathbb{C}$ with the addition of a notional point at infinity and secondly the sphere itself. We may pass from one to the other via stereographic projection. The complex plane $\mathbb{C}$ is identified with the sphere less the North Pole. The point at infinity can be thought of as both the notional point ∞ that, when adjoined to the complex plane, makes it compact or as the North Pole **N** of the sphere. At a deeper level, the sphere inherits a *complex structure* from the complex plane by means of stereographic projection. This allows us to do complex analysis on the sphere, which we can't do directly on the round surface of the sphere. The surface of the sphere is locally flat and stereographic projection flattens the surface of the sphere with the North Pole removed onto the complex plane. Also, stereographic projection sends the surface of the sphere with the South Pole removed to the extended complex plane with zero removed, and if we follow this with the inversion $z \to 1/z$, we have a bijection onto the complex plane. These flattening maps are called *charts*: our two charts together cover the entire sphere and behave well where the charts overlap, that is on the sphere with both poles removed or, from the point of view of the plane, on the complex plane with 0 removed (their composition is the inversion $z \to 1/z$, in fact). The sphere with this complex structure is called the *Riemann Sphere* and is an example of a *Riemann Surface*. Now that we have a complex structure, we can do complex analysis on the sphere: rather than working on the sphere itself, we use the charts to work on the complex plane. For example, a function f on the sphere will be meromorphic if $f \circ \varphi^{-1}$ is meromorphic in the complex plane for each chart φ.

We will study this particular Riemann surface in detail in this chapter. For example, we will be able to give a complete description of all meromorphic functions on the Riemann sphere (Theorem 3.3)—all are rational functions. This is an important result and a jumping off point for the study of compact Riemann surfaces.

Why, you may ask, such emphasis on the Riemann sphere if it is just one example of a Riemann surface? The reason is the *Uniformisation Theorem* which states that every simply connected Riemann surface is conformally equivalent to one of three special Riemann surfaces: the complex plane, the hyperbolic disk (that we will study in Chap. 4), or the Riemann sphere. The Riemann sphere, then, is the canonical compact, simply connected Riemann surface and deserves a careful study in its own right. We will see that stereographic projection has some very nice properties: it is an angle preserving map under which circles on the sphere correspond to generalised circles in the complex plane. The Riemann sphere has a natural distance function that it inherits from the sphere and this in turn induces a metric on the extended complex plane, called the *spherical metric* and for which we will calculate a simple formula. We will also be able to describe all isometries of the spherical metric and deduce that the shortest curves on the sphere are the great circles.

We begin our study of the Riemann sphere, then, with stereographic projection.

3.1 Stereographic Projection

Let $\mathscr{S}$ denote the unit sphere in three dimensional Euclidean space $\mathbb{R}^3$,

$$\mathscr{S} = \{(x_1, x_2, x_3) \subset \mathbb{R}^3 : x_1^2 + x_2^2 + x_3^2 = 1\}.$$

We think of the $x_1 x_2$-plane as the complex plane where the point $(x_1, x_2, 0)$ corresponds to the complex number $z = x_1 + ix_2$. The point $\mathbf{N} = (0, 0, 1)$ is singled out as a special point—the *North Pole* of the sphere. The sphere punctured at $\mathbf{N}$ is denoted by $\mathscr{S}'$ that is $\mathscr{S}' = \mathscr{S} \setminus \{\mathbf{N}\}$.

A line in $\mathbb{R}^3$ through $\mathbf{N}$ and through a second point $\hat{z}$ on the punctured sphere $\mathscr{S}'$ will meet the complex plane in exactly one point z. Conversely, a line in $\mathbb{R}^3$ through $\mathbf{N}$ that meets the complex plane in a point z will meet the sphere $\mathscr{S}$ in exactly one point $\hat{z}$ other than $\mathbf{N}$. The points z in $\mathbb{C}$ and $\hat{z}$ in $\mathscr{S}'$ are identified. For example, the South Pole $\hat{z} = (0, 0, -1)$ and the origin $z = 0$ are identified. It may help to think of z as the shadow of $\hat{z}$ with the sun at the North Pole $\mathbf{N}$. We can then make a function $\mathscr{P} \colon \mathscr{S}' \to \mathbb{C}$ by $\mathscr{P}(\hat{z}) = z$. This function is one-to-one and onto and, for its inverse function $\mathscr{P}^{-1} \colon \mathbb{C} \to \mathscr{S}'$, we have $\mathscr{P}^{-1}(z) = \hat{z}$. The function $\mathscr{P}$

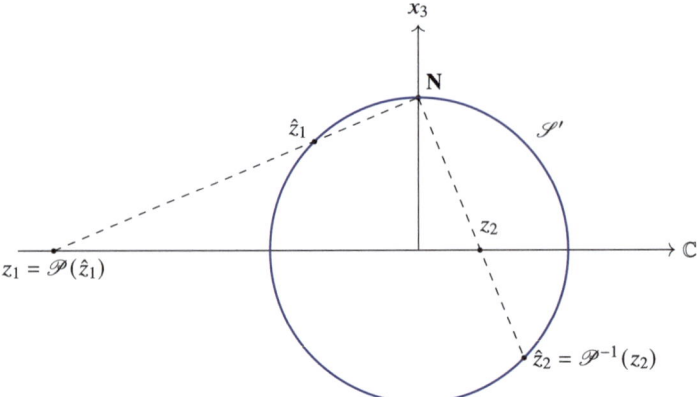

Fig. 3.2 Stereographic projection: cutaway view

is called *stereographic projection*. Figure 3.1 shows stereographic projection while Fig. 3.2 shows a cross section taken perpendicular to the $x_1 x_2$-plane.

The map $\mathscr{P}^{-1}$ takes the complex plane and wraps it onto the punctured sphere without tears or breaks. The disk $\mathbb{D} = \{z : |z| < 1\}$ is mapped onto the southern hemisphere, the unit circle stays fixed and everything outside the unit circle is folded onto the northern hemisphere. There are a myriad topologically equivalent ways[1] of achieving this continuous folding of the complex plane onto the punctured sphere, but stereographic projection has special properties. For one, stereographic projection is given by explicit formulas.

A point $\hat{z}$ on the sphere is naturally described in spherical coordinates (ρ, θ, ϕ). Here ρ is the distance of the point from the origin, (which equals 1 for each point on $\mathscr{S}$ and may be suppressed); θ is the longitude—when the point $\hat{z}$ is projected vertically onto the $x_1 x_2$-plane, then θ is the anti-clockwise angle in the $x_1 x_2$-plane between the positive x_1-axis and the radius from the origin to this projection; ϕ is the latitude, that is the angle made by the radius to $\hat{z}$ with the positive $x_1 x_2$-plane. Points on the equator of the sphere have latitude 0, points in the southern hemisphere have negative latitude with those near the South Pole having latitude close to $-\pi/2$, while points in the northern hemisphere have positive latitude.[2] In summary, any point on the sphere $\mathscr{S}$ can be represented by two angles, a longitude angle θ and a latitude angle ϕ, where $\theta \in [0, 2\pi)$ and $\phi \in [-\pi/2, \pi/2]$ (strictly speaking, the longitude angle θ is not defined at the poles; $\phi = -\pi/2$ will correspond to the

---

[1] As mentioned earlier, in many expositions the sphere $\mathscr{S}$ sits on top of the complex plane at 0 and again points on the sphere are projected down onto the complex plane from N. This too is called stereographic projection.

[2] This agrees with how we describe latitude on the earth: the equator is at latitude 0, the Tropic of Cancer is at latitude $23\frac{1}{2}$ degrees north of the equator while the Tropic of Capricorn is at $23\frac{1}{2}$ degrees south of the equator.

3.1 Stereographic Projection

South Pole and $\phi = \pi/2$ to the North Pole). For example, the equator of the sphere is $\phi = 0$. The point in the plane with cartesian coordinates $(1/\sqrt{2}, -1/\sqrt{2})$ has spherical coordinates $(7\pi/4, 0)$.

Proposition 3.1 *The point z in the punctured complex plane $\mathbb{C} \setminus \{0\}$ is the stereographic projection of the point $\hat{z} = (\theta, \phi)$ on the twice punctured sphere $\mathscr{S}' \setminus \{(0, 0, -1)\}$. Then, in polar coordinates,*

$$z = \tan\left(\frac{\phi}{2} + \frac{\pi}{4}\right) e^{i\theta}. \tag{3.1}$$

If $z = re^{i\theta} \in \mathbb{C}$ then, in spherical coordinates,

$$\hat{z} = \left(\theta, 2\arctan r - \tfrac{\pi}{2}\right). \tag{3.2}$$

Proof Stereographic projection does not involve any rotation about the x_3-axis. The stereographic projection of $\hat{z} = (\theta, \phi)$ will therefore have argument θ and can be written as $z = re^{i\theta}$ in polar form. The question is to determine $r = |z|$ in terms of ϕ (and, in principle, θ). A cross section through **N**, $\hat{z}$ and z, with $\hat{z}$ on the northern hemisphere, is shown in Fig. 3.3. The triangles $\triangle \mathbf{N}0z$ and $\triangle \hat{z}Pz$ in this figure are similar. Here P is the projection of $\hat{z}$ vertically down onto the complex plane. Hence

$$\tan(\angle 0z\mathbf{N}) = \frac{1}{r} = \frac{\sin\phi}{r - \cos\phi},$$

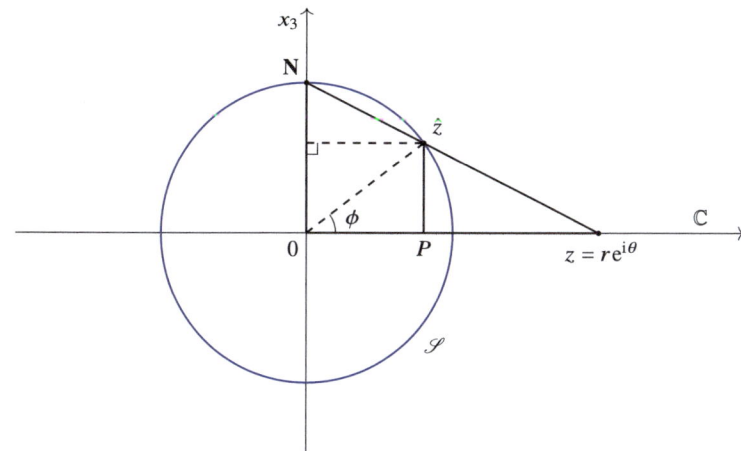

Fig. 3.3 Stereographic projection of a point on the northern hemisphere

from which we find, setting $\phi' = \phi + \pi/2$, that

$$r = \frac{\cos\phi}{1-\sin\phi} = \frac{\sin\phi'}{1+\cos\phi'} = \frac{2\sin(\phi'/2)\cos(\phi'/2)}{1+(2\cos^2(\phi'/2)-1)} = \frac{\sin(\phi'/2)}{\cos(\phi'/2)} = \tan\left(\frac{\phi'}{2}\right).$$

One may check that the formula $r = \cos\phi/(1-\sin\phi)$ also holds in the case when $\hat{z}$ lies on the southern hemisphere of the sphere $\mathscr{S}$.

Conversely, if $z = re^{i\theta}$ then $\hat{z}$, in spherical coordinates, is $\hat{z} = (\theta, \phi)$ where $r = \tan(\phi'/2)$. Then $\phi' = 2\arctan r$, giving $\phi = 2\arctan r - \pi/2$. (Remember that, by convention, arctan takes values between $-\pi/2$ and $\pi/2$.) □

Exercise 3.1 The preimage of a circle $C(0, r)$ in the complex plane under stereographic projection is the intersection of a plane $x_3 = k(r)$ with the sphere $\mathscr{S}$. Find a formula for $k(r)$ in terms of r. How does $k(r)$ behave as $r \to \infty$?

We can also derive formulas for stereographic projection in terms of Cartesian coordinates on the sphere and on the plane.

Proposition 3.2 *The point $z = x + iy$ in the complex plane $\mathbb{C}$ is the stereographic projection of the point $\hat{z} = (x_1, x_2, x_3)$ on the punctured sphere $\mathscr{S}'$. Then z and $\hat{z}$ are related by*

$$\hat{z} = \mathscr{P}^{-1}(z) = \left(\frac{2x}{|z|^2+1}, \frac{2y}{|z|^2+1}, \frac{|z|^2-1}{|z|^2+1}\right) \tag{3.3}$$

and

$$z = \mathscr{P}(\hat{z}) = \frac{x_1}{1-x_3} + i\frac{x_2}{1-x_3}. \tag{3.4}$$

Proof Let $z = x + iy = (x, y, 0)$. The line in $\mathbb{R}^3$ through z and $\mathbf{N} = (0, 0, 1)$ is given by the parametric equations

$$P(t) = t(x, y, 0) + (1-t)(0, 0, 1) = (tx, ty, 1-t), \quad -\infty < t < \infty.$$

The point $P(t)$ lies on the sphere $\mathscr{S}$ when $1 = (tx)^2 + (ty)^2 + (1-t)^2$, that is when

$$0 = t\big[(1+x^2+y^2)t - 2\big].$$

The solution $t = 0$ corresponds to the North Pole $\mathbf{N}$. The second solution is $t = 2/(1+x^2+y^2) = 2/(1+|z|^2)$. Then

$$\hat{z} = P\left(\frac{2}{1+|z|^2}\right)$$

which leads to the formula (3.3).

3.1 Stereographic Projection

To obtain the formula (3.4), start with a point $\hat{z} = (x_1, x_2, x_3)$ on the punctured sphere $\mathscr{S}'$. The line in $\mathbb{R}^3$ through $\hat{z}$ and $\mathbf{N}$ is given by the parametric equations

$$Q(t) = t(x_1, x_2, x_3) + (1-t)(0, 0, 1) = (tx_1, tx_2, tx_3 + 1 - t), \quad -\infty < t < \infty.$$

The point $Q(t)$ lies in the complex plane when the third coordinate, $tx_3 + 1 - t$, vanishes, that is when $t = 1/(1 - x_3)$ (note that $1 - x_3 \neq 0$ since $\hat{z}$ is not the North Pole). Then

$$z = Q\left(\frac{1}{1-x_3}\right) = \left(\frac{x_1}{1-x_3}, \frac{x_2}{1-x_3}, \frac{x_3}{1-x_3} + 1 - \frac{1}{1-x_3}\right)$$
$$= \left(\frac{x_1}{1-x_3}, \frac{x_2}{1-x_3}, 0\right)$$

which becomes (3.4) in complex number notation. □

Under stereographic projection, the complex plane $\mathbb{C}$ corresponds to the sphere punctured at the North Pole $\mathbf{N}$. Notice that points on the sphere near the North Pole correspond to points in $\mathbb{C}$ far away from 0. This is clear either from any of the previous figures or from the formula (3.1) where $|z| = r = \tan(\phi/2 + \pi/4)$: if ϕ is close to $\pi/2$, then $\phi/2 + \pi/4$ is close to $\pi/2$ and so, since $\lim_{t \uparrow \pi/2} \tan t = +\infty$, r becomes arbitrarily large as the latitude ϕ increases towards $\pi/2$. It is natural to fill in the puncture in the punctured sphere by adding back in the point $\mathbf{N}$. This corresponds, under stereographic projection, to adding a point to the complex plane which, by the immediately preceding discussion of the behaviour of stereographic projection near the North Pole, can serve as a realisation of the *point at infinity*, '∞', that we introduced as a notational convenience in Chap. 2. As before, the complex plane together with the point at infinity, that is $\mathbb{C} \cup \{\infty\}$, is denoted by $\mathbb{C}_\infty$ (or, in some books, by $\hat{\mathbb{C}}$ or $\overline{\mathbb{C}}$), and is called the *extended complex plane*. Stereographic projection is defined on the entire sphere $\mathscr{S}$ by setting $\mathscr{P}(\mathbf{N}) = \infty$. Then $\mathscr{P}: \mathscr{S} \to \mathbb{C}_\infty$ is a bijection. In this context, $\mathscr{S}$ is called the *Riemann sphere*. The Riemann sphere $\mathscr{S}$ and the extended complex plane $\mathbb{C}_\infty$ are two sides of the same coin, with stereographic projection being the link between them.

Exercise 3.2 Two points P_1 and P_2 on the sphere are referred to as *antipodal* if they lie diametrically opposite one another. That is, P_1 and P_2 are antipodal points if the straight line in $\mathbb{R}^3$ that joins them passes through the centre of the sphere, that is if $P_2 = -P_1$.

▷ For a point P on the sphere, write down the spherical coordinates of $-P$ in terms of the spherical coordinates of P.

Let us call two points in the extended complex plane *antipodal* if their preimages under stereographic projection are antipodal on the sphere. For example, 1 and −1 are antipodal.

▷ Find a condition for two points α and β in the extended complex plane to be antipodal.

3.1.1 Circles Under Stereographic Projection

A circle on the sphere $\mathscr{S}$ is the intersection of a plane in $\mathbb{R}^3$ with the sphere. The equation of such a plane takes the form

$$ax_1 + bx_2 + cx_3 = d, \tag{3.5}$$

and the condition on the coefficients a, b, c, d for this plane to intersect (and not just touch) the sphere is

$$a^2 + b^2 + c^2 > d^2. \tag{3.6}$$

The plane passes through the North Pole $\mathbf{N} = (0, 0, 1)$ precisely when $c = d$.

Exercise 3.3 Verify the condition (3.6) for the plane (3.5) to intersect the sphere $\mathscr{S}$. (Hint: use the method of Lagrange multipliers to minimise the function $f(x_1, x_2, x_3) = x_1^2 + x_2^2 + x_3^2$ subject to the condition $ax_1 + bx_2 + cx_3 - d = 0$.)

We now show that the stereographic image of a circle on the sphere is a generalised circle in the extended complex plane, and vice versa.

Proposition 3.3 *Circles on the sphere and generalised circles in the extended complex plane correspond under stereographic projection.*

Proof Consider a circle $\hat{\mathscr{C}}$ on the sphere $\mathscr{S}$, described as the intersection of the sphere with a plane whose equation is (3.5) and that satisfies (3.6). Let $\hat{z}$ be a point on $\hat{\mathscr{C}}$ other than the North Pole (recall that the North Pole projects to the point at infinity ∞). Then the stereographic projection $z = x + iy$ of $\hat{z}$ satisfies, by (3.3),

$$a\left(\frac{2x}{|z|^2 + 1}\right) + b\left(\frac{2y}{|z|^2 + 1}\right) + c\left(\frac{|z|^2 - 1}{|z|^2 + 1}\right) = d.$$

This is equivalent to

$$2ax + 2by + c(|z|^2 - 1) = d(|z|^2 + 1). \tag{3.7}$$

Replacing x by $(z + \bar{z})/2$ and y by $(z - \bar{z})/(2i)$, (3.7) is of the form (2.1), that is $A|z|^2 + \bar{B}z + B\bar{z} + C = 0$, with

$$A = c - d, \quad B = a + ib, \quad C = -c - d. \tag{3.8}$$

3.1 Stereographic Projection

We need to check that $|B|^2 - AC > 0$. In fact,

$$|B|^2 - AC = (a^2 + b^2) - (c-d)(-c-d) = a^2 + b^2 + c^2 - d^2,$$

which is positive by (3.6). Thus $z = \mathscr{P}(\hat{z})$ lies on the generalised circle $\mathscr{C}$ of the form (2.1) with A, B and C given by (3.8). Note that $\mathscr{C}$ is a Euclidean straight line if and only if $A = 0$, which in turn holds if and only if $c = d$ and the circle $\hat{\mathscr{C}}$ passes through the North Pole.

In the other direction, suppose that z lies on a generalised circle $\mathscr{C}$ given by (2.1) for which $|B|^2 - AC > 0$, and that z is not the point at infinity. Guided by Eqs. (3.8), we set

$$a = \Re B, \quad b = \Im B, \quad c = \frac{1}{2}(A - C), \quad d = -\frac{1}{2}(A + C). \tag{3.9}$$

Then, $a^2 + b^2 + c^2 - d^2 = |B|^2 - AC$ so that (3.6) is satisfied. Moreover, $\mathscr{P}^{-1}(z) = (x_1, x_2, x_3)$, with x_1, x_2 and x_3 given by (3.3) in terms of $z = x + iy$, satisfies

$$ax_1 + bx_2 + cx_3 - d$$

$$= \left(\frac{2ax}{|z|^2 + 1}\right) + \left(\frac{2by}{|z|^2 + 1}\right) + \frac{1}{2}(A - C)\left(\frac{|z|^2 - 1}{|z|^2 + 1}\right) + \frac{1}{2}(A + C)$$

$$= \frac{1}{|z|^2 + 1}\left[2ax + 2by + \frac{1}{2}(A - C)(|z|^2 - 1) + \frac{1}{2}(A + C)(|z|^2 + 1)\right]$$

$$= \frac{1}{|z|^2 + 1}\left[a(z + \bar{z}) + b(i\bar{z} - iz) + A|z|^2 + C\right]$$

$$= \frac{1}{|z|^2 + 1}\left[A|z|^2 + (a - ib)z + (a + ib)\bar{z} + C\right]$$

$$= \frac{1}{|z|^2 + 1}\left[A|z|^2 + \overline{B}z + B\bar{z} + C\right] = 0.$$

Thus, $\mathscr{P}^{-1}(z)$ lies on the plane $ax_1 + bx_2 + cx_3 = d$, and on the sphere of course.

We have shown that a point on $\hat{\mathscr{C}}$ projects to a point on $\mathscr{C}$ under stereographic projection and, conversely, a point on the same generalised circle $\mathscr{C}$ has preimage in $\hat{\mathscr{C}}$ under stereographic projection. The parameters of the equation of the plane defining $\hat{\mathscr{C}}$ and of the equation defining $\mathscr{C}$ are related by (3.8) or, equivalently, by (3.9). That is, $\mathscr{P}(\hat{\mathscr{C}}) \subseteq \mathscr{C}$ and $\mathscr{P}^{-1}(\mathscr{C}) \subseteq \hat{\mathscr{C}}$, and so $\mathscr{P}(\hat{\mathscr{C}}) = \mathscr{C}$ and $\mathscr{P}^{-1}(\mathscr{C}) = \hat{\mathscr{C}}$.

In the case when $A = 0$ in (3.8) or, equivalently, $c = d$ in (3.9), the generalised circle $\mathscr{C}$ is a Euclidean straight line together with the point at infinity and $\hat{\mathscr{C}}$ is a circle on the sphere passing through the North Pole **N**. □

Exercise 3.4 Find the image under stereographic projection of the circle on the sphere described by the intersection of the plane $x_1 + x_2 + x_3 = 0$ with the sphere. Find the centre and radius of the image circle in the complex plane.

Exercise 3.5 Find the image under stereographic projection of the circle on the sphere described by the intersection of the plane $x_1 = t$ with the sphere, as t varies from -1 to 1. Sketch the image for a range of values of t. Is the image always a Euclidean circle? When is it not? For those t for which it is a Euclidean circle, find formulas for the centre and radius of the image circle in the complex plane.

Exercise 3.6 A *great circle* on the sphere is the intersection with the sphere of a plane $ax_1 + bx_2 + cx_3 = 0$ that passes through the centre of the sphere. Show that a generalised circle in the extended complex plane is the image of a great circle under stereographic projection if and only if it has the form (2.1) with $C = -A$.

Show that, in this case, the generalised circle is either a straight line through the origin or a circle in the plane having the form, for some complex number z_0,

$$|z - z_0| = \sqrt{1 + |z_0|^2}. \tag{3.10}$$

3.2 The Topology of the Extended Complex Plane

The topology on the sphere $\mathscr{S}$ is the *relative topology* it inherits as a subset of three dimensional Euclidean space $\mathbb{R}^3$: that is, E is an *open subset of the sphere* precisely when E can be described as the intersection of the sphere with a set that is open in $\mathbb{R}^3$. A point x in a set $E \subseteq \mathbb{R}^3$ is an *interior point* of E if there is a ball $B(x, r)$ of positive radius r with $B(x, r) \subseteq E$. A set E is open in $\mathbb{R}^3$ if all points in E are interior points of E. By considering only balls $B(x_n, r_n)$ whose centres come from a countable dense subset of $\mathbb{R}^3$ and whose radii are rational, any open set in $\mathbb{R}^3$ can be written as

$$A = \bigcup_{n \in I} B(x_n, r_n),$$

where I is some subset of $\mathbb{N}$. That is, the topology on $\mathbb{R}^3$ has a *countable open base*. It follows that $E \subset \mathscr{S}$ is an open subset of the sphere if $E = \mathscr{S} \cap A$ where A is open in $\mathbb{R}^3$, that is if

$$E = \mathscr{S} \cap \left(\bigcup_{n \in I} B(x_n, r_n) \right) = \bigcup_{n \in I} (\mathscr{S} \cap B(x_n, r_n)).$$

A *spherical cap* on the sphere is the intersection of the sphere with an open ball in $\mathbb{R}^3$. In summary, the relative topology on the sphere has a countable open base of spherical caps.

3.2 The Topology of the Extended Complex Plane

By means of stereographic projection, via its inverse $\mathscr{P}^{-1} : \mathbb{C}_\infty \to \mathscr{S}$ in fact, this topology can be transferred from the sphere $\mathscr{S}$ to the extended complex plane $\mathbb{C}_\infty$. Following an established method, we define Ω to be open in $\mathbb{C}_\infty$ if and only if $\Omega = \left(\mathscr{P}^{-1}\right)^{-1}(E)$ for some open set E in $\mathscr{S}$. This is called either the *induced topology* or the *initial topology* on $\mathbb{C}_\infty$. Since $\mathscr{P}^{-1}$ is a bijection, Ω is open in $\mathbb{C}_\infty$ if and only if $\Omega = \mathscr{P}(E)$ for some open set in $\mathscr{S}$, that is if Ω is the union of images under $\mathscr{P}$ of spherical caps coming from some fixed countable collection of spherical caps.

The boundary of a spherical cap is a circle on the sphere since the intersection of two spheres in $\mathbb{R}^3$ is a circle. By Proposition 3.3, the image of this circle under stereographic projection is a generalised circle in the extended complex plane. If this circle passes through the North Pole, then its image under stereographic projection is a straight line in the plane together with the point at infinity. The image of the corresponding spherical cap will be one or other of the two half-planes bounded by this straight line. If the boundary of the spherical cap is a circle that doesn't pass through the North Pole, then the image of the boundary circle is a circle in the plane, and the image of the spherical cap is either the open disk in the plane bounded by this circle, or the exterior of the circle including the point at infinity. These open sets in the extended complex plane form a countable open base for the topology on $\mathbb{C}_\infty$ and are pictured in Fig. 3.4. For the record, every open set in the extended complex plane is a union of a combination of some or all of the following: open disks in the plane, open half-planes not including the point at infinity, and complements of closed disks

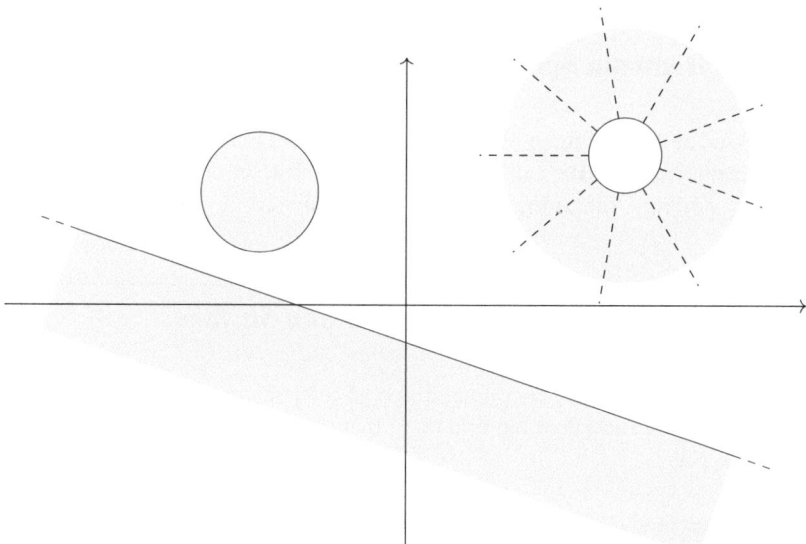

Fig. 3.4 Basic open sets in the topology of the extended complex plane: an open disk; the exterior of a closed disk including the point at infinity; a half-plane

in the plane including the point at infinity. Since a half-plane is a countable union of disks, we could drop half-planes from our list if we wished.

Both stereographic projection $\mathscr{P}\colon \mathscr{S} \to \mathbb{C}_\infty$ and its inverse are continuous since both map open sets to open sets. In other words, $\mathscr{P}$ is a topological *homeomorphism* between $\mathscr{S}$ and $\mathbb{C}_\infty$. The sphere $\mathscr{S}$ in $\mathbb{R}^3$ is compact.[3] Since the extended complex plane is the homeomorphic image of a compact topological space, it too is compact. By adding a point, the point at infinity, to the complex plane, the resulting space is compact. As mentioned previously, this is called the *Alexandroff one-point compactification* of the complex plane. Let's summarise the above discussion in the following theorem.

Theorem 3.1 *The extended complex plane $\mathbb{C}_\infty$ is equipped with the topology it inherits from that on the sphere via stereographic projection. The restriction of this topology to the complex plane $\mathbb{C}$ coincides with the Euclidean topology. Stereographic projection $\mathscr{P}\colon \mathscr{S} \to \mathbb{C}_\infty$ is a homeomorphism between the Riemann sphere $\mathscr{S}$ and the extended complex plane $\mathbb{C}_\infty$. In particular, the extended complex plane is compact.*

Exercise 3.7 Describe the action of $f(z) = 1/z$, for $z \in \mathbb{C}_\infty$, as a map on the Riemann sphere: that is, describe the map $g\colon \mathscr{S} \to \mathscr{S}$ where $g(x) = (\mathscr{P}^{-1} \circ f \circ \mathscr{P})(x)$, $x \in \mathscr{S}$. (Hint: use spherical coordinates.)

Show that the inversion $f\colon \mathbb{C}_\infty \to \mathbb{C}_\infty$ given by $f(z) = 1/z$ is continuous. (It is enough to check that the image under inversion of each basic open set in the topology of $\mathbb{C}_\infty$ is again open.)

3.3 The Riemann Sphere

As discussed at the beginning of this chapter, the Riemann Sphere is a key example of a Riemann Surface. We can get a feel for the abstract definition of a *Riemann surface* by studying this particular example in detail.

3.3.1 The Riemann Sphere as a Riemann Surface

Without entering into the general definition of a Riemann surface, we first need a collection of open subsets of the sphere $\mathscr{S}$ that cover the sphere—in this case, two will suffice. Set

$$E_0 = \{z \in \mathbb{C} \colon |z| < \sqrt{3}\} \text{ and } E_\infty = \mathbb{C}_\infty \setminus \{z \in \mathbb{C} \colon |z| \le 1/\sqrt{3}\}.$$

---

[3] This is a consequence of the Bolzano-Weierstrass Theorem: the compact subsets of n-dimensional Euclidean space $\mathbb{R}^n$ are those that are closed and bounded.

3.3 The Riemann Sphere

Take $U_0 = \mathscr{P}^{-1}(E_0)$, which is the lower hemisphere together with some of the upper hemisphere; by (3.2) and since $\arctan\sqrt{3} = \pi/3$, $U_0 = \{\hat{z}: \phi < \pi/6\}$ in spherical coordinates. Take $U_\infty = \mathscr{P}^{-1}(E_\infty)$ which is the upper hemisphere together with some of the lower hemisphere; in fact, $U_\infty = \{\hat{z}: \phi > -\pi/6\}$. Together U_0 and U_∞ cover the sphere and their overlap is $\{\hat{z}: -\pi/6 < \phi < \pi/6\}$, a band about the equator.

The pairs (U_0, ψ_0) and (U_∞, ψ_∞), where $\psi_0 = \mathscr{P}|_{U_0}$ and $\psi_\infty = 1/\mathscr{P}|_{U_\infty}$ are *charts*, homeomorphisms from open sets on the sphere onto open sets in the complex plane, in each case onto the disk $D(0, \sqrt{3})$ (stereographic projection and its inverse are continuous and, by Exercise 3.7, so is inversion). Both charts send $U_0 \cap U_\infty$ onto the annulus $\{z: 1/\sqrt{3} < |z| < \sqrt{3}\}$ in the complex plane. On this annulus, $\psi_\infty \circ \psi_0^{-1}(z) = \psi_0 \circ \psi_\infty^{-1}(z) = 1/z$, which is analytic on the annulus. Thus, the two charts work well together from the point of view of complex analysis. Formally, then, the Riemann sphere is the sphere $\mathscr{S}$ together with these two charts.

3.3.2 Analytic Functions on the Riemann Sphere

Now we extend the notion of an analytic function from the familiar setting of the complex plane to the Riemann sphere. The idea is to pass everything through the charts.

Definition 3.1 Let $\hat{\Omega}$ be an open subset of the Riemann sphere $\mathscr{S}$ and let $\hat{z}$ be in $\hat{\Omega}$. Let $f: \hat{\Omega} \to \mathbb{C}$ be a complex-valued function on $\hat{\Omega}$. We say that f *is analytic at* $\hat{z}$ if there is a chart (U, ψ) with $\hat{z}$ in U such that $f \circ \psi^{-1}$ is analytic at $\psi(\hat{z})$. We say that f *is analytic on* $\hat{\Omega}$ if f is analytic at each point of $\hat{\Omega}$.

In this definition, $f \circ \psi^{-1}$ is a complex-valued function on the open set $\psi(U \cap \hat{\Omega})$ in the complex plane which contains $\psi(\hat{z})$. When we speak of this function being 'analytic' we mean analytic in the sense of a first course in complex analysis. A further point to note, common to definitions of this type, is that it shouldn't matter which chart we choose when implementing the definition. In our case, suppose that $\hat{z}$ belongs to $U_0 \cap U_\infty \cap \hat{\Omega}$. Then, on the image of this open set under ψ_0,

$$f \circ \psi_0^{-1} = \left(f \circ \psi_\infty^{-1}\right) \circ \left(\psi_\infty \circ \psi_0^{-1}\right).$$

If $f \circ \psi_\infty^{-1}$ is analytic at $\psi_\infty(\hat{z})$ then, being a composition of analytic functions, $f \circ \psi_0^{-1}$ will be analytic at $\psi_0(\hat{z})$. Reversing the roles of ψ_0 and ψ_∞, we see that the converse holds. In short, $f \circ \psi^{-1}$ is analytic at $\psi(\hat{z})$ for every chart (U, ψ) with $\hat{z}$ in U, or for none. Our definition is therefore independent of the chart chosen.

If f is analytic on an open subset $\hat{\Omega}$ of the sphere it will be continuous there: in terms of a chart (U, ψ), we can write f locally as $(f \circ \psi^{-1}) \circ \psi$ which is a composition of the analytic, hence continuous, function $f \circ \psi^{-1}$ and the continuous function ψ.

The special case when $\hat{z}$ is ∞ (the North Pole) is of special interest, in which case the relevant chart is (U_∞, ψ_∞). Consider a function f analytic in some neighbourhood of ∞ on the sphere, which we may take to be a spherical cap $\hat{\Omega}$ centred at the North Pole. This spherical cap projects under $\mathscr{P}$ to a neighbourhood of ∞ in $\mathbb{C}_\infty$ of the form

$$\Omega = \{z \in \mathbb{C} : |z| > R\} \cup \{\infty\}.$$

(If $\hat{\Omega} = \{\phi > \phi_0\}$ on the sphere then, by (3.1), $R = \tan(\phi_0/2 + \pi/4)$.) Now, $\psi_\infty = 1/\mathscr{P}$, and so ψ_∞ maps $\hat{\Omega}$ onto the disk $D(0, 1/R)$. Moreover, $\psi_\infty(\infty) = 1/\mathscr{P}(\infty) = 1/\infty = 0$. According to Definition 3.1, the function f is analytic at ∞ if $f \circ \psi_\infty^{-1}$ is analytic at $\psi_\infty(\infty)$, that is at 0. In that case, $f \circ \psi_\infty^{-1}$ has a power series expansion about 0, say

$$(f \circ \psi_\infty^{-1})(z) = \sum_{n=0}^\infty c_n z^n, \quad |z| < 1/R.$$

Now $\psi_\infty^{-1}(z) = \mathscr{P}^{-1}(1/z)$, $|z| < 1/R$. Setting $w = 1/z$, so that $|w| > R$,

$$(f \circ \mathscr{P}^{-1})(w) = \sum_{n=0}^\infty \frac{c_n}{w^n}, \quad |w| > R.$$

Blurring the distinction between ∞ as the North Pole on the Riemann Sphere and ∞ as the point at infinity in the extended complex plane, the following definition is really a special case of Definition 3.1.

Definition 3.2 We say that a function f defined in a neighbourhood Ω of ∞ in the extended complex plane $\mathbb{C}_\infty$ is *analytic at infinity* if it has a power series expansion of the form

$$f(w) = \begin{cases} c_0, & w = \infty \\ \sum_{n=0}^\infty c_n/w^n, & |w| > R, \end{cases} \quad (3.11)$$

the series assumed to be convergent whenever $|w| > R$, for some positive R. We say that f has a *zero of order N at infinity*, $N \geq 1$, if $c_0 = c_1 = \ldots c_{N-1} = 0$.

Note that $f(w)$ is analytic at infinity if and only if $f(1/w)$ is analytic at 0. There is an abundance of functions that are analytic in any given region on the Riemann Sphere. It may come as a surprise, then, that there are no functions analytic everywhere on the Riemann Sphere except the constant functions. The determining factor here is that the Riemann Sphere is a *compact* Riemann surface.

Theorem 3.2 *Every function analytic on the Riemann sphere is constant.*

3.3 The Riemann Sphere

Proof Suppose that f is analytic on the Riemann sphere. Then $|f|$ is continuous on the sphere and, since the sphere is compact, $|f|$ achieves its maximum value on the sphere and f is therefore bounded.

Set $g = f \circ \mathscr{P}^{-1}$ on the complex plane $\mathbb{C}$. On the disk $D(0, \sqrt{3})$, g agrees with $f \circ \psi_0^{-1}$ and so is analytic there. When $|z| > 1/\sqrt{3}$, g agrees with $(f \circ \psi_\infty^{-1})(1/z)$ which is analytic there since $f \circ \psi_\infty^{-1}$ is analytic on $D(0, \sqrt{3})$. This demonstrates that g is an entire function on the complex plane. Since g is bounded, it must be constant by Liouville's Theorem, hence f is constant. □

In fact, the only analytic functions on *any* compact Riemann surface are the constant functions.

3.3.3 Meromorphic Functions on the Riemann Sphere

Complex analysis on the Riemann Sphere would be a triviality if we were to restrict ourselves to complex-valued functions analytic everywhere on the sphere—this would reduce to a study of the constant functions! It is natural, then, to consider functions taking values in the extended complex plane, and allow singularities—in this we refer back to Sect. 2.4 on isolated singularities. A function on the sphere, analytic except for removable singularities, is of no help as when the singularities are removed we once again have a function analytic on the whole sphere which will then necessarily be constant. Since we prefer to work with continuous functions, essential singularities are also ruled out since a function with an essential singularity will assume a dense set of values in any neighbourhood of the singularity and therefore won't be continuous, not even as an extended complex-valued function, that is as a function taking values in the extended complex plane. Poles, on the other hand, are ideally suited to this situation and we extend Definition 3.1 to allow for this.

Definition 3.3 Let $\hat{\Omega}$ be an open subset of the Riemann sphere $\mathscr{S}$ and let $\hat{z}$ be in $\hat{\Omega}$. Let f be an extended complex-valued function defined in $\hat{\Omega}$. We say that f *has a pole of order N at $\hat{z}$* if there is a chart (U, ψ) with $\hat{z}$ in U such that $f \circ \psi^{-1}$ has a pole of order N at $\psi(\hat{z})$. We say that f *is meromorphic on Ω* if f is analytic at each point of Ω except for poles.

As was the case for Definition 3.1, this definition is independent of the choice of chart at $\hat{z}$.

If f is meromorphic on $\hat{\Omega}$ then it is continuous on $\hat{\Omega}$ as an extended complex-valued function. We verified this after Definition 3.1 at points on the sphere where f is analytic. Now suppose that f has a pole at $\hat{z}$ and that (U, ψ) is a chart covering $\hat{z}$. Write $f = (f \circ \psi^{-1}) \circ \psi$ in a neighbourhood of $\hat{z}$, and note that ψ is continuous near $\hat{z}$. Since $g = f \circ \psi^{-1}$ has a pole at $\psi(\hat{z})$ it follows that $|g(z)| \to \infty$ as $z \to \psi(\hat{z})$. In other words, g is continuous at $\psi(\hat{z})$ as an extended complex-valued function with $g(\psi(\hat{z})) = \infty$. If in Definition 3.3 we set f to take the value ∞ at the points in $\hat{\Omega}$ where it has a pole, then $f : \hat{\Omega} \to \mathbb{C}_\infty$ becomes continuous.

By Definition 3.3, a function f has a pole of order N at ∞ if $f \circ \psi_\infty^{-1}$ has a pole of order N at 0. In that case, $f \circ \psi_\infty^{-1}$ has a Laurent series expansion about 0 of the form

$$(f \circ \psi_\infty^{-1})(z) = \sum_{n=-N}^{\infty} c_n z^n, \quad |z| < 1/R,$$

with $c_{-N} \neq 0$, this expansion being valid for some finite R. Again setting $w = 1/z$ and using $\psi_\infty^{-1}(z) = \mathscr{P}^{-1}(1/z)$, we see that

$$(f \circ \mathscr{P}^{-1})(w) = \sum_{n=-N}^{\infty} \frac{c_n}{w^n}, \quad |w| > R.$$

Now we extend Definition 3.2 to encompass poles.

Definition 3.4 We say that a function f defined in a neighbourhood Ω of ∞ in the extended complex plane $\mathbb{C}_\infty$ has a *pole of order N at infinity*, $N \geq 1$, if it has a power series expansion of the form

$$f(w) = \begin{cases} \infty, & w = \infty \\ \sum_{n=-N}^{\infty} c_n/w^n, & |w| > R, \end{cases} \quad (3.12)$$

the series being assumed convergent whenever $|w| > R$, for some positive R.

Note that $f(w)$ has a pole of order N at infinity if and only if $f(1/w)$ has a pole of order N at 0 (in which case, $g(w) = 1/f(1/w)$ has a zero of order N at 0). Note also that it follows from (3.11) and (3.12) that if f is either analytic at infinity or has a pole at infinity, then there is a finite R for which f is analytic in $\{w \in \mathbb{C} : |w| > R\}$. If, in (3.12), the expansion of f about the point at infinity is $f(w) = \sum_{n=-\infty}^{\infty} c_n/w^n$ where infinitely many c_n are non-zero for negative n, then we say that f has an *essential singularity at infinity*.

From a geometric point of view, it is most convenient to think of the Riemann Sphere as the physical sphere $\mathscr{S}$ together with its two charts. This viewpoint emphasises, for example, the fact that the point at infinity (the North Pole) on the Riemann Sphere is no more or no less special than any other point on the sphere. On the other hand, for the purpose of actually doing complex analysis on the Riemann Sphere, it is best to work directly on the extended complex plane $\mathbb{C}_\infty$. Stereographic projection renders these viewpoints equivalent. As an example of what we mean, consider a meromorphic function f on the sphere. Set $g = f \circ \mathscr{P}^{-1}$ on the extended complex plane. Working with the chart (U_0, ψ_0) for which $\psi_0 = \mathscr{P}$, we see that g is meromorphic on the disk $D(0, \sqrt{3})$ on the complex plane. Working with the chart (U_∞, ψ_∞) for which $\psi_\infty = 1/\mathscr{P}$, we see that g is meromorphic on the complement of the closed disk $\{z : |z| \leq 1/\sqrt{3}\}$ in the complex plane, and at the point at infinity in the sense of Definitions 3.2 and 3.4. That is, to the meromorphic function f on

3.3 The Riemann Sphere

the sphere we can associate the function g on the extended complex plane that is meromorphic on the complex plane $\mathbb{C}$ and at the point at infinity. We will say that g is meromorphic on the extended complex plane. Reversing the argument, it is straightforward to construct f meromorphic on the sphere from g meromorphic on the extended complex plane by $f = g \circ \mathscr{P}$. We will gloss over without comment from now on this somewhat subtle, and possibly pedantic, distinction between the sphere and the extended complex plane.

The next example shows a key difference between functions meromorphic on $\mathbb{C}$ and functions meromorphic on $\mathbb{C}_\infty$.

Example 3.1 The function $f(z) = 1/\sin z$ has a simple pole at 0. In fact,

$$\sin z = z - \frac{z^3}{3!} + \frac{z^5}{5!} - \frac{z^7}{7!} + \cdots = z\, g(z)$$

where the function

$$g(z) = 1 - \frac{z^2}{3!} + \frac{z^4}{5!} - \frac{z^6}{7!} + \cdots$$

is analytic at 0 with $g(0) \neq 0$. Thus, $1/g$ is analytic at 0, has a power series expansion about 0, and so

$$f(z) = \frac{1}{z}\left[\frac{1}{g(z)}\right] = \frac{1}{z}\left[1 + c_1 z + c_2 z^2 + c_3 z^3 + \cdots\right]$$

has a simple pole at 0. In fact, $f(z)$ has a simple pole at every zero of $\sin z$, that is at the points $z_n = n\pi$, $n \in \mathbb{Z}$, and is otherwise analytic on $\mathbb{C}$. While $f(z) = 1/\sin z$ is therefore a meromorphic function on the complex plane $\mathbb{C}$, it is not meromorphic on $\mathbb{C}_\infty$. In fact, since f has poles of arbitrarily large modulus, there is no finite R such that f is analytic everywhere in $\{z \in \mathbb{C} : |z| > R\}$.

Here are some exercises to firm up these ideas.

Exercise 3.8 Suppose that f has a pole at of order N at $a \in \mathbb{C}$. Show that there is a positive ε and a finite r such that if $|w| > r$ then there are N distinct points z_i, $i = 1, 2, \ldots, N$, in the punctured disk $D'(a, \varepsilon)$ such that $f(z_i) = w$ for each i. That is, each w of sufficiently large modulus has N distinct preimages near a under f.

Hint: choose ε so that f is defined and non-zero in $D'(a, \varepsilon)$. Then consider the analytic function $1/f(z)$, which has a zero of order N at a, and apply the result [16, Theorem 10.6].

Formulate the corresponding result in the case of a pole of order N at the point at infinity.

Exercise 3.9

▷ The function $f(z) = z$ is meromorphic on $\mathbb{C}_\infty$ with a simple zero at 0 and a simple pole at ∞.

▷ The function $f(z) = 1/z$ is meromorphic on $\mathbb{C}_\infty$ with a simple pole at 0 and a simple zero at ∞.
▷ The function $f(z) = z + 1/z$ is meromorphic on $\mathbb{C}_\infty$ with simple poles at 0 and at ∞.
▷ The function $f(z) = e^z$ is entire and has an essential singularity at ∞.
▷ Consider a polynomial $P(z) = a_0 + a_1 z + \cdots + a_n z^n$, $z \in \mathbb{C}$, with $a_n \neq 0$. Then P has a pole at ∞ of order $n = \deg(P)$.
▷ A function f has a pole of order n at $a \in \mathbb{C}_\infty$ if and only if the function $1/f$ has a zero of order n at a.
▷ If f and g both have a pole at $a \in \mathbb{C}_\infty$ then the functions $f + g$ and fg have either a pole or a removable singularity at a.

We saw in Theorem 3.2 that the only analytic functions on the Riemann Sphere are the constant functions. It will soon become apparent that the class of functions meromorphic on the sphere is also relatively limited, at least in the sense that it can be easily described. This is in stark contrast to the situation for functions meromorphic on the plane $\mathbb{C}$ as exemplified by W.K. Hayman's classic monograph *Meromorphic Functions* [14]. As a first step in describing all meromorphic functions on the sphere we show that any such meromorphic function has only finitely many poles.

Lemma 3.1 *Let $f : \mathbb{C}_\infty \to \mathbb{C}_\infty$ be meromorphic. Then f has only finitely many poles.*

Proof Since f is either analytic at infinity or has a pole at ∞, f is analytic in $\{z \in \mathbb{C} : |z| > R\}$ for some $R > 0$. If f had *infinitely* many poles in the finite complex plane $\mathbb{C}$, at the distinct points $z_1, z_2, \ldots, z_n, \ldots$ say, then all these would necessarily lie in the closed disk $\overline{D(0, R)}$. By the Heine-Borel Theorem, one version of which states that every bounded sequence of complex numbers has a convergent subsequence, there is some a with $|a| \leq R$ such that $z_{n_k} \to a$ as $k \to \infty$ for some subsequence $\{z_{n_k}\}_k$. But then a is not an isolated singularity of f as the function f is not analytic in some punctured disk about a. This contradiction (of the assumption that f is meromorphic on $\mathbb{C}_\infty$ and therefore also on $\mathbb{C}$) shows that f can have only finitely many poles in $\mathbb{C}$ and only that number plus one in $\mathbb{C}_\infty$. □

With the help of Lemma 3.1, we can now classify *all* meromorphic functions on the Riemann Sphere and show that these are the rational functions that is, quotients of polynomials. This result, and its many extensions and generalisations, has far-reaching consequences. It goes some way to explaining the origins of the field of algebraic geometry, in which the basic objects of study are the polynomials. It is clear that any rational function is meromorphic on $\mathbb{C}_\infty$. The point of Theorem 3.3 is that these are the *only* meromorphic functions on the sphere. The nature of the singularity of a rational function at ∞ is treated in Exercise 3.10.

3.3 The Riemann Sphere

Theorem 3.3 *Let $f: \mathbb{C}_\infty \to \mathbb{C}_\infty$ be meromorphic. Then there are polynomials P and Q (with no common factors) such that*

$$f(z) = \frac{P(z)}{Q(z)}, \quad z \in \mathbb{C}_\infty.$$

Proof By Lemma 3.1, f has only finitely many poles in the finite complex plane $\mathbb{C}$, say at α_i, $i = 1, 2, \ldots, n$, with orders $k_1, k_2, \ldots, k_n$ respectively. Set

$$Q(z) = \prod_{i=1}^{n}(z - \alpha_i)^{k_i}.$$

Then,

$$g(z) = Q(z) f(z)$$

is analytic in $\mathbb{C}$ (once the removable singularities at the points α_i are removed). This can be seen by expanding f as a Laurent series about α_i: the pole of f of order k_i at α_i is then cancelled out by the zero of Q of order k_i at α_i. In short, then, g is an entire function.

Accordingly, g has a power series expansion about 0, say

$$g(z) = \sum_{n=0}^{\infty} a_n z^n, \quad z \in \mathbb{C}.$$

Since $g(z)$ is either analytic at ∞ or has a pole at ∞, $g(1/z)$ is either analytic or has a pole at 0: see comments directly after Definitions 3.2 and 3.4. As a consequence, only finitely many a_n can be non-zero, and g is a polynomial. □

Exercise 3.10

▷ Let $f(z) = P(z)/Q(z)$ where P and Q are polynomials with no common factors. Classify the singularity of f at ∞ in terms of the degrees of the polynomials P and Q.

▷ Show that, counting multiplicities, the number of zeroes of f in $\mathbb{C}_\infty$ equals the number of poles of f in $\mathbb{C}_\infty$.

Exercise 3.11 Let $f: \mathbb{C}_\infty \to \mathbb{C}_\infty$ be analytic everywhere. Rework the proof of Theorem 3.3 to show (or deduce from Theorem 3.3) that f must be constant.

Exercise 3.12 Let $f: \mathbb{C}_\infty \to \mathbb{C}_\infty$ be meromorphic. Deduce from Theorem 3.3 that there is some M such that f takes every value $a \in \mathbb{C}_\infty$ at most M times.

Show that, in fact, f takes every value in $\mathbb{C}_\infty$ exactly M times counting multiplicity where, if $f = P/Q$ for polynomials P and Q with no common factors, M is the maximum of the degrees of P and Q.

Deduce that the only meromorphic bijections of the Riemann Sphere are the Möbius transformations.

3.4 The Spherical Metric

The Riemann Sphere has additional structure, over and above its structure as a Riemann surface with its atlas of charts, that comes from its embedding in $\mathbb{R}^3$ as a Euclidean sphere. Just as we measure the distance between cities on the surface of the earth, we can measure the distance between points on the surface of the sphere. This distance function or metric is transferred by stereographic projection to a metric on the extended complex plane called the *spherical metric*. It is well worth working through the details of the construction of the spherical metric as it is an explicit example of a *Riemannian metric*, a fundamental idea in differential geometry. The idea is a simple one, and essentially this: we first define the length of a general curve in the space and then set the distance between two points to be the length of the shortest curve in the space that joins them. This is familiar to us from Euclidean geometry in which the distance between two points is the length of the straight line joining them. What, then, are the 'straight lines' on the sphere?

Since the construction of the spherical metric on the extended complex plane and that of the hyperbolic metric on the disk (which we will meet in the next chapter) have so much in common, we take a general approach in the next subsection.

3.4.1 General Construction of a Riemannian Metric on a Planar Domain

We first describe how the length of a curve is defined in the setting of Riemannian geometry. We begin with a positive, continuous function ρ on a domain Ω in the complex plane. The function ρ should be thought of as a *scale factor* in that the 'ρ-length' of a vector $\mathbf{v}$ at the point z is set to be $\rho(z)|\mathbf{v}|$; this being the value of ρ at the base point z times the Euclidean length of $\mathbf{v}$.

The (Euclidean) length $L(\gamma)$ of a smooth curve γ in the complex plane, parameterised by $\gamma(t), t \in [a, b]$, is given by the integral

$$L(\gamma) = \int_a^b |\gamma'(t)| \, dt,$$

(see [16, Theorem 5.9]). Recall that γ is *smooth* if $\gamma'(t)$ is continuous and non-zero on $[a, b]$.

Let's revisit where this formula comes from, while also including the scale factor ρ to obtain a formula for the 'ρ-length' of the curve γ. We partition the parameter

3.4 The Spherical Metric

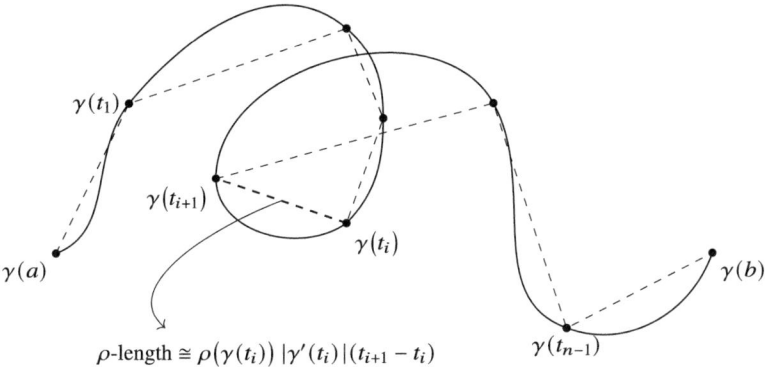

Fig. 3.5 Illustration of (3.13)

interval $[a, b]$ by points $a = t_0 < t_1 < \ldots < t_{n-1} < t_n = b$. The polygonal path through the points $\gamma(t_i)$, $i = 0 \ldots n$, approximates the curve γ (see Fig. 3.5). The segment $[\gamma(t_i), \gamma(t_{i+1})]$ has Euclidean length $|\gamma(t_{i+1}) - \gamma(t_i)|$, and this can be approximated by $|\gamma'(t_i)|(t_{i+1} - t_i)$.[4] The ρ-length of this segment of the polygonal curve is then approximately $\rho(\gamma(t_i)) |\gamma'(t_i)|(t_{i+1} - t_i)$.

The ρ-length $L_\rho(\gamma)$ of the curve γ is then approximated by the ρ-length of the approximating polygonal curve, so that

$$L_\rho(\gamma) \cong \sum_{i=0}^{n-1} \rho(\gamma(t_i)) |\gamma'(t_i)|(t_{i+1} - t_i). \qquad (3.13)$$

We recognise the right hand side of (3.13) as a Riemann sum for the integral $\int_a^b \rho(\gamma(t))|\gamma'(t)| \, dt$. Consequently, as the partition of $[a, b]$ becomes finer,[5]

$$\sum_{i=0}^{n-1} \rho(\gamma(t_i)) |\gamma'(t_i)|(t_{i+1} - t_i) \longrightarrow \int_a^b \rho(\gamma(t))|\gamma'(t)| \, dt.$$

If γ is only *piecewise smooth*, (in that $\gamma'(t)$ exists, is non-zero and is continuous for each $t \in (a, b)$, with at most finitely many exceptions) then we apply the above argument to each of the parameter intervals on which γ is smooth. These considerations motivate the following definition.

---

[4] $|\gamma'(t)|$ is the *speed* of the parameterisation at the point $\gamma(t)$ on the curve.

[5] The sum on the left is a Riemann sum for the Riemann integral of the continuous function $\rho(\gamma(t))|\gamma'(t)|$ for t in the interval $[a, b]$. The general theory of the Riemann integral tells us that Riemann sums for a continuous function *always* tend to the value of the Riemann integral of the function provided that the mesh size of the partition tends to zero.

Definition 3.5 (ρ-Length of a Curve) Let γ be a piecewise-smooth curve parameterised by $\gamma(t), t \in [a, b]$. The ρ-length of γ, denoted by $L_\rho(\gamma)$, is set to be

$$L_\rho(\gamma) = \int_a^b \rho(\gamma(t)) |\gamma'(t)| \, dt. \tag{3.14}$$

The Euclidean length of the curve is the special case $\rho \equiv 1$. It is only proper to mention that, in the general Riemannian setting, the scale factor ρ for a vector $\mathbf{v}$ at a point z may depend not only on the base position z but also on the orientation of the vector $\mathbf{v}$. In this book, ρ will depend only on position in which case it is known as a *conformal metric*. The conformal metric ρ tells us how space is scaled at a local level.

The right hand side of (3.14) is often abbreviated as $\int_\gamma \rho(w) |dw|$, which is justified by the following exercise.

Exercise 3.13 Show that the ρ-length of a curve, as defined by (3.14), does not depend on how the curve is parameterised.

Hint: The verification that the length is independent of the parameterisation is only a minor modification of the same fact for Euclidean lengths of curves. That $\tilde{\gamma}(s), s \in [c, d]$, is an equivalent parameterisation of the curve γ means that there is a function $\phi \colon [c, d] \to [a, b]$ with $\phi(c) = a$, $\phi(d) = b$, and with $\phi'(s) > 0$, $s \in [c, d]$ (in the greatest generality, ϕ continuous and increasing on $[c, d]$ is sufficient) such that

$$\tilde{\gamma}(s) = \gamma(\phi(s)), \quad s \in [c, d].$$

Exercise 3.14 Let γ be a smooth curve parameterised by $\gamma(t), t \in [a, b]$. Let $-\gamma$ be the curve γ described in the reverse direction (that is from $\gamma(b)$ to $\gamma(a)$). Write down a parameterization of $-\gamma$. Verify that $L_\rho(-\gamma) = L_\rho(\gamma)$.

Exercise 3.15 Let γ be a curve parameterised by $\gamma(t), t \in [a, b]$, and with Euclidean length $L(\gamma)$. Set

$$m = \min \{\rho(\gamma(t)) \colon t \in [a, b]\} \text{ and } M = \max \{\rho(\gamma(t)) \colon t \in [a, b]\},$$

so that m and M are, respectively, the minimum and maximum values of ρ on the curve γ. Show that

$$m L(\gamma) \leq L_\rho(\gamma) \leq M L(\gamma).$$

Now that we can compute the ρ-length of curves in the domain Ω, we would like to define the distance between two points in Ω to be the ρ-length of the shortest curve in the domain that joins them. The problem with this is that there may be no such shortest curve! Take, for example, the L-shaped open region shown in Fig. 3.6, the ordinary Euclidean length of curves (that is, $\rho \equiv 1$), and take two points, one in each arm of the L. There is no shortest curve joining these two points, rather

3.4 The Spherical Metric

Fig. 3.6 There is no shortest curve in this L-shaped domain joining the points x and y. The dashed curve has length equal to the infimum of the lengths of all curves in the domain joining x and y but leaves the domain at the corner

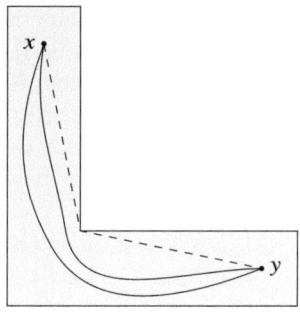

curves whose length is arbitrarily close to that of the dashed line. But this dashed line leaves the domain at the corner and isn't a curve *in the domain* that joins them. Formally, then, we need to define the distance between two points as the infimum of the lengths of all curves in the space that joins them and then prove, if we can, that this infimum is attained so that a shortest curve then exists.

Definition 3.6 (Distance Between Points) For any two points z and w in the domain Ω we set

$$d(z, w; \Omega) = \inf \{L_\rho(\gamma) : \gamma \text{ is a curve in } \Omega \text{ joining } z \text{ to } w\}. \qquad (3.15)$$

This is a perfectly natural and intuitive definition: the only difference from the Euclidean setting is how we measure the length of a curve.

Remark 3.1 It is well worth noting that the ρ-length of *any* curve γ joining z to w gives an upper estimate for $d(z, w; \Omega)$ in that $d(z, w; \Omega) \leq L_\rho(\gamma)$.

We show straightaway that (3.15) defines a genuine distance function on Ω.

Proposition 3.4 *For z and w in Ω, let $d(z, w; \Omega)$ be defined by (3.15). Then $d(z, w; \Omega)$ is a metric on Ω.*

Proof We need to verify the defining properties of a metric.[6]

(i) First, $d(z, w; \Omega) \geq 0$ for any z and w in Ω since $d(z, w; \Omega)$ is an infimum of ρ-lengths of curves and the ρ-length of any curve is non-negative.
It is clear that $d(z, z; \Omega) = 0$ whenever $z \in \Omega$ since z can be joined to itself by a curve whose ρ-length is zero (take, for example, $\gamma(t) \equiv z, 0 \leq t \leq 1$).

---

[6] A metric on a set X is a function $d : X \times X \to [0, \infty)$ with the following properties:

(i) $d(x, y) = 0$ if, and only if, $x = y$;
(ii) $d(x, y) = d(y, x)$ whenever $x \in X$ and $y \in X$;
(iii) $d(x, y) \leq d(x, z) + d(z, y)$ whenever $x \in X$, $y \in X$ and $z \in X$, (the triangle inequality).

We also need to show that $d(z, w; \Omega) = 0$ only if $z = w$. Equivalently, we need to show that $d(z, w; \Omega) > 0$ if $w \neq z$. Since $\rho(z)$ is positive by assumption and ρ is a continuous function, there is a positive δ_1 such that the disk $D(z, \delta_1)$ is contained in Ω and $\rho(\zeta) \geq \rho(z)/2$ for all ζ in $D(z, \delta_1)$. Since $w \neq z$ there is a positive δ_2 such that $w \notin D(z, \delta_2)$. Now set $\delta = \min\{\delta_1, \delta_2\}$. Any curve γ that joins z to w must leave $D(z, \delta)$ and so contains a subarc $\tilde{\gamma}$ that joins z to the circle $C(z, \delta)$ and also lies inside the disk $D(z, \delta)$. Since ρ is bounded below by $\rho(z)/2$ on $D(z, \delta)$ and since $L(\tilde{\gamma}) \geq \delta$, it follows from Exercise 3.15 that $L_\rho(\gamma) \geq L_\rho(\tilde{\gamma}) \geq \rho(z)\delta/2$. Since this inequality holds for any curve γ that joins z to w, it follows that $d(z, w; \Omega) \geq \rho(z)\delta/2$ which is positive.

(ii) Next we show that, for any z and w, the equality $d(z, w; \Omega) = d(w, z; \Omega)$ holds. Let γ be any curve joining z to w in Ω. Then $-\gamma$ joins w to z and has, by Exercise 3.14, the same ρ-length. Thus, bearing in mind Remark 3.1,

$$d(w, z; \Omega) \leq L_\rho(-\gamma) = L_\rho(\gamma).$$

Now take the infimum in the inequality $d(w, z; \Omega) \leq L_\rho(\gamma)$ over all curves γ joining z and w in Ω to obtain $d(w, z; \Omega) \leq d(z, w; \Omega)$.

Since the reverse inequality holds by interchanging the roles of z and w, we see that $d(z, w; \Omega) = d(w, z; \Omega)$.

(iii) The last property we need to verify in order to conclude that $d(\cdot, \cdot; \Omega)$ is indeed a metric on Ω is that

$$d(z, w; \Omega) \leq d(z, v; \Omega) + d(v, w; \Omega), \quad \text{for any } z, v, w \in \Omega. \tag{3.16}$$

This is the triangle inequality. To verify this property, let γ_1 be any curve joining z to v in Ω and let γ_2 be any curve joining v to w in Ω. The curve γ formed by joining γ_2 to γ_1 is a curve joining z to w in Ω. Hence, again by Remark 3.1,

$$d(z, w; \Omega) \leq L_\rho(\gamma) = L_\rho(\gamma_1) + L_\rho(\gamma_2).$$

Now fix γ_2 and take the infimum over all curves γ_1 to obtain

$$d(z, w; \Omega) \leq \left(\inf_{\gamma_1} L_\rho(\gamma_1)\right) + L_\rho(\gamma_2) = d(z, v; \Omega) + L_\rho(\gamma_2).$$

Taking the infimum over all curves γ_2 in this last estimate yields (3.16).

□

3.4.2 The Infinitesimal Spherical Metric

We now construct a special conformal metric ρ on the extended complex plane $\mathbb{C}_\infty$ so that the length of a curve on the sphere equals the ρ-length of the image curve, under stereographic projection, in the extended complex plane. We can achieve this by working locally and figuring out the infinitesimal displacement on the extended complex plane that results from an infinitesimal displacement on the sphere. One byproduct of the computations we will do is that *stereographic projection is conformal* in that it preserves angles between curves.

We begin, then, with a point $\hat{z}$ on the sphere $\mathscr{S}$, other than the North Pole, with $\hat{z} = (\theta, \phi)$ in spherical coordinates. Imagine standing on the sphere at $\hat{z}$ and then moving a very short distance away from $\hat{z}$ along the surface of the sphere in a fixed direction. It helps to think of the sphere as the earth, and of moving a short distance away from a given point in a given direction on the surface of the earth. The projected point on the plane also changes—the question is by how much?

Suppose that we move on the sphere from $\hat{z} = (\theta, \phi)$ to $\hat{z} + \Delta\hat{z} = (\theta + \Delta\theta, \phi + \Delta\phi)$. Since $\hat{z}$ lies on a horizontal circle of radius $\cos\phi$, the length of the displacement parallel to the equator is $\Delta\theta \cos\phi$. The displacement perpendicular to the equator is due to the change in latitude and is simply $\Delta\phi$ (since the sphere $\mathscr{S}$ has radius 1). These displacements are orthogonal to each other. The length of the actual displacement $\Delta\hat{z}$ is then approximately

$$|\Delta\hat{z}|^2 \cong (\Delta\theta)^2 \cos^2\phi + (\Delta\phi)^2. \tag{3.17}$$

We say 'approximately' because the displacement takes place on the curved surface of the sphere. But then the sphere can be approximated locally by a flat plane, as we well know from living on the earth. What about the corresponding displacement on the complex plane? The answer here comes from formula (3.1), in which z is the stereographic projection of $\hat{z}$. The change in latitude $\Delta\phi$ results in a change in z directly away from (or towards) the origin, that is a change in modulus but not in argument. The magnitude of this change is

$$\left|\tan\left(\frac{\phi + \Delta\phi}{2} + \frac{\pi}{4}\right) - \tan\left(\frac{\phi}{2} + \frac{\pi}{4}\right)\right| \cong |\Delta\phi| \left|\frac{d}{d\phi}\left(\tan\left(\frac{\phi}{2} + \frac{\pi}{4}\right)\right)\right|$$

$$= \frac{|\Delta\phi|}{2} \sec^2\left(\frac{\phi}{2} + \frac{\pi}{4}\right).$$

The change in longitude $\Delta\theta$ results in the same change in the argument of z, but no change in its modulus. When θ alone changes, the stereographic projection z moves on a circle of radius $\tan(\phi/2 + \pi/4)$, so the resulting displacement in z is

$$|\Delta\theta| \tan(\phi/2 + \pi/4).$$

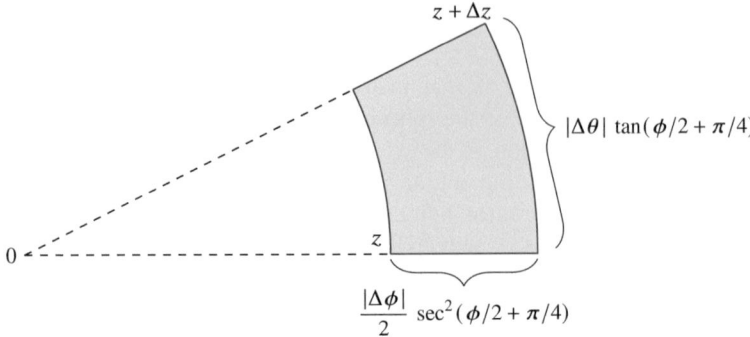

Fig. 3.7 Displacement in the plane as a result of displacement on the sphere

These two displacements in z are also roughly orthogonal (see Fig. 3.7), so that

$$|\Delta z|^2 \cong (\Delta\theta)^2 \tan^2(\phi/2 + \pi/4) + \frac{(\Delta\phi)^2}{4} \sec^4(\phi/2 + \pi/4)$$

$$= \frac{1}{4\cos^4(\phi/2 + \pi/4)}\left[4(\Delta\theta)^2 \sin^2(\phi/2 + \pi/4)\cos^2(\phi/2 + \pi/4) + (\Delta\phi)^2\right]$$

$$= \frac{1}{4\cos^4(\phi/2 + \pi/4)}\left[(\Delta\theta)^2 \sin^2(\phi + \pi/2) + (\Delta\phi)^2\right]$$

$$= \frac{1}{4\cos^4(\phi/2 + \pi/4)}\left[(\Delta\theta)^2 \cos^2\phi + (\Delta\phi)^2\right] \tag{3.18}$$

It is remarkable that the last expression in brackets is nothing other than $|\Delta\hat{z}|^2$ (see (3.17)). There is no reason to expect that the length of the displacement in the stereographic projection $|\Delta z|$ should be proportional to the length of the displacement $|\Delta\hat{z}|$ on the Riemann sphere. Yet (using the d rather than the Δ notation),

$$|dz| = \frac{1}{2\cos^2(\phi/2 + \pi/4)} |d\hat{z}|. \tag{3.19}$$

This key formula has interesting consequences. It relates distance travelled on the sphere, namely $|d\hat{z}|$, starting at the point (θ, ϕ) in spherical coordinates on the sphere, to distance travelled by the stereographic projection in the plane, namely $|dz|$. The very first and crucial thing to notice is that the displacement in the complex plane $\mathbb{C}$, (namely $|dz|$) depends only on the displacement on the sphere (namely $|d\hat{z}|$) and the position (θ, ϕ) at which the displacement takes place. It *does not* depend on the direction of travel on the sphere. This means, geometrically, that infinitesimal circles are preserved by stereographic projection. That is, an infinitesimal circle on the sphere centred at the point (θ, ϕ) in spherical coordinates will be mapped

3.4 The Spherical Metric

by stereographic projection to an infinitesimal circle in the plane but with radius $1/(2\cos^2(\phi/2 + \pi/4))$ times that of the circle on the sphere.[7] This is very much in accordance with Proposition 3.3, and is a result of the fortuitous occurrence at (3.18) that the coefficients of $(\Delta\theta)^2 \cos^2\phi$ and $(\Delta\phi)^2$ happen to be the same. In general, even with an orthogonal projection, one would not expect this to be the case: $|dz|$ could have depended differently on the sizes of $\Delta\phi$ and $(\Delta\theta)\cos\phi$, but didn't. For example, the change of coordinates $x' = x$, $y' = 2y$ in the plane is an orthogonal change of coordinates, but angles are not preserved. The following exercise teases out the magnification factors for stereographic projection in the north-south and east-west directions.

Exercise 3.16 Just as at the beginning of this section, suppose that we move in an east-west direction from $\hat{z} = (\theta, \phi)$ to $(\theta + \Delta\theta, \phi)$ on the sphere. Show that the displacement on the sphere is $|\Delta\theta|\cos\phi$ and that the displacement of the stereographic projection on the complex plane is $|\Delta\theta|\tan(\phi/2 + \pi/4)$. Compute the magnification factor $M_\theta(\hat{z})$ along the line of latitude, this being the ratio of the displacement in the complex plane to the displacement on the sphere.

Suppose that we move in a north-south direction from $\hat{z} = (\theta, \phi)$ to $(\theta, \phi + \Delta\phi)$ on the sphere. Compute the resultant displacement on the sphere and the displacement of the stereographic projection on the complex plane. Hence find the magnification factor $M_\phi(\hat{z})$ along the line of longitude at (θ, ϕ).

Verify that $M_\theta(\hat{z}) = M_\phi(\hat{z})$.

It follows from the equality of the magnification factors in two orthogonal directions that angles are preserved under stereographic projection. This property is known as *conformality* [16, Sect. 11.1]. When stereographic projection is used to make a map of part of the earth's surface, the map will be conformal in the sense that shapes of regions on the sphere will be shown correctly on the map. The map will not, however, have the other desirable property of *equal-area*: the actual area of a region on the earth will not be uniformly proportional to the corresponding area on the map: the map won't be to a uniform scale—areas will be distorted.

Theorem 3.4 *Stereographic projection is conformal.*

Proof At a point on the sphere $\hat{z} = (\theta_0, \phi_0)$, the ratio ds_1/ds_2 of a displacement ds_1 from west to east parallel to the equator and a displacement ds_2 from south to north perpendicular to the equator determine a direction/angle $\hat{A}$ at $\hat{z}$, clockwise from the due north direction say. (If $ds_2 = 0$ then one is travelling either due west or due east on the sphere, which is a special case in this proof.) The direction from west to east on the sphere corresponds to the direction in the complex plane of anticlockwise rotation about the origin, while the direction from south to north on the sphere corresponds to the radial direction in the complex plane directly away from the origin. The displacements ds_1 and ds_2 on the sphere at $\hat{z}$ therefore give

---

[7] Noting that $2\cos^2(\phi/2 + \pi/4) = 1 - \sin\phi$, does this agree with what you would expect for a displacement on the sphere at the equator, or at the South Pole?

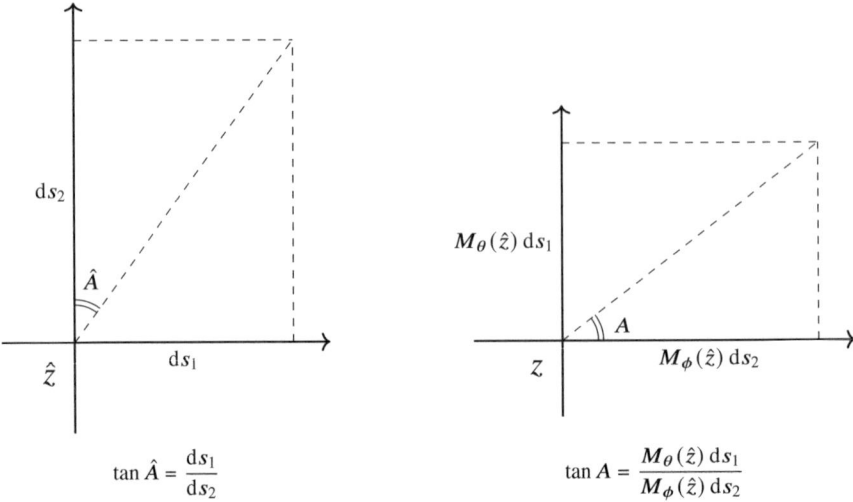

Fig. 3.8 Orthogonal displacements on the sphere and in the plane

rise to displacements $M_\theta(\hat{z}) \, ds_1$ in the anticlockwise direction at z in the plane and $M_\phi(\hat{z}) \, ds_2$ in the radial direction (see Fig. 3.8). The angle A in the complex plane (anticlockwise from the outward radial direction) is determined by the ratio

$$\frac{M_\theta(\hat{z}) \, ds_1}{M_\phi(\hat{z}) \, ds_2}.$$

This ratio equals ds_1/ds_2 by Exercise 3.16, hence the angles A and $\hat{A}$ are equal. The equality of the angle between two curves meeting at $\hat{z}$ and the angle between their projections meeting at z then follows. □

We rewrite (3.19) as

$$|d\hat{z}| = 2\cos^2(\phi/2 + \pi/4) \, |dz|. \tag{3.20}$$

To make the right hand side of this formula intrinsic to the complex plane (at the moment it is written in terms of the ϕ coordinate back on the sphere), we need to write $2\cos^2(\phi/2 + \pi/4)$ in terms of z. Now, by (3.1),

$$|z|^2 = \tan^2(\phi/2 + \pi/4) = \frac{1 - \cos^2(\phi/2 + \pi/4)}{\cos^2(\phi/2 + \pi/4)} = \frac{1}{\cos^2(\phi/2 + \pi/4)} - 1,$$

3.4 The Spherical Metric

so that

$$\cos^2(\phi/2 + \pi/4) = \frac{1}{1+|z|^2}.$$

Thus (3.20) becomes

$$|d\hat{z}| = \frac{2}{1+|z|^2} |dz|, \qquad (3.21)$$

which relates the infinitesimal distance $|d\hat{z}|$ on the curved surface of the unit sphere to the usual infinitesimal Euclidean distance $|dz|$ on the complex plane.

3.4.3 The Spherical Metric on $\mathbb{C}_\infty$

When deriving (3.18), we excluded the case when $\hat{z}$ is the North Pole, in which case z is the point at infinity. Noting that $1/(1+|z|^2) \to 0$ as $z \to \infty$, we set

$$\rho(z) = \begin{cases} \dfrac{2}{1+|z|^2} & \text{if } z \in \mathbb{C}, \\ 0 & \text{if } z = \infty; \end{cases} \qquad (3.22)$$

and rewrite (3.21) as

$$|d\hat{z}| = \rho(z)\,|dz|, \quad z \in \mathbb{C}.$$

Following the general construction in Sect. 3.4.1, we construct a Riemannian metric on the extended complex plane $\mathbb{C}_\infty$ with conformal metric ρ. We call this metric the *spherical metric* and denote the distance between points z and w in this metric by $d^\#(z, w)$. It follows from (3.21) that the ρ-length of any curve γ in $\mathbb{C}_\infty$ will be the same as the length of its preimage $\tilde{\gamma}$ on the sphere under stereographic projection. In fact, by the identity (3.21), the length on the sphere of each infinitesimal segment of the curve $\tilde{\gamma}$ is the same as the ρ-length of the corresponding segment of the projected curve γ. As a consequence, the distance between points z and w in the extended complex plane will be the same as the distance between their preimages under stereographic projection on the curved surface of the sphere.

There is a slight awkwardness, however, that we need to bear in mind. The point at infinity is not special when viewed as the North Pole on the sphere, but it is a special case from the viewpoint of the extended complex plane. One illustration of this is that $\rho(\infty) = 0$, making the point at infinity a *singularity* of the metric ρ, which is otherwise positive in the complex plane $\mathbb{C}$. A further technicality is that, in Sect. 3.4.1, we worked on a domain Ω in the plane, not in the extended complex plane. Definition 3.5 applies perfectly well to define the ρ-length of a curve lying entirely in the complex plane. For a curve that passes through the point at infinity, we split the curve into subcurves each lying entirely in the complex plane and then add their ρ-lengths.

Exercise 3.17 Write down a parameterisation $\gamma(t)$, $t \in [0, 1]$, of the positive real axis from 0 to the point at infinity.

Compute its ρ-length.

Compare your answer to the curved length on the sphere of the preimage of the positive real axis under stereographic projection.

The proof that (3.15) with ρ given by (3.22) leads to a valid distance function $d^\#(\cdot, \cdot)$ on the extended complex plane is effectively that of Proposition 3.4. The one point where a slight modification is required is in proving that $d^\#(z, w) > 0$ if $z \neq w$; here it is sufficient to realise that at least one of z or w is *not* the point at infinity and then base the argument in the proof at that point.

The function ρ is certainly continuous on $\mathbb{C}_\infty$. We also frequently make use of the observation that ρ is a radial function (that is, $\rho(z)$ depends only on $|z|$) and that ρ is a decreasing function of $|z|$. As a consequence, for each $R > 0$,

$$\rho(z) \geq \frac{2}{1 + R^2} \quad \text{if } |z| \leq R. \tag{3.23}$$

We now obtain some basic estimates for the spherical distance between points in terms of the Euclidean distance between them.

Lemma 3.2 *For any two points z and w in $\mathbb{C}$,*

$$d^\#(z, w) \leq 2|z - w|. \tag{3.24}$$

If $R > 0$ and $|z| \leq R$, $|w| \leq R$ then

$$\frac{2}{1 + 4R^2} |z - w| \leq d^\#(z, w). \tag{3.25}$$

For any non-zero z in $\mathbb{C}$,

$$d^\#(z, \infty) \leq \frac{2}{|z|}. \tag{3.26}$$

Proof The conformal factor ρ satisfies $\rho(\zeta) \leq 2$ for any $\zeta \in \mathbb{C}_\infty$. It follows from Exercise 3.15 with $M = 2$ that, for any curve γ in the extended complex plane,

$$L_\rho(\gamma) \leq 2 L(\gamma).$$

3.4 The Spherical Metric

For any two points z and w in $\mathbb{C}$, choosing γ to be the straight line segment $[z, w]$ joining them yields

$$d^{\#}(z, w) \leq L_\rho([z, w]) \leq 2 L([z, w]) = 2|z - w|.$$

This establishes (3.24). To prove (3.25), let $\gamma(t)$, $t \in [a, b]$, be any curve joining z to w where $|z| \leq R$ and $|w| \leq R$. If γ does not leave the disk $D(0, 2R)$ then $|\gamma(t)| \leq 2R$ for each t and it follows from (3.23) that

$$\rho(\gamma(t)) \geq \frac{2}{1 + (2R)^2}, \quad t \in [a, b].$$

It then follows from Exercise 3.15 with $m = 2/(1 + 4R^2)$ that

$$L_\rho(\gamma) \geq \frac{2}{1 + 4R^2} L(\gamma) \geq \frac{2}{1 + 4R^2} |z - w|.$$

On the other hand, suppose that the curve γ does leave $D(0, 2R)$. Then γ starts from z inside the disk $D(0, R)$, leaves the disk $D(0, 2R)$ and then returns to the disk $D(0, R)$ at w. Consequently, there are two distinct subarcs $\tilde{\gamma}_1$ and $\tilde{\gamma}_2$ of γ that join the circle $C(0, R)$ and the circle $C(0, 2R)$ and lie inside $D(0, 2R)$. On each subarc, $\rho \geq 2/(1 + 4R^2)$ and each subarc has Euclidean length at least R. Thus,

$$L_\rho(\gamma) \geq L_\rho(\tilde{\gamma}_1) + L_\rho(\tilde{\gamma}_2) \geq 2 \frac{2}{1 + 4R^2} R \geq \frac{2}{1 + 4R^2} |z - w|,$$

as $|z - w| \leq 2R$. Thus,

$$L_\rho(\gamma) \geq \frac{2}{1 + 4R^2} |z - w|,$$

if γ leaves the disk $D(0, 2R)$ and also if it doesn't. Since this inequality therefore holds for any curve joining z to w, the estimate (3.25) follows.

To obtain (3.26), consider the straight line curve γ parameterised by tz, $1 \leq t \leq \infty$. This curve joins z to the point at infinity. Recalling Remark 3.1, we have

$$d^{\#}(z, \infty) \leq L_\rho(\gamma) = \int_1^\infty \rho(\gamma(t)) |\gamma'(t)| \, dt$$

$$= \int_1^\infty \frac{2}{1 + t^2|z|^2} |z| \, dt \leq \int_1^\infty \frac{2}{t^2|z|} \, dt = \frac{2}{|z|},$$

which is (3.26). □

The following exercise sets out steps to obtain estimates for the spherical distance that will prove useful in Chap. 5. The crux of the exercise is to show that the spherical distance between two points in the complex plane doesn't change significantly if both points are translated or scaled.

Exercise 3.18

▷ Set $f(x) = 1/(1+x^2)$ for $x \in [0, \infty)$. Let r be a fixed positive number. Show that there is a positive constant c, depending only on r, such that $f(x+r) \geq cf(x)$ for all $x \in [0, \infty)$. Hint: note that $f(x+r)/f(x) \to 1$ as $x \to \infty$.

▷ Fix $a \in \mathbb{C}$. Show that there is a finite constant C depending only on $|a|$ such that each of the following estimates hold in turn:

$$\frac{1}{C}\rho(z) \leq \rho(z+a) \leq C\rho(z) \text{ for all } z \in \mathbb{C};$$

$$\frac{1}{C}L_\rho(\gamma) \leq L_\rho(\gamma+a) \leq CL_\rho(\gamma) \text{ for any curve } \gamma;$$

$$\frac{1}{C}d^\#(z, w) \leq d^\#(z+a, w+a) \leq Cd^\#(z, w) \text{ for all } z, w \in \mathbb{C}_\infty.$$

▷ Follow an analogous method to show that, for a fixed *and non-zero*, there is a finite constant C depending only on $|a|$ such that

$$\frac{1}{C}d^\#(z, w) \leq d^\#(az, aw) \leq Cd^\#(z, w) \text{ for all } z, w \in \mathbb{C}_\infty.$$

Even with only these preliminary estimates to hand, we can already prove that the extended complex plane is complete in the spherical metric.

Lemma 3.3 $\mathbb{C}_\infty$ *is complete in the spherical metric* $d^\#(\cdot, \cdot)$.

Proof Let $\{z_n\}_{n=1}^\infty$ be a sequence of points in $\mathbb{C}_\infty$ that is Cauchy in the spherical metric. Suppose first that the sequence is bounded (in the Euclidean metric), in that there is a finite R such that $|z_n| \leq R$ for each n. By (3.25),

$$|z_n - z_m| \leq \frac{1+4R^2}{2}d^\#(z_n, z_m), \text{ for each } n \text{ and } m.$$

From this we conclude that the sequence $\{z_n\}_{n=1}^\infty$ is a Cauchy sequence in the Euclidean metric. The sequence is therefore convergent in the Euclidean metric to some z_0 with $|z_0| \leq R$. By (3.24), $z_n \to z_0$ in the spherical metric as well.

On the other hand, suppose that there is a sequence $\{n_k\}$ such that $|z_{n_k}| \to \infty$ as $k \to \infty$. By (3.26), $d^\#(z_{n_k}, \infty) \to 0$ as $k \to \infty$, that is the subsequence $\{z_{n_k}\}_{k=1}^\infty$ converges in the spherical metric to the point at infinity as $k \to \infty$. Since the sequence $\{z_n\}_{n=1}^\infty$ is Cauchy, it follows that the sequence as a whole converges to the point at infinity. □

We would hope to show that there always is a curve of shortest spherical length joining any two points and to describe this curve of shortest length. This then helps

3.4 The Spherical Metric

us to solve the problem of describing the curves of shortest curved length on the sphere, which we show to be great circles. It turns out to be convenient to first answer the question, which is a natural one in any case, as to what the isometries of the spherical metric are. These turn out to be special Möbius transformations. With this information to hand, we can use an isometry to reduce the problem of whether there is a curve of shortest spherical length joining two arbitrary points to the special case when one of the points is the origin.

3.4.4 Isometries of the Spherical Metric

First some general facts about distance preserving maps.

Exercise 3.19 Suppose that (X, d) is a metric space and that $f : X \to X$ is an isometry of X in that $d(x, y) = d(f(x), f(y))$ for each x, y in X. Show that

▷ f is injective;
▷ f is continuous;
▷ if X is compact then f is surjective (this part requires some work);
▷ if f is a bijection then f^{-1} is an isometry and f is a homeomorphism.

Give an example of a metric space (X, d) and an isometry f that is not surjective.

In the setting of complex analysis (our setting), one should ask what meromorphic functions f on the Riemann sphere preserve the spherical distance between points, in that

$$d^\#(z, w) = d^\#(f(z), f(w)), \quad \text{for each } z, w \in \mathbb{C}_\infty. \tag{3.27}$$

By Theorem 3.3, any meromorphic function on the Riemann sphere is a rational function. If such a map is to be an isometry, it must be injective and so, by Exercise 3.12, the map must be a Möbius transformation. Our task, therefore, is to discover which Möbius transformations are spherical isometries of $\mathbb{C}_\infty$.

The key step in solving this problem is the observation that any mapping that preserves the distance between points globally will also preserve infinitesimal distances: that is, it will preserve the metric ρ as made precise in the next result. As we will use this material again in the next chapter on the hyperbolic metric, we work in the general setting of Sect. 3.4.1 with a conformal metric ρ in a domain Ω, the ρ-length of a curve as given by (3.14), and distance $d(z, w; \Omega)$ between points as given by (3.15).

Proposition 3.5 *Let $f : \Omega \to \Omega$ be analytic and injective. If f is an isometry in the sense that*

$$d(z, w; \Omega) = d(f(z), f(w); \Omega), \quad \text{for each } z, w \in \Omega, \tag{3.28}$$

then
$$\rho(f(z))|f'(z)| = \rho(z), \quad z \in \Omega. \tag{3.29}$$

Conversely, if (3.29) *holds then f preserves the ρ-length of curves in that, for any curve γ in Ω,*

$$L_\rho(f \circ \gamma) = L_\rho(\gamma), \tag{3.30}$$

and therefore f preserves the Riemannian distance between points. In other words, f is an isometry of the metric $d(\cdot, \cdot; \Omega)$.

If f is an isometry and α and β are points in Ω, then γ is a curve of shortest ρ-length joining α and β if and only if $f(\gamma)$ is a curve of shortest ρ-length joining $f(\alpha)$ and $f(\beta)$.

Remark 3.2 To be clear, the last statement in Proposition 3.5 only says that isometries preserve curves of shortest length *if* they exist, not that they *do* exist!

We use the following estimate of the local Riemannian distance in order to prove Proposition 3.5. In words, the estimate implies that, locally near a point z, distances in the Riemannian metric are comparable to Euclidean distances scaled by $\rho(z)$.

Lemma 3.4 *For $z \in \Omega$,*[8]

$$d(z, w; \Omega) = [\rho(z) + o(1)]|z - w| \quad as\ w \to z. \tag{3.31}$$

Proof Let ϵ positive be given. Since ρ is continuous at z, there is a positive number r such that

$$\rho(z) - \epsilon < \rho(\zeta) < \rho(z) + \epsilon \quad \text{whenever } |z - \zeta| < r.$$

Now suppose that $|z - w| < r$. Let $\gamma = [z, w]$ be the straight line segment from z to w. On this line segment, which has length $|z - w|$, the function ρ is bounded above by $M = \rho(z) + \epsilon$. By Exercise 3.15, therefore,

$$d(z, w; \Omega) \leq L_\rho(\gamma) \leq [\rho(z) + \epsilon]|z - w|.$$

If $\gamma(t)$, $t \in [a, b]$, is any curve from z to w that never leaves the disk $D(z, r)$ then ρ is bounded below by $m = \rho(z) - \epsilon$ on γ. Again, by Exercise 3.15,

$$L_\rho(\gamma) \geq [\rho(z) - \epsilon] L(\gamma) \geq [\rho(z) - \epsilon]|z - w|.$$

---

[8] Here $o(1)$ stands for a function of w that tends to 0 as w approaches z.

3.4 The Spherical Metric

If $\gamma(t)$, $t \in [a, b]$, is any curve from z to w that *does* leave the disk $D(z, r)$ then γ contains a subcurve $\tilde\gamma$ that lies in $D(z, r)$ and joins z to the circle $C(z, r)$. Thus $\tilde\gamma$ has length at least r and, since $\tilde\gamma$ lies in $D(z, r)$, we again have that ρ is bounded below by $m = \rho(z) - \epsilon$ on $\tilde\gamma$. Using Exercise 3.15 one final time,

$$L_\rho(\gamma) \geq L_\rho(\tilde\gamma) \geq \big[\rho(z) - \epsilon\big] r \geq \big[\rho(z) - \epsilon\big] |z - w|.$$

Since $L_\rho(\gamma) \geq \big[\rho(z) - \epsilon\big] |z - w|$ in all cases that γ is a curve joining z to w, we may conclude that $d(z, w; \Omega) \geq \big[\rho(z) - \epsilon\big] |z - w|$. Since we now have both an upper and lower estimate for $d(z, w; \Omega)$, and ϵ was arbitrary, the estimate (3.31) follows. □

Proof of Proposition 3.5 Let's first prove that (3.29) implies that f preserves the ρ-lengths of curves; it will then follow that f is distance preserving in that (3.28) holds. Let γ be any piecewise-smooth curve in Ω parameterised, say, by $\gamma(t)$, $t \in [a, b]$. Then $f \circ \gamma$ is a curve in Ω parameterized by $(f \circ \gamma)(t)$, $t \in [a, b]$. By Definition 3.5,

$$L_\rho(f \circ \gamma) = \int_a^b \rho\big((f \circ \gamma)(t)\big) \big|(f \circ \gamma)'(t)\big| \, dt$$

$$= \int_a^b \rho\big(f(\gamma(t))\big) \big|f'(\gamma(t))\big| \big|\gamma'(t)\big| \, dt.$$

But (3.29) with $z = \gamma(t)$ implies that $\rho\big(f(\gamma(t))\big) |f'(\gamma(t))| = \rho(\gamma(t))$ and so

$$L_\rho(f \circ \gamma) = \int_a^b \rho(\gamma(t)) |\gamma'(t)| \, dt = L_\rho(\gamma).$$

Thus f preserves the ρ-length of every curve, which is (3.30).

Since f is analytic and injective, its derivative never vanishes. The derivative of the inverse g of f is then given by $g'(w) = 1/f'(g(w))$. Replacing z by $g(w)$, so that $w = f(z)$, in (3.29) leads to

$$\rho(w) \frac{1}{|g'(w)|} = \rho(g(w)), \quad w \in f(\Omega).$$

Thus, (3.29) also holds for the inverse g of f. By the previous argument, $L_\rho(f^{-1} \circ \gamma) = L_\rho(\gamma)$ for any curve γ in $f(\Omega)$.

Now let z and w be any two points in Ω. If γ joins z to w then $f \circ \gamma$ joins $f(z)$ to $f(w)$ and has the same ρ-length. In words, for every curve joining z to w there is a corresponding curve of the same ρ-length joining $f(z)$ to $f(w)$. Bearing in mind that the spherical distance between points is defined as an infimum of ρ-lengths of curves, it follows that $d\big(f(z), f(w); \Omega\big) \leq d(z, w; \Omega)$.

If γ is any curve that joins $f(z)$ to $f(w)$ then $f^{-1} \circ \gamma$ joins z to w and has the same ρ-length. In this case, it follows that $d(z, w; \Omega) \leq d(f(z), f(w); \Omega)$. Thus, f preserves the spherical distance between points and (3.28) holds.

Suppose, on the other hand, that f is an isometry in that (3.28) holds. We need to show that (3.29) holds for z in Ω. From Lemma 3.4 we deduce both that

$$d(z, w; \Omega) = [\rho(z) + o(1)] |z - w| \quad \text{as } w \to z \tag{3.32}$$

and that

$$\begin{aligned} d(f(z), f(w); \Omega) &= [\rho(f(z)) + o(1)] |f(z) - f(w)| \\ &= [\rho(f(z)) + o(1)] (|f'(z)| + o(1)) |z - w|. \\ &= [\rho(f(z))|f'(z)| + o(1)] |z - w| \quad \text{as } f(w) \to f(z). \end{aligned} \tag{3.33}$$

Equating the expressions (3.32) for $d(z, w; \Omega)$ and (3.33) for $d(f(z), f(w); \Omega)$ yields

$$\rho(z) + o(1) = \rho(f(z))|f'(z)| + o(1) \quad \text{as } w \to z.$$

Letting w approach z gives (3.31).

For the last statement in Proposition 3.5, suppose that f is an isometry of the Riemannian metric. Then, f and f^{-1} both satisfy (3.29) and so preserve the lengths of curves. For α and β in Ω, suppose that γ is a curve of shortest ρ-length joining α and β. Then $f(\gamma)$ is a curve of shortest ρ-length joining $f(\alpha)$ and $f(\beta)$. Otherwise, if there was a shorter curve $\tilde{\gamma}$ joining $f(\alpha)$ and $f(\beta)$ then its preimage under f would be a curve joining α and β that would be shorter than γ. Similarly, since f preserves the ρ-lengths of curves, if $\tilde{\gamma}$ is a curve of shortest ρ-length joining $f(\alpha)$ and $f(\beta)$ then $f^{-1}(\tilde{\gamma})$ is a curve of shortest ρ-length joining α and β. □

We are now in a position to describe all (meromorphic) isometries of the spherical metric on the extended complex plane. As noted earlier, any such isometry must be a Möbius transformation M and so will not be analytic in the sense used in Proposition 3.5 at $M^{-1}(\infty)$ or at ∞. Moreover, the point at infinity is a singularity of the metric as ρ vanishes there. These technicalities mean that we can't quite apply Proposition 3.5 directly. Checking the proof, it is sufficient to replace (3.29) by

$$\rho(M(z))|M'(z)| = \rho(z), \quad z \neq \infty, \ z \neq M^{-1}(\infty), \tag{3.34}$$

as the two exceptional points ∞ and $M^{-1}(\infty)$ don't affect the spherical length of any curve that passes through them. To be explicit, the Möbius transformation M is an isometry of the spherical metric in the sense of (3.27) if and only if (3.34) holds.

3.4 The Spherical Metric

Dropping the factor '2', (3.34) is equivalent to

$$\frac{|M'(z)|}{1+|M(z)|^2} = \frac{1}{1+|z|^2}, \quad z \neq \infty, \; z \neq M^{-1}(\infty). \tag{3.35}$$

The Spherical Derivative For f meromorphic and z not a pole of f,

$$f^\#(z) := \rho(f(z))|f'(z)| = \frac{2|f'(z)|}{1+|f(z)|^2} \tag{3.36}$$

is called the *spherical derivative* of f at z.

Exercise 3.20

▷ Suppose that f has a pole of order n at z_0 in $\mathbb{C}$ so that f can be written as $f(z) = g(z)/(z-z_0)^n$ where g is analytic at z_0 with $g(z_0) \neq 0$. Show that $\lim_{z \to z_0} f^\#(z) = 0$ if $n \geq 2$.
Identify this limit if f has a simple pole at z_0.
▷ Define $f^\#(z_0) = \lim_{z \to z_0} f^\#(z)$ if f has a pole at z_0. Conclude that $f^\#$ is then a continuous, real-valued function.
▷ If f has a zero of order n at z_0, so that f can be written as $f(z) = g(z)(z-z_0)^n$ where g is analytic at z_0 with $g(z_0) \neq 0$, find $f^\#(z_0)$.
▷ Show that $(1/f)^\# = f^\#$ for f meromorphic.

The spherical derivative will play an essential role in the context of normal families of meromorphic functions in Chap. 5. Now back to our discussion of spherical isometries. With the standard normalisation of the coefficients, we write

$$M(z) = \frac{az+b}{cz+d}, \quad ad-bc = 1.$$

Exercise 3.21 Before reading any further, find all normalised Möbius transformations that satisfy (3.35).

While Exercise 3.21 is a reasonably straightforward computation, the result is so important for our story that we carry out the computation explicitly here. The derivative of M is given by $M'(z) = 1/(cz+d)^2$ (for $z \neq \infty$ and $z \neq M^{-1}(\infty) = -d/c$). Then,

$$\frac{|M'(z)|}{1+|M(z)|^2} = \frac{1}{|az+b|^2+|cz+d|^2}.$$

Thus M satisfies (3.35) if and only if

$$|az+b|^2 + |cz+d|^2 = 1+|z|^2, \quad z \in \mathbb{C} \setminus \{-d/c\},$$

that is, if and only if

$$\left(|a|^2+|c|^2\right)|z|^2+2\Re\left[(a\bar{b}+c\bar{d})z\right]+|b|^2+|d|^2=1+|z|^2. \tag{3.37}$$

Setting $a\bar{b}+c\bar{d}=Re^{it}$ and $z=re^{i\theta}$, then $\Re\left[(a\bar{b}+c\bar{d})z\right]=rR\cos(t+\theta)$. We can then rewrite (3.37) as

$$2rR\cos(t+\theta)=1+|z|^2-\left(|a|^2+|c|^2\right)|z|^2-|b|^2-|d|^2.$$

As z travels around a circle $C(0,r)$, the modulus of z and therefore the right hand side of this equation stays fixed but the left hand side varies with the argument θ of z, unless $R=0$. Thus, in order that (3.37) holds we require that $a\bar{b}+c\bar{d}=0$. In that case, (3.37) becomes

$$\left(|a|^2+|c|^2\right)|z|^2+|b|^2+|d|^2=1+|z|^2,$$

for z complex. This leads to

$$\begin{cases} |a|^2+|c|^2=1 \\ |b|^2+|d|^2=1. \end{cases}$$

Taking account of the normalisation, we also have

$$\begin{cases} a\bar{b}+c\bar{d}=0 \\ ad-bc=1 \end{cases}$$

Multiplying the first equation by d and the second by $\bar{b}$ and subtracting (and using $|b|^2+|d|^2=1$) leads to $c=-\bar{b}$. The equation $ad-bc=1$ then leads to

$$ad=1+bc=1-|b|^2=|d|^2=\bar{d}d,$$

and so either $d=0$ or $d=\bar{a}$. If $d=0$ then $|b|=1$ and $0=a\bar{b}$ forces $a=0$ as well, in which case $d=\bar{a}$ trivially. The work in this subsection is summarised in the following result.

Theorem 3.5 *The meromorphic isometries of the spherical metric on the extended complex plane $\mathbb{C}_\infty$ are precisely those Möbius transformations that can be written in the form*

$$M(z)=\frac{az+b}{\bar{a}-\bar{b}z}, \text{ where } |a|^2+|b|^2=1. \tag{3.38}$$

3.4 The Spherical Metric

Exercise 3.22

▷ Check that a Möbius transformation of the form (3.38) is indeed an isometry of the spherical metric on $\mathbb{C}_\infty$.
▷ Do these isometries form a subgroup of the full group of Möbius transformations under composition?

Setting $b = 0$ in (3.38) results in $M(z) = (a/\bar{a})z$ which has the general form $M(z) = e^{i\theta}z$, $\theta \in \mathbb{R}$, and so is a rotation. Setting $a = 0$ in (3.38) results in $M(z) = -b/(\bar{b}z)$ which has the general form $M(z) = e^{i\theta}/z$, $\theta \in \mathbb{R}$; this is the inversion $z \to 1/z$ that interchanges 0 and ∞, followed by a rotation.

The expression for M in (3.38) can be multiplied top and bottom by any non-zero *real* number. This shows that the condition $|a|^2 + |b|^2 = 1$ in (3.38) can be replaced by $|a|^2 + |b|^2 > 0$ at one's convenience.

We now give an alternative description of the isometries of the spherical metric. Suppose that M is such an isometry and therefore, by Theorem 3.5, is of the form (3.38). Set $\alpha = M^{-1}(0)$ (remember that M is surjective) so that $M(\alpha) = 0$. If $\alpha = \infty$ then $a = 0$ in (3.38) and so $M(z) = e^{i\theta}/z$, for some $\theta \in \mathbb{R}$. If $\alpha \in \mathbb{C}$ then, writing $a = ie^{i\theta/2}$, $\theta \in \mathbb{R}$, (which is the most general form up to multiplication by a positive real number), we have $a\alpha + b = 0$ so that $b = -i\alpha e^{i\theta/2}$. Then

$$M(z) = \frac{ie^{i\theta/2}(z - \alpha)}{-ie^{-i\theta/2} - i\bar{\alpha}e^{-i\theta/2}z} = e^{i\theta}\frac{\alpha - z}{1 + \bar{\alpha}z}.$$

Since the case $\alpha = \infty$ covers all isometries of the form $M(z) = e^{i\theta}/z$, we therefore have the following re-statement of Theorem 3.5.

Theorem 3.6 *The meromorphic isometries of the spherical metric on the extended complex plane $\mathbb{C}_\infty$ are precisely those Möbius transformations that can be written in the form*

$$M(z) = e^{i\theta}\frac{\alpha - z}{1 + \bar{\alpha}z} \tag{3.39}$$

for some $\theta \in \mathbb{R}$ and some $\alpha \in \mathbb{C}_\infty$.

Exercise 3.23 Show that both great circles (see Exercise 3.6) and antipodal points (see Exercise 3.2) are preserved under isometries of the spherical metric.

3.4.5 Geodesics for the Spherical Metric

Next, we're going to show that there is always a curve of shortest spherical length joining any two points in the extended complex plane. Such a curve will be called a *geodesic arc*. With this terminology, Proposition 3.5 includes the statement that isometries map geodesic arcs to geodesic arcs. A *geodesic* is a locally length-minimising curve, in that any sufficiently short section of the geodesic is the shortest curve between its endpoints. Formally, we make the following definition.

Definition 3.7 A *geodesic* γ is a maximal locally length-minimising curve. Let γ be parameterised by $\gamma(t)$, $t \in [a, b]$. That γ is locally length minimising means that, to each $t_0 \in (a, b)$ there corresponds a positive ε such that, for any choice of t_1 and t_2 with $t_0 - \varepsilon \leq t_1 \leq t_2 \leq t_0 + \varepsilon$, the curve $\gamma(t)$, $t \in [t_1, t_2]$, is a geodesic arc joining $\gamma(t_1)$ and $\gamma(t_2)$. If γ is a closed curve, so that $\gamma(a) = \gamma(b)$, then we also require that there is a positive ε such that, if $b - \varepsilon \leq t_1 \leq b$ and $a \leq t_2 \leq a + \varepsilon$ then the curve $\gamma(t)$, $t_1 \leq t \leq b$, followed by the curve $\gamma(t)$, $a \leq t \leq t_2$ is a geodesic arc joining $\gamma(t_1)$ and $\gamma(t_2)$. That γ is maximal means that γ is not a proper subcurve of a locally length minimising curve.

For the moment, we concentrate on describing geodesic arcs and show that they exist! Since we have a full description of the isometries of the spherical metric, we first describe geodesic arcs with one endpoint at the origin. The general case will follow by mapping with an isometry.

Exercise 3.24

▷ Let γ be any curve in the extended complex plane, parameterised by $\gamma(t)$, $t \in [a, b]$, say and let $\theta \in \mathbb{R}$. Let $\tilde{\gamma}$ be the rotated curve $\tilde{\gamma}(t) = e^{i\theta}\gamma(t)$, $t \in [a, b]$. Show that the curves γ and $\tilde{\gamma}$ have the same ρ-length.
▷ Deduce that if Γ is a geodesic in the spherical metric then so is the curve $e^{i\theta}\Gamma$, any $\theta \in \mathbb{R}$.
▷ Deduce that $d^{\#}(0, z) = d^{\#}(0, w)$ if $|z| = |w|$, z, w in $\mathbb{C}$.
▷ Deduce also that $d^{\#}(\infty, z) = d^{\#}(\infty, w)$ if $|z| = |w|$, z, w in $\mathbb{C}$.

To begin with, we will show that a curve of shortest spherical length exists between 0 and any $z \in \mathbb{C}$. In this case there is a unique geodesic arc from 0 to z and it coincides with the Euclidean geodesic arc, that is with the Euclidean straight line segment from 0 to z. On the other hand, *any* half-ray from 0 is a geodesic arc joining 0 to the point at infinity. There are, therefore, infinitely many curves of shortest spherical length joining 0 and ∞.

Proposition 3.6 *For $z \in \mathbb{C}$, the Euclidean line segment $[0, z]$ alone has the shortest spherical length among all curves joining 0 to z, and*

$$d^{\#}(0, z) = 2 \arctan |z|. \tag{3.40}$$

In the case that z is the point at infinity,

$$d^{\#}(0, \infty) = \pi.$$

Any curve of shortest spherical length joining 0 to the point at infinity is a Euclidean half-line originating at 0, and conversely.

3.4 The Spherical Metric

Proof Let $z \in \mathbb{C}$. Let γ be any curve joining 0 to z, other than the straight line segment $[0, z]$. We will show that its ρ-length is strictly greater than that of the line segment $[0, z]$, that is that $L_\rho(\gamma) > L_\rho([0, z])$. It will then follow that

$$d^{\#}(0, z) = \inf_\gamma L_\rho(\gamma) = L_\rho([0, z])$$

and that $d^{\#}(0, z) = L_\rho(\gamma)$ if and only if $\gamma = [0, z]$.

Let $L = L(\gamma)$ be the Euclidean length of γ. Then $L > |z|$. Parameterise γ on $[0, L]$ by arclength so that $|\gamma'(t)| = 1$ for $t \in [0, L]$. Then, by (3.14),

$$L_\rho(\gamma) = \int_0^L \rho(\gamma(t)) |\gamma'(t)| \, dt = \int_0^L \rho(\gamma(t)) \, dt. \tag{3.41}$$

Since $\gamma(0) = 0$,

$$|\gamma(t)| = \left| \int_0^t \gamma'(s) \, ds \right| \leq \int_0^t |\gamma'(s)| \, ds = t, \text{ for } t \in [0, L].$$

Since $\rho(w)$ is a decreasing function of $|w|$, it follows that $\rho(\gamma(t)) \geq \rho(t)$. Returning to (3.41) we find that

$$L_\rho(\gamma) = \int_0^L \rho(\gamma(t)) \, dt \geq \int_0^L \rho(t) \, dt > \int_0^{|z|} \rho(t) \, dt = L_\rho([0, z]),$$

as required.

The last equality comes from the following calculation. Writing $z = |z|e^{i\theta}$, the line segment $[0, z]$ may be parameterised by $\gamma(t) = te^{i\theta}$, for $t \in [0, |z|]$, in which case $\rho(\gamma(te^{i\theta})) = \rho(t)$ and $|\gamma'(t)| \equiv 1$. Thus, again by (3.14),

$$L_\rho([0, z]) = \int_0^{|z|} \rho(\gamma(t)) |\gamma'(t)| \, dt$$

$$= \int_0^{|z|} \rho(t) \, dt = \int_0^{|z|} \frac{2 \, dt}{1 + t^2} = 2 \arctan |z|.$$

This establishes (3.40).

Finally, we discuss the case when z is the point at infinity. Any infinite half-line starting from 0 is a curve joining 0 to the point at infinity. By Exercise 3.24, all such half-lines have the same spherical length. All other curves joining 0 to ∞ have strictly longer spherical length. To see this, take any curve γ joining 0 to the point at infinity. For successive values of $n \in \mathbb{N}$, choose a point z_n on γ with $|z_n| = n$, replace the subcurve γ_n of γ joining 0 to z_n by the line segment $[0, z_n]$. By the previous part of this proof, $L_\rho([0, z_n]) \leq L_\rho(\gamma_n)$ and so

$$L_\rho([0, n]) \leq L_\rho(\gamma_n) \leq L_\rho(\gamma).$$

Since this holds for each n, we deduce that $L_\rho([0,\infty]) \le L_\rho(\gamma)$. If γ is not an infinite half-line, we produce a strictly shorter curve by replacing the subcurve of γ joining 0 to z by the line segment $[0, z]$ for some point z on γ, and this second curve also has length at least that of the half-line $[0, \infty]$.

The spherical length of an infinite half-line γ emanating from 0 with angle θ, which can be parameterised by $\gamma(t) = te^{i\theta}, 0 \le t \le \infty$, is

$$L_\rho(\gamma) = \int_0^\infty \rho(t)\,dt = \int_0^\infty \frac{2\,dt}{1+t^2} = \pi.$$

As a consequence, $d^{\#}(0, \infty) = \pi$. □

We can now begin to describe all geodesics in the spherical metric. Recall that a 'Euclidean line through 0' is understood to include the point at infinity in $\mathbb{C}_\infty$ and that $\mathbb{R}_\infty$ denotes the real line with the point at infinity included. To be precise, $\mathbb{R}_\infty$ denotes the curve parameterised on $[-1, 1]$ by $\gamma(t) = t/(1-t^2)$. Note that $\mathbb{R}_\infty$ is a closed curve passing through ∞, and is a generalised circle.

Proposition 3.7 *Each Euclidean line through 0 is a geodesic in the spherical metric on the extended complex plane $\mathbb{C}_\infty$.*

Proof Since, by Exercise 3.24, the rotation of any geodesic is again a geodesic, it is enough to show that the line $\mathbb{R}_\infty$ is a geodesic. To show that $\mathbb{R}_\infty$ is a geodesic, we need to show that this curve satisfies Definition 3.7.

Suppose that x_0 lies on the positive real line, and that x and y are both near x_0, both positive and with $x < y$ say. The line segment $[x, y]$ then has to be the curve of shortest spherical length joining x to y. For if there was a shorter curve, we could adjoin it to the straight line segment $[0, x]$ to produce a curve joining 0 to y that is shorter than the line segment $[0, y]$, contradicting Proposition 3.6. A similar argument works if x_0 lies on the negative real line and x and y are near x_0 and both lie on the negative real line.

Next we deal with the case when we are working in a neighbourhood of 0 on the curve $\mathbb{R}_\infty$. Suppose that x and y are close to 0. If both x and y are positive, or if both are negative, then the line segment joining them is, as we have just seen, the curve of shortest spherical length joining them. Now suppose that $x < 0$ and $y > 0$ and that x and y are close to 0 in the sense that $1 + xy > 0$, (this is certainly the case if, say, $-1/2 < x < y < 1/2$). The map

$$M(z) = \frac{z-x}{1+xz}, \quad z \in \mathbb{C}_\infty,$$

is, by Theorem 3.6, an isometry of the spherical metric. Thus,

$$d^{\#}(x, y) = d^{\#}(M(x), M(y)) = d^{\#}\left(0, \frac{y-x}{1+xy}\right) = 2\arctan\left(\frac{y-x}{1+xy}\right)$$

3.4 The Spherical Metric

by Proposition 3.6, since $(y-x)/(1+xy) > 0$. The double angle formula for arctan then shows that

$$d^{\#}(x, y) = 2\arctan(-x) + 2\arctan y = d^{\#}(x, 0) + d^{\#}(0, y).$$

But, again by Proposition 3.6, $d^{\#}(x, 0) = L_\rho([x, 0])$ and $d^{\#}(0, y) = L_\rho([0, y])$ so that, by joining these two curves to form the line segment $[x, y]$, we obtain a curve whose length is

$$L_\rho([x, y]) = L_\rho([x, 0]) + L_\rho([0, y]) = d^{\#}(x, 0) + d^{\#}(0, y) = d^{\#}(x, y).$$

Thus, the line segment $[x, y]$ is seen to be a geodesic arc joining x and y.

Finally, we deal with the case when we are working in a neighbourhood of ∞ on the curve $\mathbb{R}_\infty$; we need to consider this case since $\mathbb{R}_\infty$ is a closed curve. The inversion map $M(z) = 1/z$ is an isometry of the spherical metric and so sends geodesic arcs to geodesic arcs. The map M fixes $\mathbb{R}_\infty$ and sends ∞ to 0. Suppose that x and y are close to ∞ with, say, $x > 0$ and $y < 0$ and $1 + xy < 0$ (this is certainly the case if, say, $x > 2$ and $y < -2$). Then $M(x) = 1/x > 0$, $M(y) = 1/y < 0$, and $1 + (1/x)(1/y) > 0$ so that, by the previous paragraph, the line segment $[1/y, 1/x]$ is a geodesic arc joining them. Mapping by the isometry $M(z) = 1/z$, we see that the subcurve $[x, \infty] \cup [\infty, y]$ of $\mathbb{R}_\infty$ is a geodesic arc joining x and y.

Since all cases have now been covered, we see that the extended real line is indeed a geodesic in the spherical metric. □

We will shortly learn that the Euclidean straight lines through 0 are the only geodesics that pass through 0. But first, using an isometry, we reduce the computation of the spherical distance between two general points to the special case (Proposition 3.6) when one point is the origin. For α and β in $\mathbb{C}$, set

$$\delta^{\#}(\alpha, \beta) = \left|\frac{\alpha - \beta}{1 + \overline{\alpha}\beta}\right|, \tag{3.42}$$

and set

$$\delta^{\#}(\infty, \infty) = 0, \quad \delta^{\#}(\infty, \beta) = 1/|\beta|, \quad \delta^{\#}(\alpha, \infty) = 1/|\alpha|. \tag{3.43}$$

The quantity $\delta^{\#}(\alpha, \beta)$ is sometimes called the pseudo-spherical distance between α and β in $\mathbb{C}_\infty$.

Theorem 3.7 *Let $\alpha \in \mathbb{C}_\infty$. If $\beta \neq -1/\overline{\alpha}$, the spherical distance between α and β in the extended complex plane is given by*

$$d^{\#}(\alpha, \beta) = 2\arctan\left(\delta^{\#}(\alpha, \beta)\right). \tag{3.44}$$

Two arcs join α to β on the generalised circle that also passes though the point $-1/\overline{\alpha}$. The shorter of these two arcs is the unique geodesic arc joining α to β.

If $\beta = -1/\overline{\alpha}$, then

$$d^{\#}(\alpha, -1/\overline{\alpha}) = \pi.$$

Geodesic arcs are images of rays from the origin under the isometry $M(z) = (\alpha - z)/(1 + \overline{\alpha}z)$ and all have the same spherical length π.

In particular, $d^{\#}(\alpha, \beta) \leq \pi$ for any α and β in $\mathbb{C}_{\infty}$.

Proof Let $\alpha \in \mathbb{C}_{\infty}$. By (3.39), the Möbius mapping

$$M(z) = \frac{\alpha - z}{1 + \overline{\alpha}z}$$

is an isometry of the spherical metric if $\alpha \neq \infty$ and, if $\alpha = \infty$, then $M(z) = 1/z$ performs this role. In either case, $M(\alpha) = 0$. Suppose that $M(\beta) \neq \infty$, that is that $\beta \neq -1/\overline{\alpha}$ if $\alpha \in \mathbb{C}$ or $\beta \neq 0$ if $\alpha = \infty$. Using the fact that M is an isometry of the spherical metric and Proposition 3.6,

$$d^{\#}(\alpha, \beta) = d^{\#}(M(\alpha), M(\beta)) = d^{\#}(0, M(\beta)) = 2\arctan\left(\delta^{\#}(\alpha, \beta)\right)$$

where $\delta^{\#}(\alpha, \beta)$ is given by (3.42) and (3.43), which is (3.44). At the last step we used Proposition 3.6.

By Proposition 3.5, a curve of shortest spherical length joining $M(\alpha)$ to $M(\beta)$ is mapped by M to a curve of shortest spherical length joining α to β and conversely; this is because $M^{-1} = M$. Since $M(\beta)$ is not the point at infinity then, by Proposition 3.6, there is a unique geodesic arc from $M(\alpha) = 0$ to this point, namely the straight line segment joining 0 to $M(\beta)$. The image of this line segment under M is therefore the unique geodesic arc joining α to β. Being the image of a line segment under a Möbius mapping, it is an arc of a generalised circle. This circle must pass through $M(\infty)$ which is $-1/\overline{\alpha}$, and so is the unique generalised circle that passes through α, β and $-1/\overline{\alpha}$. The geodesic arc joining α and β is then the shorter of the two arcs of this generalised circle that join α and β. The generic situation is pictured in Fig. 3.9. We now consider the case when $\beta = -1/\overline{\alpha}$, that is when α and β are antipodal (see Exercise 3.2). With M as before, we have $M(\alpha) = 0$ and $M(\beta) = \infty$. In this case there are infinitely many spherical geodesics joining α to $-1/\overline{\alpha}$, each of which is the image of a Euclidean half-line emanating from 0 under the map M. All have spherical length π. □

Exercise 3.25 Describe all spherical geodesic arcs that join the antipodal points 1 and -1.

3.4 The Spherical Metric

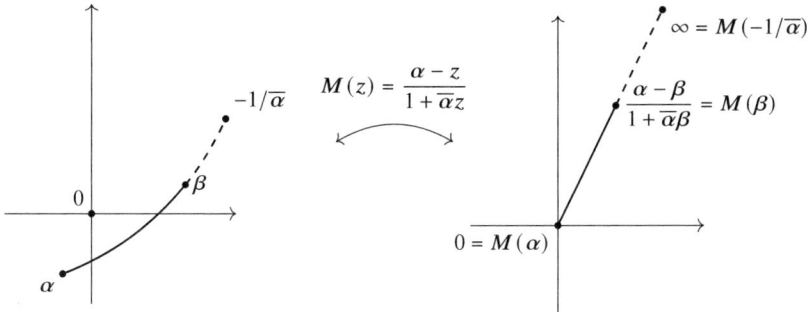

Fig. 3.9 Geodesic arc between α and β

Exercise 3.26 Let x and y be points on the real axis with $x < y$, and let u be a point on the real axis between x and y. Show that u is equidistant from x and y in the spherical metric if and only if $u = 0$ if $y = -x$ and, if $y \neq -x$,

$$u = \frac{-1 + xy + \sqrt{(1+x^2)(1+y^2)}}{x+y}.$$

We can now prove the reverse implication of Proposition 3.7, namely that the only geodesics passing through 0 are Euclidean lines. What we effectively want to rule out is that a geodesic might split, for example by having a Y-shape. To do so, we bear in mind that it is a consequence of Theorem 3.7 that any geodesic arc between points α and β in $\mathbb{C}_\infty$ is necessarily an arc of a generalised circle.

Proposition 3.8 *The only spherical geodesics that pass through 0 are Euclidean straight lines with the point at infinity added.*

Proof Let Γ be any geodesic that passes through 0. For points z on Γ sufficiently close to 0, the arc of Γ from 0 to z is the geodesic arc from 0 to z. This is, in turn, the line segment $[0, z]$. Since Γ locally looks like the arc of a generalised circle, Γ coincides with some Euclidean straight line L in some neighbourhood of 0.

Starting from 0 we head along Γ (in either direction) following L until, perhaps, L and Γ diverge at $z_0 \in \mathbb{C}$, say. Choose z_1 on $\Gamma \cap L$ just before z_0. Choose z_2 on Γ just after Γ diverges from L. By choosing z_1 and z_2 sufficiently close to z_0, the arc γ of Γ from z_1 to z_2 will be the geodesic arc joining z_1 to z_2. Now γ, being a geodesic arc between its endpoints, is necessarily an arc of a generalised circle. However, the part of γ from z_1 to z_0 is a (Euclidean) straight line segment and so the whole of the curve γ from z_1 to z_2 has to be a straight line segment (see Fig. 3.10). Thus, Γ coincides with the Euclidean straight line L from 0 as far as z_2 and not just as far as z_0. This contradicts the property of z_0 that it is the point where Γ and L diverge. As a consequence, Γ entirely coincides with the straight line L. □

Fig. 3.10 Γ and L in the proof of Proposition 3.8

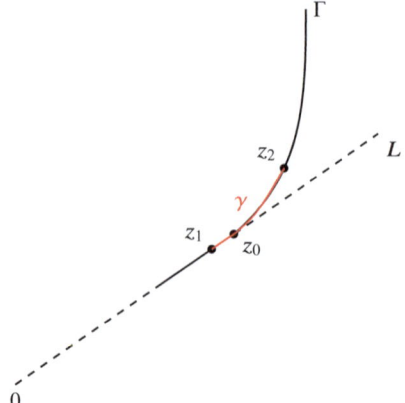

Theorem 3.8 *The geodesics in the spherical metric on the extended complex plane $\mathbb{C}_\infty$ comprise the extended real line $\mathbb{R}_\infty$ together with its images under the spherical isometries*

$$M(z) = e^{i\theta} \frac{\alpha - z}{1 + \overline{\alpha} z}, \qquad (3.45)$$

with $\theta \in [0, 2\pi)$ and as α ranges over $\mathbb{C}$.

Proof Since a spherical isometry preserves distances between points (by definition), and keeping in mind Proposition 3.5, it is clear that the image of a geodesic under an isometry is again a geodesic. Since we saw in Proposition 3.7 that the generalised circle $\mathbb{R}_\infty$ is a geodesic, it follows that its image under the spherical isometry (3.45) is also a geodesic.

On the other hand, suppose that Γ is a geodesic in the spherical metric. Choose any point α (but not the point at infinity) on Γ. Then, $M(z) = (\alpha - z)/(1 + \overline{\alpha} z)$, $z \in \mathbb{C}_\infty$, is a spherical isometry. The image of Γ under M is again a geodesic, and this geodesic passes through $M(\alpha) = 0$. By Proposition 3.8, $M(\Gamma)$ must be a Euclidean straight line through 0. Then $\Gamma = M(M(\Gamma))$ (remember that M is its own inverse) is the image of a Euclidean straight line that passes through 0 under a spherical isometry. By precomposing M with a suitable rotation, we may assume that this straight line is $\mathbb{R}_\infty$. □

3.5 Some Spherical Geometry

The spherical metric on the extended complex plane $\mathbb{C}_\infty$ was constructed in such a way that the length of a curve on the surface of the sphere $\mathscr{S}$ and the length of its stereographic projection in $\mathbb{C}_\infty$ are the same (this is encapsulated by (3.20)). Consequently, the distance between two points on the curved surface of the sphere

3.5 Some Spherical Geometry

$\mathscr{S}$ is the same as the spherical distance between their stereographic projections in $\mathbb{C}_\infty$. Not only that, but the geodesics on the curved surface of the sphere $\mathscr{S}$ are precisely those curves on the sphere that project to geodesics in the spherical metric on the extended complex plane. We've just characterised these geodesics in the extended complex plane. But what are the corresponding geodesics on the sphere $\mathscr{S}$?

Recall from Exercise 3.6 that a generalised circle in the extended complex plane is the image of a great circle on the sphere under stereographic projection precisely when it can be written as the locus of

$$A|z|^2 + \overline{B}z + B\overline{z} - A = 0, \text{ where } B \in \mathbb{C}, A \in \mathbb{R}. \tag{3.46}$$

Theorem 3.9 *A curve in the extended complex plane $\mathbb{C}_\infty$ is a geodesic in the spherical metric if and only if it is the image of a great circle on the sphere $\mathscr{S}$ under stereographic projection.*

In particular, the geodesics on the sphere $\mathscr{S}$ are precisely the great circles on the sphere.

That the shortest distance between two points on the sphere is the length of the shorter arc of the great circle joining them is a classical result from geometry. The Earth is close to being a sphere. Then, for example, the shortest flight path from Dubai to Los Angeles passes close to the North Pole.

Proof Suppose that Γ is a geodesic in the spherical metric on $\mathbb{C}_\infty$. It is therefore, by Theorem 3.8, the image of $\mathbb{R}_\infty$ under a spherical isometry M of the form (3.45). We wish to show that $M(\mathbb{R}_\infty)$ is of the form (3.46). It is enough to do so when $M(z) = M_\alpha(z) = (\alpha - z)/(1 + \overline{\alpha}z)$ for some complex α as the image of the locus of (3.46) under a rotation $z \to e^{i\theta}z$ is again of the form (3.46). If $-1/\overline{\alpha}$ is real, that is if α is real, then $M_\alpha(\mathbb{R}_\infty) = \mathbb{R}_\infty$, which is of the form (3.46). Suppose, then, that α is not real so that $M_\alpha(\mathbb{R}_\infty)$ is a Euclidean circle. Since $M_\alpha^{-1} = M_\alpha$ and treating $z = -1/\overline{\alpha}$, $w = \infty$ as a special case, a point z lies on $\Gamma = M_\alpha(\mathbb{R}_\infty)$ if and only if $w = M_\alpha(z)$ lies on $\mathbb{R}$, that is if and only if

$$0 = w - \overline{w} = M_\alpha(z) - \overline{M_\alpha(z)} = \frac{\alpha - z}{1 + \overline{\alpha}z} - \overline{\left(\frac{\alpha - z}{1 + \overline{\alpha}z}\right)},$$

that is if and only if

$$0 = (\alpha - z)(1 + \alpha\overline{z}) - (\overline{\alpha} - \overline{z})(1 + \overline{\alpha}z)$$
$$= (\overline{\alpha} - \alpha)|z|^2 - (1 + \overline{\alpha}^2)z + (1 + \alpha^2)\overline{z} - (\overline{\alpha} - \alpha).$$

Diving by $-2i$, we see that $z \in \Gamma = M_\alpha(\mathbb{R}_\infty)$ if and only if

$$\Im\alpha |z|^2 - \tfrac{1}{2}i(1 + \overline{\alpha}^2)z + \tfrac{1}{2}i(1 + \alpha^2)\overline{z} - \Im\alpha = 0,$$

which is of the form (3.46) with $A = \Im\alpha$ and $B = \tfrac{1}{2}i(1 + \alpha^2)$.

In the other direction, the locus of (3.46) when $A = 0$ is a straight line through the origin and so is a spherical geodesic. If $A \neq 0$ in (3.46) then we can write (3.46) equivalently as the circle

$$\left| z + \frac{B}{A} \right| = \sqrt{1 + \frac{|B|^2}{A^2}}. \qquad (3.47)$$

Since, by Exercise 3.24, the rotation of any geodesic is again a geodesic, it is enough to show that such a circle is a geodesic when its centre, $-B/A$ is real. With B real, we choose

$$\alpha = -\frac{B}{A} - \sqrt{1 + \frac{B^2}{A^2}}.$$

Note that α is also real and that M_α sends the point $z = -\frac{B}{A} + \sqrt{1 + \frac{B^2}{A^2}}$, that lies on the circle (3.47), to the point at infinity. Hence the image of the circle (3.47) under M_α is a Euclidean straight line. Moreover, the point α also lies on the circle (3.47) and is sent to 0 by M_α. Thus, the image of the circle (3.47) under M_α is a Euclidean straight line that passes through the origin, in fact $\mathbb{R}_\infty$. Hence, by Theorem 3.8, the circle (3.47) is the spherical geodesic $\Gamma = M_\alpha(\mathbb{R}_\infty)$ for this choice of α. □

By Exercise 3.6, the images under stereographic projection of the great circles on the sphere are precisely those curves that are either (i) Euclidean straight lines through the origin (and the point at infinity) or (ii) Euclidean circles in the complex plane with equation, for some $z_0 \in \mathbb{C}$,

$$|z - z_0| = \sqrt{1 + |z_0|^2}. \qquad (3.48)$$

The first case occurs precisely when the great circle passes through the North Pole and the second case when it doesn't.

There are key differences between spherical geometry, that is geometry on the sphere, and the Euclidean geometry of the plane. Euclid (about 300 BCE) set down five 'postulates', five 'obvious' facts about the geometry of the plane, which was the geometry he had in mind, on which proofs of all subsequent theorems, such as Pythagoras' Theorem, could be based. The first four of these were:

1. To draw a straight line from any point to any other.
2. To produce a finite straight line continuously in a straight line.
3. To describe a circle with any centre and radius.
4. That all right angles are equal to each other.

None of these was in any way controversial. Euclid's Fifth Postulate, however, gave rise to centuries of controversy and many failed attempts to deduce it from the first

3.5 Some Spherical Geometry

four: it just wasn't as neat and straightforward as the others. Playfair's equivalent statement of Euclid's Fifth Postulate, or Parallel Postulate, is

5. Given a line and a point not on the line, it is possible to draw exactly one line through the given point parallel to the given line.

Here, 'parallel' is to mean 'non-intersecting'.

Spherical geometry is not a true *non-Euclidean* geometry as it doesn't satisfy Postulates 1 and 3. We will meet a genuine non-Euclidean geometry in the next chapter, namely the hyperbolic geometry of the disk. A 'line' in spherical geometry is an arc of a great circle. It is implicit in Postulate 1 as used by Euclid that the straight line is unique. This fails in spherical geometry since, as we have seen, there are infinitely many lines between any two antipodal points. Postulate 3 doesn't hold as there are no circles with arbitrarily large radius. The Parallel Postulate fails as well since it is clear that any two great circles on the sphere intersect (or see Exercise 3.27 to follow), and so there are no parallel lines in this geometry.

A statement that is equivalent to the Parallel Postulate is that the sum of the angles in any triangle is $180°$. This fails on the sphere where the angle sum in any triangle is strictly greater than $180°$. To see this consider, for example, three points $\hat{a}, \hat{b}$ and $\hat{c}$ on the sphere forming a spherical triangle that doesn't include the North Pole **N**. The image of this spherical triangle under stereographic projection is a 'triangle' in the complex plane with vertices at points a, b and c and sides that are geodesic arcs. We write A, B and C for the angles these geodesic arcs make at the vertices. These angles are the same as the corresponding angles back on the sphere since stereographic projection is conformal (Theorem 3.4). We can map this 'triangle' by a spherical isometry M so that a is sent to the origin 0, and let us continue to denote the other two vertices by b and c. Since the spherical isometry is a Möbius transformation, it too preserves angles. This means that the angles at the vertices are still A, B and C (see Fig. 3.11).

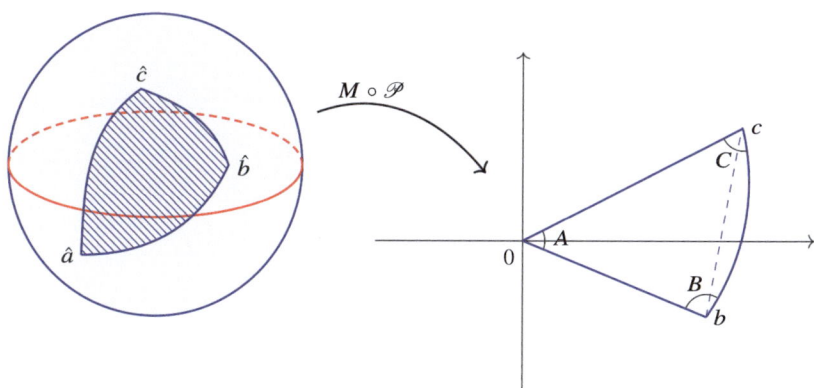

Fig. 3.11 A spherical triangle and its image in the complex plane

The geodesic arc joining 0 to b and the geodesic arc joining 0 to c are, by Proposition 3.6, the Euclidean line segments $[0, b]$ and $[0, c]$ respectively. The geodesic arc joining b and c is an arc of a circle of the form (3.48). The origin lies inside this circle and so the Euclidean triangle $\Delta 0bc$ lies inside this circle. The angles B at b and C at c in the *spherical* triangle $\Delta 0bc$ are therefore greater than the angles at b and at c, respectively, in the *Euclidean* triangle $\Delta 0bc$. The angle at the origin in both triangles is the same. Since the angles in the Euclidean triangle add to 180°, the angles in the spherical triangle add to more than 180°.

Exercise 3.27 Show that any two great circles on the sphere intersect in two antipodal points.

Hint: Show that their images under stereographic projection intersect by using Exercise 3.6.

Greenberg [12] is a readable, authoritative account of the history of Euclidean and non-Euclidean geometry and, in particular, of the fascinating story behind the parallel postulate.

Chapter 4
The Hyperbolic Disk

Complex analysis, the branch of mathematics of which this book is a partial account, interacts with other branches of mathematics in many rich and extraordinary ways. One such connection, namely that with geometry, is a central theme of this book. We already came across important connections between geometry and complex analysis in Chap. 3 on the Riemann sphere. We continue to develop this connection, under the heading of 'geometric function theory', by constructing the hyperbolic metric first in the disk and, in Chap. 6, in a general simply connected domain.

The hyperbolic metric on the disk is a genuine example of a non-Euclidean geometry in which the first four of Euclid's postulates hold but not the fifth Parallel Postulate. The 'straight lines' in this geometry are all arcs of circles, including diameters, that meet the unit circle at right angles. The geometry is non-Euclidean in that given any line and any point not on the line there are *infinitely many* lines through the given point that are parallel to the given line.

In the hyperbolic metric in the disk, also known as the Poincaré metric, space is stretched as one approaches (in the Euclidean sense) the boundary circle of the disk. In fact, this boundary circle is hyperbolically infinitely far away. This is beautifully illustrated in several drawings by the Dutch artist, M. C. Escher[1] including a drawing, Circle Limit IV, of angels and demons that tile the disk and that appear smaller and smaller closer to the boundary circle: yet, hyperbolically, they are all the same size!

The hyperbolic metric in the disk is a Riemannian metric, as was the spherical metric on the extended complex plane, and we make use of the general construction set out in Sect. 3.4.1. It is worth bearing in mind an important distinction between our constructions of spherical geometry and of hyperbolic geometry. In the case of spherical geometry, the intrinsic distance on the sphere guided us to the particular form (3.22) of the conformal metric ρ. We subsequently set ourselves the task of classifying the isometries of the resulting metric. In the case of hyperbolic geometry,

---

[1] https://mcescher.com.

we first classify all conformal mappings of the unit disk onto itself—this is the group of conformal automorphisms of the disk—and then set out to find a Riemannian metric for which these are the isometries.

The classification of the conformal automorphisms of the disk comes ultimately from Schwarz's Lemma, familiar to us from a first course in complex analysis. We later derive more general forms of Schwarz's Lemma, that generally go by the name 'Schwarz-Pick Lemma' and that state that analytic maps are contractions in the hyperbolic metric.

However, we are getting ahead of ourselves. Let's go back to the beginning of the story with, simply for the sake of completeness, a statement and proof of Schwarz's Lemma. Recall the notation $\mathbb{D}$ for the unit disk, so that $\mathbb{D} = \{z \in \mathbb{C} : |z| < 1\}$.

4.1 The Basic Schwarz Lemma

Theorem 4.1 (Schwarz's Lemma) *Suppose that f is an analytic map from the unit disk $\mathbb{D}$ into itself for which $f(0) = 0$. Then*

$$|f'(0)| \leq 1, \tag{4.1a}$$

$$|f(z)| \leq |z| \quad \text{for each } z \in \mathbb{D}. \tag{4.1b}$$

If equality holds in (4.1a), or if equality holds for some non-zero z in (4.1b), then f is a rotation of the disk, that is there is a real number θ for which f may be written as $f(z) = e^{i\theta}z$, $z \in \mathbb{D}$.

Remark 4.1 It is helpful to think of (4.1b) in Schwarz's Lemma pictorially. In the case of an analytic map f of the disk $\mathbb{D}$ into itself that fixes the centre of the disk, (4.1b) says that f then maps each smaller disk $D(0, r)$ into itself as shown in Fig. 4.1.

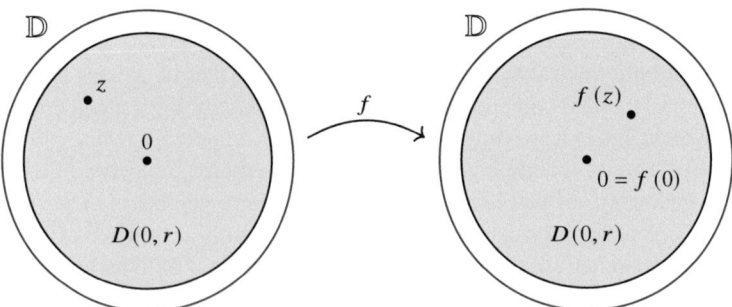

Fig. 4.1 Pictorial representation of Schwarz's Lemma

4.1 The Basic Schwarz Lemma

Proof Suppose that f is an analytic map of the unit disk into itself that fixes the origin. This function can be written in the form of a power series,

$$f(z) = \sum_{n=0}^{\infty} a_n z^n = a_0 + a_1 z + a_2 z^2 + \ldots,$$

that is convergent at least when $z \in \mathbb{D}$. The power series

$$g(z) = \sum_{n=1}^{\infty} a_n z^{n-1} = a_1 + a_2 z + a_3 z^2 + \ldots$$

has the same radius of convergence[2] as the power series for f, and so it too is convergent whenever $z \in \mathbb{D}$.

We can write the function g in terms of the function f. First, $g(0) = a_1 = f'(0)$. We haven't used the assumption that $f(0) = 0$ as yet. This assumption leads to $a_0 = 0$, since $f(0) = a_0$. It then follows, crucially, that for $z \neq 0$, $f(z)/z = a_1 + a_2 z + a_3 z^2 + \ldots = g(z)$. Thus

$$g(z) = \begin{cases} f'(0), & \text{if } z = 0; \\ f(z)/z, & \text{if } 0 < |z| < 1. \end{cases} \quad (4.2)$$

The next, and final, step is to apply the Maximum Principle[3] to g (see [16, Theorem 10.8]). Fix z in $\mathbb{D}$, and choose r with $|z| < r < 1$. Then

$$|g(z)| \leq \max\{|g(w)| : |w| \leq r\}$$
$$= \max\{|g(w)| : |w| = r\} \quad \text{(by the maximum principle)}$$
$$= \max\{|f(w)/w| : |w| = r\} \quad \text{(since } r > 0\text{)}$$
$$= \frac{1}{r} \max\{|f(w)| : |w| = r\} \leq \frac{1}{r}.$$

---

[2] If a power series $P(z) = \sum_{n=0}^{\infty} a_n z^n$ has positive radius of convergence R and if $|z| < R$ then there are $M > 0$ and $\rho < 1$ such that $|a_n z^n| \leq M\rho^n$, each n. (To see this, set $x = (R+|z|)/2$ so that $|z| < x < R$ and note that $\sum_{n=0}^{\infty} a_n x^n$ is convergent. Set $M = \max\{|a_n|x^n : n \in \mathbb{N}\}$, which is finite, and $\rho = |z|/x$. Then, $|a_n z^n| = |a_n|x^n (|z|/x)^n \leq M\rho^n$.) The power series P can be compared, term by term, to a geometric series. Then, for $z \neq 0$, $|a_n z^{n-1}| \leq (M/|z|)\rho^n$.

[3] The Maximum Principle states that the modulus of a function f, analytic on some open set D, cannot have a local maximum at an interior point of D unless f is a constant function. It is a consequence of the mean value property of analytic functions: the value of f at the centre of a circle equals the average value of f on the circle. In the proof of the Schwarz Lemma, $|g|$ is a continuous function on the compact set $\{w : |w| \leq r\}$ and so this function *must* assume its maximum value at some point of this closed disk. By the maximum principle, a point at which $|g|$ assumes its maximum cannot be in the disk $D(0, r)$, unless g is constant, and so must be on the circle $C(0, r)$. In either case, the maximum value of $|g|$ on $\{|w| \leq r\}$ is assumed on the circle $C(0, r)$.

The last inequality comes from the fact that $f : \mathbb{D} \to \mathbb{D}$ (that is, $|f(w)| < 1$ for each w in $\mathbb{D}$). We have now used all the assumptions on f (including the assumption that f was analytic).

In short, $|g(z)| \leq 1/r$ for each r with $|z| < r < 1$. Though $1/r$ is greater than 1, it approaches 1 as r approaches 1. It follows that $|g(z)| \leq 1$, and this inequality holds for each $z \in \mathbb{D}$. Writing g in terms of f using (4.2) yields the main part of Schwarz's Lemma.

It remains to decide when equality may hold in (4.1a), or in (4.1b) for some non-zero z in $\mathbb{D}$. By (4.2), this is equivalent to $|g(z)| = 1$ for some z in $\mathbb{D}$, and so the modulus of g has assumed its maximum value at an interior point of $\mathbb{D}$. By the maximum principle, g is constant, and this constant value must be a complex number of modulus 1, that is $e^{i\theta}$ for some real θ. $\square$

Exercise 4.1 Suppose that f is an analytic map from the disk $D(0, r_1)$ into the disk $D(0, r_2)$, with $f(0) = 0$. Prove that

$$|f'(0)| \leq \frac{r_2}{r_1},$$

$$|f(z)| \leq \frac{r_2}{r_1}|z|, \text{ for } z \in D(0, r_1).$$

(There is no need to start over and rework the *proof* of Schwarz's Lemma in this slightly more general context. Schwarz's Lemma has now been established and is a tool in our arsenal, so it may be applied to a suitable auxiliary function to solve this exercise.)

Definition 4.1 A *self-map* of a domain D is an analytic function $f : D \to D$, that is a map f that is analytic in D and takes values in D.

Note that a self-map need not be onto D, in that it need not be surjective, nor need it be one-to-one. For example, each of $f(z) = z/2$, $f(z) = z^2$ and $f(z) = (1+z^2)/2$ is a self-map of $\mathbb{D}$. The map $f(z) = \bar{z}$ does map $\mathbb{D}$ into $\mathbb{D}$ but is not a self-map of $\mathbb{D}$ as it is not analytic.

The Schwarz Lemma, as stated above, applies only to self-maps of the unit disk that fix the centre 0 of the disk. Suppose that the information we are given is that f is a self-map of the disk $\mathbb{D}$ and that $f(a) = b$, for specified points a and b in $\mathbb{D}$. What can we then say about $|f'(a)|$, for example? Faced with this problem, one approach might be to adapt the above proof of the Schwarz Lemma to work in this more general setting. However, mathematicians should be slow to start again from scratch, discarding any progress that has already been made. In the present setting, this means that we should attempt to deduce the more general theorem (in which $f(a) = b$) from the special case in which $a = b = 0$ (cf. Exercise 4.1).

The goal, then, is to modify f so that 0 becomes a fixed point. Suppose that we have a self-map g of $\mathbb{D}$ for which $g(0) = a$, and a second self-map h of $\mathbb{D}$ for which $h(b) = 0$. Then

$$F = h \circ f \circ g$$

is, in turn, analytic and is a self-map of the disk for which $F(0) = 0$. Schwarz's Lemma therefore applies to F and we deduce that $|F'(0)| \leq 1$. But, by the chain rule for derivatives,

$$F'(z) = h'\big((f \circ g)(z)\big) \times f'\big(g(z)\big) \times g'(z).$$

Since $g(0) = a$ and $(f \circ g)(0) = f(a) = b$, we have

$$|F'(0)| = |h'(b)|\,|f'(a)|\,|g'(0)| \leq 1,$$

and find that

$$|f'(a)| \leq \frac{1}{|h'(b)|\,|g'(0)|}. \tag{4.3}$$

For the moment, this is but a partial solution to the problem of finding a more general version of Schwarz's Lemma. To proceed further, we need to make an explicit choice of the maps g and h. There are many possible choices of g and h, and each gives rise to an inequality (4.3). To optimise the inequality (4.3), we should choose g and h so that $|g'(0)|$ and $|h'(b)|$ are as large as possible. The derivative $|g'(0)|$ is the expansion factor of g at 0 so this should, in principle, be maximised if g maps $\mathbb{D}$ onto the largest possible region (subject to $g \colon \mathbb{D} \to \mathbb{D}$ and $g(0) = a$), that is if g maps $\mathbb{D}$ *onto* $\mathbb{D}$. Similar considerations suggest that we should probably aim to have h also map $\mathbb{D}$ onto $\mathbb{D}$. With this in mind, we study surjective self-maps of $\mathbb{D}$ in more detail in the next section.

4.2 Automorphisms of the Disk

Definition 4.2 An *automorphism* of a domain D is a one-to-one, onto, analytic function $f \colon D \to D$. The collection of all automorphisms of a domain D is denoted by $\mathrm{Aut}(D)$.

The term *conformal map of D onto D* is also used. A *homeomorphism* is a continuous bijection with a continuous inverse. $\mathrm{Aut}(D)$ may be thought of as the subgroup of the full group of homeomorphisms from D to D consisting of those homeomorphisms that are also analytic. Note that though a self-map of a domain need be neither injective nor surjective, an automorphism is a self-map that is both.

Exercise 4.2

▷ Given two functions $f : A \to B$ and $g : B \to C$, their composition $h = g \circ f$ is given by $h(a) = g(f(a))$, $a \in A$, so that $h : A \to C$. Prove that if f and g are both injective maps then so is h.
Prove that if f and g are both surjective maps then so is h.

▷ Prove that $\text{Aut}(D)$, as defined above, forms a group under composition of functions.[4]

The natural place to look for a conformal[5] mapping of the disk $\mathbb{D}$ onto itself is the class of Möbius transformations since these map generalised circles to generalised circles. What conditions will guarantee that a Möbius transformation maps the disk $\mathbb{D}$ onto itself (recall that Möbius maps are automatically injective)?

Suppose that M is a Möbius transformation that maps $\mathbb{D}$ onto $\mathbb{D}$ and let's see if we can discover the form of M. We would expect that M maps the unit circle onto itself since it maps the disk $\mathbb{D}$ onto itself. Also, since $M : \mathbb{D} \to \mathbb{D}$ is onto, there is a point α in $\mathbb{D}$ with $M(\alpha) = 0$. If $\alpha = 0$ then we can simply set $M(z) = \beta z$ where β has modulus 1. This Möbius transformation is a rotation and maps $\mathbb{D}$ onto $\mathbb{D}$. Next suppose that α is non-zero and write

$$M(z) = \frac{\beta(\alpha - z)}{cz + d},$$

so that $M(\alpha) = 0$. It cannot be that $d = 0$, as then (since $\alpha \neq 0$) M would have a pole at 0. Let's normalise so that $d = 1$ and then $M(z) = \beta(\alpha - z)/(1 + cz)$. Imposing the condition that $|M(z)| = 1$ if $|z| = 1$ leads to

$$|\beta|^2 |\alpha - z|^2 = |1 + cz|^2 \quad \text{if } |z| = 1,$$

---

[4] One of the many attractive properties of analytic functions of a complex variable is that an analytic function has a local analytic inverse once its derivative isn't zero.

[5] 'Conformal' in the context of complex analysis means 'angle-preserving'. That is, f is conformal at a if any two curves γ_1 and γ_2 through a that meet at an angle α are mapped, by f, to two curves $\Gamma_1 = f(\gamma_1)$, $\Gamma_2 = f(\gamma_2)$ through $b = f(a)$ that also meet in the same angle α at b. This is a local property of the map. It transpires that a map f is conformal at a point a in its domain precisely when $f'(a) \neq 0$ [16, Theorem 11.1]. If $f'(a) = 0$ then f does *not* preserve angles at a and neither is f one-to-one in *any* neighbourhood of a: in any disk $D(a, r)$ there are guaranteed to be two distinct points z_1 and z_2 for which $f(z_1) = f(z_2)$. Hence, if f is one-to-one in a domain D then f' cannot vanish anywhere in D and consequently f is conformal at each point in D. Technically speaking, the converse is not true in that a map f may be conformal in D but not one-to-one in D (a classic example is $f(z) = \exp z$ in the strip $\{z : |\Im z| < 2\pi\}$). Even so, it is common to describe a map as 'conformal' when what is really meant is 'one-to-one'; this is the convention adopted in this book. To end this note, there are some synonyms of 'conformal' in the sense of 'one-to-one, analytic function' of which the reader should be aware; these include 'univalent', which is used in this book, and the older German term 'schlicht'.

4.2 Automorphisms of the Disk

that is

$$|\beta|^2 \left[|\alpha|^2 + 1 - 2\Re(\bar{\alpha}z)\right] = 1 + |c|^2 + 2\Re(cz), \quad \text{if } |z| = 1,$$

or

$$|\beta|^2 |\alpha|^2 + |\beta|^2 - 1 - |c|^2 = 2\Re\left[(|\beta|^2 \bar{\alpha} + c)z\right], \quad \text{if } |z| = 1. \tag{4.4}$$

As z travels around the unit circle, the right hand side of (4.4) changes while the left hand side stays constant unless, that is, $|\beta|^2 \bar{\alpha} + c = 0$. We conclude that $c = -|\beta|^2 \bar{\alpha}$. Returning to (4.4), in which the right hand side is now zero, we find that

$$\begin{aligned} 0 &= |\beta|^2 |\alpha|^2 + |\beta|^2 - 1 - |c|^2 \\ &= |\beta|^2 |\alpha|^2 + |\beta|^2 - 1 - |\beta|^4 |\alpha|^2 \\ &= \left(|\beta|^2 - 1\right)\left(1 - |\alpha|^2 |\beta|^2\right). \end{aligned}$$

Now, $|M(0)| = |\alpha\beta| < 1$ and so the only possibility is that $|\beta| = 1$. Then, $c = -\bar{\alpha}$ and we find that

$$M(z) = \beta \frac{\alpha - z}{1 - \bar{\alpha}z}, \quad z \in \mathbb{D}.$$

This suggests the following result, formally stated.

Proposition 4.1 *Each Möbius transformation of the form*

$$M(z) = \beta \frac{\alpha - z}{1 - \bar{\alpha}z}, \quad \text{where } |\beta| = 1 \text{ and } |\alpha| < 1, \tag{4.5}$$

is an automorphism of the unit disk $\mathbb{D}$.

Exercise 4.3 Prove Proposition 4.1 as follows. Suppose that M is of the form (4.5). Being a Möbius transformation, M is certainly injective, in fact a homeomorphism of $\mathbb{C}_\infty$. The function M is also analytic in $\mathbb{D}$ (its only pole is at $z = 1/\bar{\alpha}$, which lies outside $\mathbb{D}$). Verify that $|M(z)| < 1$ if $|z| < 1$ so that M maps $\mathbb{D}$ into $\mathbb{D}$. Finally, to show that $M: \mathbb{D} \to \mathbb{D}$ is onto, compute M^{-1} and verify that $M^{-1}: \mathbb{D} \to \mathbb{D}$. (Note that $M^{-1} = M$ in the case that $\beta = 1$. We will often make use of this observation.)

Note 4.1 Once Proposition 4.1 is proven, the work preceding it that led us to the Möbius transformations (4.5) becomes redundant and can be discarded as rough work.

Note 4.2 A quick way to verify that $|M(z)| < 1$ if $|z| < 1$ is to verify that $|M(z)| = 1$ if $|z| = 1$ and then apply the maximum principle.

Exercise 4.4 For the Möbius map M given by (4.5), show that

$$M'(0) = \beta(|\alpha|^2 - 1) \quad \text{and} \quad M'(\alpha) = \frac{\beta}{|\alpha|^2 - 1}.$$

Exercise 4.5 Compute the power series expansion of $M(z) = \dfrac{\alpha - z}{1 - \bar{\alpha}z}$ about 0.

Exercise 4.6 Suppose that f is a self-map of the unit disk $\mathbb{D}$ with $f(z) = a_1 z + a_2 z^2 + \ldots$, so that $f(0) = 0$. Set $g(z) = a_1 + a_2 z + a_3 z^2 + \ldots$. By the proof of Schwarz's Lemma, g is either a constant of modulus 1 or g is a self-map of $\mathbb{D}$. In the latter case, verify that h satisfies the conditions of Schwarz's Lemma where

$$h(z) = \frac{a_1 - g(z)}{1 - \overline{a_1} g(z)} = (M_{a_1} \circ g)(z)$$

and $M_{a_1}(z) = (a_1 - z)/(1 - \overline{a_1} z)$. Deduce that

$$|a_2| \leq 1 - |a_1|^2. \tag{4.6}$$

Determine when equality holds in (4.6).

From now on, for $a \in \mathbb{D}$, we write M_a for the Möbius transformation

$$M_a(z) = \frac{a - z}{1 - \bar{a}z}, \quad z \in \mathbb{D}. \tag{4.7}$$

While Proposition 4.1 may seem modest, in that we have found some automorphisms of the disk and they are all Möbius transformations, it is an important consequence of Schwarz's Lemma that we have actually found *all* automorphisms.

Theorem 4.2 *Every automorphism of the unit disk $\mathbb{D}$ is a Möbius map of the form* (4.5). *Thus,*

$$\operatorname{Aut}(\mathbb{D}) = \left\{ \phi(z) = \beta \frac{\alpha - z}{1 - \bar{\alpha}z} : |\beta| = 1 \text{ and } |\alpha| < 1 \right\}.$$

Proof We first show that an automorphism f of the disk that fixes 0 is a rotation. In fact, Schwarz's Lemma applies to such an f, since $f(0) = 0$, and implies that

$$|f'(0)| \leq 1.$$

The inverse function f^{-1} is likewise an automorphism of $\mathbb{D}$, with $f^{-1}(0) = 0$, and so this function too satisfies the conditions of Schwarz's Lemma. Applying Schwarz's Lemma to f^{-1} leads to

$$|(f^{-1})'(0)| \leq 1.$$

But $|(f^{-1})'(0)| = 1/|f'(0)|$ (bearing in mind that $f'(0) \neq 0$ since f is conformal), so that $|f'(0)| \geq 1$. Taken together, these two inequalities for $|f'(0)|$ lead to $|f'(0)| = 1$. By the equality statement in Schwarz's Lemma, there is a β of modulus 1 for which $f(z) = \beta z$. This is of the form (4.5) with 'α' = 0 and 'β' = $-\beta$.

Let us now suppose that g is any automorphism of $\mathbb{D}$. Set $\alpha = g^{-1}(0)$. Since, by Proposition 4.1, M_α is an automorphism of $\mathbb{D}$, the composition $f = g \circ M_\alpha$ is also an automorphism of the disk for which

$$f(0) = g(M_\alpha(0)) = g(\alpha) = 0.$$

Hence, by the immediately preceding discussion, f is a rotation in that there is a β of modulus 1 such that

$$g(M_\alpha(z)) = \beta z, \quad z \in \mathbb{D}. \tag{4.8}$$

Given $z \in \mathbb{D}$, then $M_\alpha(z) \in \mathbb{D}$ and (4.8) implies that

$$g(M_\alpha(M_\alpha(z))) = \beta M_\alpha(z), \quad z \in \mathbb{D}.$$

The result follows since $M_\alpha(M_\alpha(z)) = z$ (case $\beta = 1$ of Exercise 4.3). $\square$

Exercise 4.7 Find all automorphisms of the disk $D(0, R)$, with centre at 0 and radius R. Hint: use a simple conformal map of $\mathbb{D}$ to $D(0, R)$ together with Theorem 4.2.

Exercise 4.8 Verify that

$$\mathrm{Aut}(\mathbb{D}) = \left\{ \phi(z) = \beta \frac{az + b}{\bar{b}z + \bar{a}} : |a|^2 - |b|^2 = 1, \ |\beta| = 1 \right\}$$

is an alternative description of the automorphism group of the unit disk.

4.3 Automorphisms of a Half-Plane

It is natural to ask for descriptions of the groups of automorphisms of regions other than the unit disk. We did this for a disk of arbitrary radius in Exercise 4.7. Another region that comes naturally to mind is the upper half-plane $\mathbb{H} = \{z : \Im z > 0\}$. In proving the Riemann Mapping Theorem in Chap. 6 we will see that all simply connected domains other than $\mathbb{C}$ are conformally equivalent. This implies that any simply connected domain can be chosen as the model simply connected domain. The most natural model domains are the disk $\mathbb{D}$ and the half-plane $\mathbb{H}$; occasionally the infinite strip is a convenient choice of model domain.

We would like to describe the automorphism group $\mathrm{Aut}(\mathbb{H})$ of the half-plane $\mathbb{H}$. By pre-composing with a conformal map of the disk to the half-plane, and post-

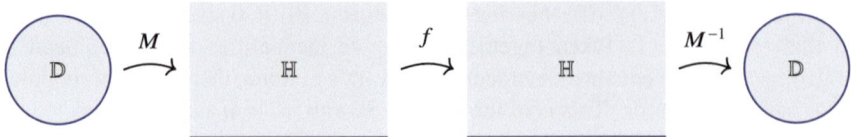

Fig. 4.2 Illustration of the composition of maps $g = M^{-1} \circ f \circ M$

composing with its inverse, we can reduce the problem to one we have already solved, that of describing the automorphisms of the disk. The map

$$M(z) = i\frac{1+z}{1-z}$$

is a conformal map of the disk $\mathbb{D}$ onto $\mathbb{H}$.

Exercise 4.9 Verify this last statement. (That is, argue that M is analytic and univalent in $\mathbb{D}$, that M maps $\mathbb{D}$ into $\mathbb{H}$, and that M maps $\mathbb{D}$ onto $\mathbb{H}$.)

Find an explicit formula for the inverse of the map M.

Suppose that $f \in \text{Aut}(\mathbb{H})$ so that f is a conformal map of $\mathbb{H}$ onto $\mathbb{H}$. Then

$$g = M^{-1} \circ f \circ M$$

is an automorphism of the disk, that is $g \in \text{Aut}(\mathbb{D})$. You should be sure that you understand why this is the case: the above diagram—Fig. 4.2—helps (cf. Exercise 4.2). We may then write f in terms of g as[6]

$$f = M \circ g \circ M^{-1}. \tag{4.9}$$

Conversely, if we start with $g \in \text{Aut}(\mathbb{D})$ and define f by (4.9) then $f \in \text{Aut}(\mathbb{H})$ (it may help to the draw the version of Fig. 4.2 that corresponds to (4.9)). With this simple argument, we have established the following.

Proposition 4.2 *A map f is an automorphism of the upper half-plane $\mathbb{H}$ if and only if it can be written in the form $f = M \circ g \circ M^{-1}$ where g is an automorphism of the disk $\mathbb{D}$ and $M(z) = i(1+z)/(1-z)$.*

Since every automorphism of $\mathbb{D}$ is a Möbius transformation (see Theorem 4.2) and M is also a Möbius transformation, this proposition exhibits each automorphism f of the half-plane $\mathbb{H}$ as a composition of three Möbius transformations. With very little work, we have found out that every automorphism of the half-plane $\mathbb{H}$ is a Möbius transformation.

---

[6] Note the similarity here to typical computations in a group: if $x = a^{-1}ya$ then $xa^{-1} = a^{-1}y(aa^{-1}) = a^{-1}y$, and then $axa^{-1} = a(a^{-1}y) = y$. In fact, we're already seen that $\text{Aut}(D)$ forms a group under composition of functions.

4.3 Automorphisms of a Half-Plane

Working directly with (4.9) to find an explicit form for an automorphism f of $\mathbb{H}$ is a little messy, so we take an indirect approach and first find an explicit description of all automorphisms of $\mathbb{H}$ that fix the point i, since $i = M(0)$.

Proposition 4.3 *Suppose that $h \in \mathrm{Aut}(\mathbb{H})$ and that $h(i) = i$. Then either $h(z) = -1/z$ or there is a real number k such that*

$$h(z) = \frac{z-k}{kz+1}, \quad z \in \mathbb{H}. \tag{4.10}$$

Conversely, $h(z) = -1/z$ and any map h given by (4.10) is an automorphism of $\mathbb{H}$ that fixes i.

Proof As noted above, if $h \in \mathrm{Aut}(\mathbb{H})$ then the map $g = M^{-1} \circ h \circ M$ is an automorphism of the disk $\mathbb{D}$. Not only that but, if $h(i) = i$, then

$$g(0) = M^{-1}(h(M(0))) = M^{-1}(h(i)) = M^{-1}(i) = 0.$$

Hence, by Theorem 4.2, g is a rotation in that $g(z) = \beta z$, $|\beta| = 1$, for $z \in \mathbb{D}$. For $z \in \mathbb{H}$, by (4.9),

$$h(z) = (M \circ g \circ M^{-1})(z) = (M \circ g)\left(\frac{z-i}{z+i}\right)$$

$$= M\left(\beta \frac{z-i}{z+i}\right) = i \frac{1 + \beta \frac{z-i}{z+i}}{1 - \beta \frac{z-i}{z+i}}$$

$$= i \frac{(1+\beta)z + i(1-\beta)}{(1-\beta)z + i(1+\beta)}$$

$$= \frac{z + i\frac{1-\beta}{1+\beta}}{-i\frac{1-\beta}{1+\beta}z + 1}.$$

The case $\beta = -1$ is a special case and gives rise to the automorphism $h(z) = -1/z$ of $\mathbb{H}$. Otherwise, since $|\beta| = 1$, we can write $\beta = e^{2i\theta}$ where $-\pi/2 < \theta < \pi/2$. Then,

$$k := -i\frac{1-\beta}{1+\beta} = -\tan\theta. \tag{4.11}$$

In particular, k is real and we see that h can be written in the form (4.10).

For the converse, $h(z) = -1/z$ is an automorphism of $\mathbb{H}$ and is its own inverse. Next, note that $-\tan\theta$ is a bijection from $(-\pi/2, \pi/2)$ to $\mathbb{R}$, and so any real k can be written in the form (4.11) for some β of modulus 1 with $\beta \neq -1$. Then, if h has the form (4.10), we reverse the above steps to see that $h = M \circ g \circ M^{-1}$ where g is a rotation of the disk. Hence h is an automorphism of $\mathbb{H}$ that fixes i. Alternatively, one may check directly that h as in (4.10) is an automorphism of the upper half-plane that fixes i. □

Now suppose that f is *any* automorphism of $\mathbb{H}$. Set $w_0 = f(i)$ with $w_0 = u_0 + iv_0$, $v_0 > 0$. Next set

$$h(z) = \frac{1}{v_0}\left[f(z) - u_0\right].$$

Then, h is also an automorphism of $\mathbb{H}$ (this should be checked!). Moreover, $h(i) = i$ and $f(z) = u_0 + v_0 h(z)$. Conversely, any automorphism h of the half-plane that fixes i gives rise to an automorphism f that sends i to $w_0 = u_0 + iv_0$, $v_0 > 0$, via

$$f(z) = u_0 + v_0 h(z), \quad z \in \mathbb{H}. \tag{4.12}$$

Thus, (4.12) is a complete description of those elements in $\mathrm{Aut}(\mathbb{H})$ that map i to w_0. Proposition 4.3 is a complete description of the maps h that arise in the context of (4.12). The choice $h(z) = -1/z$ leads to

$$f(z) = u_0 - \frac{v_0}{z} = \frac{u_0 z - v_0}{z}, \quad z \in \mathbb{H}. \tag{4.13}$$

Making use of (4.10), leads to

$$f(z) = u_0 + v_0 \frac{z-k}{kz+1} = \frac{(ku_0 + v_0)z + (u_0 - kv_0)}{kz+1}, \quad z \in \mathbb{H}. \tag{4.14}$$

The maps in (4.13) and (4.14) are both of the form

$$f(z) = \frac{az+b}{cz+d}$$

where a, b, c and d are real and $ad - bc > 0$. In the case of (4.13), $ad - bc = v_0 > 0$. In the case of (4.14),

$$ad - bc = (ku_0 + v_0) - k(u_0 - kv_0) = (1+k^2)v_0 > 0.$$

Together, (4.13) and (4.14) give a complete description of the automorphisms of the upper half-plane as u_0 varies in $\mathbb{R}$, v_0 varies in $(0, \infty)$, and k in (4.14) varies in $\mathbb{R}$. We may alternatively state this result in the following form

4.4 The Schwarz-Pick Lemma: First Version

Theorem 4.3 *The automorphisms of the upper half-plane $\mathbb{H}$ are precisely those Möbius maps that can be written in the form*

$$f(z) = \frac{az+b}{cz+d}, \quad z \in \mathbb{H}, \qquad (4.15)$$

where a, b, c and d are real and $ad - bc$ is positive.

Proof We have just seen that each map of the form (4.13) or (4.14) is of the form (4.15). Conversely, if f is a Möbius map of the form (4.15) with $d = 0$, we normalise the coefficients so that $c = 1$. Then, $f(z) = (az+b)/z$ with $-b > 0$, which is of the form (4.13) with $u_0 = a$ and $v_0 = -b$. If f is a Möbius map of the form (4.15) with $d \neq 0$ then we normalise the coefficients so that $d = 1$. Then, $f(z) = (az+b)/(cz+1)$ and $a - bc > 0$. This is of the form (4.14) with

$$k = c, \qquad u_0 = \frac{ca+b}{1+c^2}, \qquad v_0 = \frac{a-bc}{1+c^2}.$$

This shows that any map of the form (4.15) is an automorphism of $\mathbb{H}$. □

Exercise 4.10 Verify directly that every map of the form (4.15) is an automorphism of the half-plane $\mathbb{H}$.

4.4 The Schwarz-Pick Lemma: First Version

As promised at the end of Sect. 4.1, we will now be able to use the complete description of the automorphisms of the unit disk set out in Theorem 4.2 to prove a more general version of Schwarz's Lemma, one without the restriction that 0 is a fixed point of the self-map f of the disk $\mathbb{D}$.

Theorem 4.4 (The Schwarz-Pick Lemma—First Version) *Suppose that f is a self-map of the unit disk $\mathbb{D}$. Then, for points α and β in $\mathbb{D}$,*

$$|f'(\alpha)| \leq \frac{1-|f(\alpha)|^2}{1-|\alpha|^2} \qquad (4.16\text{a})$$

$$\left|\frac{f(\alpha)-f(\beta)}{1-\overline{f(\alpha)}f(\beta)}\right| \leq \left|\frac{\alpha-\beta}{1-\overline{\alpha}\beta}\right| \qquad (4.16\text{b})$$

Equality holds in (4.16a), or in (4.16b) for distinct α and β, if any only if f is an automorphism of the unit disk $\mathbb{D}$.

Proof We use the ideas outlined at the end of Sect. 4.1, in particular (4.3). For $a \in \mathbb{D}$, we consider the Möbius map given by (4.7). By Proposition 4.1, each function M_a is an automorphism of $\mathbb{D}$. For $\alpha \in \mathbb{D}$, we consider

$$g(z) = \bigl(M_{f(\alpha)} \circ f \circ M_\alpha\bigr)(z)$$

which, being a composition of three self-maps of $\mathbb{D}$, is itself a self-map of $\mathbb{D}$. Note that

$$g(0) = \bigl(M_{f(\alpha)} \circ f\bigr)\bigl(M_\alpha(0)\bigr) = \bigl(M_{f(\alpha)} \circ f\bigr)(\alpha) = M_{f(\alpha)}\bigl(f(\alpha)\bigr) = 0.$$

Thus g satisfies the conditions of Schwarz's Lemma. Applying Schwarz's Lemma to g, we deduce from (4.1a) that $|g'(0)| \leq 1$. But

$$\bigl|g'(0)\bigr| = \bigl|M'_{f(\alpha)}\bigl(f(\alpha)\bigr)\bigr|\bigl|f'(\alpha)\bigr|\bigl|M'_\alpha(0)\bigr|$$

so that, by Exercise 4.4,

$$|f'(\alpha)| \leq \frac{1}{\bigl|M'_{f(\alpha)}\bigl(f(\alpha)\bigr)\bigr|\bigl|M'_\alpha(0)\bigr|} = \frac{1 - |f(\alpha)|^2}{1 - |\alpha|^2},$$

which is (4.16a).

The inequality (4.1b) from Schwarz's Lemma applied to g at $z = M_\alpha(\beta)$ gives

$$\bigl|g\bigl(M_\alpha(\beta)\bigr)\bigr| \leq \bigl|M_\alpha(\beta)\bigr| = \left|\frac{\alpha - \beta}{1 - \overline{\alpha}\beta}\right|.$$

But, since $M_\alpha\bigl(M_\alpha(\beta)\bigr) = \beta$ (see Exercise 4.3),

$$\bigl|g\bigl(M_\alpha(\beta)\bigr)\bigr| = \bigl|M_{f(\alpha)}\bigl(f(\beta)\bigr)\bigr| = \left|\frac{f(\alpha) - f(\beta)}{1 - \overline{f(\alpha)}f(\beta)}\right|$$

and (4.16b) follows.

Equality holds in either (4.16a), or in (4.16b) for distinct α and β, precisely when equality holds in Schwarz's Lemma for the function g. This, in turn, is the case precisely when g is a rotation, that is $g(z) = az$ where $|a| = 1$. Since $f = M_{f(\alpha)} \circ g \circ M_\alpha$, the function f is an automorphism of $\mathbb{D}$ in this case. Conversely, if f is an automorphism of $\mathbb{D}$ then so is g. Since g fixes 0, g must then be a rotation and so equality holds in (4.16a), and in (4.16b) for any choice of α and β. □

Exercise 4.11 Verify by direct computation that equality holds in (4.16a) and (4.16b) if ϕ is an automorphism of the unit disk $\mathbb{D}$. In other words verify that, for $z \in \mathbb{D}$,

$$\frac{|\phi'(z)|}{1 - |\phi(z)|^2} = \frac{1}{1 - |z|^2}, \qquad (4.17)$$

and that, for z and w in $\mathbb{D}$,

$$\left|\frac{\phi(z) - \phi(w)}{1 - \overline{\phi(z)}\phi(w)}\right| = \left|\frac{z-w}{1-\bar{z}w}\right|. \tag{4.18}$$

4.5 The Hyperbolic Metric in the Disk

It is natural to draw a parallel between an automorphism of a domain D on the one hand and a rotation or translation of the Euclidean plane on the other, in that both are bijective transformations of their underlying space. Automorphisms are the natural transformations of a domain from the point of view of complex analysis, whereas the rigid motions of Euclidean space are natural from the point of view of Euclidean geometry—they are, in fact, the isometries of Euclidean space. From this perspective, it is natural to ask whether there is a geometry on a domain D in the complex plane such that the group of transformations $\mathrm{Aut}(D)$ are isometries in this geometry.

The perspective here differs from that adopted when we discussed the Riemann sphere in the following sense. In that case, we had a metric on the extended complex plane and set about finding the (holomorphic) isometries of that metric. In the current setting, we have a group of transformations $\mathrm{Aut}(\mathbb{D})$ of the underlying space $\mathbb{D}$ and wish to find a metric for which these are the isometries.

4.5.1 Construction of the Hyperbolic Metric

Recall the general construction of a Riemannian metric on a domain Ω with conformal metric ρ as developed in Sect. 3.4.1. With this in mind, we look for a scale function $\rho(z)$, $z \in \mathbb{D}$, such that each ϕ in $\mathrm{Aut}(\mathbb{D})$ is an isometry in the resulting Riemannian metric. By Proposition 3.5, ρ should satisfy

$$\rho(\phi(z))|\phi'(z)| = \rho(z), \text{ for } z \in \mathbb{D}, \ \phi \in \mathrm{Aut}(\mathbb{D}). \tag{4.19}$$

One look at (4.17) and we see that the choice

$$\rho(z) = \frac{2}{1-|z|^2}, \ z \in \mathbb{D}, \tag{4.20}$$

will work. Following the construction in Sect. 3.4.1, we make a Riemannian metric in the disk $\mathbb{D}$ with this choice of conformal metric ρ. This is the *hyperbolic metric* or *Poincaré metric* on the unit disk $\mathbb{D}$ and $d(z, w; \mathbb{D})$ is referred to as the *hyperbolic distance* between z and w in $\mathbb{D}$. By Proposition 3.4, $d(z, w; \mathbb{D})$ is a metric on the unit disk $\mathbb{D}$. By (4.19) and Proposition 3.5, each automorphism ϕ of the disk is an isometry of this metric in that, if z and w are in the disk $\mathbb{D}$ and $\phi \in \text{Aut}(\mathbb{D})$, then

$$d(\phi(z), \phi(w); \mathbb{D}) = d(z, w; \mathbb{D}). \qquad (4.21)$$

Notice that the conformal metric ρ in (4.20) is a radial function (that is, ρ is constant on circles about the origin). In contrast to the case of the spherical metric, ρ is an increasing radial function (rather than a decreasing radial function) and, in fact, $\rho(z)$ 'blows up' as z approaches the boundary of the unit disk. In particular, $\rho(z) \geq \rho(0) = 2$ everywhere in $\mathbb{D}$.

Exercise 4.12 Suppose that we require ρ to be chosen so that every ϕ in $\text{Aut}(\mathbb{D})$ is to be an isometry, equivalently so that

$$\rho(\phi(z)) |\phi'(z)| = \rho(z), \quad z \in \mathbb{D}, \ \phi \in \text{Aut}(\mathbb{D}).$$

Show that

$$\rho(z) = \frac{\rho(0)}{1 - |z|^2}, \quad z \in \mathbb{D}.$$

4.5.2 Geodesics, and a Formula for Hyperbolic Distance

Just as in the case of the spherical metric in Chap. 3, we should decide whether there is a curve of shortest hyperbolic length joining any two points in the disk and whether we can compute its length explicitly. Again in analogy with spherical distance, it is simplest to first look for curves of shortest length starting from the origin and then use isometries of the metric to move (literally) to the case of a general base point. Since the scale function $\rho(z) = 2/(1 - |z|^2)$ is rotationally symmetric (as was the scale factor $\rho(z) = 2/(1 + |z|^2)$ for the spherical metric on $\mathbb{C}_\infty$), it should not come as a surprise that the curve of shortest hyperbolic length from 0 to z in $\mathbb{D}$ is the Euclidean line segment $[0, z]$. The following is the counterpart of Proposition 3.6. The main difference between their proofs is that in Proposition 3.6 we worked forwards along the curve γ starting at 0 whereas in Proposition 4.4 we work backwards along the curve towards 0. The reason for this is that the conformal factor ρ is a decreasing function of $|z|$ in the case of the spherical metric and is an increasing function of $|z|$ in the case of the hyperbolic metric.

Proposition 4.4 *For $z \in \mathbb{D}$, the Euclidean line segment $[0, z]$ has the shortest hyperbolic length among all curves joining 0 to z. That is,*

$$d(0, z; \mathbb{D}) = L_\rho([0, z]) = \log\left(\frac{1 + |z|}{1 - |z|}\right). \qquad (4.22)$$

4.5 The Hyperbolic Metric in the Disk

Proof Let $z \in \mathbb{D}$. Let γ be any curve joining z to 0 in $\mathbb{D}$, which we parameterize by arclength $\gamma(t)$, $t \in [0, L]$, (so that $|\gamma'(t)| = 1$ for $t \in [0, L]$ and L is the Euclidean length of γ). Note that $L \geq |z|$ with equality if and only if L is the straight line segment $[z, 0]$. For $t \in [0, L]$,

$$|\gamma(t) - z| = |\gamma(t) - \gamma(0)| = \left|\int_0^t \gamma'(s)\,ds\right| \leq \int_0^t |\gamma'(s)|\,ds = \int_0^t ds = t.$$

Hence

$$|\gamma(t)| = |z + (\gamma(t) - z)| \geq |z| - |\gamma(t) - z| \geq |z| - t.$$

Since ρ decreases with decreasing modulus, it follows that

$$\rho(\gamma(t)) \geq \rho(|z| - t), \quad 0 \leq t \leq |z|.$$

Hence,

$$L_\rho(\gamma) = \int_0^L \rho(\gamma(t))\,|\gamma'(t)|\,dt = \int_0^L \rho(\gamma(t))\,dt$$

$$\geq \int_0^{|z|} \rho(\gamma(t))\,dt \qquad (4.23)$$

$$\geq \int_0^{|z|} \rho(|z| - t)\,dt = \int_0^{|z|} \frac{2\,dt}{1 - (|z| - t)^2} = L_\rho([z, 0]).$$

The last equality follows from the fact that the line segment $[z, 0]$ has the arclength parametrisation $\tilde{\gamma}(t) = (1 - t/|z|)z$, $0 \leq t \leq |z|$. Since $\rho \geq 2$, the inequality (4.23) is strict if $L > |z|$, that is if γ is not the Euclidean line segment $[z, 0]$. Hence, the line segment $[z, 0]$ or, by Exercise 3.14, the line segment $[0, z]$, has the shortest hyperbolic length of all curves joining 0 and z, and is the unique shortest curve. Its length is $\int_0^{|z|} 2\,dt/(1 - t^2)$ which evaluates to $\log\big((1 + |z|)/(1 - |z|)\big)$. □

We next wish to obtain the general version of Proposition 4.4 in which we find the hyperbolic distance between any two points in the disk and the curve of shortest hyperbolic length joining them.

Theorem 4.5 *Let α and β be any two points in the unit disk $\mathbb{D}$. Then*

$$d(\alpha, \beta; \mathbb{D}) = \log\left(\frac{1 + \delta(\alpha, \beta)}{1 - \delta(\alpha, \beta)}\right) \text{ where } \delta(\alpha, \beta) = \left|\frac{\alpha - \beta}{1 - \overline{\alpha}\beta}\right|. \qquad (4.24)$$

The unique curve of shortest hyperbolic length joining α to β lies on the generalised circle that passes through α and β and that meets the unit circle at right angles.

Moreover, the unit disk $\mathbb{D}$ is complete in the hyperbolic metric.

Note that $\delta(\alpha, \beta)$ is sometimes referred to as the *pseudo-hyperbolic distance* between α and β.

Proof Let α be a point in the unit disk $\mathbb{D}$, and consider the automorphism M_α (see (4.7)) of $\mathbb{D}$. Since M_α satisfies (4.19), M_α preserves the lengths of curves in the sense of (3.30) and is an isometry of the hyperbolic metric in the sense of (4.21).

Let β be another point in $\mathbb{D}$ and let γ be any curve joining α to β in $\mathbb{D}$. Then $M_\alpha \circ \gamma$ is a curve joining 0 to $M_\alpha(\beta)$ in $\mathbb{D}$ and, by (3.30) and Proposition 4.4,

$$L_\rho(\gamma) = L_\rho(M_\alpha \circ \gamma) \geq L_\rho\bigl([0, M_\alpha(\beta)]\bigr) = \log\left(\frac{1 + |M_\alpha(\beta)|}{1 - |M_\alpha(\beta)|}\right) \quad (4.24)$$

with equality if and only if $M_\alpha \circ \gamma$ is the line segment $[0, M_\alpha(\beta)]$. This is since $|M_\alpha(\beta)| = \delta(\alpha, \beta)$.

Now $M_\alpha \circ \gamma$ is the line segment $[0, M_\alpha(\beta)]$ precisely when $\gamma = M_\alpha\bigl([0, M_\alpha(\beta)]\bigr)$ (recall that $M_\alpha^{-1} = M_\alpha$). Since M_α is a Möbius transformation, the image of the line segment $[0, M_\alpha(\beta)]$ is an arc of a generalised circle C. This generalised circle C is the image under M_α of the Euclidean line L passing through 0 and $M_\alpha(\beta)$. This line L meets the unit circle $\mathbb{T}$ at right angles. Since M_α preserves angles and fixes the unit circle, $C = M_\alpha(L)$ and $\mathbb{T} = M_\alpha(\mathbb{T})$ also meet at right angles. The curve of shortest hyperbolic length joining α and β is therefore that arc of C that joins α and β as illustrated in Fig. 4.3.

To establish the completeness of $\mathbb{D}$ in the hyperbolic metric, suppose that $\{z_n\}_{n=1}^\infty$ is a sequence of points in $\mathbb{D}$ that is Cauchy in the hyperbolic metric. Since, in general, $|\alpha - \beta| \leq 2\delta(\alpha, \beta)$ for α and β in $\mathbb{D}$, the sequence $\{z_n\}_{n=1}^\infty$ is also Cauchy in the Euclidean metric. Since the sequence is Cauchy in the hyperbolic metric, there is some N such that $d(z_N, z_n; \mathbb{D}) \leq 1$ if $n > N$. Then, by the triangle inequality, $d(0, z_n; \mathbb{D}) \leq C$ if $n \geq N$ where $C = d(0, z_N; \mathbb{D}) + 1$. By (4.22),

$$\frac{1 + |z_n|}{1 - |z_n|} \leq e^C, \quad \text{that is} \quad |z_n| \leq \frac{e^C - 1}{e^C + 1} = r.$$

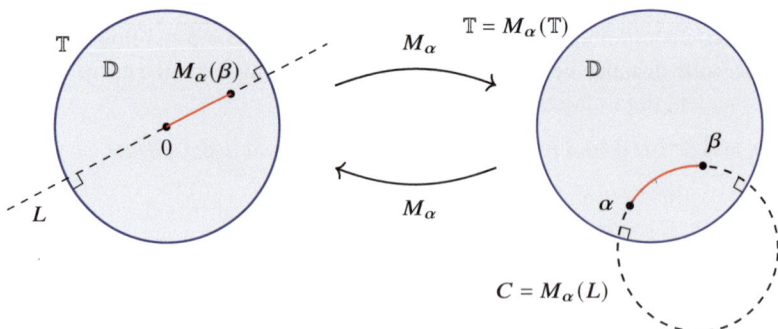

Fig. 4.3 A geodesic for the hyperbolic metric in the disk $\mathbb{D}$ as the image of a diameter under M_α

4.5 The Hyperbolic Metric in the Disk

Since $r < 1$, we conclude that $\{z_n\}_{n=N}^{\infty}$ is a sequence of points in the closed disk $\overline{D(0,r)}$ that is Cauchy in the Euclidean metric. There is therefore a point z with $|z| \leq r$ such that $z_n \to z$ as $n \to \infty$ in the Euclidean metric. Then,

$$\delta(z_n, z) = \frac{|z_n - z|}{|1 - \overline{z_n} z|} \leq \frac{|z_n - z|}{1 - r^2}$$

and so $\delta(z_n, z) \to 0$ as $n \to \infty$. It then follows from (4.24) that $d(z_n, z; \mathbb{D}) \to 0$ as $n \to \infty$, so that the hyperbolically Cauchy sequence $\{z_n\}_{n=1}^{\infty}$ is hyperbolically convergent. □

Next we turn to a complete description of the geodesics in the hyperbolic geometry in the disk. We again rely on the fact that the image of a geodesic under an isometry is again a geodesic. First we show that the diameter of the disk, $\Gamma = (-1, 1)$, is a geodesic. To this end, suppose that α and β both lie on this diameter. Since this diameter meets the unit circle at right angles, it follows from the statement of Theorem 4.5 that the subarc of Γ that joins them, in other words the line segment $[\alpha, \beta]$, is the curve of shortest length joining α and β. Thus, $\Gamma = (-1, 1)$ satisfies the condition required to be a geodesic. Conversely, arguing exactly as in the proof of Proposition 3.8 for geodesics passing through 0 in the spherical metric, we can show that if Γ is any hyperbolic geodesic that passes through the centre 0 of the disk $\mathbb{D}$ then Γ is a diameter of $\mathbb{D}$.

Finally, suppose that Γ is any hyperbolic geodesic. Let α be any point on Γ and apply the isometry M_α given by (4.7) so that $M_\alpha(\Gamma)$ is a geodesic that passes through 0 and hence is a diameter of the disk. Then, $\Gamma = M_\alpha(M_\alpha(\Gamma))$ is the image of this diameter under an automorphism of the disk $\mathbb{D}$. A typical situation is shown in Fig. 4.3. Conversely, the image of a diameter of the disk under such an automorphism will be a hyperbolic geodesic. We have found the following description of geodesics.

Theorem 4.6 *The hyperbolic geodesics in the disk $\mathbb{D}$ are precisely those curves which are images of the diameter $(-1, 1)$ of the disk under automorphisms of the disk.*

Alternatively, each geodesic is the intersection with the disk $\mathbb{D}$ of a generalised circle that intersects the unit circle $\mathbb{T}$ at right angles.

The second statement in Theorem 4.6 follows from the observation that a diameter of the disk meets the circle bounding the disk at right angles, and this angle of contact is preserved under any automorphism (all of which are *conformal* maps). Conversely, you should convince yourself that if C is a circle that meets the unit circle at right angles then there is an automorphism of the disk that maps C to the real line. In fact, a generalised circle that meets the unit circle at right angles and passes through 0 must be a Euclidean straight line. These geodesics are the 'straight lines' in Poincaré's non-Euclidean geometry. This is a genuine non-Euclidean geometry as the first four of Euclid's Postulates hold but the fifth, the Parallel Postulate does not. The Parallel Postulate fails because, given a 'line' L

Fig. 4.4 The Parallel Postulate fails for the hyperbolic geometry of the disk

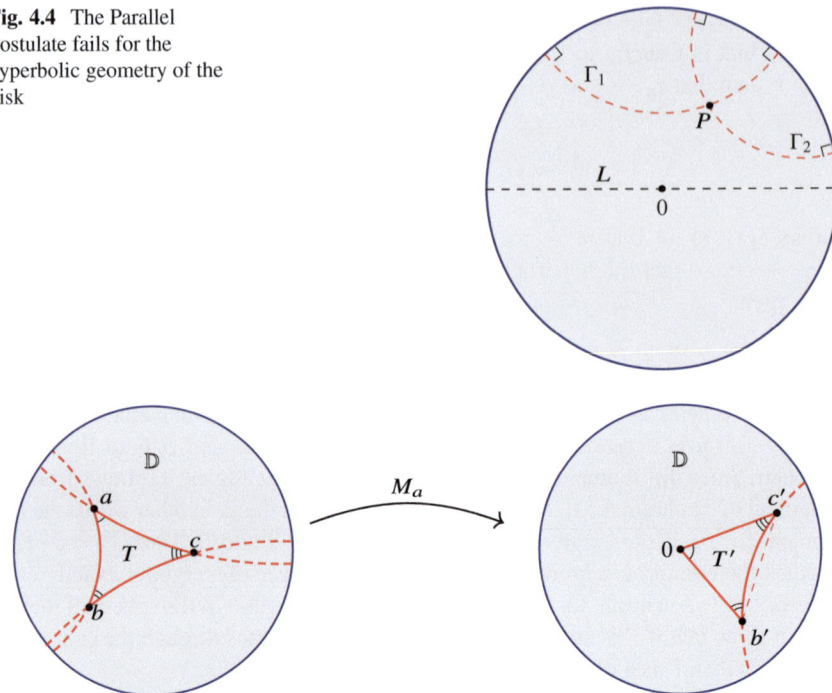

Fig. 4.5 A hyperbolic triangle: its angle sum is less than 180°

and a point P not on L, there are infinitely many 'lines' parallel to (that is, not intersecting) L and passing through P. By mapping with an isometry, we may arrange that L passes through 0 (and so is a diameter of the disk), in fact the diameter $(-1, 1)$. Figure 4.4 shows two of the infinitely many 'lines' that pass through P and are parallel to L. It was this geometry that finally resolved the debate, lasting centuries, as to whether the Parallel Postulate was logically independent of the first four postulates or not. A potted history of this fascinating story is to be found under 'Non-Euclidean Geometry' on the MacTutor website of the University of St. Andrews. See also the wonderfully readable and comprehensive account in Greenberg's book [12].

The Parallel Postulate is equivalent to the statement that the sum of the angles in any triangle is 180°, and so this cannot be the case in hyperbolic geometry. Consider a hyperbolic triangle T with vertices $a, b\ c$ in $\mathbb{D}$ as shown in Fig. 4.5, and angles A, B, C at these vertices. We map this triangle by the hyperbolic isometry M_a which, being conformal, preserves the angles at the vertices. The image triangle T' has vertices $M_a(a) = 0$ and then b' and c' say. The hyperbolic geodesic from b' to c' that forms one side of the hyperbolic triangle T' lies inside the Euclidean triangle with vertices at 0, b' and c'. Thus the angles at b' and c' in the hyperbolic triangle T' are smaller than the angles at b' and c' in the Euclidean triangle $\Delta 0b'c'$. The angles at 0 in both triangles are the same since the sides $0b'$ and $0c'$ are the Euclidean line segments $[0, b']$ and $[0, c']$ in the case of each triangle. As a consequence, the angle sum in the hyperbolic triangle is strictly less than the angle sum in the Euclidean triangle, hence strictly less than $180°$.

Exercise 4.13 Show that if the circle $C(a, r)$ meets the unit circle $\mathbb{T}$ at right angles at the points $e^{i\theta}$ and $e^{-i\theta}$ ($0 < \theta < \pi/2$) then the circle has centre $a = 1/\cos\theta$ and radius $r = \tan\theta$.

Hence find the centre and radius of a circle that meets the unit circle at right angles at the points $e^{i\theta_1}$ and $e^{i\theta_2}$ ($0 < \theta_2 - \theta_1 < \pi$).

4.6 The Schwarz-Pick Lemma: Second Version

The Schwarz-Pick Lemma, Theorem 4.4, states that if f is a self-map of the unit disk $\mathbb{D}$ and if α and β are points in $\mathbb{D}$, then

$$\left| \frac{f(\alpha) - f(\beta)}{1 - \overline{f(\alpha)} f(\beta)} \right| \leq \left| \frac{\alpha - \beta}{1 - \overline{\alpha}\beta} \right|.$$

In terms of the notation

$$\delta(\alpha, \beta) = \left| \frac{\alpha - \beta}{1 - \overline{\alpha}\beta} \right|$$

this inequality can be rephrased as

$$\delta\big(f(\alpha), f(\beta)\big) \leq \delta(\alpha, \beta).$$

Now the function

$$r \to \log\left(\frac{1+r}{1-r}\right)$$

is a strictly increasing function of r on $[0, 1)$ (its derivative is positive). Thus,

$$\log\left(\frac{1 + \delta\big(f(\alpha), f(\beta)\big)}{1 - \delta\big(f(\alpha), f(\beta)\big)}\right) \leq \log\left(\frac{1 + \delta(\alpha, \beta)}{1 - \delta(\alpha, \beta)}\right).$$

Comparing this to (4.24), this is an inequality for hyperbolic distances! Moreover, by the equality statement in the Schwarz-Pick Lemma [Theorem 4.4], equality holds for distinct points α and β if any only if f is a disk automorphism. To sum up, we have the following equivalent statement of Theorem 4.4.

Theorem 4.7 (The Schwarz-Pick Lemma—Second Version) *Each self-map f of the unit disk is a contraction in the hyperbolic metric: for any points α and β in $\mathbb{D}$,*

$$d\big(f(\alpha), f(\beta); \mathbb{D}\big) \leq d(\alpha, \beta; \mathbb{D}).$$

If equality holds for some pair of distinct points α and β then f is a disk automorphism, in which case equality then holds for every choice of α and β.

Exercise 4.14 Part of the statement of the Schwarz-Pick Lemma, Theorem 4.4, is the inequality

$$|f'(\alpha)| \leq \frac{1 - |f(\alpha)|^2}{1 - |\alpha|^2},$$

valid for any self-map f of the disk $\mathbb{D}$ and any point α in $\mathbb{D}$. Interpret this inequality in terms of hyperbolic geometry.

4.7 The Hyperbolic Metric in a Half-Plane

The question arises whether regions other than a disk have an associated 'hyperbolic metric' for which the automorphisms of the region are isometries. If so, is there a corresponding Schwarz-Pick Lemma? We will be able to give a reasonable answer to this question in due course. For now, we will work on a specific case, that of the upper half-plane $\mathbb{H}$ for which we computed the corresponding group of automorphisms in Sect. 4.3.

There are two possible approaches, both of which need to be kept in mind. The simplest approach is to transfer the hyperbolic distance directly from the disk to the half-plane via a conformal map. So choose a conformal map g from $\mathbb{H}$ to $\mathbb{D}$ (cf. Exercise 4.9). Given any two points α and β in $\mathbb{H}$, set

$$d(\alpha, \beta; \mathbb{H}) := d\big(g(\alpha), g(\beta); \mathbb{D}\big). \tag{4.25}$$

For this to be a valid definition of the distance between points α and β relative to the half-plane $\mathbb{H}$, the right hand side of (4.25) must not depend on the choice of conformal map g. Suppose, then, that $g_1 : \mathbb{H} \to \mathbb{D}$ is a second conformal map of $\mathbb{H}$ onto the disk $\mathbb{D}$ (see Fig. 4.6).

Then $\phi = g_1 \circ g^{-1}$ is an automorphism of $\mathbb{D}$, and so is an isometry of the hyperbolic metric on $\mathbb{D}$. In particular,

$$d\big(g(\alpha), g(\beta); \mathbb{D}\big) = d\big(\phi(g(\alpha)), \phi(g(\beta)); \mathbb{D}\big) = d\big(g_1(\alpha), g_1(\beta); \mathbb{D}\big).$$

Thus the right hand side of (4.25) is indeed independent of the choice of conformal map g.

4.7 The Hyperbolic Metric in a Half-Plane

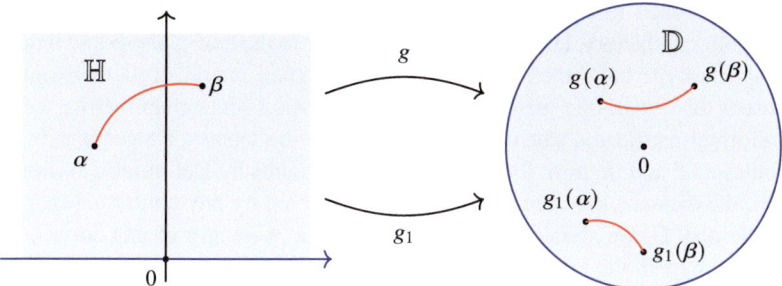

Fig. 4.6 The definition (4.25) of hyperbolic distance in the half-plane $\mathbb{H}$ does not depend on the choice of map g

Exercise 4.15 Verify that $d(\alpha, \beta; \mathbb{H})$, as defined by (4.25) in terms of the hyperbolic metric on the disk, satisfies the three properties of a metric.

Now suppose that $f: \mathbb{H} \to \mathbb{H}$ is a self-map of the half-plane $\mathbb{H}$. Again choose a conformal map $g: \mathbb{H} \to \mathbb{D}$ (for example, $g(w) = (w - i)/(w + i)$, $w \in \mathbb{H}$, as you will have found in Exercise 4.9). Then,

$$h = g \circ f \circ g^{-1}$$

is a self-map of $\mathbb{D}$ and so, by the Schwarz-Pick Lemma, Theorem 4.7,

$$d(h(z), h(w); \mathbb{D}) \leq d(z, w; \mathbb{D}) \tag{4.26}$$

for any choice of points z and w in $\mathbb{D}$. Given α and β in the half-plane $\mathbb{H}$, we apply (4.26) with $z = g(\alpha)$ and $w = g(\beta)$. This leads to

$$d\big(g(f(\alpha)), g(f(\beta)); \mathbb{D}\big) \leq d\big(g(\alpha), g(\beta); \mathbb{D}\big).$$

By (4.25), this becomes

$$d\big(f(\alpha), f(\beta); \mathbb{H}\big) \leq d(\alpha, \beta; \mathbb{H}).$$

Equality holds (for $\alpha \neq \beta$) if and only if equality holds in (4.26) (for $z \neq w$), which is the case if and only if h is a disk automorphism. Since $f = g^{-1} \circ h \circ g$, the map h is a disk automorphism if and only if f is an automorphism of the half-plane $\mathbb{H}$.

In summary, we have established the following. Let $g: \mathbb{H} \to \mathbb{D}$ be any conformal map of the half-plane $\mathbb{H}$ onto the disk $\mathbb{D}$. For α and β in $\mathbb{H}$, define the hyperbolic distance between α and β relative to $\mathbb{H}$ by (4.25). Then $d(\alpha, \beta; \mathbb{H})$ is a well-defined distance function on $\mathbb{H}$ for which the Schwarz-Pick Lemma holds: that is, every self-map of $\mathbb{H}$ is a contraction of the hyperbolic metric on $\mathbb{H}$ with the isometries being precisely the automorphisms Aut($\mathbb{H}$).

This approach to constructing a hyperbolic metric on the plane is quick and reasonably satisfactory. However, it is not intrinsic to the half-plane $\mathbb{H}$ and it doesn't display the hyperbolic metric on $\mathbb{H}$ as a Riemannian metric in its own right. To construct the hyperbolic metric on $\mathbb{H}$ as an intrinsic Riemannian metric, we need a conformal metric $\rho_\mathbb{H}$, say, on $\mathbb{H}$, to then define the length of a curve γ in $\mathbb{H}$ by Definition 3.5 and, in turn, the distance between points by Definition 3.6. Since, in (4.25), the distance between points in $\mathbb{H}$ is preserved by any conformal map of $\mathbb{H}$ onto the disk $\mathbb{D}$, we should choose $\rho_\mathbb{H}$ so that the $\rho_\mathbb{H}$-length of any curve γ in $\mathbb{H}$ equals the hyperbolic length of its image under g in $\mathbb{D}$. Taking a curve γ in $\mathbb{H}$ to be parameterised by $\gamma(t)$, $t \in [a, b]$, its $\rho_\mathbb{H}$-length would be

$$L_{\rho_\mathbb{H}}(\gamma) = \int_a^b \rho_\mathbb{H}(\gamma(t)) |\gamma'(t)| \, dt. \tag{4.27}$$

The image of γ under a conformal map g of $\mathbb{H}$ onto $\mathbb{D}$ is $g \circ \gamma$, and the hyperbolic length of this curve in $\mathbb{D}$ is

$$L_{\rho_\mathbb{D}}(g \circ \gamma) = \int_a^b \rho_\mathbb{D}\big((g \circ \gamma)(t)\big) |(g \circ \gamma)'(t)| \, dt$$

$$= \int_a^b \rho_\mathbb{D}\big(g(\gamma(t))\big) |g'(\gamma(t))| |\gamma'(t)| \, dt. \tag{4.28}$$

Here we have written, to avoid confusion, $\rho_\mathbb{D}(z)$ for $\rho(z) = 2/(1 - |z|^2)$. The simplest way in which the integrals (4.27) and (4.28) can be made equal is if the integrands are equal. Since $\gamma(t)$ could effectively be any point in $\mathbb{H}$, we would want (setting $\gamma(t) = w$)

$$\rho_\mathbb{H}(w) := \rho_\mathbb{D}(g(w)) |g'(w)|, \quad w \in \mathbb{H}. \tag{4.29}$$

If we wish to use (4.29) as the definition of a conformal metric in the half-plane $\mathbb{H}$, we need to check that its value does not depend on the choice of conformal map g from $\mathbb{H}$ to $\mathbb{D}$. Again, let $g_1 : \mathbb{H} \to \mathbb{D}$ be any other conformal map of $\mathbb{H}$ onto $\mathbb{D}$. As before, $\phi = g_1 \circ g^{-1}$ is an automorphism of $\mathbb{D}$. The identity (4.19) leads to

$$\rho_\mathbb{D}(\phi(z)) |\phi'(z)| = \rho_\mathbb{D}(z), \quad z \in \mathbb{D}.$$

This becomes, when expanded,

$$\rho_\mathbb{D}(g_1 \circ g^{-1}(z)) |g_1'(g^{-1}(z))| |(g^{-1})'(z)| = \rho_\mathbb{D}(z), \quad z \in \mathbb{D}. \tag{4.30}$$

Set $w = g^{-1}(z)$ (so that w is an arbitrary point in $\mathbb{H}$ since z is an arbitrary point in $\mathbb{D}$) and use

$$|(g^{-1})'(z)| = 1/|g'(g^{-1}(z))| = 1/|g'(w)|.$$

4.7 The Hyperbolic Metric in a Half-Plane

Since $z = g(w)$, (4.30) becomes

$$\rho_\mathbb{D}(g_1(w)) |g_1'(w)|/|g'(w)| = \rho_\mathbb{D}(g(w)), \quad w \in \mathbb{H},$$

or

$$\rho_\mathbb{D}(g_1(w)) |g_1'(w)| = \rho_\mathbb{D}(g(w)) |g'(w)|, \quad w \in \mathbb{H}.$$

This shows that the definition (4.29) of $\rho_\mathbb{H}$ does not depend on the choice of conformal map g.

We may therefore define the conformal metric $\rho_\mathbb{H}$ on the half-plane $\mathbb{H}$ by (4.29), and define the length of curves in $\mathbb{H}$ by Definition 3.5 and the distance $d(\alpha, \beta; \mathbb{H})$ between points α and β in $\mathbb{H}$ by Definition 3.6. From (4.29), and comparing (4.27) and (4.28), we see that

$$L_{\rho_\mathbb{D}}(g(\gamma)) = L_{\rho_\mathbb{H}}(\gamma) \tag{4.31}$$

for any curve γ in $\mathbb{H}$ and any conformal map g of $\mathbb{H}$ onto the disk $\mathbb{D}$. Definition 3.6 for points α and β in $\mathbb{H}$ becomes

$$d(\alpha, \beta; \mathbb{H}) = \inf \{ L_{\rho_\mathbb{H}}(\gamma) : \gamma \text{ is a curve in } \mathbb{H} \text{ joining } \alpha \text{ to } \beta \}. \tag{4.32}$$

It then follows from (4.31) that

$$d(\alpha, \beta; \mathbb{H}) = d(g(\alpha), g(\beta); \mathbb{D}),$$

and we're back to (4.25).

Notice that we didn't use an *explicit* conformal map $g: \mathbb{H} \to \mathbb{D}$ at any point. In fact, the above construction of the hyperbolic metric in $\mathbb{H}$, based on the conformal metric $\rho_\mathbb{H}$ defined by (4.29), will work in a general region D once we know, in theory at least, that *there is* a conformal map $g: D \to \mathbb{D}$. We will revisit this in more detail in Chap. 6.

That said, in the case of the half-plane $\mathbb{H}$, we can take

$$g(w) = \frac{w-i}{w+i}, \quad w \in \mathbb{H}, \tag{4.33}$$

as an explicit conformal map of $\mathbb{H}$ onto $\mathbb{D}$ in (4.29). Then, for $w \in \mathbb{H}$,

$$\rho_\mathbb{H}(w) = \rho_\mathbb{D}(g(w)) |g'(w)| = \frac{2}{1 - |g(w)|^2} |g'(w)|$$

$$= \frac{2}{1 - \left|\frac{w-i}{w+i}\right|^2} \frac{2}{|w+i|^2} = \frac{4}{|w+i|^2 - |w-i|^2}$$

$$= \frac{4}{\left(|w|^2 + 1 + i\overline{w} - iw\right) - \left(|w|^2 + 1 - i\overline{w} + iw\right)}$$

$$= \frac{2}{i\overline{w} - iw}$$

This gives the conformal metric in the half-plane $\mathbb{H}$ explicitly as

$$\rho_\mathbb{H}(w) = \frac{1}{\Im w}, \quad w \in \mathbb{H}. \tag{4.34}$$

Finally, we turn to a description of the geodesics in the hyperbolic metric of the half-plane $\mathbb{H}$. Geodesics in the half-plane and geodesics in the disk correspond under conformal maps between the half-plane and the disk. By this we mean that if g is any conformal mapping of the half-plane $\mathbb{H}$ onto the disk $\mathbb{D}$ then γ is a geodesic in $\mathbb{H}$ if and only if $g(\gamma)$ is a geodesic in $\mathbb{D}$. In fact, there is a one-to-one correspondence between curves in $\mathbb{H}$ joining α to β in $\mathbb{H}$ and curves in $\mathbb{D}$ joining $g(\alpha)$ to $g(\beta)$ in $\mathbb{D}$, by which corresponding curves have the same length. It follows that a curve of shortest length joining α to β in $\mathbb{H}$ does exist and is the preimage under g of the curve of shortest length joining $g(\alpha)$ and $g(\beta)$ in $\mathbb{D}$. With g given by (4.33), for example, and its inverse $g^{-1}(z) = i(1+z)/(1-z)$, $z \in \mathbb{D}$, this conformal map is also a Möbius map and so maps generalised circles to generalised circles. The unit circle and the real line correspond under g and g^{-1}. The hyperbolic geodesics in $\mathbb{D}$ are precisely those arcs of circles in $\mathbb{D}$ that meet the unit circle at right angles. Hence, their images under g^{-1}, that is the geodesics in $\mathbb{H}$, are precisely those arcs of generalised circles in $\mathbb{H}$ that meet the real line at right angles. These will include half-lines to infinity perpendicular to the real line, as shown in Fig. 4.7.

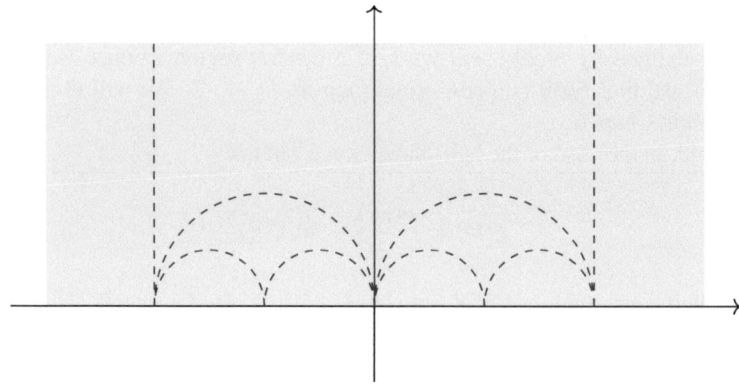

Fig. 4.7 Geodesics in the hyperbolic metric for the half-plane $\mathbb{H}$

Chapter 5
Normal Families and Value Distribution

One of the most striking results from a first course in complex analysis is the Casorati-Weierstrass Theorem [16, Theorem 8.10]. It states that if f has an essential singularity at the point a, say, then the set of values of f in any neighbourhood of a is dense in the complex plane. In this chapter, we are going to prove the much stronger result that, in fact, f assumes every complex value with at most one exception in any neighbourhood of an essential singularity. This result is known as the *Great Picard Theorem*. The example $f(z) = e^{1/z}$, which has an essential singularity at 0 and yet never assumes the value 0, shows that it is necessary to allow for one omitted value.

Another result on value distribution is the *Little Picard Theorem* which states that an entire function takes on every complex value with at most one exception. The example $f(z) = e^z$ is an entire function that never assumes the value 0, again showing that an omitted value is required. The Little Picard Theorem is a powerful extension of Liouville's Theorem which states that a bounded entire function is constant. It is easy to do a little better than the standard statement of Liouville's Theorem: if f is entire and never takes any values in the disk $D(a, r)$ (for some $a \subset \mathbb{C}$ and positive r) then $g(z) = 1/(f(z) - a)$ is entire and is bounded (by $1/r$); hence g is constant and so the original function f is also constant. The Little Picard Theorem says that an entire function is necessarily constant if, rather than a whole disk, it omits only two values.

We approach the theorems of Picard via the notion of a *normal family* which is the somewhat archaic, yet useful, terminology for a relatively compact set of functions. Before even discussing compactness in this context, we first need to put a metric space structure on the space of analytic functions on a domain D in the plane. The metric space structure will be chosen with convergence in mind: specifically, limits of convergent sequences of analytic functions will be analytic—the convergence that guarantees this is uniform convergence on compact subsets of D, that is locally uniform convergence. We will repeat this for meromorphic functions as opposed to analytic functions alone: in the latter case it is natural to view the target space as the complex plane $\mathbb{C}$ and in the former as the Riemann sphere $\mathbb{C}_\infty$.

To begin with, we will work more generally with continuous functions on a compact set and look for conditions that imply that a set of continuous functions is compact—this is the famous Arzelà-Ascoli Theorem. Starting from the Arzelà-Ascoli Theorem, Montel's Theorem gives a simpler criterion for a set of analytic functions to be relatively compact while Marty's Theorem gives a corresponding simpler criterion for a set of meromorphic functions to be relatively compact. We will also prove Zalcman's Lemma which is a criterion for a set of meromorphic functions to be non-normal. All these ingredients together will combine to provide a proof of Montel's Second Theorem and then the theorems of Picard.

The theorems of Picard are but one instance of the usefulness of the notion of a normal family. Normal families are a key tool in the arsenal of a complex analyst as compactness allows one to show that certain extremal problems have a solution, namely a function that realises the extremum. In general, a subset of a metric space is compact if and only if it is sequentially compact, that is if and only if every infinite sequence of points in the subset has a convergent subsequence. Montel's Theorem can then be used to produce analytic functions with certain desired properties. One sets up a suitable set of analytic functions, uses Montel's Theorem to show that it is compact, and the required function then arises as the limit of a subsequence of functions in the set. This general method will be used in the next chapter in the proof of the Riemann Mapping Theorem.

5.1 The Arzelà-Ascoli Theorem

For now we work in a general metric space setting with the goal of obtaining a criterion for a set of continuous functions to be compact. This is (one version of) the Arzelà-Ascoli Theorem.

In this section, (X, d_X) and (Y, d_Y) will denote metric spaces with X being compact. The space Y is assumed to be complete in some of the results, in which case this will be stated explicitly. Relevant examples of the target space Y are $\mathbb{C}$ with the Euclidean metric, or $\mathbb{C}_\infty$ with the spherical metric or, indeed, the unit disk $\mathbb{D}$ with the hyperbolic metric.

Let $\mathscr{C}(X, Y)$ be the space of continuous functions $f \colon X \to Y$. Note that, since X is compact, each function f in $\mathscr{C}(X, Y)$ is uniformly continuous. For f and g in $\mathscr{C}(X, Y)$, set

$$d(f, g) = \sup \{d_Y(f(x), g(x)) : x \in X\}. \tag{5.1}$$

Since $x \to d_Y(f(x), g(x))$ is a continuous function on X and since continuous real-valued functions on a compact set always achieve their supremum and infimum

5.1 The Arzelà-Ascoli Theorem

values, the 'supremum' in this definition can be replaced by 'maximum'. A trivial, yet often used estimate, one that is valid for each x_0 in X, is

$$d_Y\big(f(x_0), g(x_0)\big) \leq \sup\big\{d_Y\big(f(x), g(x)\big) : x \in X\big\} = d(f, g).$$

Exercise 5.1

▷ Show that, for f, g in $\mathscr{C}(X, Y)$, the function $x \to d_Y\big(f(x), g(x)\big)$ is a continuous function on X.
▷ Verify that (5.1) defines a metric on $\mathscr{C}(X, Y)$.

Note that a sequence of continuous functions $\{f_n\}_{n=1}^\infty$ in $\mathscr{C}(X, Y)$ converges to f in this metric if and only if f_n converges to f uniformly on X.

We will now see that, if Y is assumed to be complete, then $\mathscr{C}(X, Y)$ is complete in the supremum metric $d(\cdot, \cdot)$. Suppose that $\{f_n\}_{n=1}^\infty$ is a Cauchy sequence in $\mathscr{C}(X, Y)$ so that $d(f_n, f_m) \to 0$ as $n, m \to \infty$. Let ϵ positive be given. Then there is $N \in \mathbb{N}$ such that, for each $x \in X$,

$$d_Y\big(f_n(x), f_m(x)\big) \leq d\big(f_n, f_m\big) \leq \epsilon, \text{ for } n \geq N \text{ and } m \geq N. \tag{5.2}$$

This shows that the sequence of values $\{f_k(x)\}_{k=1}^\infty$ is a Cauchy sequence in Y, this for each $x \in X$, and hence convergent (since Y was taken to be complete). Say $f_k(x) \to f(x)$ as $k \to \infty$. This gives us a function $f: X \to Y$. Letting $m \to \infty$ in (5.2) we see, again for each individual x in X, that

$$d_Y\big(f_n(x), f(x)\big) \leq \epsilon \quad \text{if } n \geq N,$$

that is $d(f_n, f) \leq \epsilon$ if $n \geq N$. From this we can draw two conclusions: first that f_n converges to f uniformly on X and hence[1] the limit function f is continuous and belongs to $\mathscr{C}(X, Y)$; second that $f_n \to f$ in the metric $d(\cdot, \cdot)$. Thus, $\mathscr{C}(X, Y)$ is, indeed, complete.

Proposition 5.1 *Let (X, d_X) and (Y, d_Y) be metric spaces with X compact. Let $\mathscr{C}(X, Y)$ be the space of continuous functions from X to Y with metric given by (5.1). Then $\mathscr{C}(X, Y)$ is a metric space, and is complete if Y is complete.*

It is natural to ask what the compact subsets of our new metric space might look like. We first recall some basic properties of compact subsets of a metric space.

Proposition 5.2 *Let K be a subset of a metric space (M, d). Then the following are equivalent:*

(a) *K is a compact subset of M (that is, every cover of K by open subsets of M has a finite subcover);*
(b) *Every infinite set in K has a limit point in K;*

---

[1] The uniform limit of a sequence of continuous functions is itself continuous.

(c) K is sequentially compact (that is, every sequence in K has a subsequence that converges to a point in K);
(d) K is both complete (every Cauchy sequence in K converges to a point in K) and totally bounded, meaning that to each positive ϵ there correspond points $x_1, x_2, \ldots x_n$ in K such that

$$K \subseteq \bigcup_{i=1}^{n} B(x_i, \epsilon).$$

Here $B(x, \epsilon) = \{w \in M : d(x, w) < \epsilon\}$ is the ball with centre x and radius ϵ.

Exercise 5.2 Prove Proposition 5.1 (either yourself, or write out the details from your favourite analysis textbook).

Exercise 5.3

▷ Prove that a subset E of a metric space (M, d) is totally bounded if and only if its closure $\mathrm{cl}(E)$ is totally bounded.
▷ If (M, d) is also complete, prove that E is totally bounded if and only if $\mathrm{cl}(E)$ is compact.

It is useful to have the notion of a relatively compact (sometimes called a precompact) subset of a metric space, as well of that of a compact subset. A set F in a metric space (M, d) is said to be *relatively compact* if its closure $\mathrm{cl}(F)$ in (M, d) is compact. For example, the relatively compact subsets of $\mathbb{R}^n$ are the bounded sets. In analogy with Proposition 5.2, Part (c), we have the following.

Lemma 5.1 *A subset F of a metric space (M, d) is relatively compact if and only if every sequence in F has a convergent subsequence.*

Proof Suppose that every sequence in F has a convergent subsequence. We will show that $\mathrm{cl}(F)$ is sequentially compact and hence, by Proposition 5.2, is compact. Let $\{g_n\}_1^\infty$ be a sequence in $\mathrm{cl}(F)$. To each g_n there corresponds an f_n in F for which $d(g_n, f_n) < 2^{-n}$. The sequence $\{f_n\}_1^\infty$ in F has a convergent subsequence, say $\{f_{n_k}\}_{k=1}^\infty$, that converges to, say, f in M. Since

$$d(g_{n_k}, f) \leq d(g_{n_k}, f_{n_k}) + d(f_{n_k}, f) \leq 2^{-n_k} + d(f_{n_k}, f),$$

it follows that the subsequence $\{g_{n_k}\}_{k=1}^\infty$ of $\{g_n\}_1^\infty$ also converges to f. Since f is a limit point of F, we see that f lies in $\mathrm{cl}(F)$. Thus, $\mathrm{cl}(F)$ is sequentially compact and so F is relatively compact, by definition.

Suppose, conversely, that F is relatively compact and that $\{f_n\}_1^\infty$ is a sequence in F. Then $\{f_n\}_1^\infty$ is a sequence in the compact set $\mathrm{cl}(F)$ whence it has a convergent subsequence. □

Exercise 5.4 Show that each subset of a relatively compact set is itself relatively compact. (Bear in mind that a closed subset of a compact set is itself compact.)

5.1 The Arzelà-Ascoli Theorem

Our goal in this section is to find a way of recognising a relatively compact subset of $\mathscr{C}(X, Y)$ in terms of the behaviour of the functions in the family. This is achieved in the Arzelà-Ascoli Theorem, but we still have some work to do before reaching this goal. The first step is a general result about the convergence of functions defined on a *countable* set. The proof uses Cantor's diagonalization argument.

Proposition 5.3 *Let E be a countable set, let (M, d) be a metric space, and let $\{f_n\}_{n=1}^{\infty}$ a sequence of functions from E to M. Suppose that, for each $e \in E$, the set of values $\{f_n(e) : n \in \mathbb{N}\}$ is relatively compact in M. Then some subsequence $\{f_{n_k}\}_{k=1}^{\infty}$ is convergent pointwise at each point of E.*

Note that if $M = \mathbb{C}$ or $\mathbb{R}^n$ then the condition that 'the set $\{f_n(e) : n \in \mathbb{N}\}$ is relatively compact' can be rephrased as the condition that 'the set $\{f_n(e) : n \in \mathbb{N}\}$ is bounded'. If M is the Riemann sphere $\mathbb{C}_\infty$ then this condition is automatically satisfied as $\mathbb{C}_\infty$ itself is compact. Note also that there are no regularity assumptions on the functions in this result.

Proof Let $\{e_n\}_{n=1}^{\infty}$ be an ordering of the elements in E. The sequence of values $\{f_n(e_1)\}_{n=1}^{\infty}$ lies in a compact subset of M and therefore has a convergent subsequence, which we will write as $\{f_{1,n}(e_1)\}_{n=1}^{\infty}$. Now consider the values of this subsequence of functions at the second point e_2, namely the sequence of values $\{f_{1,n}(e_2)\}_{n=1}^{\infty}$. Again, by assumption and Exercise 5.4, this set of values is relatively compact in M and so it too has a convergent subsequence, say $\{f_{2,n}(e_2)\}_{n=1}^{\infty}$. Note that $\{f_{2,n}\}_{n=1}^{\infty}$ is a subsequence of a subsequence of the original sequence. Being a subsequence of $\{f_{1,n}\}_{n=1}^{\infty}$, it converges not just at e_2 but also at e_1.

Continuing in this fashion (we would next consider the sequence of values $\{f_{2,n}(e_3)\}_{n=1}^{\infty}$), we produce for each k a sequence of functions $\{f_{k,n}\}_{n=1}^{\infty}$ that converge at e_k. Moreover, each sequence $\{f_{k,n}\}_{n=1}^{\infty}$ is a subsequence of the previous sequence $\{f_{k-1,n}\}_{n=1}^{\infty}$, and hence of all previous sequences, and so also converges at $e_{k-1}, e_{k-2}, \ldots e_2, e_1$.

Now consider the sequence of functions $\{f_{i,i}\}_{i=1}^{\infty}$ (this is the diagonal sequence underlying Cantor's diagonalization argument). For each fixed k, and for each i bigger than k, the term $f_{i,i}$ is in the subsequence $\{f_{i,n}\}_{n=1}^{\infty}$, which is itself a subsequence of $\{f_{k,n}\}_{n=1}^{\infty}$ (since $i \geq k$). Thus, $f_{i,i}$ for $i \geq k$ is a subsequence of $\{f_{k,n}\}_{n=1}^{\infty}$, and so $\{f_{i,i}(e_k)\}_{i=1}^{\infty}$ is convergent. Since this holds for every k, we see that the sequence of functions $\{f_{i,i}\}_{i=1}^{\infty}$, a subsequence of the original sequence $\{f_n\}_{n=1}^{\infty}$, is convergent at each point of E. □

We will use Proposition 5.3 in the context of functions defined on a compact metric space X, in which case we want to argue that X has a countable dense[2] subset. This is not too difficult to see. Since X is totally bounded (by Proposition 5.2), for each $n \in \mathbb{N}$ there are finitely many $x_{n,1}, x_{n,2}, \ldots, x_{n,k_n}$ such that

$$X \subseteq \bigcup_{k=1}^{k_n} B_X(x_{n,k}, 1/n).$$

---

[2] A subset of X is dense if its closure is X.

It follows that

$$E = \bigcup_{n=1}^{\infty} \{x_{n,k} : k = 1 \ldots k_n\}$$

is a countable dense subset of X.

In order to understand relatively compact subsets of the metric space $\mathscr{C}(X, Y)$ we need, by Proposition 5.2 (c), to have a better understanding of uniformly convergent sequences of functions on X. Suppose that $\{f_n\}_{n=1}^{\infty}$ is a convergent sequence of functions in $\mathscr{C}(X, Y)$. Consequently, it is also a Cauchy sequence of functions. Let ϵ positive be given. Then, since $\{f_n\}_{n=1}^{\infty}$ is a Cauchy sequence of functions, there is a natural number N such that

$$d_Y\big(f_m(x), f_n(x)\big) \leq \frac{\epsilon}{3} \text{ whenever } x \in X \text{ and } m \geq N,\ n \geq N. \tag{5.3}$$

Since each function f_k is uniformly continuous on X, there is a positive number δ_k such that

$$d_Y\big(f_k(x_1), f_k(x_2)\big) \leq \frac{\epsilon}{3} \text{ whenever } d_X(x_1, x_2) < \delta_k. \tag{5.4}$$

Fix $m = N$ in (5.3), and apply (5.4) with $k = N$. Then, if $n \geq N$ and $d_X(x_1, x_2) < \delta_N$,

$$d_Y\big(f_n(x_1), f_n(x_2)\big) \leq d_Y\big(f_n(x_1), f_N(x_1)\big) + d_Y\big(f_N(x_1), f_N(x_2)\big)$$
$$+ d_Y\big(f_N(x_2), f_n(x_2)\big)$$
$$\leq \frac{\epsilon}{3} + \frac{\epsilon}{3} + \frac{\epsilon}{3} = \epsilon.$$

Here we used (5.3), then (5.4), then (5.3) again to estimate the terms on the right. It follows that,

$$d_Y\big(f_n(x_1), f_n(x_2)\big) \leq \epsilon \text{ whenever } d_X(x_1, x_2) < \delta_N \text{ and } n \geq N.$$

Using the uniform continuity of the finitely many remaining functions $f_1, f_2, \ldots, f_{N-1}$, we see that there is a positive δ such that, for each n,

$$d_Y\big(f_n(x_1), f_n(x_2)\big) \leq \epsilon \text{ whenever } d_X(x_1, x_2) < \delta.$$

This shows that a convergent sequence of functions in $\mathscr{C}(X, Y)$ is *equicontinuous* in the sense of the following definition.

5.1 The Arzelà-Ascoli Theorem

Definition 5.1 Let $\mathscr{F}$ be a family of functions in $\mathscr{C}(X, Y)$. We say that $\mathscr{F}$ is *equicontinuous* on X if to each positive number ϵ there corresponds a positive number δ such that

$$d_Y\big(f(x_1), f(x_2)\big) \leq \epsilon \text{ whenever } d_X(x_1, x_2) \leq \delta \text{ and } f \in \mathscr{F}. \tag{5.5}$$

As we have already pointed out, a continuous function f on X is uniformly continuous, this because X is compact: given ϵ positive there corresponds a positive number δ_f such that

$$d_Y\big(f(x_1), f(x_2)\big) \leq \epsilon \quad \text{whenever } d_X(x_1, x_2) \leq \delta_f. \tag{5.6}$$

The family $\mathscr{F}$ is equicontinuous if there is a positive δ for which it is possible to choose δ_f so that $\delta_f \geq \delta$ for every f in $\mathscr{F}$. Equicontinuity can be thought of as 'uniform, uniform continuity' in that the same 'δ' works not only for x_1 and x_2 sufficiently close in X but also for every function f in the family $\mathscr{F}$. A finite family of functions is clearly equicontinuous: in fact, if $\mathscr{F} = \{f_1, f_2, \ldots, f_n\}$ then (5.6) holds separately for each f_i, $1 \leq i \leq n$, with $\delta = \delta_i$ say. Then, (5.5) holds with

$$\delta = \min\{\delta_1, \delta_2, \ldots, \delta_n\},$$

which is positive.

The discussion preceding the definition of equicontinuity shows that uniform convergence of a sequence of functions implies that the sequence of functions is equicontinuous (by equicontinuity of a sequence of functions $\{f_n\}_{n=1}^\infty$ we mean that the set of functions occurring in the sequence forms an equicontinuous family). In the other direction, equicontinuity implies convergence—almost! For a trivial yet illustrative example, set $f_n \equiv (-1)^n$ on $X = [0, 1]$. The sequence $\{f_n\}_{n=1}^\infty$ is certainly equicontinuous as each function in the sequence is constant, but it is certainly not convergent. The correct result, which avoids this anomaly, is the following.

Proposition 5.4 *Let $\{f_n\}_{n=1}^\infty$ be a sequence of functions in $\mathscr{C}(X, Y)$. If $f_n \to f$ in $\mathscr{C}(X, Y)$, that is if $f_n \to f$ uniformly on X, then the sequence of functions is equicontinuous. Conversely, if the sequence $\{f_n\}_{n=1}^\infty$ is equicontinuous and if there is a function $f : X \to Y$ such that $f_n(x) \to f(x)$ pointwise for each $x \in X$ then $f_n \to f$ in $\mathscr{C}(X, Y)$.*

Proof That $f_n \to f$ in $\mathscr{C}(X, Y)$ implies that the sequence $\{f_n\}_{n=1}^\infty$ is equicontinuous has already been shown in the discussion preceding Definition 5.1.

In the other direction, suppose that the sequence $\{f_n\}_{n=1}^\infty$ is equicontinuous and that there is a function $f : X \to Y$ such that $f_n \to f$ pointwise on X. Note that we make no assumption on the pointwise limit function f, not even that it is continuous. Suppose to the contrary that the sequence $\{f_n\}_{n=1}^\infty$ does not converge to this function f *uniformly* on X. Then there is some positive ϵ, some sequence $\{n_k\}_{k=1}^\infty$ with $n_k \to \infty$, and points x_{n_k} in X such that

$$d_Y\left(f_{n_k}(x_{n_k}), f(x_{n_k})\right) \geq \epsilon \text{ for each } k.$$

Since X is compact, the sequence of points $\{x_{n_k}\}_{k=1}^\infty$ has a convergent subsequence. To simplify notation, let's denote the convergent subsequence of points in X by $\{a_k\}_{k=1}^\infty$ with its limit being a, and rename the corresponding subsequence of $\{f_{n_k}\}_{k=1}^\infty$ as $\{g_n\}_{n=1}^\infty$. Thus, $\{g_n\}_{n=1}^\infty$ is an equicontinuous sequence of functions that converges pointwise to f, and there is a sequence of points $\{a_k\}_{k=1}^\infty$ converging to a in X such that

$$d_Y\left(g_n(a_n), f(a_n)\right) \geq \epsilon \text{ for each } n. \tag{5.7}$$

Now the assumption that the sequence of functions $\{g_n\}_{n=1}^\infty$ is equicontinuous implies that there is a positive number δ such that

$$d_Y\left(g_n(x_1), g_n(x_2)\right) \leq \frac{\epsilon}{6} \text{ for every } n, \text{ once } d_X(x_1, x_2) \leq \delta. \tag{5.8}$$

Since $a_k \to a$, we can choose K large enough so that $d_X(a_k, a) \leq \delta$ whenever $k \geq K$. By (5.8), then,

$$d_Y\left(g_n(a_k), g_n(a)\right) \leq \frac{\epsilon}{6} \text{ for } k \geq K \text{ and for every } n. \tag{5.9}$$

(Do keep in mind the relatively subtle distinction between n and k here and in the remainder of the proof!). Since $g_n(a) \to f(a)$ as $n \to \infty$, there is a positive integer N such that

$$d_Y\left(g_n(a), g_m(a)\right) \leq \frac{\epsilon}{6} \text{ for } n, m \geq N. \tag{5.10}$$

For $n \geq \max\{K, N\}$, and for $m \geq N$,

$$d_Y\left(g_n(a_n), g_m(a_n)\right)$$
$$\leq d_Y\left(g_n(a_n), g_n(a)\right) + d_Y\left(g_n(a), g_m(a)\right) + d_Y\left(g_m(a), g_m(a_n)\right)$$
$$\leq \frac{\epsilon}{6} + \frac{\epsilon}{6} + \frac{\epsilon}{6} = \frac{\epsilon}{2},$$

where we used (5.9), then (5.10), and then (5.9) again. Letting $m \to \infty$ in this last estimate we find, since $g_m(a_n) \to f(a_n)$ as $m \to \infty$, that

$$d_Y\left(g_n(a_n), f(a_n)\right) \leq \frac{\epsilon}{2} \text{ for } n \geq \max\{K, N\}.$$

5.1 The Arzelà-Ascoli Theorem

But this contradicts (5.7), thereby establishing the uniform convergence of the sequence of functions $\{f_n\}_{n=1}^{\infty}$ to its pointwise limit f. Moreover, being the uniform limit of a sequence of continuous functions, f is itself continuous, and $f_n \to f$ in $\mathscr{C}(X, Y)$. □

We are now ready to state and prove the Arzelà-Ascoli Theorem.

Theorem 5.1 (Arzelà-Ascoli Theorem—Compact Space Setting) *Suppose that (X, d_X) and (Y, d_Y) are metric spaces with X compact and Y complete. A subset $\mathscr{F}$ of $\mathscr{C}(X, Y)$ is relatively compact if and only if*

(a) *for every x in X, the set of function values $\mathscr{F}(x) = \{f(x) : f \in \mathscr{F}\}$ is a relatively compact subset of Y;*
(b) *$\mathscr{F}$ is an equicontinuous family of functions.*

Proof We will first show that (a) and (b) hold if $\mathscr{F}$ is a relatively compact subset of $\mathscr{C}(X, Y)$. First (a): fix x in X. Let $\{y_n\}_{n=1}^{\infty}$ be a sequence in the set $\{f(x) : f \in \mathscr{F}\}$. Then, $y_n = f_n(x)$ for some sequence of functions $\{f_n\}_{n=1}^{\infty}$ in $\mathscr{F}$. Since $\mathscr{F}$ is relatively compact, this sequence of functions has a convergent subsequence by Lemma 5.1, say $\{f_{n_k}\}_{k=1}^{\infty}$, and this subsequence converges, in particular, at x. That is, $\{f_{n_k}(x)\}_{k=1}^{\infty} = \{y_{n_k}\}_{k=1}^{\infty}$ is a convergent subsequence of the original sequence $\{y_n\}_{n=1}^{\infty}$. Again by Lemma 5.1, the set $\{f(x) : f \in \mathscr{F}\}$ is relatively compact in Y.

To show that (b) holds if $\mathscr{F}$ is relatively compact, we prove the contrapositive: if $\mathscr{F}$ is not an equicontinuous family of functions then it is not a relatively compact subset of $\mathscr{C}(X, Y)$. Suppose, then, that $\mathscr{F}$ is not equicontinuous. Then there is some positive ϵ for which it is not possible to find any positive δ such that (5.5) holds, so that (5.5) fails to hold with $\delta = 1/n$, this for every $n \in \mathbb{N}$. That is, for each $n \in \mathbb{N}$, there is some function $f_n \in \mathscr{F}$ and points x_{n_1}, x_{n_2} in X such that

$$d_Y(f_n(x_{n_1}), f_n(x_{n_2})) \geq \epsilon \text{ and } d_X(x_{n_1}, x_{n_2}) \leq \frac{1}{n}.$$

In particular, the sequence of functions $\{f_n\}_{n=1}^{\infty}$ is not equicontinuous, and neither is any subsequence of this sequence of functions. It follows from Proposition 5.4 that no subsequence of $\{f_n\}_{n=1}^{\infty}$ is convergent in $\mathscr{C}(X, Y)$ which in turn shows, by Lemma 5.1 again, that $\mathscr{F}$ is not relatively compact.

Conversely, we now show that if $\mathscr{F}$ is a family of functions in $\mathscr{C}(X, Y)$ for which (a) and (b) hold then $\mathscr{F}$ is relatively compact in $\mathscr{C}(X, Y)$. Let $\{f_n\}_{n=1}^{\infty}$ be any sequence of functions in $\mathscr{F}$. Let E be a countable dense subset of X. By Assumption (a), the conditions of Proposition 5.3 are satisfied and we may conclude that there is a subsequence $\{f_{n_k}\}_{k=1}^{\infty}$ that converges pointwise at each point of E. For convenience, we write g_k for f_{n_k}. From the assumption (b) that the whole family $\mathscr{F}$ is equicontinuous, we see that the sequence of functions $\{g_k\}_{k=1}^{\infty}$ is equicontinuous. We use this equicontinuity to show that the sequence $\{g_k\}_{k=1}^{\infty}$ converges pointwise not just at each point of E but on all of X. In fact, let ϵ positive be given. By equicontinuity, there is a positive δ such that

$$d_Y(g_k(x_1), g_k(x_2)) \leq \frac{\epsilon}{3} \text{ whenever } d_X(x_1, x_2) \leq \delta, \text{ each } k \in \mathbb{N}. \tag{5.11}$$

Let $x \in X$. Chose $x_1 \in E$ with $d_X(x, x_1) \leq \delta$. Then choose N such that

$$d_Y(g_k(x_1), g_j(x_1)) \leq \frac{\epsilon}{3} \text{ whenever } k, j \geq N. \tag{5.12}$$

We can do this because $\{g_k(x_1)\}$ is a convergent sequence in Y. Putting this together shows that, for $k, j \geq N$,

$$d_Y(g_k(x), g_j(x)) \leq d_Y(g_k(x), g_k(x_1)) + d_Y(g_k(x_1), g_j(x_1)) + d_Y(g_j(x_1), g_j(x))$$
$$\leq \frac{\epsilon}{3} + \frac{\epsilon}{3} + \frac{\epsilon}{3} = \epsilon,$$

where we used (5.11), then (5.12), then (5.11) again. This shows that the sequence $\{g_k(x)\}_{k=1}^\infty$ is a Cauchy sequence in the complete space Y and therefore convergent. Hence there is some function g on X such that $g_k(x) \to g(x)$ for every $x \in X$. We conclude from Proposition 5.4 that $g_k \to g$ uniformly on X, that is $\{g_k\}_{k=1}^\infty = \{f_{n_k}\}_{k=1}^\infty$ is a convergent subsequence of the original sequence $\{f_n\}_{n=1}^\infty$ in $\mathscr{C}(X, Y)$. By Lemma 5.1, the family $\mathscr{F}$ is relatively compact in $\mathscr{C}(X, Y)$. □

Exercise 5.5 Let $\mathscr{F}$ be a subset of $\mathscr{C}(X, Y)$. A subset $\mathscr{F}$ of $\mathscr{C}(X, Y)$ is said to be *equicontinuous at the point* x_0 *in* X if to each positive number ϵ there corresponds a positive number δ such that

$$d_Y(f(x), f(x_0)) \leq \epsilon \text{ whenever } d_X(x, x_0) \leq \delta \text{ and } f \in \mathscr{F}.$$

Prove that $\mathscr{F}$ is equicontinuous on X (in the sense of Definition 5.1) if and only if $\mathscr{F}$ is equicontinuous at each point of X. (Note that one direction is trivial and the other is essentially the proof that a continuous function on a compact set is uniformly continuous.)

Exercise 5.6 Prove that the continuous image of a relatively compact set is relatively compact. (Recall that the continuous image of a compact set is compact.)

Exercise 5.7 Suppose that $\{f_n\}_1^\infty$ is a sequence in $\mathscr{C}(X, Y)$ that converges to f in $\mathscr{C}(X, Y)$. Suppose that $\{x_n\}_1^\infty$ is a sequence in X that converges to x in X. Show that

$$\lim_{n \to \infty} f_n(x_n) = f(x).$$

5.2 The Space of Continuous Functions on a Planar Domain

In the case when the target space is $\mathbb{R}^d$ (or, more generally, a normed space or a Banach space) we write $\mathscr{C}(X)$ for $\mathscr{C}(X, \mathbb{R}^d)$ and, for $f \in \mathscr{C}(X)$, we set

$$\|f\|_{\mathscr{C}(X)} = \max\left\{\|f(x)\|_{\mathbb{R}^d} : x \in X\right\}.$$

Then, $\mathscr{C}(X)$ becomes a Banach space and the metric $d(f, g)$ defined by (5.1) can be recast as

$$d(f, g) = \|f - g\|_{\mathscr{C}(X)}, \text{ for } f, g \in \mathscr{C}(X).$$

A subset $\mathscr{F}$ of $\mathscr{C}(X)$ is *bounded* if there is a finite M such that $\|f\|_{\mathscr{C}(X)} \leq M$ for each $f \in M$. Equivalently,

$$\|f(x)\|_{\mathbb{R}^d} \leq M \text{ for every } x \in X \text{ and every } f \in \mathscr{F}.$$

Exercise 5.8 Verify the following form of the Arzelà-Ascoli Theorem: a subset $\mathscr{F}$ of $\mathscr{C}(X)$ is relatively compact if and only if it is bounded and equicontinuous, and is compact if and only if it is closed, bounded and equicontinuous. (Note that a subset of $\mathbb{R}^n$ is compact if and only if it is closed and bounded and is relatively compact if and only if it is bounded; this is the Heine-Borel Theorem.)

5.2 The Space of Continuous Functions on a Planar Domain

In the last section on the Arzelà-Ascoli Theorem, we worked in a reasonably general setting. We now move to the more concrete setting of continuous, Y-valued functions on a domain (an open connected set) in the complex plane and write $\mathscr{C}(D, Y)$ for the space of such functions. We make the standing assumption that (Y, d_Y) is a complete metric space. In reality, though, we will only ever use $Y = \mathbb{C}$ when working with analytic functions and $Y = \mathbb{C}_\infty$ when working with meromorphic functions. The domain D is open and therefore not compact, so the results of the previous section do not apply directly. The trick is to 'exhaust' D by compact sets, meaning an increasing sequence of compact sets that fill out all of D.

Our primary interest is in $\mathscr{H}(D) \subset \mathscr{C}(D, \mathbb{C})$ consisting of the analytic functions on D and in $\mathscr{M}(D) \subset \mathscr{C}(D, \mathbb{C}_\infty)$ consisting of the meromorphic functions on D. For this reason, we give $\mathscr{C}(D, Y)$ a metric space structure corresponding to that of *uniform convergence on compact subsets*. We will show later (Theorem 5.5) that, with this definition of convergence, the limit of a sequence of analytic functions is analytic and (Theorem 5.6) that the limit of a sequence of meromorphic functions is either the constant function ∞ or meromorphic. The first thing we need, then, is an increasing sequence of compact subsets whose union is D. It's worth bearing in mind that the results in this section go through with only minor changes in notation if we wish to deal with continuous functions from a metric space (X, d_X) to a complete metric space (Y, d_Y) so long as the space X can be exhausted by compact sets. In the next lemma, we show that any domain in the plane satisfies this assumption. It is fine to skip the proof, though not the statement, at first reading.

Lemma 5.2 *Suppose that D is a domain in the plane. There is a sequence $\{K_n\}_{n=1}^{\infty}$ of compact subsets of D with the property that*

(i) $K_n \subseteq \operatorname{int}(K_{n+1})$ *for* $n \geq 1$;

(ii) $\bigcup_{n=1}^{\infty} K_n = D$.

If K is any compact subset of D then there is an n for which $K \subseteq K_n$.

Such a sequence $\{K_n\}$ is called an *exhaustion of D by compact subsets*. Note that, as a consequence of (i), $K_n \subseteq \operatorname{int}(K_{n+1}) \subseteq K_{n+1}$, so that the sequence of compact sets $\{K_n\}$ is increasing.

Proof We write down an explicit sequence of compact sets $\{K_n\}_{n=1}^{\infty}$ and prove that it has the required properties. We set, for each positive integer n,

$$K_n = \{z \in D : \operatorname{dist}(z; \mathbb{C} \setminus D) \geq 1/n\} \cap \{z : |z| \leq n\}. \tag{5.13}$$

Here $\operatorname{dist}(z; \mathbb{C} \setminus D)$ refers to the distance between the point z and the set $\mathbb{C} \setminus D$, this being the infimum of $|z - w|$ as w ranges over $\mathbb{C} \setminus D$.

If $z \in D$ then $\operatorname{dist}(z; \mathbb{C} \setminus D)$ is positive, and so z will belong to K_n for all sufficiently large n (in fact, once $|z| \leq n$ and $n \times \operatorname{dist}(z; \mathbb{C} \setminus D) \geq 1$. Thus, $\bigcup_{n=1}^{\infty} K_n = D$, which is (ii).

We need to show that K_n is compact that is, by the Heine-Borel Theorem, that K_n is closed and bounded. Since $K_n \subseteq \{z : |z| \leq n\}$, it is bounded. Suppose that $\{z_m\}_{m=1}^{\infty}$ is a sequence of points in K_n that converges to z. Since $|z_m| \leq n$ for each m, then $|z| \leq n$ as well. Moreover, if $\operatorname{dist}(z; \mathbb{C} \setminus D)$ was strictly less than $1/n$ there would be some point ζ in $\mathbb{C} \setminus D$ with $|z - \zeta| < 1/n$. Put $\epsilon = \frac{1}{n} - |z - \zeta|$. Since $\{z_m\}_1^{\infty}$ converges to z, there is an integer k such that $|z - z_k| \leq \epsilon/2$. Then,

$$|z_k - \zeta| \leq |z_k - z| + |z - \zeta| \leq \epsilon/2 + (1/n - \epsilon) < \frac{1}{n},$$

which contradicts the fact that $\operatorname{dist}(z_k; \mathbb{C} \setminus D) \geq 1/n$. Hence, $\operatorname{dist}(z; \mathbb{C} \setminus D) \geq 1/n$ and $z \in K_n$. In summary, K_n is closed and bounded, hence compact.

Next, we prove that K_n lies in the interior of K_{n+1}. In order to show this, given $z \in K_n$, we need to produce a disk $D(z, \delta)$ with $D(z, \delta) \subseteq K_{n+1}$. Since $\operatorname{dist}(z; \mathbb{C} \setminus D) \geq 1/n$, if w is a point with $|z - w| \leq (\frac{1}{n} - \frac{1}{n+1})$ then, for any $\zeta \in \mathbb{C} \setminus D$,

$$|w - \zeta| \geq |z - \zeta| - |w - z| \geq \frac{1}{n} - \left(\frac{1}{n} - \frac{1}{n+1}\right) = \frac{1}{n+1}. \tag{5.14}$$

5.2 The Space of Continuous Functions on a Planar Domain

Set $\delta = 1/n - 1/(n+1)$. If $w \in D(z, \delta)$ then $|w| \leq |z| + \delta \leq n+1$ and consequently, taking account of (5.14), $w \in K_{n+1}$.

Finally, we prove that if K is any compact subset of D then K is contained in one of the sets K_n. First, by (i) and (ii),

$$D = \bigcup_{n=1}^{\infty} K_n = \bigcup_{n=1}^{\infty} \text{int}(K_{n+1}).$$

Thus, $\{\text{int}(K_n)\}_{n=2}^{\infty}$ is an open cover of K and, since K is compact, a finite number of these open sets are sufficient to cover K. Hence, there is a positive integer m such that

$$K \subseteq \bigcup_{n=2}^{m} \text{int}(K_n) \subseteq \bigcup_{n=2}^{m} K_n = K_m.$$

This proves the last assertion in the lemma. □

Exercise 5.9 Describe the compact sets $\{K_n\}$ constructed in the proof of Lemma 5.2 in the particular cases $D = \mathbb{D}$ and $D = \mathbb{H}$.

We are now in a position to construct the required metric on $\mathscr{C}(D, Y)$. For each f and g in $\mathscr{C}(D, Y)$, we set

$$\rho_n(f, g) = \max \{d_Y(f(z), g(z)) : z \in K_n\}.$$

In words, $\rho_n(f, g)$ is the maximum distance in Y between the values of $f(z)$ and $g(z)$ as z ranges over K_n—compare with (5.1). Again, since $d_Y(f(z), g(z))$ is a continuous real-valued function on the compact set K_n, it achieves its supremum value on K_n—thus, we may confidently write 'max' rather than 'sup' here. If $\rho_n(f, g) \leq \epsilon$ then $d_Y(f(z), g(z)) \leq \epsilon$ for *every* z in K_n. Consequently, if $\{f_m\}_1^{\infty}$ is a sequence in $\mathscr{C}(D, Y)$ then $\rho_n(f_m, f) \to 0$ as $m \to \infty$ if and only if $\{f_m\}$ converges to f uniformly on the compact set K_n.

It is not difficult to see that ρ_n satisfies each of the properties of a metric bar one: $\rho_n(f, g) = 0$ does not necessarily imply that $f = g$. Thus, $\rho_n(\cdot, \cdot)$ is a pseudometric on $\mathscr{C}(D, Y)$.

Exercise 5.10 Deduce from Proposition 5.1 that $\rho_n : \mathscr{C}(D, Y) \times \mathscr{C}(D, Y) \to [0, \infty)$ is a pseudo-metric on $\mathscr{C}(D, Y)$ in that it satisfies all properties of a metric except, possibly, that $\rho_n(f, g) = 0$ does not necessarily imply that $f = g$.

Give an example of a domain D and functions f and g in $\mathscr{C}(D, \mathbb{C})$ such that $\rho_n(f, g) = 0$ but $f \neq g$. (Urysohn's Lemma can give a general construction).

Since we require uniform convergence on *all* compact subsets of D, we need to take into account all of the compacta K_n, $n \geq 1$. There is a standard method of achieving this (of constructing a metric from a sequence of pseudometrics, or a norm from a sequence of seminorms): one sets, for f and g in $\mathscr{C}(D, Y)$,

$$\rho(f, g) = \sum_{n=1}^{\infty} \frac{1}{2^n} \frac{\rho_n(f, g)}{1 + \rho_n(f, g)}. \tag{5.15}$$

This may seem somewhat complicated, but the issue is that of convergence. If we simply had, say, $\sum_n \rho_n(f, g)$ then this series would not converge unless $f = g$ since otherwise $\rho_n(f, g)$ will be non-zero for some n and $\rho_m(f, g) \geq \rho_n(f, g)$ for $m \geq n$. However, since $x/(1 + x) \leq 1$ for $x \in [0, \infty)$,

$$\tilde{\rho}_n(f, g) = \frac{\rho_n(f, g)}{1 + \rho_n(f, g)} \leq 1,$$

irrespective of the value of $\rho_n(f, g)$. The term $1/2^n$ guarantees convergence since $\sum_n 1/2^n < \infty$; any other convergent series of positive terms would work just as well.

That this construction works hinges on the following lemma, worth knowing in its own right.

Lemma 5.3 *Suppose that d is a metric, or a pseudometric, on a set X. Then so is $\tilde{d}$ where*

$$\tilde{d}(x, y) = \frac{d(x, y)}{1 + d(x, y)}, \quad \text{for } x, y \in X.$$

Exercise 5.11 Prove Lemma 5.3.

Hint: the only non-trivial aspect of the proof is verification of the triangle inequality

$$\tilde{d}(x, y) \leq \tilde{d}(x, z) + \tilde{d}(z, y), \quad x, y, z \in X.$$

In the denominator of the expression for $\tilde{d}(x, z)$, replace $1 + d(x, z)$ by $1 + d(x, z) + d(z, y)$ and similarly for the denominator of $\tilde{d}(z, y)$. Then use the triangle inequality for d and the fact that $x \to x/(1 + x)$ is an increasing function on $[0, \infty)$.

Theorem 5.2 *$(\mathscr{C}(D, Y), \rho)$, with ρ given by (5.15), is a metric space.*

Proof It is clear that $\rho: \mathscr{C}(D, Y) \times \mathscr{C}(D, Y) \to [0, \infty)$ and that $\rho(f, f) = 0$ for $f \in \mathscr{C}(D, Y)$. Suppose that $f, g \in \mathscr{C}(D, Y)$ and that $\rho(f, g) = 0$. Then, each term $\rho_n(f, g)/(1 + \rho_n(f, g))$ must be zero, whence $\rho_n(f, g) = 0$ for each n. This implies that $f(z) = g(z)$ for each $z \in K_n$, and this for each n. Since $\bigcup_{n=1}^{\infty} K_n = D$, we see that $f = g$. Thus, $\rho(f, g) = 0$ if and only if $f = g$.

It is clear that ρ is symmetric, in that $\rho(f, g) = \rho(g, f)$, since this is the case for each ρ_n.

5.2 The Space of Continuous Functions on a Planar Domain

Finally, since ρ_n satisfies the triangle inequality, so does $\rho_n/(1 + \rho_n)$ by Lemma 5.3. It then follows that ρ itself satisfies the triangle inequality. □

Let us pause for a moment and summarise the present position and where we might want to go from here. We have a metric ρ on the space $\mathscr{C}(D, Y)$ of continuous Y-valued functions on a domain D. This metric is constructed from the sequence of compacta $\{K_n\}$ described in Lemma 5.2. If we have a second sequence of compacta $\{K'_n\}$ that also has the properties (i) and (ii) in Lemma 5.2 we then end up with a different metric ρ' instead of ρ. We would hope that these two metrics, though different, would be equivalent in some sense. Here, equivalent means that both metric spaces induce the same topology on $\mathscr{C}(D, Y)$, that is that they have the same open subsets, and moreover have the same convergent sequences. In the next exercise, we firm up these concepts.

Exercise 5.12 Two metrics d_1 and d_2 on a set X are said to be *topologically equivalent* if they induce the same topology on X, in that any subset E of X is open in the metric space (X, d_1) if and only if it is open in the metric space (X, d_2).[3]

▷ Prove that the metrics d_1 and d_2 on X are topologically equivalent if and only if

   (i) to each x in X and to each positive r, there corresponds a positive r_2 such that $B_{d_2}(x, r_2) \subseteq B_{d_1}(x, r)$,
   and
   (ii) to each x in X and to each positive r, there corresponds a positive r_1 such that $B_{d_1}(x, r_1) \subseteq B_{d_2}(x, r)$.

   Here, $B_{d_1}(x, r)$ stands for the ball $\{y \in X : d_1(x, y) < r\}$, and similarly for $B_{d_2}(x, r)$.

▷ Suppose that d_1 and d_2 are topologically equivalent metrics on X. Let $\{x_n\}_{n=1}^{\infty}$ be a sequence of points in X. Prove that $\{x_n\}_{n=1}^{\infty}$ is convergent to x in the metric space (X, d_1) if and only if it is convergent to x in the metric space (X, d_2). Thus, topologically equivalent metrics have the same convergent sequences.

▷ Suppose that d_1 and d_2 are metrics on a set X, and that a sequence $\{x_n\}_{n=1}^{\infty}$ is convergent to x in the metric space (X, d_1) if and only if it is convergent to x in the metric space (X, d_2). Prove that d_1 and d_2 are topologically equivalent metrics on X.

We see from this last exercise that the convergent sequences determine the topology in a metric space. We are led to characterise the convergent sequences

---

[3] There is another stronger notion of equivalence of metrics, namely that d_1 and d_2 are equivalent metrics on X if there is a constant C, with $0 < C < \infty$, such that

$$\frac{1}{C} d_1(x, y) \leq d_2(x, y) \leq C d_1(x, y) \text{ for every } x, y \in X.$$

Metrics that are equivalent in this sense are automatically topologically equivalent.

in the metric space $(\mathscr{C}(D, Y), \rho)$ (where ρ is given by (5.15)), and expect these to be the sequences that converge uniformly on each compact subset of D. This is, in fact, why ρ was defined in this way.

Definition 5.2 We say that a sequence of functions $\{f_n\}_{n=1}^{\infty}$ in $\mathscr{C}(D, Y)$ converges *locally uniformly* to f on D if and only if $\{f_n\}_{n=1}^{\infty}$ converges uniformly to f on each compact subset of D, that is if and only if

$$\rho_m(f_n, f) \to 0 \text{ as } n \to \infty, \text{ for each } m.$$

Note that, since the uniform limit of continuous functions is continuous, the limit function f mentioned in Definition 5.2 is automatically in $\mathscr{C}(D, Y)$.

Exercise 5.13 Show that a sequence of functions $\{f_n\}_{n=1}^{\infty}$ converges locally uniformly to f on D if and only if to each $z \in D$ there corresponds a positive r such that $f_n \to f$ uniformly on the closed (and compact) disk $\overline{D(z, r)}$.

Theorem 5.3 *A sequence of functions $\{f_n\}_{n=1}^{\infty}$ in $(\mathscr{C}(D, Y), \rho)$ converges to f in $(\mathscr{C}(D, Y), \rho)$ if and only if $\{f_n\}_{n=1}^{\infty}$ converges to f locally uniformly on D. Moreover, $(\mathscr{C}(D, Y), \rho)$ is complete.*

Proof Suppose that $f_n \to f$ in $(\mathscr{C}(D, Y), \rho)$; that is $\rho(f_n, f) \to 0$ as $n \to \infty$. Let K be a compact subset of D. By Lemma 5.2, there is a positive integer m such that $K \subseteq K_m$. Since

$$0 \leq \frac{1}{2^m} \frac{\rho_m(f_n, f)}{1 + \rho_m(f_n, f)} \leq \rho(f_n, f),$$

it follows that $\rho_m(f_n, f)/(1 + \rho_m(f_n, f))$, and hence $\rho_m(f_n, f)$, tends to 0 as $n \to \infty$. Thus, $\{f_n\}_{n=1}^{\infty}$ converges to f uniformly on K_m and hence uniformly on K.

Conversely, suppose that $\{f_n\}_{n=1}^{\infty}$ is a sequence in $(\mathscr{C}(D, Y), \rho)$ that converges to f uniformly on each compact subset of D. Then, $\{f_n\}_{n=1}^{\infty}$ converges to f uniformly on K_m for each m, that is $\rho_m(f_n, f) \to 0$ as $n \to \infty$. Let a positive number ϵ be given. Choose M such that

$$\sum_{m=M+1}^{\infty} \frac{1}{2^m} = \frac{1}{2^M} \leq \frac{\epsilon}{2}.$$

Now,

$$\sum_{m=1}^{M} \frac{1}{2^m} \frac{\rho_m(f_n, f)}{1 + \rho_m(f_n, f)} \to 0 \quad \text{as } n \to \infty,$$

5.2 The Space of Continuous Functions on a Planar Domain

(since, for each m, $\rho_m(f_n, f) \to 0$ as $n \to \infty$ and only finitely many terms are involved). Thus, there is a positive integer N such that

$$\sum_{m=1}^{M} \frac{1}{2^m} \frac{\rho_m(f_n, f)}{1 + \rho_m(f_n, f)} \leq \frac{\epsilon}{2} \quad \text{if } n \geq N.$$

Hence, if $n \geq N$,

$$\rho(f_n, f) \leq \sum_{m=1}^{M} \frac{1}{2^m} \frac{\rho_m(f_n, f)}{1 + \rho_m(f_n, f)} + \sum_{m=M+1}^{\infty} \frac{1}{2^m} \leq \frac{\epsilon}{2} + \frac{\epsilon}{2} = \epsilon,$$

whence $\rho(f_n, f) \to 0$ as $n \to \infty$.

To establish completeness of the metric space $(\mathscr{C}(D, Y), \rho)$, suppose that $\{f_j\}_{j=1}^{\infty}$ is a Cauchy sequence in $\mathscr{C}(D, Y)$. Fix any particular compact set K_n in the exhaustion of D by compact sets, and restrict each function f_j to K_n. The restricted functions then form a Cauchy sequence in the metric space $\mathscr{C}(K_n, Y)$, and this space in turn is complete by Proposition 5.1. Thus, $\{f_j\}_{j=1}^{\infty}$ converges uniformly on K_n to some function in $\mathscr{C}(K_n, Y)$. Since this is the case for each K_n, and the union of the K_n is all of D, we have a continuous function f defined on all of D for which the original sequence $\{f_j\}$ converges uniformly to f on each compact subset of D. $\square$

Irrespective of the choice of sequence of compact sets $\{K_n\}_{n=1}^{\infty}$ that exhaust D, so long as they satisfy the conditions (i) and (ii) of Lemma 5.2, the convergent sequences in $(\mathscr{C}(D, Y), \rho)$ are precisely those sequences of functions that converge uniformly on every compact subset of D. Thus, the convergent sequences in $(\mathscr{C}(D, Y), \rho)$ don't depend on the exhaustion by compact sets $\{K_n\}_{n=1}^{\infty}$ and so, by the last part of Exercise 5.12, the metrics on $\mathscr{C}(D, Y)$ arising from any two such exhaustions of D are topologically equivalent. That is, the topology on $\mathscr{C}(D, Y)$ induced by the metric ρ in (5.15) is independent of the particular exhaustion of D by compact sets.

We now need the version of the Arzelà-Ascoli Theorem appropriate to the setting of $\mathscr{C}(D, Y)$ in order to characterise the relatively compact subsets of $\mathscr{C}(D, Y)$. We again refer to this result as 'the' Arzelà-Ascoli Theorem with the justification that there are many versions of this result characterising relatively compact sets of functions that all come under the general heading of 'Arzelà-Ascoli'.

Theorem 5.4 (Arzelà-Ascoli Theorem) *A subset $\mathscr{F}$ of $\mathscr{C}(D, Y)$ is relatively compact if and only if*

(a) *for every z in D, the set of function values $\mathscr{F}(z) = \{f(z) : f \in \mathscr{F}\}$ is a relatively compact subset of Y;*
(b) *$\mathscr{F}$ is equicontinuous on each compact subset of D.*

To be precise, possibly pedantic, the statement that '$\mathscr{F}$ is equicontinuous on each compact subset of D' means that if K is a compact subset of D and $\mathscr{F}_K$ is the family of continuous functions on K obtained by restricting each of the functions in $\mathscr{F}$ to K, then $\mathscr{F}_K$ is an equicontinuous family of functions on K. We use this notation in the proof of Theorem 5.4.

Proof We will establish that a family of functions $\mathscr{F}$ in $\mathscr{C}(D, Y)$ is a relatively compact subset of $\mathscr{C}(D, Y)$ if and only if $\mathscr{F}_K$ is a relatively compact subset of $\mathscr{C}(K, Y)$ for each compact subset K of D. Once this is done, the conditions (a) and (b) of Theorem 5.4 are simply a restatement of the corresponding conditions (a) and (b) of the Arzelà-Ascoli Theorem 5.1 in the case $X = K$.

Suppose, then, that $\mathscr{F}$ is a relatively compact subset of $\mathscr{C}(D, Y)$ and that K is a compact subset of D. Then $\mathscr{F}_K$ is a relatively compact subset of $\mathscr{C}(K, Y)$. In fact, if $\{f_n^K\}_{n=1}^{\infty}$ is a sequence of functions from $\mathscr{F}_K$ then there is a sequence of functions $\{f_n\}_{n=1}^{\infty}$ from $\mathscr{F}$ such that $f_n^K = f_n|_K$ for each n. Since $\mathscr{F}$ is a relatively compact subset of $\mathscr{C}(D, Y)$, $\{f_n\}_{n=1}^{\infty}$ has a subsequence, say $\{f_{n_k}\}_{k=1}^{\infty}$, that converges uniformly on each compact subset of D. In particular, $\{f_{n_k}\}_{k=1}^{\infty}$ converges uniformly on K, that is the subsequence $\{f_{n_k}^K\}_{k=1}^{\infty}$ of $\{f_n^K\}_{n=1}^{\infty}$ converges uniformly on K. This shows, by Lemma 5.1, that $\mathscr{F}_K$ is a relatively compact subset of $\mathscr{C}(K, Y)$.

Conversely, suppose that $\mathscr{F}$ is a subset of $\mathscr{C}(D, Y)$ and that $\mathscr{F}_K$ is a relatively compact subset of $\mathscr{C}(K, Y)$ for each compact subset K of D. Now we need to use a diagonalisation argument as in the proof of Proposition 5.3. Choose an exhaustion $\{K_j\}_{j=1}^{\infty}$ of D by compact sets K_j as constructed in Lemma 5.2. Let $\{f_n\}_{n=1}^{\infty}$ be a sequence of functions from $\mathscr{F}$. Since $\mathscr{F}_{K_1}$ is a relatively compact subset of $\mathscr{C}(K_1, Y)$, there is a subsequence of $\{f_n\}_{n=1}^{\infty}$ that converges uniformly on K_1. This subsequence, in turn, has a subsequence that converges uniformly on K_2, this because $\mathscr{F}_{K_2}$ is a relatively compact subset of $\mathscr{C}(K_2, Y)$ by assumption. Continuing in this way, we produce a sequence of subsequences $\{f_{j,k}\}_{k=1}^{\infty}$, say, of $\{f_n\}_{n=1}^{\infty}$ such that the $(j + 1)$st subsequence is itself a subsequence of the jth subsequence, and the jth subsequence converges uniformly on the compact set K_j. The diagonal sequence $\{f_{m,m}\}_{m=1}^{\infty}$ then converges uniformly on K_j for each j and is a subsequence of the original sequence $\{f_n\}_{n=1}^{\infty}$. Since any compact subset K of D is contained in some set K_j of the exhaustion of D by compact subsets, $\{f_{m,m}\}_{m=1}^{\infty}$ converges uniformly on K since it converges uniformly on K_j. This shows, by Lemma 5.1 and Theorem 5.3, that $\mathscr{F}$ is a relatively compact subset of $\mathscr{C}(D, Y)$. □

Now suppose that Y is the complex plane $\mathbb{C}$ and let $\mathscr{F}$ be a relatively compact subset of $\mathscr{C}(D, \mathbb{C})$, so that conditions (a) and (b) of Theorem 5.4 hold. Let z_0 be an arbitrary point in D. Choose $r_0 > 0$ so that the closed disk $K = \overline{D(z_0, r_0)}$, which is compact, is contained in D. Then $\mathscr{F}$ is equicontinuous on K and therefore there is a positive r (less than r_0) such that

$$|f(z_1) - f(z_2)| \leq 1 \text{ whenever } z_1, z_2 \in K, \ |z_1 - z_2| \leq r \text{ and } f \in \mathscr{F}. \quad (5.16)$$

By (a), there is a finite number M such that $|f(z_0)| \leq M$ for each $f \in \mathscr{F}$. From (5.16) we deduce that

$$|f(z)| \leq |f(z) - f(z_0)| + |f(z_0)| \leq 1 + M,$$

whenever $f \in \mathscr{F}$ and $z \in D(z_0, r)$. This shows that a relatively compact subset $\mathscr{F}$ of $\mathscr{C}(D, \mathbb{C})$ is *locally bounded* in the sense of the following definition.

Definition 5.3 A family of functions $\mathscr{F}$ in $\mathscr{C}(D, \mathbb{C})$ is said to be *locally bounded* if to each point z_0 in D there corresponds a number M and a positive number r such that $D(z_0, r) \subseteq D$ and

$$|f(z)| \leq M \text{ whenever } z \in D(z_0, r) \text{ and } f \in \mathscr{F}.$$

Exercise 5.14 Show that a family of functions $\mathscr{F}$ in $\mathscr{C}(D, \mathbb{C})$ is locally bounded if and only if the family $\mathscr{F}$ is uniformly bounded on each compact subset K of D, that is to each compact subset K of D there corresponds a finite number M such that

$$|f(z)| \leq M \text{ whenever } z \in K \text{ and } f \in \mathscr{F}.$$

We will see that the converse holds if all the functions in the family are analytic, that local boundedness of a family of analytic functions implies relative compactness of the family. This is Montel's Theorem. But first we take a little time to discuss convergence of analytic and meromorphic functions.

5.3 Convergence of Sequences of Analytic and Meromorphic Functions

When discussing analytic functions on a domain D it is natural to regard them as belonging to the metric space $\mathscr{C}(D, \mathbb{C})$, in other words as continuous functions on D taking values in $\mathbb{C}$ with the standard Euclidean metric. In the case of meromorphic functions on D, it is natural to regard these as belonging to the metric space $\mathscr{C}(D, \mathbb{C}_\infty)$, that is as taking values in the extended complex plane $\mathbb{C}_\infty$ with the spherical metric. In each case, D is equipped with the usual Euclidean metric. To distinguish convergence in these two distinct settings we refer to convergence in $\mathscr{C}(D, \mathbb{C})$ as *locally uniform convergence* and to convergence in $\mathscr{C}(D, \mathbb{C}_\infty)$ as *locally uniform spherical convergence*. We denote the collection of analytic functions on D by $\mathscr{H}(D)$ and denote the collection of meromorphic functions on D by $\mathscr{M}(D)$. To avoid any misunderstanding, a meromorphic function is one that is analytic apart from poles. It is spherically continuous, that is continuous as a mapping to the Riemann sphere. If f has a pole at z_0 then $1/f$ has an isolated zero at z_0. Conversely, if f is analytic at z_0 and has an isolated zero there then $1/f$ has a pole at z_0. However, it is not quite the case that f is meromorphic on D if and only if $1/f$ is meromorphic on D. If f is identically zero on D it is meromorphic, in fact analytic, on D but $1/f$ has constant value ∞ on D and is *not* meromorphic. What is true is that if f is meromorphic on D and not identically zero then $1/f$ is meromorphic on D as, in this case, all zeroes of f are isolated.

We will show (Theorem 5.5) that $\mathscr{H}(D)$ is a closed subset of $\mathscr{C}(D, \mathbb{C})$ and (Theorem 5.6) that $\mathscr{M}(D)$ is almost(!) a closed subset of $\mathscr{C}(D, \mathbb{C}_\infty)$. The spherical limit of a sequence of meromorphic functions could be identically infinite. For example, if $f_n(z) = n$, $z \in D$, then by (3.26) f_n converges spherically locally uniformly to the function ∞_D, this being the function on D with constant value ∞. The constant function ∞_D belongs to $\mathscr{C}(D, \mathbb{C}_\infty)$ but is not meromorphic. This is the only way, though, in which a sequence of meromorphic functions can converge spherically locally uniformly to a non-meromorphic function. Thus, $\mathscr{M}(D) \cup \{\infty_D\}$ is a closed subset of $\mathscr{C}(D, \mathbb{C}_\infty)$. Being closed subsets of complete metric spaces (cf. Theorem 5.3), both $\mathscr{H}(D)$ and $\mathscr{M}(D) \cup \{\infty_D\}$ are complete metric spaces in their own right. For ease of notation, we write $\mathscr{M}^*(D)$ for $\mathscr{M}(D) \cup \{\infty_D\}$. First we show that $\mathscr{H}(D)$ is closed in $\mathscr{C}(D, \mathbb{C})$.

Theorem 5.5 *Suppose that $\{f_n\}_{n=1}^\infty$ is a sequence of analytic functions on D that converges locally uniformly to a function f on D. Then f is analytic on D. Moreover, for each k, the sequence of kth derivatives $\{f_n^{(k)}\}_{n=1}^\infty$ converges to $f^{(k)}$ locally uniformly on D.*

Proof We will need Morera's Theorem [16, Theorem 7.8]: suppose that $f : D \to \mathbb{C}$ is continuous and that, whenever T is a triangle that is contained in a disk that is, in turn, contained in D, we have $\int_T f(z)\,dz = 0$. Then f is analytic in D.

Suppose that $\{f_n\}_{n=1}^\infty$ is a sequence of functions in $\mathscr{H}(D)$ that converges to f locally uniformly on D. Then f is continuous on D, being locally a uniform limit of continuous functions. Suppose that T is a triangle in a disk that is contained in D. Then T is a compact subset of D and so, by assumption, $f_n \to f$ uniformly on T. Moreover, by Cauchy's Theorem, $\int_T f_n(z)\,dz = 0$ for each n. As a consequence,

$$\left| \int_T f(z)\,dz \right| = \left| \int_T (f(z) - f_n(z))\,dz \right|$$
$$\leq \text{length}(T) \times \left(\max \{ |f(z) - f_n(z)| : z \in T \} \right),$$

by the ML-inequality.[4] Since this last maximum will be arbitrarily small for large n while the length of the boundary of the triangle is fixed, $\int_T f(z)\,dz = 0$. By Morera's Theorem, f is analytic in D.

---
[4] If f is continuous and bounded by M on a curve γ of length L then $|\int_\gamma f(z)\,dz| \leq ML$.

5.3 Convergence of Sequences of Analytic and Meromorphic Functions

Now we turn to kth derivatives. Let $D(a, r)$ be a disk such that $\overline{D(a, 2r)}$ lies in D. By the Cauchy Integral Formula for derivatives,

$$f_n^{(k)}(z) = \frac{k!}{2\pi i} \int_{C(a,2r)} \frac{f_n(w)}{(w-z)^{k+1}} \, dw \text{ and}$$

$$f^{(k)}(z) = \frac{k!}{2\pi i} \int_{C(a,2r)} \frac{f(w)}{(w-z)^{k+1}} \, dw,$$

this for $z \in D(a, 2r)$. For $z \in D(a, r)$ we have $|w - z| \geq r$ if $w \in C(a, 2r)$. Thus,

$$\left| f_n^{(k)}(z) - f^{(k)}(z) \right| = \frac{k!}{2\pi} \left| \int_{C(a,2r)} \frac{f_n(w) - f(w)}{(w-z)^{k+1}} \, dw \right|$$

$$\leq \frac{2k!}{r^k} \max \left\{ |f_n(w) - f(w)| : w \in C(a, 2r) \right\}.$$

Since $f_n \to f$ uniformly on compact subsets of D, in particular $f_n \to f$ uniformly on the circle $C(a, 2r)$. Hence, $f_n^{(k)}(z) \to f^{(k)}(z)$ uniformly on $D(a, r)$. It follows from Exercise 5.13 that $f_n^{(k)} \to f^{(k)}$ uniformly on each compact subset of D. □

Remark 5.1 For $k = 0, 1, 2, \ldots$, and for fixed $a \in D$, the linear functional $\mathscr{D}_a^{(k)} : \mathscr{H}(D) \to \mathbb{C}$ defined by $\mathscr{D}_a^{(k)}(f) = f^{(k)}(a)$ returns the kth derivative of f at a. The functional $\mathscr{D}_a^{(k)}$ is continuous. In fact, the continuity of $\mathscr{D}_a^{(k)}$ on $\mathscr{H}(D)$ is equivalent to the statement that if $f_n \to f$ in $\mathscr{H}(D)$ then $\mathscr{D}_a^{(k)}(f_n) \to \mathscr{D}_a^{(k)}(f)$ in $\mathbb{C}$, and this follows from Theorem 5.5 since the singleton set $\{a\}$ is a compact subset of D.

Next we establish the corresponding convergence result for meromorphic functions, showing that $\mathscr{M}^*(D)$ is closed in $\mathscr{C}(D, \mathbb{C}_\infty)$. This is a little more involved than the result for analytic functions. It helps to first firm up the relationship between locally uniform convergence and locally uniform spherical convergence—as long as the point at infinity is not in the picture, the two notions are effectively equivalent.

Lemma 5.4 *Let K be a compact subset of the complex plane. Suppose that f_n, $n \geq 1$, are continuous complex-valued functions on K.*

(i) *If $f_n \to f$ uniformly on K then $f_n \to f$ spherically uniformly on K.*
(ii) *If $f_n \to f$ spherically uniformly on K and $f(K)$ is a bounded subset of the complex plane then $f_n \to f$ uniformly on K.*

Proof

(i) Given ε positive there is $N \in \mathbb{N}$ such that $|f_n(z) - f(z)| \leq \varepsilon/2$ for $z \in K$ and $n \geq N$. By (3.24), $d^\#(f_n(z), f(z)) \leq 2|f_n(z) - f(z)| \leq \varepsilon$ for $z \in K$ and $n \geq N$ so that $f_n \to f$ spherically uniformly on K.
(ii) Suppose that $|f(z)| \leq M$, $z \in K$. Since $f_n \to f$ spherically uniformly on K, there is some $N \in \mathbb{N}$ such that $d^\#(f_n(z), f(z)) \leq d^\#(M, 2M)$, $z \in K$ and

$n \geq N$. Then, by the triangle inequality and the rotational invariance of the spherical metric,

$$d^{\#}(0, f_n(z)) \leq d^{\#}(0, M) + d^{\#}(M, 2M) = d^{\#}(0, 2M),$$

and so $|f_n(z)| \leq 2M$ for $z \in K$ and $n \geq N$. By (3.25),

$$|f_n(z) - f(z)| \leq \frac{1 + 4(2M)^2}{2} d^{\#}(f_n(z), f(z)),$$

again for $z \in K$ and $n \geq N$. Since $d^{\#}(f_n(z), f(z)) \to 0$ uniformly on K as $n \to \infty$, it follows that $f_n \to f$ uniformly on K. □

The next step is the following proposition in which locally uniform spherical convergence of a sequence of meromorphic functions is interpreted in terms of the usual Euclidean metric. As an example where (5.17b) holds, take $f(z) = 1/z$, $f_n(z) = 1/(z - 1/n)$, $z \in \mathbb{D}$.

Proposition 5.5 *A sequence of functions $\{f_n\}_{n=1}^{\infty}$ in $\mathscr{C}(D, \mathbb{C}_\infty)$ converges spherically locally uniformly to a function f on D if and only if there is, for each z_0 in D, a closed disk $\overline{D(z_0, r)}$ of positive radius in D on which at least one of the following holds:*

$$|f_n(z) - f(z)| \to 0 \text{ as } n \to \infty, \tag{5.17a}$$

$$\left|\frac{1}{f_n(z)} - \frac{1}{f(z)}\right| \to 0 \text{ as } n \to \infty, \tag{5.17b}$$

uniformly for $z \in \overline{D(z_0, r)}$.

Suppose that f_n converges spherically locally uniformly to f and that each f_n is meromorphic. If $z_0 \in D$ and $f(z_0) \neq \infty$ then there is some disk about z_0 in which f, and each f_n for n sufficiently large, is analytic.

Suppose that f_n converges spherically locally uniformly to f and each f_n is analytic. Then either f is identically infinite or f is analytic in D. If the limit function f is analytic then f_n converges to f locally uniformly on D.

Proof Suppose that for each z_0 in D there is a closed disk $\overline{D(z_0, r)}$ in D on which (at least) one of (5.17a) or (5.17b) holds. If (5.17a) holds then, by (3.24),

$$d^{\#}(f_n(z), f(z)) \leq 2|f_n(z) - f(z)|, \text{ for } z \in \overline{D(z_0, r)}.$$

5.3 Convergence of Sequences of Analytic and Meromorphic Functions

If (5.17b) holds then, since the inversion $z \to 1/z$ is an isometry of the spherical metric (Theorem 3.6),

$$d^{\#}(f_n(z), f(z)) = d^{\#}(1/f_n(z), 1/f(z)) \leq 2 \left| \frac{1}{f_n(z)} - \frac{1}{f(z)} \right|, \text{ for } z \in \overline{D(z_0, r)}.$$

Hence, each of (5.17a) and (5.17b) implies that $f_n \to f$ spherically uniformly on $\overline{D(z_0, r)}$, and hence spherically uniformly on each compact subset of D.

Conversely, suppose that $d^{\#}(f_n(z), f(z)) \to 0$ uniformly on compact subsets of D. Suppose that z_0 is in D and that, in the first instance, $f(z_0) \in \mathbb{C}$ (that is, $f(z_0) \neq \infty$). Since $f : D \to \mathbb{C}_\infty$ is spherically continuous, there is a closed disk $K = \overline{D(z_0, r)}$ in D on which f is bounded, say by $2|f(z_0)|$. By Lemma 5.4 (ii), $f_n \to f$ uniformly on K, which is (5.17a). In the case that $f(z_0) = \infty$ we have $1/f(z_0) = 0$. Since $z \to 1/z$ is an isometry of the spherical metric, $1/f_n \to 1/f$ spherically locally uniformly as $n \to \infty$. The previous case therefore applies to the sequence $1/f_n$ and shows that (5.17b) holds.

Returning to the case where $f(z_0) \neq \infty$, suppose also that each f_n is meromorphic. With f bounded by M, say, on the closed disk $K = \overline{D(z_0, r)}$, we see from the proof of Lemma 5.4 (ii) that f_n is bounded by $2M$ on K for all sufficiently large n. As a consequence, f_n doesn't have any poles in K and is therefore analytic in $D(z_0, r)$. Since, by Lemma 5.4 (ii), f is then a uniform limit of analytic functions on $D(z_0, r)$, it is itself analytic on $D(z_0, r)$ by Theorem 5.5.

Now suppose that each f_n is analytic and converges spherically locally uniformly to f on D. Denote the set of all points z_0 in D for which $f(z_0) = \infty$ by E. Clearly, E is closed as $f : D \to \mathbb{C}_\infty$ is continuous. Now we argue that E is also open. If $z_0 \in E$ then, for sufficiently large n, the function f_n is not identically zero and so $1/f_n$ is meromorphic and converges spherically locally uniformly to $1/f$. Since $1/f(z_0) = 0$, it follows from the previous paragraph that $1/f$ is analytic at z_0. Either $1/f$ is identically zero in a disk about z_0, in which case f is identically infinite there and z_0 is an interior point of E, or $1/f$ has an isolated zero at z_0 in which case f has a pole at z_0. We will show that this last cannot occur—the limit function f can only have a pole in D if the converging functions f_n have poles. Write n for the order of the pole of f at z_0. Then, $f(z) = (z - z_0)^{-n} g(z)$ in some disk about z_0 where g is analytic at z_0 with $g(z_0) \neq 0$. There is a positive r such that f is analytic in the annulus $A = \{z : r/2 \leq |z - z_0| \leq 3r/2\}$. Then f is bounded on A and, by assumption, f_n converges to f spherically uniformly on A. By Lemma 5.4 (ii), f_n converges to f uniformly on A. As a consequence,

$$2\pi i g(z_0) = \int_{C(z_0, r)} \frac{g(z)}{z - z_0} \, dz$$

$$= \int_{C(z_0, r)} (z - z_0)^{n-1} f(z) \, dz$$

$$= \lim_{n \to \infty} \int_{C(z_0, r)} (z - z_0)^{n-1} f_n(z) \, dz.$$

But $(z-z_0)^{n-1} f_n(z)$ is analytic in D and so the integral $\int_{C(z_0,r)} (z-z_0)^{n-1} f_n(z)\, dz$ vanishes. This contradicts $g(z_0) \neq 0$ and so f does not have a pole at z_0. Hence, E is open. Being both open and closed, E is either empty or all of D, that is the limit function f is either identically infinite or analytic. Finally, if f is analytic in D (that is, not identically infinite) and K is a compact subset of D then f is bounded on K. Again it follows from Lemma 5.4 (ii) that f_n converges to f uniformly on K, not just spherically uniformly. □

The proof of this last proposition suggests the following exercise.

Exercise 5.15 Show that a sequence of functions $\{f_n\}_{n=1}^{\infty}$ in $\mathscr{C}(D, \mathbb{C}_\infty)$ converges to f in $\mathscr{C}(D, \mathbb{C}_\infty)$ if and only if the sequence $\{1/f_n\}_{n=1}^{\infty}$ converges to $1/f$ in $\mathscr{C}(D, \mathbb{C}_\infty)$.

With this proposition to hand, we can now describe the spherically locally uniform limit of a sequence of meromorphic functions.

Theorem 5.6 *Suppose that $\{f_n\}_{n=1}^{\infty}$ is a sequence of meromorphic functions on D that converges spherically locally uniformly to a function f on D. Then, either f is identically infinite in D or else f is meromorphic in D.*

Proof Suppose that $\{f_n\}_{n=1}^{\infty}$ is a sequence of meromorphic functions on D that converges spherically locally uniformly to a function f in $\mathscr{C}(D, \mathbb{C}_\infty)$. At any point z_0 in D for which $f(z_0) \neq \infty$, f is analytic in some disk $D(z_0, r)$ about z_0; this is part of the statement of Proposition 5.5.

Suppose that $z_0 \in D$ and $f(z_0) = \infty$. Then $f_n(z_0) \to \infty$ as $n \to \infty$ so that f_n cannot be identically zero in D for sufficiently large n and $1/f_n$ is meromorphic. By Exercise 5.15, $1/f_n$ converges spherically locally uniformly to $1/f$ on D. Since $1/f(z_0) = 0$, Proposition 5.5 again applies and we find that $1/f$ is analytic in some disk about z_0. If the zero of $1/f$ at z_0 is not isolated then, by the Uniqueness Theorem [16, Theorem 10.5], $1/f$ is identically equal to zero at least in this disk about z_0. If the zero of $1/f$ at z_0 is, in fact, isolated, then f itself has a pole at z_0.

Now write E for the set of all points z_0 in D for which there is some disk about z_0 on which $1/f$ is identically 0. Obviously, E is an open subset of D. It is also a closed subset of D. In fact, if points z_n lie in E and have limit z in D then, since $1/f$ is spherically continuous and $1/f(z_n) = 0$ for each n, we have $1/f(z_0) = 0$ as well. By the argument in the previous paragraph, $1/f$ is analytic in a disk about z_0. But then z_0 is a limit point of zeroes of $1/f$ and so, by the Uniqueness Theorem again, $1/f$ must be identically zero in this disk. That is, E is closed. Since D is connected, it now follows that E is either empty or E is all of the domain D. In the latter case, $1/f$ is identically zero on D so that f is identically ∞ on D. If E is empty then, by previous analysis, at each point z_0 in D either $f(z_0) \neq \infty$, in which case f is analytic in a disk about z_0, or $f(z_0) = \infty$, in which case $1/f$ is analytic in a disk about z_0 and has an isolated zero at z_0 so that f itself has a pole at z_0. That is, if E is empty the limit function f is meromorphic on D. □

5.3 Convergence of Sequences of Analytic and Meromorphic Functions

Before we move on, we prove a classical result on convergence of a sequence of one-to-one analytic (that is *univalent*) functions, one that will be particularly useful in the proof of the Riemann Mapping Theorem. We already know from Theorem 5.5 that $\mathscr{H}(D)$ is a closed subset of $\mathscr{C}(D, \mathbb{C})$ in that the limit f of a locally uniformly convergent sequence of analytic functions is itself analytic. If, in addition, the functions f_n are univalent then so is the limit function f, with one exception.

Theorem 5.7 (Hurwitz's Theorem) *Suppose that a sequence of analytic functions $\{f_n\}_{n=1}^\infty$ converges to f in $\mathscr{H}(D)$ and that the closed disk $\overline{D(a, r)}$ is contained in D. If f is non-zero on the circle $C(a, r)$ then there is an integer N such that f_n and f have the same number of zeros in $D(a, r)$ whenever $n \geq N$.*

If each function f_n is univalent in D then the limit function f is either constant or univalent.

Proof We need Rouché's Theorem: suppose that g and h are analytic on the disk $D(a, R)$ and that $r < R$. If

$$|g(z) + h(z)| < |g(z)| + |h(z)| \text{ for } z \in C(a, r),$$

(so that, in particular, neither has any zeros on the circle $C(a, r)$) then g and h have the same number of zeros in $D(a, r)$.

With $\{f_n\}_{n=1}^\infty$, f, a and r as in the statement in the theorem, we have

$$\delta = \min \{|f(z)| : z \in C(a, r)\} > 0,$$

since the continuous function $|f|$ does not vanish on $C(a, r)$. Moreover, since $f_n \to f$ uniformly on the compact set $C(a, r)$, there is a positive integer N such that

$$|f_n(z) - f(z)| \leq \delta/2 \text{ whenever } z \in C(a, r) \text{ and } n \geq N.$$

Thus, for $z \in C(a, r)$ and $n \geq N$,

$$|f_n(z) - f(z)| \leq \delta/2 < \delta \leq |f(z)| \leq |f_n(z)| + |f(z)|,$$

and it follows from Rouché's Theorem that f_n and f have the same number of zeros in the disk $D(a, r)$.

For the last statement, suppose that each function f_n is univalent in D but that the limit function f is neither constant nor univalent. Since f is not univalent, there are z_1 and z_2 in D, $z_1 \neq z_2$, with $f(z_1) = f(z_2) = w$, say. The sequence of functions $\{f_n - w\}_{n=1}^\infty$ converges to $f - w$ in $\mathscr{H}(D)$. Choose a closed disk $\overline{D(z_1, r_1)}$ contained in D such that $f - w$ doesn't vanish on $C(z_1, r_1)$. If such a choice was not possible then the set where $f(z) = w$ would have $z_1 \in D$ as a limit point, forcing f to be constant. Then, for all sufficiently large n, $f_n - w$ has the same number of zeros as $f - w$ in $D(z_1, r_1)$, that is at least one. Next, choose a closed disk $\overline{D(z_2, r_2)}$ contained in D and not meeting $\overline{D(z_1, r_1)}$ such that $f - w$ doesn't vanish on $C(z_2, r_2)$. Again, for all sufficiently large n, $f_n - w$ has the same number of zeros as $f - w$ in $D(z_2, r_2)$, that is at least one. This means that f_n (for all sufficiently large n) takes the value w at least once in $D(z_1, r_1)$ and at least once in $D(z_2, r_2)$, which is at least twice in total—as these two disks don't meet—this being a contradiction of the assumption that f_n is univalent. □

Note the case of a constant limit function can arise: for example, $f_n(z) = z/n$ is univalent in the disk $\mathbb{D}$ for each n, while this sequence of functions has the constant function $f \equiv 0$ as its limit in $\mathscr{H}(\mathbb{D})$.

Exercise 5.16 Suppose that the sequence of analytic functions $\{f_n\}_{n=1}^{\infty}$ converges to f in $\mathscr{H}(D)$ and that each function f_n is never zero in D. Prove that either f is identically zero in D or f is never zero in D.

5.4 The Theorems of Montel, Marty and Zalcman

The conditions in the Arzelà-Ascoli Theorem are not always easy to check. In the case of analytic or meromorphic functions, however, there are much simpler conditions for normality: Montel's Theorem in the case of analytic functions and Marty's Theorem in the case of meromorphic functions. Zalcman has a condition for a family of meromorphic functions *not* to be normal. We will be able to put these together in the final section of this chapter to prove the Great Picard Theorem which is a landmark result in complex analysis.

That a family of meromorphic functions $\mathscr{F}$ is a relatively compact subset of $\mathscr{M}^*(D)$ means that every sequence of functions from $\mathscr{F}$ has a subsequence that converges spherically locally uniformly to some function f in $\mathscr{M}^*(D)$. Since an analytic function is also meromorphic, a family of analytic functions can be considered as belonging to either $\mathscr{H}(D)$ or to $\mathscr{M}^*(D)$, there being a subtle difference between the two viewpoints. Both viewpoints are legitimate and are covered by Theorems 5.5 and 5.6. Take, for example, D to be the unit disk $\mathbb{D}$ and set, for each $n \geq 1$, $f_n(z) = n$, $z \in \mathbb{D}$. This sequence of functions does not converge in the space $\mathscr{C}(\mathbb{D}, \mathbb{C})$. But in the space $\mathscr{C}(\mathbb{D}, \mathbb{C}_\infty)$ it converges to the constant function ∞_D (which is in agreement with Theorem 5.6). With this in mind, we have the classical complex analysis definition of a normal family.

Definition 5.4 (Normal Family) Let $\mathscr{F}$ be a family of meromorphic functions on a domain D. We say that $\mathscr{F}$ is a *normal family* if it is a relatively compact subset of $\mathscr{M}^*(D)$ or, equivalently, if each sequence of functions from $\mathscr{F}$ has a subsequence that converges spherically locally uniformly to some function f in $\mathscr{M}^*(D)$.

If all functions in $\mathscr{F}$ are analytic then, by Proposition 5.5, $\mathscr{F}$ is a normal family if and only if every sequence of functions from $\mathscr{F}$ has a subsequence that converges locally uniformly to an analytic function f or spherically locally uniformly to the constant function ∞_D. The point is that locally uniform spherical convergence of a sequence of analytic functions to a limit analytic function is equivalent to locally

5.4 The Theorems of Montel, Marty and Zalcman

uniform convergence of the sequence. The only difference when a family $\mathscr{F}$ of analytic functions is viewed as a normal family in the sense of the above definition rather than as a relatively compact subset of $\mathscr{H}(D)$ is that, in the first case, spherical convergence to the constant function ∞_D is allowed.

It is clear that if a family $\mathscr{F}$ of analytic functions in $\mathscr{H}(D)$ is relatively compact then, for any non-zero complex a, the translated family $\mathscr{F} + a = \{f + a : f \in \mathscr{F}\}$ is relatively compact and the scaled family $a\mathscr{F} = \{af : f \in \mathscr{F}\}$ is relatively compact. The same is true for locally uniform spherical convergence, in which case first Exercise 5.15 and then Exercise 3.18 prove useful.

Exercise 5.17 Let $\mathscr{F}$ be a family of functions in $\mathscr{M}^*(D)$. Let a be a non-zero complex number.

▷ Show that $\mathscr{F}$ is relatively compact in $\mathscr{M}^*(D)$, in other words a normal family, if and only if $1/\mathscr{F} := \{1/f : f \in \mathscr{F}\}$ is a normal family.
▷ Show that $\mathscr{F}$ is a normal family if and only if $\mathscr{F} + a$ is a normal family.
▷ Show that $\mathscr{F}$ is a normal family if and only if $a\mathscr{F}$ is a normal family.
▷ Let M be a Möbius transformation. Show that $\mathscr{F}$ is a normal family if and only if $M(\mathscr{F}) = \{M \circ f : f \in \mathscr{F}\}$ is a normal family. (Make use of the decomposition (2.8) of a general Möbius transformation as a composition of translations, dilations and inversions.)

At the end of Sect. 5.2, we saw that if a family $\mathscr{F}$ of functions is relatively compact in $\mathscr{C}(D, \mathbb{C})$ then $\mathscr{F}$ is locally bounded in the sense of Definition 5.3. Montel's remarkable result is that the converse holds if the functions in the family are analytic: local boundedness is enough, and equicontinuity comes for free.

Theorem 5.8 (Montel's First Theorem) *A family $\mathscr{F}$ of analytic functions is a relatively compact subset of $\mathscr{H}(D)$ if and only if $\mathscr{F}$ is locally bounded on D.*

Proof As noted at the end of Sect. 5.2, if a family of functions in $\mathscr{C}(D, \mathbb{C})$ is relatively compact then it is locally bounded. In this direction, analyticity of the functions in the family is not needed.

Now suppose that $\mathscr{F}$ is a locally bounded family of analytic functions. We need to show that conditions (a) and (b) of the Arzelà-Ascoli Theorem, Theorem 5.4, hold with $Y = \mathbb{C}$. Condition (a) is an automatic consequence of local boundedness. In order to establish the equicontinuity of $\mathscr{F}$ on each compact subset of D it is enough, by Exercise 5.5, to show that $\mathscr{F}$ is equicontinuous at each point z_0 of D. Let $z_0 \in D$. Since $\mathscr{F}$ is locally bounded, there is a number M and a positive number r such that $\overline{(D(z_0, r)} \subseteq D$ and)

$$|f(z)| \leq M \text{ whenever } z \in \overline{D(z_0, r)} \text{ and } f \in \mathscr{F}.$$

If $z \in D(z_0, r)$ and $f \in \mathscr{F}$ then, by the Cauchy Integral Formula,

$$f(z) - f(z_0) = \frac{1}{2\pi i} \int_{C(z_0,r)} \frac{f(w)}{w-z} dw - \frac{1}{2\pi i} \int_{C(z_0,r)} \frac{f(w)}{w-z_0} dw$$

$$= \frac{1}{2\pi i} \int_{C(z_0,r)} \frac{(z-z_0)f(w)}{(w-z)(w-z_0)} dw.$$

By the *ML*-inequality, and for $|z - z_0| \leq r/2$,

$$|f(z) - f(z_0)| \leq \frac{1}{2\pi} \frac{|z-z_0| M}{r/2 \times r} 2\pi r = \frac{2M}{r} |z - z_0|.$$

Since this inequality holds for $z \in D(z_0, r/2)$ independently of $f \in \mathscr{F}$, it follows that $\mathscr{F}$ is equicontinuous at z_0. □

As a consequence of Exercise 5.14, Montel's Theorem can be stated as 'A subset $\mathscr{F}$ of $\mathscr{H}(D)$ is relatively compact if and only if $\mathscr{F}$ is uniformly bounded on compact subsets of D'.

Exercise 5.18 Prove that a family of analytic functions $\mathscr{F}$ is a relatively compact subset of $\mathscr{H}(\mathbb{D})$ if and only if there is a sequence of positive numbers M_n, $n = 0, 1, 2, \ldots$, such that

$$\limsup_{n \to \infty} \sqrt[n]{M_n} \leq 1$$

and, if $f(z) = \sum_{n=0}^{\infty} a_n z^n \in \mathscr{F}$, then $|a_n| \leq M_n$.

Exercise 5.19 Suppose that $\mathscr{F}$ is a family of analytic functions defined on a disk $D = D(z_0, r)$ and that $\mathscr{F}$ is *not* a normal family. Show that $\bigcup_{f \in \mathscr{F}} f(D)$ is dense in the complex plane.

In the case of meromorphic functions, the result corresponding to Montel's Theorem is known as Marty's Theorem. The condition that guarantees we have a normal family of meromorphic functions is that their spherical derivatives are uniformly bounded on compact subsets. Recall that the spherical derivative of a meromorphic function is defined by (3.36) in Chap. 3. We'll use the following exercise in the proof of Marty's Theorem.

Exercise 5.20 Suppose that $\{f_n\}_{n=1}^{\infty}$ is a sequence of meromorphic functions that converges to a meromorphic function f spherically locally uniformly on a domain D. Then the sequence of spherical derivatives $\{f_n^{\#}\}_{n=1}^{\infty}$ converges locally uniformly to $f^{\#}$ on D.

(Hint: consider two cases (i) $z_0 \in D$ and z_0 is not a pole of f and (ii) $z_0 \in D$ and z_0 is a pole of f. In the second case, consider the sequence $\{1/f_n\}_{n=1}^{\infty}$.)

5.4 The Theorems of Montel, Marty and Zalcman

Theorem 5.9 (Marty's Theorem) *A family $\mathscr{F}$ of meromorphic functions is a normal family if and only if the family of spherical derivatives $\mathscr{F}^\# = \{f^\# : f \in \mathscr{F}\}$ is locally bounded on D.*

Proof First we show that if the family of spherical derivatives of the meromorphic functions in $\mathscr{F}$ is uniformly bounded on compact subsets of D then $\mathscr{F}$ is a normal family. To do this we use, of course, the Arzelà-Ascoli Theorem, Theorem 5.4, and verify conditions (a) and (b) of that theorem. We are working in the space $\mathscr{C}(D, \mathbb{C}_\infty)$ with the spherical metric on the target space $\mathbb{C}_\infty$. The condition (a) that 'for every z in D, the set of function values $\mathscr{F}(z) = \{f(z) : f \in \mathscr{F}\}$ is a bounded subset of $\mathbb{C}_\infty$' is automatically satisfied since the Riemann sphere is bounded as a metric space.

To establish equicontinuity, let $z_0 \in D$ and choose a positive r for which the closed disk $K = \overline{D(z_0, r)}$ is contained in D. Since K is a compact subset of D, there is a finite M such that $f^\#(z) \leq M$ whenever $f \in \mathscr{F}$ and $z \in K$. Now, for $z \in D(z_0, r)$ and $f \in \mathscr{F}$, consider the spherical length of the image under f of the line segment $[z_0, z]$. This curve can be parameterised by

$$\gamma(t) = f(\tilde{\gamma}(t)) \text{ where } \tilde{\gamma}(t) = (1-t)z_0 + tz, \ 0 \leq t \leq 1.$$

Bearing the mind Definition 3.6, Remark 3.1 following it, and the definition (3.36) of the spherical derivative,

$$d^\#(f(z_0), f(z)) \leq L_\rho(f([z_0, z]))$$

$$= \int_0^1 \rho(\gamma(t)) |\gamma'(t)| \, dt$$

$$= \int_0^1 \frac{2}{1 + |f(\tilde{\gamma}(t))|^2} |f'(\tilde{\gamma}(t))| |z - z_0| \, dt$$

$$= \int_0^1 f^\#(\tilde{\gamma}(t)) |z - z_0| \, dt \leq M |z - z_0|.$$

Equicontinuity of the family $\mathscr{F}$ at z_0 follows. Hence, by Exercise 5.5, $\mathscr{F}$ is equicontinuous on D and, in turn, is a relatively compact subset of $\mathscr{M}^*(D)$.

For the converse, suppose that the family of meromorphic functions $\mathscr{F}$ is a normal family but that the corresponding family of spherical derivatives $\mathscr{F}^\#$ is not locally bounded on D. In that case, there is a compact subset K of D and a sequence of functions $\{f_n\}_{n=1}^\infty$ from the family $\mathscr{F}$ for which $M_n = \max\{f_n^\#(z) : z \in K\}$ tends to ∞ with n. Since $\mathscr{F}$ is a normal family, there is a subsequence of $\{f_n\}_{n=1}^\infty$, which we again call $\{f_n\}_{n=1}^\infty$, that converges spherically locally uniformly to either a meromorphic function f on D or to ∞_D on D. In the first case, it follows from Exercise 5.20 that the sequence of spherical derivatives $\{f_n^\#\}_{n=1}^\infty$ converges locally uniformly to $f^\#$ on D, and hence uniformly on K. But $f^\#$, being continuous (see Exercise 3.20), is bounded on K say $f^\#(z) \leq C$, $z \in K$. Since $f_n^\#$ converges uniformly to $f^{\#}$ on K, there is an N such that $|f_n^{\#}(z) - f^{\#}(z)| \leq 1$ for all $n \geq N$ and all z in K. This implies that $f_n^{\#} \leq C+1$ on K for all $n \geq N$, a contradiction once $M_n > C + 1$. If $f = \infty_D$ then, since f_n is not identically zero, the functions $1/f_n$ are meromorphic and converge spherically locally uniformly to $1/f$ on D, that is to the function that is identically zero. Again, by Exercise 5.20, $(1/f_n)^{\#}$ converges to $(1/f)^{\#}$, that is to zero, locally uniformly. Since $(1/f_n)^{\#}(z) = f_n^{\#}(z)$ we again have a contradiction as $\max\{(1/f_n)^{\#}(z) : z \in K\}$ tends to ∞ with n whereas $(1/f)^{\#} = 0$ on K. □

The third classic result in this section is known as Zalcman's Lemma (see [28, 29]). If a family $\mathscr{F}$ of meromorphic functions is *not* normal then we can find a sequence of functions in $\mathscr{F}$ that, when appropriately scaled, converges to a meromorphic function on the whole complex plane. Though it takes a little while to digest, Zalcman's Lemma provides an easy route to the Great Picard Theorem.

Theorem 5.10 (Zalcman's Lemma) *A family $\mathscr{F}$ of meromorphic functions on a domain D is not a normal family if and only if there are*

- *a disk $D(z_0, r)$ for which the disk $D(z_0, 2r)$ lies in D;*
- *a sequence of points $\{z_n\}_{n=1}^{\infty}$ in $D(z_0, r)$;*
- *a sequence $\{r_n\}_{n=1}^{\infty}$ of positive numbers with $r_n \to 0$ as $n \to \infty$;*
- *a sequence of functions $\{f_n\}_{n=1}^{\infty}$ in $\mathscr{F}$;*
- *a non-constant function g, meromorphic in the complex plane $\mathbb{C}$, for which*

$$f_n(z_n + r_n z) \to g(z) \text{ as } n \to \infty$$

spherically locally uniformly on $\mathbb{C}$. In addition, $g^{\#}(z) \leq g^{\#}(0) = 1$, $z \in \mathbb{C}$. If the functions in $\mathscr{F}$ are analytic then g is entire and the convergence is locally uniform.

Proof Suppose that $\mathscr{F}$ is not a normal family. Then, by Marty's Theorem, the family of spherical derivatives of functions in $\mathscr{F}$ is not locally uniformly bounded on D. There is therefore a disk $D(z_0, 2r)$ contained in D such that the family $\mathscr{F}^{\#}$ of spherical derivatives of functions in $\mathscr{F}$ is not uniformly bounded on the closed disk $K = \overline{D(z_0, r/2)}$. (Otherwise, since any compact subset of D can be covered by finitely many such disks, the family $\mathscr{F}^{\#}$ would be bounded on each compact subset of D.) That is, there is a sequence $\{f_n\}_{n=1}^{\infty}$ of functions in $\mathscr{F}$ and points w_n in K such that $f_n^{\#}(w_n) \to \infty$ as $n \to \infty$.

Now set, for each n,

$$M_n = \max\left\{\left(1 - \frac{|z - z_0|^2}{r^2}\right) f_n^{\#}(z) : z \in D(z_0, r)\right\}.$$

The maximum exists since $f_n^{\#}$ is a continuous function. The effect of the 'cut-off' factor $1 - |z - z_0|^2/r^2$, which vanishes on the circle $C(z_0, r)$, is to ensure that the maximum occurs at a point away from this circle. Moreover, since $|w_n - z_0| \leq r/2$, we have the estimate $M_n \geq 3f_n^\#(w_n)/4$, and so $M_n \to \infty$ as $n \to \infty$. We write z_n for a point in the disk $D(z_0, r)$ at which the maximum value occurs, so that

$$M_n = \left(1 - \frac{|z_n - z_0|^2}{r^2}\right) f_n^\#(z_n),$$

and $f_n^\#(z_n) \geq M_n$. Next, we set

$$r_n = \frac{1}{M_n}\left(1 - \frac{|z_n - z_0|^2}{r^2}\right) = \frac{1}{f_n^\#(z_n)}. \tag{5.18}$$

The numbers r_n tend to 0 as $n \to \infty$ since $f_n^\#(z_n) \to \infty$. Next, since

$$1 - \frac{|z_n - z_0|^2}{r^2} = \frac{(r - |z_n - z_0|)(r + |z_n - z_0|)}{r^2}$$
$$\leq \frac{(r - |z_n - z_0|)(2r)}{r^2} = \frac{2(r - |z_n - z_0|)}{r},$$

we find that

$$r_n M_n \leq \frac{2(r - |z_n - z_0|)}{r}.$$

From this we conclude that $R_n := (r - |z_n - z_0|)/r_n \to \infty$ as $n \to \infty$.

We now set

$$g_n(z) = f_n(z_n + r_n z).$$

The function f_n is certainly defined in the disk $D(z_0, r)$. If $|z| < R_n$ then

$$|(z_n + r_n z) - z_0| < |z_n - z_0| + r_n R_n = |z_n - z_0| + (r - |z_n - z_0|) = r.$$

We see that, in turn, the function g_n is defined and meromorphic in the disk $D(0, R_n)$. At a point z in $D(0, R_n)$ that isn't a pole of g_n, we calculate the spherical derivative of g_n to be

$$g_n^\#(z) = \frac{2|g_n'(z)|}{1 + |g_n(z)|^2} = \frac{2r_n|f_n'(z_n + r_n z)|}{1 + |f_n(z_n + r_n z)|^2} = r_n f_n^\#(z_n + r_n z). \tag{5.19}$$

By continuity of the spherical derivative (see Exercise 3.20), $g_n^\#(z) = r_n f_n^\#(z_n + r_n z)$ holds throughout $D(0, R_n)$. Putting $z = 0$ and using (5.18), we see that $g_n^\#(0) = r_n f_n^\#(z_n) = 1$.

We use Marty's Theorem to show that the sequence of functions $\{g_n\}_{n=1}^\infty$ is a spherically normal family. We would ideally wish to apply Marty's Theorem on the whole complex plane $\mathbb{C}$. However, the functions g_n are only defined on disks $D(0, R_n)$. This technicality means that we first work on a fixed disk $D(0, R)$ and apply Marty's Theorem on this disk to find a subsequence $\{g_{R,n}\}_{n=1}^{\infty}$ of $\{g_n\}_{n=1}^{\infty}$ that converges on $D(0, R)$. We therefore need to estimate the spherical derivative $g_n^\#$ of g_n on $D(0, R)$ where $R \leq R_n$. Recall that $z_n + r_n z \in D(z_0, r)$ if $z \in D(0, R_n)$. For $z \in D(0, R)$ with $R \leq R_n$ and z not a pole of g_n, we first use the definition of M_n and then (5.18) to estimate

$$g_n^\#(z) = r_n f_n^\#(z_n + r_n z) \leq r_n M_n \frac{1}{1 - \frac{|(z_n + r_n z) - z_0|^2}{r^2}} = \frac{1 - \frac{|z_n - z_0|^2}{r^2}}{1 - \frac{|(z_n + r_n z) - z_0|^2}{r^2}}$$

$$= \frac{r^2 - |z_n - z_0|^2}{r^2 - |(z_n + r_n z) - z_0|^2}$$

$$= \frac{r - |z_n - z_0|}{r - |(z_n + r_n z) - z_0|} \times \frac{r + |z_n - z_0|}{r + |(z_n + r_n z) - z_0|}.$$

Since $r_n R_n = r - |z_n - z_0|$, we can estimate the first term for sufficiently large n by

$$\frac{r - |z_n - z_0|}{r - |(z_n + r_n z) - z_0|} \leq \frac{r_n R_n}{r_n R_n - r_n |z|} \leq \frac{R_n}{R_n - R}.$$

For fixed R and since $R_n \to \infty$, the expression $R_n/(R_n - R) \to 1$ as $n \to \infty$. For the second term,

$$\frac{r + |z_n - z_0|}{r + |(z_n + r_n z) - z_0|} \leq \frac{r + |z_n - z_0|}{r + |z_n - z_0| - r_n R},$$

which also tends to 1 as $n \to \infty$ since $r_n \to 0$. Thus, we can bound $g_n^\#(z)$ uniformly on $D(0, R)$ by a number as close to 1 as we wish by taking n sufficiently large. Initially this bound holds only where g_n doesn't have a pole, but then holds everywhere in the disk since $g_n^\#$ is continuous. By Marty's Theorem, therefore, there is some large $N = N(R)$ such that the meromorphic functions $\{g_n\}_{n=N(R)}^{\infty}$ form a normal family on $D(0, R)$. As a consequence, we can find a subsequence that converges spherically locally uniformly on $D(0, R)$ to a function $\tilde{g}_R$, say. The possibility remains that $\tilde{g}_R$ is the constant function infinity. If that was the case, the meromorphic functions $1/g_n$ would converge spherically locally uniformly to $1/\tilde{g}_R$ (see Exercise 5.15), which is identically 0. Hence, by Exercise 5.20, $(1/g_n)^\# = g_n^\#$ converges locally uniformly to 0 as well. But $g_n^\#(0) = 1$ for each n, which is a contradiction. We have shown that $\tilde{g}_R$ is meromorphic on $D(0, R)$ and that $\tilde{g}_R^\#(z) \leq 1$ there.

To complete the argument we choose an increasing sequence of radii and find a limit function on the whole complex plane using a diagonalisation argument as in the proof of Proposition 5.3. Here are the details. Setting $R = 1$, we find a subsequence $\{g_{1,n}\}_{n=1}^{\infty}$ of $\{g_n\}_{n=1}^{\infty}$ that converges to a meromorphic function $\tilde{g}_1$ on $D(0, 1)$. Since

$\{g_{1,n}\}_{n=n_1}^{\infty}$ forms, for sufficiently large n_1, a normal family on $D(0, 2)$, we can find a subsequence $\{g_{2,n}\}_{n=1}^{\infty}$ of $\{g_{1,n}\}_{n=1}^{\infty}$ that converges to a meromorphic function $\tilde{g}_2$ on $D(0, 2)$. The function $\tilde{g}_2$ must agree with $\tilde{g}_1$ on $D(0, 1)$ since it is the limit of a subsequence of a sequence that converges to $\tilde{g}_1$. We continue to take subsequences of subsequences at infinitum to produce nested sequences $\{g_{k,n}\}_{n=1}^{\infty}$ that converge to meromorphic functions $\tilde{g}_k$ on the disks $D(0, k)$, $k = 1, 2, \ldots$. Moreover, if $k_1 < k_2$ then $\tilde{g}_{k_2}$ coincides with $\tilde{g}_{k_1}$ on the disk $D(0, k_1)$.

Finally, we set

$$g(z) = \tilde{g}_n(z) \text{ if } n > |z|.$$

This is a consistent definition of a meromorphic function on $\mathbb{C}$. Moreover, the diagonal sequence $\{g_{n,n}\}_{n=1}^{\infty}$ converges to g uniformly on each disk $D(0, k)$, $k \geq 1$, as $\{g_{n,n}\}_{n=k}^{\infty}$ is, for each k, a subsequence of $\{g_{k,n}\}_{n=k}^{\infty}$. Of course, $g_{n,n}$ is a term in the original sequence $\{g_n\}_{n=1}^{\infty}$, say $g_{n,n} = g_{n_k}$. Let f_{n_k} be the corresponding function in the sequence $\{f_n\}_{n=1}^{\infty}$. Then, $\{f_{n_k}\}_{k=1}^{\infty}$ and the corresponding sequences $\{z_{n_k}\}$ and $\{r_{n_k}\}$ satisfy the conditions in the statement of the theorem. Moreover, for each fixed z, once $n_k > |z|$ we have $g^{\#}(z) = \lim_{k \to \infty} g_{n_k}^{\#}(z) \leq 1$. Since $g^{\#}(0) = 1$, g is not constant. Finally, if the functions in $\mathscr{F}$ are analytic then so is each g_{n_k} and then so is their limit g, so that g is entire. By Proposition 5.5, the convergence of $f_{n_k}(z_{n_k} + r_{n_k}z)$ to $g(z)$ is uniform on compact sets.

For the converse, suppose that $\mathscr{F}$ is a relatively compact family of meromorphic functions in $\mathscr{M}^*(D)$. Suppose that there is a disk $D(z_0, 2r)$ in D, points $\{z_n\}_{n=1}^{\infty}$ in the disk $D(z_0, r)$, a sequence of positive numbers $\{r_n\}_{n=1}^{\infty}$ that converges to 0, and a sequence of functions $\{f_n\}_{n=1}^{\infty}$ from $\mathscr{F}$ such that $g_n(z) = f_n(z_n + r_n z) \to g(z)$ as $n \to \infty$ spherically uniformly on $\mathbb{C}$. Suppose that g is not identically equal to infinity.

Set $R = 3r/2$. By Marty's Theorem, there is a finite M such that $f^{\#}(z) \leq M$ if $|z - z_0| \leq R$ and $f \in \mathscr{F}$. Fix $z \in \mathbb{C}$. For all large n, we have

$$|(z_n + r_n z) - z_0| \leq |z_n - z_0| + r_n|z| \leq R$$

since $|z_n - z_0| < r$ and $r_n \to 0$ as $n \to \infty$. Then, by (5.19).

$$g^{\#}(z) = \lim_{n \to \infty} g_n^{\#}(z) = \lim_{n \to \infty} r_n f_n^{\#}(z_n + r_n z) \leq \lim_{n \to \infty} r_n M = 0.$$

Hence g is constant. □

5.5 Montel's Second Theorem and the Great Picard Theorem

Montel's First Theorem, Theorem 5.8, states that local boundedness of a family of analytic functions in a domain is equivalent to the property that every sequence in the family has a subsequence that converges locally uniformly in the domain. Montel's Second Theorem is a deeper and even more striking result.

Theorem 5.11 (Montel's Second Theorem) *Suppose that $\mathscr{F}$ is a family of analytic functions in a domain D all of which omit the same two values. Then $\mathscr{F}$ is a normal family.*

Remark 5.2 Being a normal family is a local property in the sense that a family $\mathscr{F}$ of meromorphic functions on D is normal if and only if the restrictions of the functions in $\mathscr{F}$ to each disk $D(z_0, r)$ in D is a normal family. The 'only if' direction is clear. The 'if' direction follows from Marty's Theorem. If $\mathscr{F}^{\#}$ is locally bounded on each disk $D(z_0, r)$ in D, hence bounded on each disk $D(z_0, r/2)$ with $D(z_0, r)$ in D, then $\mathscr{F}^{\#}$ is bounded on each compact subset of D. This last statement follows since any compact subset of D can be covered by finitely many such disks $D(z_0, r/2)$.

Proof of Montel's Second Theorem Suppose that each analytic function in $\mathscr{F}$ never takes the distinct values a and b in $\mathbb{C}$. Replacing each function f in $\mathscr{F}$ by $(f - a)/(b - a)$ we make a new family of analytic functions on D none of which ever takes the value 0 or 1. By Exercise 5.17, this new family is normal if and only if the original family is normal. We may therefore assume that the omitted values are 0 and 1.

Taking account of the remark above, we may further assume that D is a disk $D(z_0, r)$. By scaling and translating, that is by considering the family of functions $\tilde{f}$ in $\mathbb{D}$ defined by $\tilde{f}(z) = f(z_0 + rz)$, $z \in \mathbb{D}$, we may assume that D is the unit disk $\mathbb{D}$.

So, let $\mathscr{F}$ be the family of functions analytic on the unit disk each of which omits both 0 and 1. For each natural number n, let $\mathscr{F}_n$ be the family of analytic functions on $\mathbb{D}$ that omit the value 0 as well as all nth roots of unity (that is, $\exp(2k\pi i/n)$ for $k = 1, 2, \ldots, n$). If f belongs to $\mathscr{F}_n$ then f^n belongs to $\mathscr{F}$. If f belongs to $\mathscr{F}$ then, since $\mathbb{D}$ is simply connected and f omits the value 0, f has an analytic nth-root (for this we need Theorem 6.6, which we take for granted for now). This analytic function $f^{1/n}$ omits 0 and all nth-roots of unity, as otherwise $f = (f^{1/n})^n$ would take the value 1. In this sense, f belongs to $\mathscr{F}_n$ if and only if f^n belongs to $\mathscr{F}$. In fact, there is an n-to-1 correspondence between functions in $\mathscr{F}_n$ and functions in $\mathscr{F}$.[5]

Next we show that if $\mathscr{F}_n$ is a normal family then so is $\mathscr{F}$. Let $\{f_k\}_{k=1}^{\infty}$ be a sequence of functions from $\mathscr{F}$. The sequence of functions $\{f_k^{1/n}\}_{k=1}^{\infty}$ (in each case, we choose any one of the n distinct analytic nth roots of f_k) belongs to $\mathscr{F}_n$ and, therefore, has a subsequence $\{f_{k_j}^{1/n}\}_{j=1}^{\infty}$ that converges spherically locally uniformly. Then $\{f_{k_j}\}_{j=1}^{\infty}$ is a subsequence of $\{f_k\}_{k=1}^{\infty}$ that converges spherically locally uniformly.

---

[5] All analytic nth-roots of f differ by a multiplicative nth-root of unity. If $f_1^n = f_2^n$ then $(f_1/f_2)^n = 1$ and so, since f_1/f_2 is continuous, it must be a constant nth-root of unity.

5.5 Montel's Second Theorem and the Great Picard Theorem

Now suppose that $\mathscr{F}$ is not normal so that no $\mathscr{F}_{2^n}$ is normal either—we work with powers of 2 since the sets of 2^n-roots of unity are nested. By Zalcman's Lemma, therefore, for each n there is a non-constant entire function g_n with

$$g_n(z) = \lim_{k \to \infty} f_k(z_k + r_k z), \quad z \in \mathbb{C},$$

where z_k belongs to a disk $D(z_0, r)$ for which $D(z_0, 2r) \subseteq \mathbb{D}$, $r_k \to 0$, and f_k are functions in $\mathscr{F}_{2^n}$. Also, $g_n^\#(0) = 1$ and $g_n^\#(z) \leq 1$, $z \in \mathbb{C}$. Since each f_k never assumes the value 0 or any 2^n-root of unity, the same can be said for g_n by Hurwitz's Theorem.

By Marty's Theorem, the functions $\{g_n\}_{n=1}^\infty$ form a normal family. A subsequence $\{g_{n_k}\}_{k=1}^\infty$ therefore converges spherically locally uniformly to a function G, say. If G was identically infinite then the sequence of entire functions $\{1/g_{n_k}\}_{k=1}^\infty$ would converge spherically locally uniformly to 0 which is not possible since $(1/g_{n_k})^\#(0) = g_{n_k}^\#(0) = 1$. As a consequence, G must be an entire function, and cannot be constant since $G^\#(0) = \lim_{k \to \infty} g_{n_k}^\#(0) = 1$.

For each fixed n, the entire functions g_k never assume any 2^n-root of unity if $k \geq n$—this is because any 2^n-root of unity is a 2^k-root of unity if $k \geq n$. By Hurwitz's Theorem again, G never assumes any 2^n-root of unity either, and this is true for every n. Taken all together, these roots of unity are dense in the unit circle. By the Open Mapping Theorem, $G(\mathbb{C})$ is a connected open set that does not meet the unit circle, so either $G(\mathbb{C}) \subset \mathbb{D}$ or $G(\mathbb{C}) \subset \mathbb{C}\setminus\overline{\mathbb{D}}$. In either case, Liouville's Theorem implies that G must be constant, which is a contradiction of the assumption that $\mathscr{F}$ was not a normal family. □

A corollary of what we have just proved is the following version of Montel's Second Theorem: *A family $\mathscr{F}$ of functions meromorphic in a domain D, none of which assumes any of three distinct values, is normal.* If one of the omitted values is ∞, then $\mathscr{F}$ consists of analytic functions that omit the same two values so that Theorem 5.11 applies. Otherwise, the omitted values are a, b and c in $\mathbb{C}$. For $f \in \mathscr{F}$,

$$g_f(z) = \frac{b-c}{b-a}\left(\frac{f(z)-a}{f(z)-c}\right), \quad z \in D,$$

is meromorphic and omits the values 0, 1 and ∞. Since it omits the value ∞, g_f is analytic and never takes the values 0 and 1. Hence, by Theorem 5.11, $\mathscr{G} = \{g_f : f \in \mathscr{F}\}$ is a normal family and then, by Exercise 5.17, so is $\mathscr{F}$.

We now come to the highlight of this section, even of this chapter, namely the Great Picard Theorem.

Theorem 5.12 (Great Picard Theorem) *Suppose that f has an essential singularity at $a \in \mathbb{C}$. Then, in each punctured neighbourhood of a, f assumes all complex values with at most one possible exception.*

Before we move to the proof, let's see why it might be called 'great'. There is also a *Little Picard Theorem* which states that an entire function assumes every complex value with at most one possible exception. This result is a celebrated improvement on Liouville's Theorem—by Liouville's Theorem, an entire function that omits a disk must be constant; by the Little Picard Theorem, if an entire function omits just two values it must be constant. The Little Picard Theorem is a consequence of the Great Picard Theorem by the following argument. Suppose that f is an entire function. If f is a polynomial then, by the Fundamental Theorem of Algebra [16, Theorem 7.11], f takes every complex value (in fact, f takes every complex value n times counting multiplicity where n is the degree of the polynomial). Otherwise, f is a *transcendental entire function* with a power series expansion about the origin

$$f(z) = \sum_{n=0}^{\infty} a_n z^n$$

convergent for every complex z and where infinitely many of the coefficients a_n are non-zero (if only finitely many coefficients are non-zero then f is a polynomial). Then,

$$g(z) = f(1/z) = \sum_{n=0}^{\infty} \frac{a_n}{z^n}$$

has an essential singularity at the origin. By the Great Picard Theorem, g assumes every complex value with at most one possible exception in every neighbourhood of the origin, from which it follows that the entire function f takes on every value with at most one exception.

Actually, we can deduce a little more, namely that the transcendental entire function f takes on every complex value, with at most one exception, infinitely often. To see this, let c be any complex number other than this possible exceptional value for g, and let z_1 be a non-zero complex number such that $g(z_1) = c$ (so that $f(1/z_1) = c$). By the Great Picard Theorem, g will take on the value c again somewhere in the punctured neighbourhood $D'(0, |z_1|)$ of 0, say at z_2. Then, $f(1/z_2) = c$ as well. Continuing in this way (the next neighbourhood of the origin we consider is $D'(0, |z_2|)$ and produce z_3 with $g(z_3) = c$ and $0 < |z_3| < |z_2|$), we produce a sequence $\{z_n\}_{n=1}^{\infty}$ of distinct, non-zero complex numbers for which $g(z_n) = f(1/z_n) = c$. Of course, $1/z_n \to \infty$ as $n \to \infty$.

The Great Picard Theorem is a major extension of the Casorati-Weierstrass Theorem [16, Theorem 8.10] which simply states that the image of any neighbourhood of a under f, where f has an essential singularity at a, is dense in the complex plane. It is a major step to go from 'dense' to 'everywhere with at most one exception'. With this in mind, we use Montel's Second Theorem to prove the Great Picard Theorem.

Proof of the Great Picard Theorem By considering $\tilde{f}(z) = f(z + a)$, we may assume that f has an essential singularity at the origin. Suppose that the Laurent series for f is convergent in the punctured disk $D'(0, R)$ and fix r with $r \leq R$. Suppose to the contrary that f omits two distinct values in the punctured disk

5.5 Montel's Second Theorem and the Great Picard Theorem

$D'(0, r)$, say α and β. By replacing f by $\tilde{f}(z) = (f(z) - \alpha)/(\beta - \alpha)$, we may assume that the omitted values are 0 and 1.

For each n, set $f_n(z) = f(z/n)$ for $z \in D'(0, r)$. By Montel's Second Theorem, the family of analytic functions $\mathscr{F} = \{f_n : n \in \mathbb{N}\}$ is a normal family in the domain $D = D'(0, r)$ as all functions in $\mathscr{F}$ omit the same two values. There therefore exists a subsequence $\{f_{n_k}\}_{k=1}^{\infty}$ that converges, say to g, spherically locally uniformly on D where either g is analytic or g is identically infinite.

Suppose that g is analytic in D. Then g is bounded on the circle $C(0, r/2)$ and $f_{n_k} \to g$ uniformly on $C(0, r/2)$ since this circle is a compact subset of D. It follows that there is a finite M such that $|f_{n_k}| \leq M$ on $C(0, r/2)$ for each k, that is $|f(z/n_k)| \leq M$ for $|z| = r/2$. Thus, for each $k \in \mathbb{N}$, f is bounded by M on the two circles bounding the annulus

$$A_k = \left\{z : \frac{r}{2n_{k+1}} < |z| < \frac{r}{2n_k}\right\}$$

By the Maximum Principle, f is bounded by M inside the annulus A_k. Since the union of all such annuli A_k is a punctured disk about the origin, we conclude that f is bounded in a punctured neighbourhood of 0 and so has a removable singularity there and not an essential singularity.

The other possibility is that g is identically infinite in $D'(0, r)$, in which case the sequence of analytic functions $1/f_{n_k}$ (remember that f never takes the value 0) converges to 0 locally uniformly on D. As above, there will then be a finite M such that $|1/f_{n_k}|$ is bounded by M on the circle $C(0, r/2)$, so that $|1/f(z)| \leq M$ for $|z| = r/(2n_k)$. By the Maximum Principle, $1/f$ is bounded by M on each annulus A_k and therefore $1/f$ is bounded in a punctured neighbourhood of 0. It follows that $1/f$ has a removable singularity at 0. Then, $(1/f)(z) = z^m g(z)$ in $D(0, r)$ where $m \geq 0$ and $g(0) \neq 0$ so that $f(z) = z^{-m}/g(z)$. In this case, f itself has either a removable singularity at 0 (if $m = 0$) or has a pole at 0 (if $m > 0$), which in either case contradicts the assumption that f has an essential singularity at 0. □

There are various approaches to the circle of theorems and ideas that include the Picard Theorems, Montel's First and Second Theorems, Zalcman's Lemma, and also the theorems of Bloch and Schottky. A particularly nice account is to be found in Chap. 10 of Remmert's book [23].

Chapter 6
Simply Connected Domains, the Riemann Mapping Theorem and Conformal Mapping

One of the main goals of this chapter is to set up the machinery needed to prove the Riemann Mapping Theorem. This classic result states that, in the case of each proper 'simply connected' domain, there is a conformal map between the domain and a disk. A simply connected domain is one that, loosely speaking, 'has no holes', and is proper if it isn't the whole complex plane $\mathbb{C}$. The case of the complex plane has to be ruled out since any map of the complex plane into a disk would be a bounded entire function and therefore constant by Liouville's Theorem. With the existence of a conformal mapping of any proper simply connected domain onto the unit disk being guaranteed by this result, the hyperbolic geometry of the disk transfers naturally to the simply connected domain, just as we saw in Sect. 4.7 in the case of the upper half-plane $\mathbb{H}$. In Sect. 6.5 we work out some further explicit conformal maps and the induced hyperbolic geometry in the mapped domains.

There are several equivalent formulations of the notion of 'simply connected'. We will meet some of these in due course. A planar domain is simply connected precisely when its complement relative to the extended complex plane is connected, as will be established in the next chapter. As a starting point, we adopt the standard definition that a domain is simply connected if every closed curve in the domain can be shrunk to a point without leaving the domain. In general, if there is a conformal map of one planar domain onto another then (by Proposition 6.1) this map is a homeomorphism and the two domains are homeomorphic. By the Riemann Mapping Theorem, the converse holds in the case of simply connected domains: if a domain is homeomorphic to a disk and is not the complex plane then there is a conformal map of the domain onto a disk. That topological equivalence implies conformal equivalence in the case of simply connected domains is surprising. Indeed, this fails dramatically in general, even in the case of the simplest 'doubly connected' domains, namely annuli. Even though all annuli are homeomorphic there is a continuum of different annuli from the point of view of conformal equivalence.

This latter result requires some information on how conformal maps behave on the boundary. It is quite extraordinary that, even though the boundary of a simply connected domain can be highly complicated (it can even have Hausdorff dimension 2), the domain itself can be mapped conformally onto a disk whose boundary is a smooth circle. Clearly, a general conformal mapping will therefore not necessarily behave well on the boundary. On the other hand, Carathéodory showed that a conformal map from a simply connected domain onto a disk will extend to a homeomorphism of the closure of the domain onto the closure of the disk if and only if the domain is bounded by a Jordan curve. Boundary behaviour of general conformal maps is formalised in the theory of 'prime ends' [6]. It is sufficient for our purposes to show (in Sect. 6.6) that a conformal map between domains whose boundaries are 'nice' (in a manner to be described precisely later) extends continuously to the boundary. With this result to hand, we are able to analyse conformal maps between annuli.

To begin this programme of work, we first need a general form of Cauchy's Theorem. The statement of this result is, loosely speaking, that the integrals of an analytic function around two closed curves are the same if one curve can be continuously deformed into the other without leaving the domain of analyticity of the function.

6.1 The Homotopic Form of Cauchy's Theorem

This is a second course on complex analysis, so we will assume Cauchy's Theorem for a disk or, with essentially the same proof, Cauchy's Theorem for a star-shaped domain. When it comes to computing specific integrals, one can go a long way with just this version of Cauchy's Theorem together with crosscuts or, perhaps, the more formalised Residue Theorem.

Theorem 6.1 (Cauchy's Theorem for a Disk) *Suppose that f is analytic in a disk $D(a, r)$ and that γ is a piecewise-smooth closed curve in $D(a, r)$. Then,*

$$\int_\gamma f(z)\, dz = 0.$$

For a proof see [16, Theorem 6.1]. The proof works perfectly well in a *star-shaped* domain, not just in a disk. A domain D is said to be star-shaped about a point a if the line segment $[a, z]$ lies in D whenever z lies in D. A domain is said to be star-shaped if it is star-shaped about some point.

In a standard first course on complex analysis, the Cauchy Integral formula is obtained as an easy consequence of Cauchy's Theorem for star-shaped domains. See, for example, [16, Theorem 7.1].

6.1 The Homotopic Form of Cauchy's Theorem

Theorem 6.2 (Cauchy's Integral Formula) *Suppose that f is analytic in a disk $D(a, r)$ and that $0 < \rho < r$. Then, for each z in $D(a, \rho)$,*

$$f(z) = \frac{1}{2\pi i} \int_{C(a,\rho)} \frac{f(w)}{w - z} \, dw. \tag{6.1}$$

The proof, though relatively easy, introduces a new idea. One considers the difference quotient

$$g(w) = \frac{f(w) - f(z)}{w - z}.$$

The function g is an analytic function of w in $D(a, r) \setminus \{z\}$. One proceeds to prove, using suitably chosen crosscuts, that

$$\int_{C(a,\rho)} g(w) \, dw = \int_{C(z,\delta)} g(w) \, dw \tag{6.2}$$

for all sufficiently small, positive δ. The integrand in the second integral approaches $f'(z)$ as δ approaches 0 and so, once δ is sufficiently small, the integrand is bounded above by $2|f'(z)|$, say. It then follows from the ML-inequality that $\left| \int_{C(z,\delta)} g(w) \, dw \right| \leq 4\pi |f'(z)| \delta$. Since this inequality holds for all small positive δ it must be, by (6.2), that $\int_{C(a,\rho)} g(w) \, dw$ equals 0. One completes the proof of (6.1) by showing that

$$\int_{C(a,\rho)} \frac{dw}{w - z} = \int_{C(z,\delta)} \frac{dw}{w - z} = 2\pi i. \tag{6.3}$$

The integrals in (6.2), and again in (6.3), are equal so that, though the path has changed, the value of the integrals has not. The reason is, loosely speaking, that neither g nor $1/(w - z)$ has any singularity lying in the region between the circles $C(z, \delta)$ and $C(a, \rho)$.

The homotopic form of Cauchy's Theorem is a significant formalization of this idea. To begin with, we explain what we mean when we say that two curves are homotopic relative to a domain. Again, loosely speaking, this means that we can continuously deform one curve into the other curve without leaving the domain. For example, the circles $C(3, 1)$ and $C(0, 1)$ are not homotopic relative to the domain $\mathbb{C} \setminus \{0\}$—see Fig. 6.1. If one tries to continuously deform the circle $C(0, 1)$ to the circle $C(3, 1)$ without leaving the domain $\mathbb{C} \setminus \{0\}$, the 'hole' at 0 prevents this. This is the reason why the integrals $\int_{C(0,1)} dz/z$ and $\int_{C(3,1)} dz/z$ have different values even if $f(z) = 1/z$ is an analytic function in $\mathbb{C} \setminus \{0\}$: the first integral equals $2\pi i$ by direct computation and the second equals 0 by applying, for example, Cauchy's Theorem to $f(z) = 1/z$ in the disk $D(3, 2)$. To proceed, we need a formal, somewhat technical, definition of homotopy of curves.

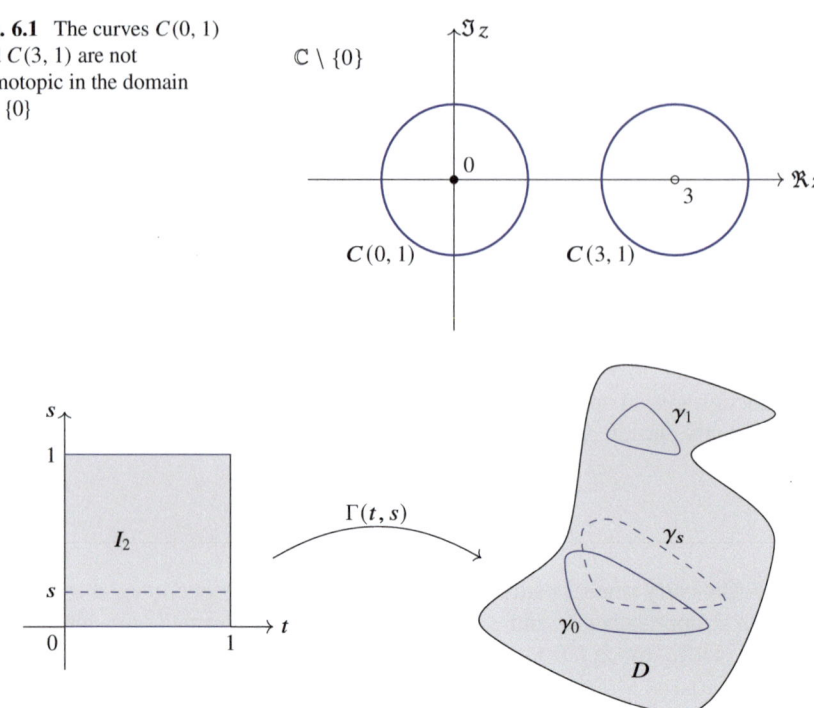

Fig. 6.1 The curves $C(0, 1)$ and $C(3, 1)$ are not homotopic in the domain $\mathbb{C} \setminus \{0\}$

Fig. 6.2 The curve γ_0 is homotopic to the curve γ_1 relative to the domain D

Definition 6.1 (Homotopy of Closed Curves) Suppose that γ_0 and γ_1 are two closed piecewise-smooth curves lying in a domain D. Assume that the curves are parameterised on $[0, 1]$ by $\gamma_0(t)$ and $\gamma_1(t)$, respectively. The curve γ_0 is said to be homotopic to the curve γ_1 relative to the domain D if there is a function $\Gamma(t, s)$ on the closed square $I_2 = [0, 1] \times [0, 1]$ with the following properties:

(i) $\Gamma(t, s)$ is continuous on I_2;
(ii) $\Gamma(t, s)$ lies in D for every (t, s) in I_2;
(iii) $\Gamma(t, 0) = \gamma_0(t)$ and $\Gamma(t, 1) = \gamma_1(t)$ for $0 \leq t \leq 1$;
(iv) $\Gamma(0, s) = \Gamma(1, s)$ for $0 \leq s \leq 1$.

For each s in $[0, 1]$, these properties imply that $\gamma_s(t) = \Gamma(t, s)$, $0 \leq t \leq 1$, is a continuous closed curve lying in D. That γ_s is a curve comes from the continuity condition (i). That γ_s lies in D comes from (ii), and that the curve γ_s is closed follows from (iv). When $s = 0$ we obtain the original curve γ_0 and when $s = 1$ we obtain the original curve γ_1; this is what (iii) in the definition says. As s changes from 0 to 1, the closed curve γ_s changes continuously—because of condition (i) in the definition—from γ_0 to γ_1, without leaving the domain D. In this sense, $\Gamma(t, s)$ provides a continuous deformation of γ_0 into γ_1 through a series of closed curves γ_s, $0 \leq s \leq 1$, without ever leaving the domain D. This deformation, which is not by any means unique, is illustrated in Fig. 6.2.

6.1 The Homotopic Form of Cauchy's Theorem

Example 6.1 The curves $\gamma_0 = C(2, 1)$ and $\gamma_1 = C(4, 2)$ are homotopic in the slit-plane $D = \mathbb{C} \setminus \{z \colon \Im z = 0, \Re z \leq 0\}$. In fact, $\Gamma(t, s)$ will work as a homotopy where

$$\Gamma(t, s) = 2 + 2s + (1 + s)e^{2\pi i t}, \quad (t, s) \in I_2.$$

Exercise 6.1 Show that homotopy is an equivalence relation on the collection of all piecewise-smooth closed curves lying in a domain D.

All constant curves in a domain are homotopic. In fact, suppose that γ_0 is the constant curve $\gamma_0(t) = z_0$, $0 \leq t \leq 1$, and that γ_1 is the constant curve $\gamma_1(t) = z_1$, $0 \leq t \leq 1$, where z_0 and z_1 are in D. Since D is a domain, it is pathwise connected and so there is a curve, say $\tilde{\gamma}(s)$, $0 \leq s \leq 1$, joining z_0 to z_1 in D. Then,

$$\Gamma(t, s) = \tilde{\gamma}(s), \quad (t, s) \in I_2,$$

is a homotopy between γ_0 and γ_1. Thus all constant curves in a domain lie in the same equivalence class which will, of course, contain lots of other closed curves as well.

We say that a closed curve γ is *homotopic to zero in a domain* D if γ is homotopic to some constant curve relative to D. A curve that is homotopic to zero in a domain can, therefore, be continuously shrunk to a point without leaving the domain.

Exercise 6.2 Show that any closed curve in a star-shaped domain is homotopic to zero.

A domain in which every closed curve is homotopic to zero is called *simply connected*. Equivalently, a simply connected domain is one with a single equivalence class of homotopic closed curves.

Exercise 6.3 Show that if h is a homeomorphism from a domain D_1 onto a domain D_2 and if γ_0 and γ_1 are homotopic closed curves in D_1 then $h(\gamma_0)$ and $h(\gamma_1)$ are homotopic closed curves in D_2.

In particular, the image of a simply connected domain under a homeomorphism is again simply connected.

We are now ready to state and prove the homotopic form of Cauchy's Theorem. Even if the idea behind the proof is straightforward, the proof itself is a little technical mainly due to the notation involved. You could be forgiven for omitting it at first reading.

Theorem 6.3 (Homotopic Form of Cauchy's Theorem) *Suppose that f is analytic in a domain D. Suppose that γ_0 and γ_1 are homotopic piecewise-smooth closed curves in D. Then*

$$\int_{\gamma_0} f(z)\,dz = \int_{\gamma_1} f(z)\,dz.$$

Proof Let $\Gamma(t, s)$, $(t, s) \in I_2$, be a homotopy of γ_0 and γ_1 in D. We divide the square $I_2 = [0, 1] \times [0, 1]$ into n^2 similar squares $I_{j,k}$ where

$$I_{j,k} = \left[\frac{j}{n}, \frac{j+1}{n}\right] \times \left[\frac{k}{n}, \frac{k+1}{n}\right], \quad 0 \le j \le n-1, \ 0 \le k \le n-1.$$

Since $\Gamma(I_2) \subset D$ by Property (ii) of homotopy, $\Gamma(I_{j,k})$ is also contained in D for each pair j, k. We wish to choose n sufficiently large so that each $\Gamma(I_{j,k})$ lies in a disk that lies, in turn, inside D. We will then be in a position to apply Cauchy's Theorem for a disk.

Being the image of the compact set I_2 under a continuous function, $\Gamma(I_2)$ is a compact subset of D. Hence, the distance r between $\Gamma(I_2)$ and the closed set $D^c = \mathbb{C} \setminus D$ is positive. Moreover, since I_2 is compact and Γ is continuous on I_2, the function Γ is, in fact, uniformly continuous on I_2. Hence, it is possible to find a positive integer n such that

$$|\Gamma(t, s) - \Gamma(t', s')| < r \quad \text{whenever } |(t, s) - (t', s')| < 2/n.$$

With this choice of n, $\Gamma(I_{j,k}) \subseteq D(z_{j,k}, r) = D_{j,k}$ where $z_{j,k} = \Gamma(\frac{j}{n}, \frac{k}{n})$. In fact,

$$|(t, s) - (\tfrac{j}{n}, \tfrac{k}{n})| < \sqrt{\left(\tfrac{1}{n}\right)^2 + \left(\tfrac{1}{n}\right)^2} = \tfrac{\sqrt{2}}{n} < \tfrac{2}{n} \quad \text{if } (t, s) \in I_{j,k}.$$

Since $z_{j,k} \in \Gamma(I_2)$ and the distance from any point in $\Gamma(I_2)$ to the complement of D is at least r, we see that $D_{j,k} \subseteq D$. In particular, f is analytic on the disk $D_{j,k}$. For $0 \le j \le n-1$ and $0 \le k \le n-1$, let $P_{j,k}$ be the closed polygonal path

$$P_{j,k} = [z_{j,k}, z_{j+1,k}] \cup [z_{j+1,k}, z_{j+1,k+1}] \cup [z_{j+1,k+1}, z_{j,k+1}] \cup [z_{j,k+1}, z_{j,k}].$$

See Fig. 6.3. Since the four vertices of the path all lie in the disk $D_{j,k}$, the path $P_{j,k}$ itself lies in $D_{j,k}$. It follows from Cauchy's Theorem for a disk that $\int_{P_{j,k}} f(z)\, dz = 0$. Hence, for each k with $0 \le k \le n-1$,

$$0 = \sum_{j=0}^{n-1} \int_{P_{j,k}} f(z)\, dz. \tag{6.4}$$

For $0 \le k \le n$, let Q_k be the closed polygonal path

$$Q_k = [z_{0,k}, z_{1,k}] \cup [z_{1,k}, z_{2,k}] \cup \ldots \cup [z_{n-1,k}, z_{n,k}]$$

6.1 The Homotopic Form of Cauchy's Theorem

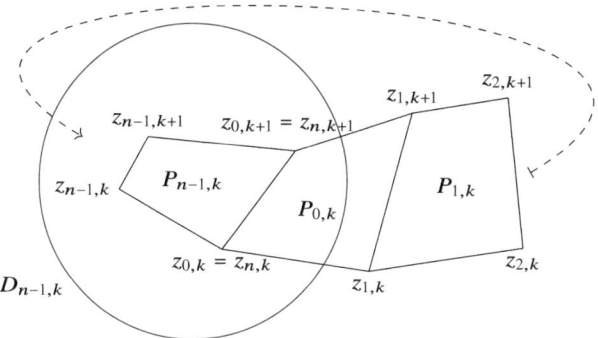

Fig. 6.3 The closed polygonal paths $P_{j,k}$ in the proof of the homotopic form of Cauchy's Theorem

(noting that $z_{0,k} = z_{n,k}$ by Property (iv) of homotopy). In (6.4), the integrals along the sides of the rectangles $P_{j,k}$ that are described twice, but in opposite directions, cancel as shown in Fig. 6.3. It follows that, for $0 \leq k \leq n-1$,

$$\sum_{j=0}^{n-1} \int_{P_{j,k}} f(z)\,dz = \int_{Q_k} f(z)\,dz - \int_{Q_{k+1}} f(z)\,dz,$$

whence, by (6.4),

$$\int_{Q_k} f(z)\,dz = \int_{Q_{k+1}} f(z)\,dz.$$

Hence, the integrals of f over Q_0 and Q_n are equal. We complete the proof by showing that

$$\int_{Q_0} f(z)\,dz = \int_{\gamma_0} f(z)\,dz \quad \text{and} \quad \int_{Q_n} f(z)\,dz = \int_{\gamma_1} f(z)\,dz. \tag{6.5}$$

We argue the case for Q_0 and γ_0, see Fig. 6.4, the case for Q_1 and γ_1 being entirely analogous. For $0 \leq j \leq n-1$, let C_j be the closed path

$$C_j = \left(\gamma_0(t) : t \in \left[\tfrac{j}{n}, \tfrac{j+1}{n}\right]\right) \cup [z_{j+1,0}, z_{j,0}],$$

as shown in Fig. 6.5. Since C_j is a piecewise-smooth closed curve and lies in the disk $D_{j,0}$, Cauchy's Theorem for a disk again leads to

$$\int_{C_j} f(z)\,dz = 0.$$

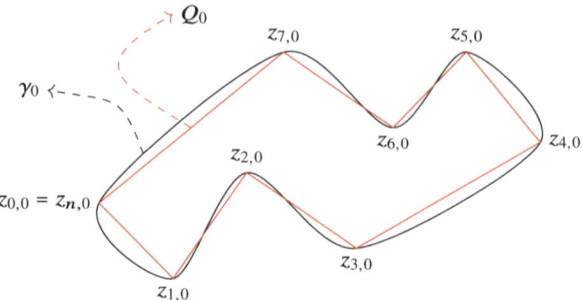

Fig. 6.4 The closed curve γ_0 and the closed polygonal path Q_0

Fig. 6.5 The curve C_j in the proof of the homotopic form of Cauchy's Theorem

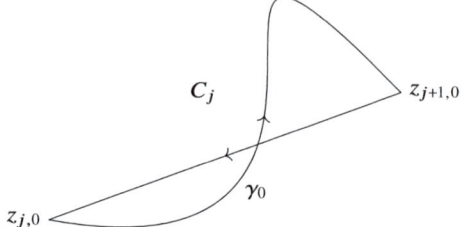

Thus,

$$0 = \sum_{j=0}^{n-1} \int_{C_j} f(z)\,dz = \int_{\gamma_0} f(z)\,dz - \int_{Q_0} f(z)\,dz.$$

The equality of the integrals of f along γ_1 and Q_1 is established similarly. Thus (6.5) is established and the result is proved. □

6.2 Simply Connected Domains

A simply connected domain is one in which every closed curve in the domain is homotopic to zero, that is every closed curve in the domain can be continuously shrunk to a point without leaving the domain. A corollary of the homotopic form of Cauchy's Theorem, of sufficient importance to be called a theorem in its own right, is the following.

Theorem 6.4 *Suppose that γ is a piecewise-smooth closed curve in a simply connected domain D and that f is analytic in D. Then*

$$\int_\gamma f(z)\,dz = 0.$$

6.2 Simply Connected Domains

Fig. 6.6 A simply connected domain

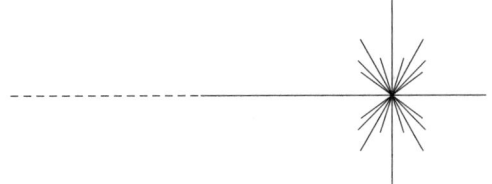

In fact, since the curve γ in the statement of this theorem is homotopic to a constant curve, and the integral of f over a constant curve is most certainly equal to 0, the result is a direct consequence of Theorem 6.3.

By Exercise 6.2, every star-shaped domain is simply connected. On the other hand, simply connected domains can be quite complicated. They can be characterised in several equivalent ways, some of which are listed in Theorem 6.8 to follow. One further characterisation, to be obtained as Theorem 7.4 in the next chapter, is that a planar domain is simply connected if and only if its complement relative to $\mathbb{C}_\infty$ is connected. For example, the domain D shown in Fig. 6.6 is simply connected.

Theorem 6.4 effectively says that a simply connected domain has what we will (unimaginatively) call **Property I** ('I' for 'integral'). A domain D will be said to have Property I if for every piecewise-smooth closed curve γ in D, and for every function f analytic in D, we have

$$\int_\gamma f(z)\,dz = 0.$$

We will say that a domain D has **Property P** ('P' for 'primitive') if every function f analytic in D has a primitive (an antiderivative), that is there is a function F analytic in D for which $F' = f$ in D. We will show that these two properties are equivalent so that, by Theorem 6.4, any simply connected domain has both properties.

Theorem 6.5 *A domain D has Property I if and only if it has Property P.*

Proof Suppose that a domain D has Property I and that f is analytic in D. We need to produce a primitive for f. We first fix some point z_0 in D. Suppose that z lies in D and that γ_1 and γ_2 are two curves in D that join z_0 to z. The curve $\gamma = \gamma_1 \cup -\gamma_2$ is a closed curve in the simply connected domain D. Thus, by Property I,

$$0 = \int_\gamma f(w)\,dw = \int_{\gamma_1} f(w)\,dw - \int_{\gamma_2} f(w)\,dw.$$

Thus, the value of

$$F(z) = \int_{\gamma_z} f(w)\,dw,$$

where γ_z is a piecewise-smooth curve in D joining z_0 to z, does not depend on the particular choice of curve γ_z. Thus F is a bona fide function and is our candidate for an antiderivative of f.

We will show that $F'(z) = f(z)$ for each z in D. We first choose r positive so that the disk $D(z, r)$ is contained in D. Let γ_z be some curve joining z_0 to z. If $|h| < r$, the curve $\gamma_z \cup [z, z+h]$ lies in D and joins z_0 to $z+h$. Then,

$$F(z+h) - F(z) = \int_{\gamma_z \cup [z,z+h]} f(w)\,dw - \int_{\gamma_z} f(w)\,dw = \int_{[z,z+h]} f(w)\,dw.$$

Since $\int_{[z,z+h]} f(z)\,dw = h f(z)$, if $h \neq 0$ we have

$$\left| \frac{F(z+h) - F(z)}{h} - f(z) \right| = \left| \left(\frac{1}{h} \int_{[z,z+h]} f(w)\,dw \right) - f(z) \right|$$

$$= \left| \frac{1}{h} \int_{[z,z+h]} [f(w) - f(z)]\,dw \right|$$

$$= \frac{1}{|h|} \left| \int_{[z,z+h]} [f(w) - f(z)]\,dw \right|$$

$$\leq \frac{1}{|h|} |h| \sup\{|f(w) - f(z)| : w \in [z, z+h]\}$$

$$\leq \sup\{|f(w) - f(z)| : w \in D(z, |h|)\}$$

where the ML-inequality was used at the penultimate step. Since f is continuous at z, given ϵ positive there is a positive number δ such that $|f(w) - f(z)| \leq \epsilon$ if $w \in D(z, \delta)$. It then follows that, if $0 < |h| \leq \delta$,

$$\left| \frac{F(z+h) - F(z)}{h} - f(z) \right| \leq \epsilon,$$

whence

$$F'(z) = \lim_{h \to 0} \frac{F(z+h) - F(z)}{h}$$

exists and equals $f(z)$ as required.

Conversely, suppose that D has Property P, that γ is a piecewise-smooth closed curve in D and that f is analytic in D. By Property P, f has a primitive F, so that

$$\int_\gamma f(z)\,dz = \int_\gamma F'(z)\,dz = 0.$$

6.2 Simply Connected Domains

This establishes that D has Property I. That this last integral vanishes follows from the general fact that if $\gamma(t)$, $t \in [a,b]$, is a curve and F is analytic, then $\int_\gamma F'(z)\,dz = F(\gamma(b)) - F(\gamma(a))$. In our case, $\gamma(b) = \gamma(a)$ as the curve in closed. $\square$

A further defining property of simply connected domains is the existence of an analytic branch of $\log f$ if f is analytic and non-zero. We say that a domain D has **Property L** ('L' for 'log') if whenever f is analytic and never vanishes in D there is an analytic function g in D such that $f = \exp g$. Then, $g(z)$ is a logarithm of $f(z)$ for each z in D.

In general, it is possible to define $\log f(z)$ if f is non-zero in D at each individual point z in D. Of course, there are infinitely many choices of the value of $\log f(z)$ that differ from each other by integer multiples of $2\pi i$; the issue is whether the choice of $\log f(z)$ at each z in D can be made in such a way that the resulting function of z is analytic. The issue, in fact, is whether it is possible to make a continuous choice of $\arg f(z)$ in D. As we will now see, this can be done if the domain D has Property P. The idea behind the proof is that if f has an analytic logarithm, say g, then $g'(z) = \frac{d}{dz} \log f(z) = f'(z)/f(z)$. Thus, we can recover $\log f$ as an antiderivative of f'/f.

Theorem 6.6 *Suppose that a domain D has Property P. Then D has Property L. That is, if f is analytic and never takes the value 0 in D then there exists a function g that is analytic in D for which $f(z) = \exp(g(z))$, z in D.*

Proof Since f is analytic and non-zero in D, the function f'/f is also analytic in D. Since D has Property P, there is a function h that is analytic in D and for which

$$h'(z) = \frac{f'(z)}{f(z)}.$$

Now we consider the analytic function e^h/f. Its derivative is

$$\frac{d}{dz}\left(\frac{\exp h(z)}{f(z)}\right) = \frac{f(z)h'(z)\exp h(z) - f'(z)\exp h(z)}{f(z)^2}$$

$$= \frac{\exp h(z)}{f(z)^2}\left[f(z)h'(z) - f'(z)\right] = 0.$$

Thus, $(\exp h(z))/f(z)$ is constant in D, say equal to K where K must, of course, be non-zero. Choose $C = \log(1/K)$ (any choice of logarithm is fine) and set $g(z) = h(z) + C$. Then,

$$e^{g(z)} = e^{h(z)} \times e^C = (K f(z))\frac{1}{K} = f(z),$$

as required. $\square$

We write $\log f(z)$ for the function $g(z)$ appearing in Theorem 6.6. If an analytic function has an analytic logarithm then we can define analytic powers of the function according to the formula

$$f(z)^\alpha = \exp\left(\alpha \log f(z)\right).$$

Here $\alpha \in \mathbb{C}$. In particular, taking α to be $1/2$ we can define an analytic branch of $\sqrt{f(z)}$. We say that a domain D has **Property S** ('S' for 'square root') if whenever f is analytic and never vanishes in D there is an analytic function g in D such that $f = g^2$.

At this point we have the following implications for a domain D:

Simply connected $\implies$ Property I $\iff$ Property P
$\implies$ Property L $\implies$ Property S.

When it comes to proving the Riemann Mapping Theorem in the next section, we will make the apparently weaker assumption that the domain in question has Property S rather than assuming it is simply connected.

6.3 The Riemann Mapping Theorem

Our aim in this section is to prove the Riemann Mapping Theorem.

Theorem 6.7 (Riemann Mapping Theorem) *Suppose that D is a simply connected domain other than the complex plane $\mathbb{C}$, and that the point a lies in D. Then there exists a unique conformal mapping f of D onto the unit disk $\mathbb{D}$ for which $f(a) = 0$ and $f'(a) > 0$.*

The Riemann Mapping Theorem is illustrated in Fig. 6.7. We sometimes refer to a conformal map of a simply connected domain D onto the unit disk $\mathbb{D}$ as a 'Riemann map'. Before we continue, we make precise the earlier statement that a conformal

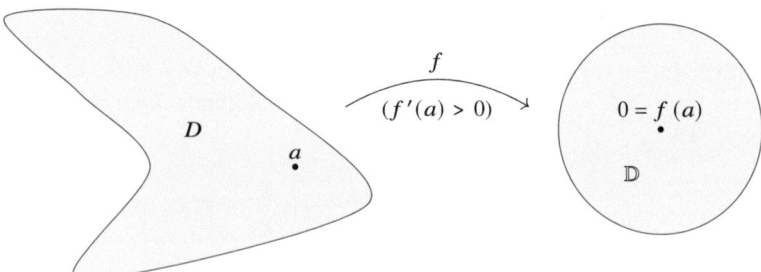

Fig. 6.7 The Riemann mapping of a simply connected domain

6.3 The Riemann Mapping Theorem

map between two domains is a homeomorphism by showing that the inverse of a conformal mapping is itself a conformal mapping.

Proposition 6.1 *Suppose that D_1 is a domain in the complex plane and that f is analytic and univalent in D_1. Then the function $g = f^{-1}$ is a conformal mapping of $D_2 = f(D_1)$ onto D_1.*

Proof First, $D_2 = f(D_1)$ is a domain since the continuous image of a connected set is connected and, by the Open Mapping Theorem [16, Thm. 10.7], analytic functions map open sets to open sets.

We need to show that g is analytic on D_2; this is sufficient since, being the inverse function of f, the function g is automatically one-to-one and onto. We begin by showing that $g = f^{-1}$ is continuous on the domain D_2. Suppose that $w_0 \in D_2$. We write $z_0 = g(w_0)$ so that $f(z_0) = w_0$. Since f is (locally) univalent, it follows from [16, Theorem 10.6] that, for each sufficiently small positive number ε, a positive number δ can be found such that if w is in the disk $D(w_0, \delta) \subset D_2$ there is precisely one z in the disk $D(z_0, \varepsilon) \subset D_1$ for which $f(z) = w$. Thus, since f is univalent in D_1 with inverse g, we have $g(w) \in D(z_0, \varepsilon)$ whenever $w \in D(w_0, \delta)$, that is $g(D(w_0, \delta)) \subseteq D(z_0, \varepsilon)$. This is exactly what is required for g to be continuous at w_0.

Finally, we show that g is not only continuous but analytic. In general, a function $h(z)$ is differentiable at a if and only if there is a continuous function $p(z)$ that is defined in $D(a, r)$ for some positive r and for which

$$h(z) = h(a) + (z - a) p(z), \quad z \in D(a, r). \tag{6.6}$$

If this holds, then $h'(a) = p(a)$.

Again, suppose that $w_0 \in D_2$ with $g(w_0) = z_0$. Since f is analytic at z_0, there is a function $p(z)$ that is continuous at z_0 and for which

$$f(z) = f(z_0) + (z - z_0) p(z), \quad z \in D(z_0, r). \tag{6.7}$$

Here r is some positive number and $D(z_0, r) \subset D_1$. Note that $p(z_0) = f'(z_0) \neq 0$. By the continuity of g there is a positive number s such that $g(D(w_0, s)) \subseteq D(z_0, r)$. We may therefore rewrite (6.7) replacing z by $g(w)$ with $w \in D(w_0, s)$. We obtain

$$w = w_0 + \big(g(w) - g(w_0)\big) p\big(g(w)\big)$$

and hence

$$g(w) = g(w_0) + (w - w_0) \frac{1}{p(g(w))}, \quad w \in D(w_0, s).$$

Since g is continuous at w_0 and p is continuous at $z_0 = g(w_0)$ with $p(z_0) \neq 0$, the function $1/(p \circ g)(w)$ is continuous at w_0. Consequently, by comparison with (6.6), g is differentiable at w_0 with $g'(w_0) = 1/f'(z_0)$. □

Before we embark on a proof of the Riemann Mapping Theorem, here is an exercise that can motivate the proof. Expressed somewhat loosely, the size of the derivative of a conformal map increases if the image domain expands. If we begin with the family of *all* conformal maps of D into the unit disk that map a to 0, and the conformal map that we want maps D onto the full unit disk, it will have the largest image of all maps in the family. Accordingly, its derivative at a should be the largest in the family.

Exercise 6.4 Let D be a simply connected domain and $a \in D$. Let $f_1 \colon D \to D_1$ and $f_2 \colon D \to D_2$ be conformal maps of D for which $D_1 \subseteq D_2$ and $f_1(a) = f_2(a)$. Prove that $|f_1'(a)| \leq |f_2'(a)|$, with equality if and only if $D_1 = D_2$.

(You may assume, for the purpose of the exercise, the Riemann Mapping Theorem and that there exists a conformal map $f \colon \mathbb{D} \to D$ with $f(0) = a$. Set

$$g(z) = (f^{-1} \circ f_2^{-1} \circ f_1 \circ f)(z), \quad z \in \mathbb{D}.$$

Why does the function g make sense? Check that g is a self-map of $\mathbb{D}$ that fixes 0. What does Schwarz's Lemma tell us in this context?)

Proof of the Riemann Mapping Theorem

Uniqueness We first prove uniqueness of the conformal mapping f, this being the easiest part. Suppose that both f and g are conformal mappings of D onto the unit disk $\mathbb{D}$ for which $f(a) = 0 = g(a)$ and both $f'(a)$ and $g'(a)$ are positive. Consider the conformal mapping $h(z) = (f \circ g^{-1})(z)$, $z \in \mathbb{D}$, where we note that g^{-1} is indeed a conformal mapping by Proposition 6.1. The map h is a conformal map of the unit disk $\mathbb{D}$ onto itself and also satisfies $h(0) = 0$ since $h(0) = f(g^{-1}(0)) = f(a) = 0$. By Theorem 4.2 (see the very first sentence in its proof), the only conformal self maps of the disk $\mathbb{D}$ that fix the origin are the rotations, and so $h(z) = \beta z$ for some β of modulus 1. Moreover, $\beta = h'(0) > 0$ since $h'(0) = f'(a)/g'(a) > 0$, and then the only possibility is $\beta = 1$, giving $h(z) = z$, the identity map. It follows that $f = g$ (given $w \in D$, we have $g(w) \in \mathbb{D}$ and so $g(w) = h(g(w)) = f(w)$).

Existence We now turn to the significant part of the proof, the existence of the conformal mapping of D onto $\mathbb{D}$. We make the apparently weaker assumption that D has Property S, rather than assuming that D is simply connected, and consider the set $\mathscr{F}$ of conformal mappings f of D *into* the unit disk $\mathbb{D}$ for which $f(a) = 0$ and $f'(a) > 0$.

6.3 The Riemann Mapping Theorem

Existence 1—$\mathscr{F}$ Is Non-empty We need to show that $\mathscr{F}$ is not empty by constructing at least one function in $\mathscr{F}$.[1] Since D is not the whole complex plane, there is a point α that does not lie in D. The function $f(z) = z - \alpha$ is analytic and non-zero in the domain D. Since D is assumed to have Property S, there is an analytic function g in D such that $g^2(z) = z - \alpha$. This function g is univalent in D, since if $g(z_1) = g(z_2)$ then $g(z_1)^2 = g(z_2)^2$, that is $z_1 - \alpha = z_2 - \alpha$, so that $z_1 = z_2$.

Next, suppose that $g(z_1) = w$. Then, $g(z_2) = -w$ has no solution $z_2 \in D$. If there were such a solution then $z_1 - \alpha = g(z_1)^2 = g(z_2)^2 = z_2 - \alpha$ forcing $z_1 = z_2$ and hence $w = -w$, that is $w = 0$. This is not possible as g is non-zero in D. That is, g does not assume both the values w and $-w$. Now g is certainly not a constant function and so, by the Open Mapping Theorem, we can find a disk $D(-\beta, r)$ in $g(D)$ with positive radius r. The disk $D(\beta, r)$ then has to lie outside $g(D)$ since if $w \in D(-\beta, r) \subseteq g(D)$ then $-w \notin g(D)$. The function

$$h(z) = \frac{r}{g(z) - \beta}$$

is analytic and univalent in D and satisfies $|h(z)| < 1$ for $z \in D$ (since $|w - \beta| > r$ if $w \in g(D)$), so that h maps D into $\mathbb{D}$. Finally, in order to satisfy the requirement on functions in $\mathscr{F}$ that they map a to 0 and have positive derivative at a, we post-compose h with an automorphism M of the disk $\mathbb{D}$. The map

$$M(z) = e^{i\theta} \frac{h(a) - z}{1 - \overline{h(a)}z},$$

where $\theta = \pi - \arg h'(a)$, serves this purpose. Using Exercise 4.4, it is straightforward to check that $f = M \circ h$ satisfies $f(a) = 0$ and $f'(a) > 0$ so that f belongs to $\mathscr{F}$.

Existence 2—The Normal Families Argument Since $\mathscr{F}$ is a non-empty family of bounded analytic functions on the domain D, Montel's Theorem implies that $\mathscr{F}$ is a normal family. Evaluation of the derivative at a is a continuous function on $\mathscr{H}(D)$ (see Remark 5.1), hence the real-valued function

$$\chi_a : f \longrightarrow |f'(a)|$$

---

[1] If the complement of D contains a disk $D(z_0, r)$ then the mapping $g(z) = r/(z - z_0)$ is univalent on D and maps D into $\mathbb{D}$ (since $|z - z_0| > r$ for $z \in D$, we have $|g(z)| < 1$ in D). The point $g(a) = r/(a - z_0)$ lies in $\mathbb{D}$, so if we now use an automorphism M of the disk to send $g(a)$ to 0 then, perhaps with a final rotation, $f = M \circ g \in \mathscr{F}$. The problem is that the complement of D need not contain such a disk, for instance if D is the complex plane slit along the negative real axis. But then the square root function converts the slit-plane to a half-plane whose complement certainly contains a disk. This is the idea behind the argument that follows.

is continuous on $\mathcal{H}(D)$. Since $\overline{\mathcal{F}}$ is a compact subset of $\mathcal{H}(D)$, the continuous function χ_a assumes its maximum value on $\overline{\mathcal{F}}$. That is, there is a function F in $\overline{\mathcal{F}}$ with the largest, in modulus, derivative at a so that

$$|f'(a)| \leq |F'(a)|, \text{ for each } f \in \overline{\mathcal{F}}.$$

In particular, by comparing with the derivative at a of any one function in $\mathcal{F}$, we find that $|F'(a)| > 0$. This function F will be our candidate for the conformal mapping of D onto the unit disk $\mathbb{D}$—see Exercise 6.4.

If F is to be a conformal mapping it needs to belong to $\mathcal{F}$, not just $\overline{\mathcal{F}}$. Since $F \in \overline{\mathcal{F}}$, there is a sequence of functions $\{f_n\}_1^\infty$ in $\mathcal{F}$ that converges to F in $\mathcal{H}(D)$. The continuity of point evaluations and derivative evaluations on $\mathcal{H}(D)$ shows that $F(a) = 0$ and $F'(a) \geq 0$, in fact $F'(a) > 0$ since it isn't zero. By Hurwitz's Theorem (see Theorem 5.7), F is either univalent in D or identically zero. Since, $F'(a) \neq 0$, the latter is certainly not the case and so $F \in \mathcal{F}$. (In fact, this argument shows that $\overline{\mathcal{F}} = \mathcal{F} \cup \{0\}$.)

Existence 3—The Mapping F Does the Job We now have a conformal map F of the simply connected domain D into the unit disk $\mathbb{D}$ that satisfies the conditions $F(a) = 0$ and $F'(a) > 0$, and that has the largest derivative at a among all such functions. It remains to show that F maps D onto $\mathbb{D}$. Suppose not, and that there is some point, say α, in $\mathbb{D}$ but $\alpha \notin F(D)$. This should lead to a contradiction, the most obvious contradiction being that it should enable us to produce a function f in $\mathcal{F}$ with a derivative at a *larger* than that of F.

To produce such as f, we first postcompose F with a conformal self-map M_α of the disk so that the omitted value becomes 0 rather than α. Thus, we set

$$h(z) = \frac{\alpha - F(z)}{1 - \overline{\alpha}F(z)} = (M_\alpha \circ F)(z), \quad z \in D,$$

so that h is univalent (since both F and M_α are) and *non-zero* in D, and so that h maps D into $\mathbb{D}$. Since we assume that the domain D has Property S, we deduce that there is an analytic function g in D for which $h(z) = g(z)^2$, $z \in D$. Moreover g, like h, must be univalent in D and map D into $\mathbb{D}$.

We don't, as yet, have control over $g(a)$ and $g'(a)$, so we set

$$f(z) = e^{i\theta} \frac{g(a) - g(z)}{1 - \overline{g(a)}g(z)} = e^{i\theta}(M_{g(a)} \circ g)(z), \quad z \in D,$$

which postcomposes g with an automorphism $M_{g(a)}$ of the disk $\mathbb{D}$, and choose $\theta = \pi - \arg(g'(a))$. This function f is again univalent in D, maps D into $\mathbb{D}$, and satisfies $f(a) = 0$ and, using Exercise 4.4 again, $f'(a) > 0$. In short, $f \in \mathcal{F}$.

6.3 The Riemann Mapping Theorem

Finally, we compute $f'(a)$ and compare it to $F'(a)$. We have

$$f'(a) = \frac{|g'(a)|}{1-|g(a)|^2} = \frac{|g'(a)|}{1-|h(a)|} = \frac{|g'(a)|}{1-|\alpha|},$$

since $F(a) = 0$. Next, since $g(z)^2 = h(z)$ we obtain $2g(z)g'(z) = h'(z)$. Setting $z = a$ and using Exercise 4.4 again leads to

$$|g'(a)| = \frac{|h'(a)|}{2|g(a)|} = \frac{(1-|\alpha|^2)F'(a)}{2|g(a)|} = \frac{1-|\alpha|^2}{2\sqrt{|\alpha|}} F'(a).$$

At the last step, we used $g(a)^2 = h(a) = \alpha$. Putting these computations together gives

$$f'(a) = \frac{|g'(a)|}{1-|\alpha|} = \frac{1+|\alpha|}{2\sqrt{|\alpha|}} F'(a).$$

The function $(1+x)/(2\sqrt{x})$ is strictly decreasing on $[0, 1]$ and takes the value 1 at 1. Thus, $f'(a) > F'(a)$, which is the desired contradiction and completes the proof of the Riemann Mapping Theorem. □

At this point, we can write down several conditions on a planar domain D each of which is equivalent to the domain being simply connected. There is one further equivalent condition, that $\mathbb{C}_\infty \setminus D$ is connected, that must wait until the next chapter and Runge's Theorem. For now we have the following result.

Theorem 6.8 *The following statements are equivalent for a domain D in the plane:*

(a) *D is simply connected;*
(b) *for every analytic function f in D and every piecewise smooth closed curve γ in D,*

$$\int_\gamma f(z)\, dz = 0;$$

(c) *to every analytic function f in D there corresponds an analytic function F in D such that $F' = f$;*
(d) *to every non-vanishing analytic function f in D there corresponds an analytic function g in D such that $f(z) = \exp(g(z))$;*
(e) *to every non-vanishing analytic function f in D there corresponds an analytic function g in D such that $f(z) = g(z)^2$;*
(f) *D is homeomorphic to the unit disk $\mathbb{D}$.*

Proof That (a) implies (b) is Theorem 6.4. That (b) implies (c) is Theorem 6.5 (in fact, (b) and (c) are already known to be equivalent). That (c) implies (d) is

Theorem 6.6, and that (d) implies (e) is mentioned in the remarks following the proof of Theorem 6.6.

Next, we show that (e) implies (f). Suppose that every non-zero analytic function in D has an analytic square root. If D is the whole complex plane $\mathbb{C}$ then D is certainly homeomorphic to the disk (an explicit homeomorphism is $h(z) = z/(1+|z|)$, $z \in \mathbb{C}$). If D is not the whole complex plane then the Riemann Mapping Theorem (whose *proof* only used that D had Property S rather than the assumption in the statement that D was simply connected) guarantees the existence of a conformal mapping of D onto the unit disk $\mathbb{D}$, and this conformal map provides a homeomorphism from D onto $\mathbb{D}$.

Finally, (f) implies (a) by Exercise 6.3 since D is then the image of the simply connected domain $\mathbb{D}$ under a homeomorphism. □

Exercise 6.5 Suppose that D_1 and D_2 are simply connected domains with non-empty and *connected* intersection. Show that $D = D_1 \cup D_2$ is a simply connected domain. (Hint: Show that D satisfies Property (c) of Theorem 6.8.)

Give an example to show that the result may fail if $D_1 \cap D_2$ is not connected.

We say that the domain D_1 is *conformally equivalent* to the domain D_2 if there is a conformal map from D_1 onto D_2. Every domain is clearly conformally equivalent to itself via the identity map. By Proposition 6.1, if D_1 is conformally equivalent to D_2 via a conformal mapping f then D_2 is conformally equivalent to D_1 via the conformal mapping f^{-1}. Since the composition of conformal mappings is again conformal, we conclude that conformal equivalence is an equivalence relation on the collection of all planar domains, thereby dividing planar domains into conformal equivalence classes.

By the Riemann Mapping Theorem, all proper simply connected domains belong to the same conformal equivalence class of domains (recall that a proper simply connected domain is any simply connected domain other than the complex plane $\mathbb{C}$). The complex plane has an equivalence class all to itself. If D is conformally equivalent to the complex plane, then D must be simply connected. But then D cannot be proper as if it were it would be conformally equivalent to the unit disk $\mathbb{D}$ by the Riemann Mapping Theorem, and hence so would the complex plane. Since, by Liouville's Theorem, the complex plane is not conformally equivalent to the disk $\mathbb{D}$, the only possibility is that $D = \mathbb{C}$. Another way to see this is via a characterisation of the possible conformal maps, a complete description of Aut($\mathbb{C}$).

Proposition 6.2 *All entire, univalent functions are linear of the form* $f(z) = az+b$, $a \neq 0$.

Proof Suppose that f is an entire, univalent function. First we show that if f is not a polynomial (that is, if f is a transcendental entire function) then f is not univalent in $\mathbb{C}$. In this case, $g(z) = f(1/z)$ has an essential singularity at 0 and so, by the Casorati-Weierstrass Theorem, the image of $\mathbb{D} \setminus \{0\}$ under g will be dense in $\mathbb{C}$. That is, the image of $\mathbb{C} \setminus \overline{\mathbb{D}}$ under f will be dense in $\mathbb{C}$ and, in particular, be dense in any disk about $f(0)$. However, by the Open Mapping Theorem, $f(\mathbb{D})$ contains some open disk about $f(0)$ and then some points in this disk will be taken on more

6.3 The Riemann Mapping Theorem

than once by f, (at least) once in $\mathbb{D}$ and again in $\mathbb{C} \setminus \overline{\mathbb{D}}$. If f is an entire, univalent function then it must be a polynomial. By the Fundamental Theorem of Algebra [16, Thm. 7.11], if f is not linear then it will definitely not be univalent. Thus, the only entire, univalent functions are the non-constant linear functions $f(z) = az + b$, $a \neq 0$, in which case $f(\mathbb{C}) = \mathbb{C}$. □

Before we continue, we introduce a slightly different version of homotopy of curves that will be needed in Chap. 10. This version is known as *fixed endpoint homotopy* and applies to curves that are not necessarily closed.

Definition 6.2 (Fixed Endpoint Homotopy) Suppose that γ_0 and γ_1 are piecewise-smooth curves lying in a domain D. Assume that both curves have initial point a and both have final point b. Assume that the curves are parameterised on $[0, 1]$ by $\gamma_0(t)$ and $\gamma_1(t)$, respectively. The curve γ_0 is said to be *fixed endpoint homotopic* to the curve γ_1 relative to the domain D if there is a function $\Gamma(t, s)$ on the closed square $I_2 = [0, 1] \times [0, 1]$ with the following properties:

(i) $\Gamma(t, s)$ is continuous on I_2;
(ii) $\Gamma(t, s)$ lies in D for every (t, s) in I_2;
(iii) $\Gamma(t, 0) = \gamma_0(t)$ and $\Gamma(t, 1) = \gamma_1(t)$ for $0 \leq t \leq 1$;
(iv) $\Gamma(0, s) = a$ and $\Gamma(1, s) = b$ for $0 \leq s \leq 1$.

The only difference between Conditions (i) to (iv) in Definitions 6.1 and 6.2 is Condition (iv). In the definition of fixed endpoint homotopy, we continuously deform the curve γ_0 into the curve γ_1 through a family of curves γ_s, $0 \leq s \leq 1$, where $\gamma_s(t) = \Gamma(t, s)$, $0 \leq t \leq 1$, insisting that each curve γ_s has the prescribed endpoints a and b.

As in the case of homotopy of closed curves, fixed endpoint homotopy is an equivalence relation on the collection of all curves in the domain D with fixed endpoints a and b. For ease of notation, we abbreviate 'fixed endpoint homotopic' to 'FEP-homotopic'.

The following lemma gives an equivalent definition of a simply connected domain in terms of FEP-homotopy rather than homotopy of closed curves. One direction is straightforward. The other is proved using the Riemann Mapping Theorem, which is certainly excessive.

Lemma 6.1 *A domain D is simply connected if and only if it has the property that any two curves in D with the same initial point and the same final point are FEP-homotopic relative to D.*

Proof First suppose that the domain D has the property that whenever γ_0 and γ_1 are curves in D with the same initial point and the same final point then γ_0 and γ_1 are FEP-homotopic. Suppose that γ is a closed curve in D, starting and ending at a in D, say. The constant curve $\gamma_1 \equiv a$ has the same endpoints as γ, hence γ and γ_1 are FEP-homotopic. Let $\Gamma(t, s)$ be the corresponding homotopy in the sense of Definition 6.2 of FEP-homotopy. This homotopy also satisfies the conditions of Definition 6.1 of homotopy of closed curves: only Condition (iv) needs checking and this holds since $\Gamma(0, s) = \gamma(0) = a$ and $\Gamma(1, s) = \gamma(1) = a$, $0 \le s \le 1$, so that all intermediate curves in the homotopy are closed. In short, every closed curve in D is homotopic to zero and so D is simply connected.

For the converse, suppose that D is simply connected and that the curves γ_0 and γ_1, both parameterised on $[0, 1]$ and lying in D, share the same initial point a and the same final point b. If $D = \mathbb{C}$ then $\Gamma(t, s) = (1 - s)\gamma_0(t) + s\gamma_1(t), t \in I_2$, is a suitable FEP-homotopy. If D is proper, choose a conformal mapping f of D onto the unit disk $\mathbb{D}$—such a mapping is guaranteed by the Riemann Mapping Theorem. Then,

$$\Gamma(t, s) = f^{-1}\big[(1 - s)f(\gamma_0(t)) + sf(\gamma_1(t))\big], \quad t \in I_2,$$

is a FEP-homotopy from γ_0 to γ_1. Since $\mathbb{D}$ is convex and $(1-s)f(\gamma_0(t))+sf(\gamma_1(t))$ is a convex combination of points in the unit disk, f^{-1} is defined at $(1-s)f(\gamma_0(t))+sf(\gamma_1(t))$ and $\Gamma(t, s)$ takes values in D. It is then clear that Γ is continuous on I_2. Conditions (iii) and (iv) of Definition 6.2 are immediate. For example, $\Gamma(1, s) = f^{-1}[(1-s)f(\gamma_0(1))+sf(\gamma_1(1))] = f^{-1}[(1-s)f(b)+sf(b)] = f^{-1}[f(b)] = b$. That is, γ_0 and γ_1 are FEP-homotopic relative to D. □

6.4 The Hyperbolic Metric in a Simply Connected Domain

We have already asked whether regions other than the disk have an associated 'hyperbolic metric' for which the automorphisms of the region are isometries and, if so, if there is a corresponding Schwarz-Pick Lemma? We answered this in Sect. 4.7 in the case of the half-plane $\mathbb{H}$, making use of an explicit conformal map from the half-plane $\mathbb{H}$ to the disk $\mathbb{D}$. Now that we are assured of the existence of conformal maps from a general, proper simply connected domain to the disk, we can rework Sect. 4.7 in this much greater generality, albeit that the Riemann mappings we work with will not be explicit except in special cases.

Before we begin, we need to describe the automorphisms of a general, proper simply connected domain D—bearing in mind that these should be the isometries of any hyperbolic metric we construct on D. The description is in terms of the Riemann map from the domain to the disk and the automorphisms of the disk as already characterised in Theorem 4.2.

Theorem 6.9 *Let D be a proper simply connected domain. Then $f : D \to D$ is an automorphism of D, that is a conformal self-map of D, if and only if f has the form*

$$f = g^{-1} \circ h \circ g, \tag{6.8}$$

where g is a conformal map of D onto the unit disk $\mathbb{D}$ and h is an automorphism of the unit disk $\mathbb{D}$.

6.4 The Hyperbolic Metric in a Simply Connected Domain

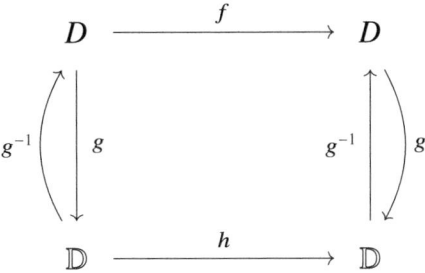

Fig. 6.8 Graphical representation of the mappings in (6.8)

Proof If f is given by (6.8) where g is a conformal map from D onto $\mathbb{D}$ and h is an automorphism of $\mathbb{D}$ then f is a composition of conformal maps and so is conformal, and f maps D onto D. This is illustrated in Fig. 6.8. Conversely, suppose that f is an automorphism of D, that is f is a conformal map of D onto D. Choose any conformal map g of D onto $\mathbb{D}$ (whose existence is guaranteed by the Riemann Mapping Theorem). Then $h = g \circ f \circ g^{-1}$, being a composition of conformal maps, is itself a conformal map, and maps the unit disk $\mathbb{D}$ onto itself. That is, h is an automorphism of $\mathbb{D}$. Since $g^{-1} \circ h \circ g$ simplifies to f, we see that f can be written in the form (6.8). □

This description of the group of automorphisms $\text{Aut}(D)$ of a proper simply connected domain D can written symbolically as

$$\text{Aut}(D) = g^{-1} \circ \text{Aut}(\mathbb{D}) \circ g,$$

where g is any Riemann map from D onto the disk $\mathbb{D}$.

6.4.1 Construction of the Hyperbolic Metric

Taking our cue from the work in Sect. 4.7 where we constructed the hyperbolic metric on the half-plane $\mathbb{H}$, we implement the same construction but now for a general proper simply connected domain D. We define the density of the hyperbolic metric (often called simply the 'hyperbolic metric') in D, the length of a curve with respect to this density, and finally the Riemannian metric on D associated with this density.

Definition 6.3 Let D be a proper simply connected domain. Let f be a conformal mapping of D onto the unit disk $\mathbb{D}$. For $z \in D$, we define the (density of) the hyperbolic metric by

$$\rho_D(z) := \rho_{\mathbb{D}}\big(f(z)\big) |f'(z)| = \frac{2|f'(z)|}{1 - |f(z)|^2}. \tag{6.9}$$

For a curve γ in D, parameterised by $\gamma(t)$, $t \in [a, b]$, its ρ_D-length is defined to be

$$L_{\rho_D}(\gamma) = \int_a^b \rho_D(\gamma(t)) |\gamma'(t)| \, dt. \tag{6.10}$$

We then define, for α and β in D,

$$d(\alpha, \beta; D) = \inf \{L_{\rho_D}(\gamma) : \gamma \text{ is a curve in } D \text{ joining } \alpha \text{ to } \beta\}. \tag{6.11}$$

Note that (6.9) generalises (4.29), that (6.10) generalises (4.27), and that (6.11) generalises (4.32) from the half-plane $\mathbb{H}$ to an arbitrary proper simply connected domain.

Before we go any further, we must check that (6.9) and (6.10) are legitimate definitions. We have already checked in a few different contexts that the definition (6.10) of the ρ-length $L_\rho(\gamma)$ of a curve γ doesn't depend on the particular parameterisation. The same argument suffices here.

Now we check that the definition (6.9) of the hyperbolic metric at a point z doesn't depend on the particular choice of mapping f. Let f_1 be a second conformal map of D onto $\mathbb{D}$. Then, $\phi = f \circ f_1^{-1}$ is an automorphism of $\mathbb{D}$ to which Exercise 4.11 applies. Thus, for any $w \in \mathbb{D}$,

$$\frac{1}{1 - |w|^2} = \frac{|\phi'(w)|}{1 - |\phi(w)|^2} = \frac{|f'(f_1^{-1}(w))| \, |(f_1^{-1})'(w)|}{1 - |f(f_1^{-1}(w))|^2}$$

Given $z \in D$, set $w = f_1(z)$ so that $(f_1^{-1})'(w) = 1/f_1'(f_1^{-1}(w)) = 1/f_1'(z)$. We obtain that

$$\frac{1}{1 - |f_1(z)|^2} = \frac{|f'(z)| \, |1/f_1'(z)|}{1 - |f(z)|^2} = \frac{|f'(z)|}{1 - |f(z)|^2} \frac{1}{|f_1'(z)|},$$

and so

$$\frac{|f'(z)|}{1 - |f(z)|^2} = \frac{|f_1'(z)|}{1 - |f_1(z)|^2}.$$

This shows that the definition (6.9) of the hyperbolic metric doesn't depend on the choice of conformal map f.

Exercise 6.6 Use Theorem 4.4—the Schwarz-Pick Lemma, First Version—to show that, for D proper and simply connected and for $z \in D$,

$$\rho_D(z) = \sup \left\{ \frac{2|f'(z)|}{1 - |f(z)|^2} : f \text{ is analytic in } D \text{ and } f(D) \subseteq \mathbb{D} \right\}$$

$$= \sup \{2|f'(z)| : f \text{ is analytic in } D, \ f(z) = 0, \text{ and } f(D) \subseteq \mathbb{D}\}.$$

6.4 The Hyperbolic Metric in a Simply Connected Domain

That is, if f is analytic in D taking values in $\mathbb{D}$ then

$$\rho_\mathbb{D}(f(z))|f'(z)| \leq \rho_D(z), \quad z \in D,$$

with equality at any point, and then at every point, if and only if f is a conformal map of D onto $\mathbb{D}$.

Much as we have argued before, (6.11) defines a valid metric on the simply connected domain D: certainly $d(\alpha, \beta; D)$ is non-negative; that $d(\alpha, \beta; D)$ equals $d(\beta, \alpha; D)$ follows from $L_{\rho_D}(\gamma) = L_{\rho_D}(-\gamma)$; the triangle inequality $d(\alpha, \beta; D) \leq d(\alpha, z; D) + d(z, \beta; D)$, α, β, z in D, derives from the fact that the family of curves joining α to β in D and passing through z is a subset of the full collection of curves joining α to β in D. The last thing to check is that $d(\alpha, \beta; D) > 0$ if $\alpha \neq \beta$. First and foremost, ρ_D is a positive continuous function on D since, for any given conformal map f of D onto $\mathbb{D}$, the function $|f'(z)|/(1 - |f(z)|^2)$ is positive and continuous on D. If $\beta \neq \alpha$ in D, choose $r > 0$ so that $D(\alpha, 2r) \subset D$ and $\beta \notin D(\alpha, 2r)$. By continuity, ρ_D is bounded below by some positive constant c on $D(\alpha, r)$, and so any curve joining α to β in D has to have a subcurve lying in $D(\alpha, r)$ of Euclidean length at least r and so of ρ_D-length at least cr. This then shows that $d(\alpha, \beta; D) \geq cr > 0$.

Conformal Radius As in (6.9), let D be a proper simply connected domain, let z be a point in D, and let f be a conformal map of D onto the unit disk $\mathbb{D}$. As in the statement of the Riemann Mapping Theorem, let's choose f so that $f(z) = 0$ (we can even have $f'(z)$ positive as well if we wish). Since $\rho_\mathbb{D}(0) = 2$, (6.9) simplifies to

$$\rho_D(z) = 2|f'(z)|.$$

The *conformal radius* $r(z; D)$ of D at z is defined to be

$$r(z; D) := 1/|f'(z)|$$

for this conformal map f. Thus, the conformal radius is no more than twice the reciprocal of the hyperbolic metric. The conformal radius has a nice geometric interpretation as illustrated in Fig. 6.9. For $r > 0$, the map $w \to rw$ sends the unit disk $\mathbb{D}$ to the disk $D(0, r)$ of radius r. The composition $g = rf$ maps D onto $r\mathbb{D}$, maps z to 0, and

$$|g'(z)| = r|f'(z)| = \frac{r}{r(z; D)}.$$

If we choose $r = r(z; D)$ then $|g'(z)| = 1$. The conformal radius is therefore the radius r of the disk $D(0, r)$ such that the conformal map g of D onto $D(0, r)$ with $g(z) = 0$ satisfies $|g'(z)| = 1$.

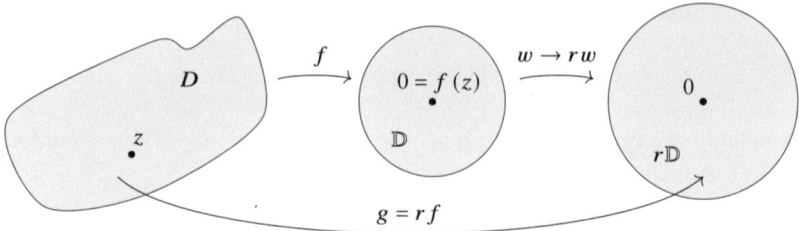

Fig. 6.9 Graphical representation of the conformal radius $r(z; D)$

Exercise 6.7 Let D be a proper simply connected domain and z_0 a point in D. Let r be positive. The domain D scaled by r is $rD = \{rz : z \in D\}$. Show that

$$\rho_{rD}(rz_0) = \frac{1}{r}\rho_D(z_0).$$

Show that, if D is the disk $D(0, r)$, then $\rho_D(0) = 2/r$.

6.4.2 The Schwarz-Pick Lemma: Third Version

A key result now follows: it is the third (and final for us) version of the Schwarz-Pick Lemma! We see that analytic maps between simply connected domains are contractions in the hyperbolic metric and that the isometries are the conformal maps.

We begin with the preliminary statement of the result for analytic functions defined on a proper simply connected domain and taking values in the unit disk—the full version (Theorem 6.10) is for analytic maps between any two proper simply connected domains. In this preliminary statement we see that, just as the identity (4.25) holds for hyperbolic distance in the half-plane $\mathbb{H}$, hyperbolic distance is preserved under conformal maps of D onto $\mathbb{D}$.

Proposition 6.3 *Let D be a proper simply connected domain. Any function f analytic in D and taking values in the unit disk is a contraction in the hyperbolic metric in that, for any points α and β in D,*

$$d\bigl(f(\alpha), f(\beta); \mathbb{D}\bigr) \leq d(\alpha, \beta; D). \tag{6.12}$$

If equality holds for some pair of distinct points α and β then f is a conformal mapping from D onto the disk $\mathbb{D}$, in which case equality then holds for every choice of α and β.

Moreover, there is a unique curve of shortest hyperbolic length, a geodesic arc, joining any two points α and β in D. This geodesic arc is the preimage, under any conformal map f of D onto $\mathbb{D}$, of the geodesic arc in $\mathbb{D}$ joining $f(\alpha)$ and $f(\beta)$.

6.4 The Hyperbolic Metric in a Simply Connected Domain

Proof We first work with a Riemann map g of D onto $\mathbb{D}$. Let γ be any curve in D and $g(\gamma)$ be its image in the disk $\mathbb{D}$ under g. We will show that

$$L_{\rho_D}(\gamma) = L_{\rho_\mathbb{D}}(g \circ \gamma).$$

As usual, we parameterise γ by $\gamma(t)$, $t \in [a, b]$. From definition (6.9) of the hyperbolic metric ρ_D and definition (6.10) of the hyperbolic length of a curve,

$$\begin{aligned} L_{\rho_\mathbb{D}}(g \circ \gamma) &= \int_a^b \rho_\mathbb{D}((g \circ \gamma)(t)) \, |(g \circ \gamma)'(t)| \, dt \\ &= \int_a^b \rho_\mathbb{D}(g(\gamma(t))) \, |g'(\gamma(t))| \, |\gamma'(t)| \, dt \\ &= \int_a^b \rho_D(\gamma(t)) \, |\gamma'(t)| \, dt = L_{\rho_D}(\gamma). \end{aligned}$$

The fact that hyperbolic lengths of curves are preserved allows us to conclude that hyperbolic distances are preserved. Let α and β be distinct points in D. To each curve γ joining α to β in D we can find a curve $\tilde{\gamma}$ joining $g(\alpha)$ to $g(\beta)$ in $\mathbb{D}$ with the same hyperbolic length—we can take $\tilde{\gamma}$ to be $g(\gamma)$. Conversely, given a curve $\tilde{\gamma}$ joining $g(\alpha)$ to $g(\beta)$ in $\mathbb{D}$, we can find a curve γ joining α to β in D with the same hyperbolic length—we can take $\gamma = g^{-1}(\tilde{\gamma})$. The sets of real numbers

$$\left\{ L_{\rho_D}(\gamma) : \gamma \text{ is a curve in } D \text{ joining } \alpha \text{ to } \beta \right\}$$

and

$$\left\{ L_{\rho_\mathbb{D}}(\tilde{\gamma}) : \tilde{\gamma} \text{ is a curve in } \mathbb{D} \text{ joining } g(\alpha) \text{ to } g(\beta) \right\}$$

are therefore the same, hence they have the same infimum and so

$$d(\alpha, \beta; D) = d(g(\alpha), g(\beta); \mathbb{D}). \tag{6.13}$$

As regards the existence of a curve of shortest hyperbolic length joining α and β in D, we know from Theorem 4.5 that there is a unique curve $\tilde{\gamma}$ of shortest hyperbolic length joining $g(\alpha)$ and $g(\beta)$ in $\mathbb{D}$. Then, again using the invariance of hyperbolic lengths of curves under g, we see that $\gamma = g^{-1}(\tilde{\gamma})$ is the unique curve of shortest hyperbolic length joining α and β in D.

Finally, and more generally, let f be an analytic map of D into $\mathbb{D}$ (not necessarily one-to-one or onto). Let g be a Riemann map of D onto $\mathbb{D}$ and set $h = f \circ g^{-1}$, which is analytic in $\mathbb{D}$ and takes values in $\mathbb{D}$. Therefore, by the Schwarz-Pick Lemma, Second Version (Theorem 4.7), for any distinct points z and w in $\mathbb{D}$,

$$d(h(z), h(w); \mathbb{D}) \leq d(z, w; \mathbb{D}), \tag{6.14}$$

with equality if and only if h is an automorphism of $\mathbb{D}$, which in turn is the case if and only if f is a conformal map of D onto $\mathbb{D}$. For distinct points α and β in D, set $z = g(\alpha)$ and $w = g(\beta)$ so that z and w are distinct points in $\mathbb{D}$ and apply (6.14). We obtain, since $h \circ g = f$,

$$d(f(\alpha), f(\beta); \mathbb{D}) = d(h(g(\alpha)), h(g(\beta)); \mathbb{D}) \leq d(g(\alpha), g(\beta); \mathbb{D}).$$

Taking account of (6.13), this proves (6.12) together with the case of equality. □

We can use this special case of analytic maps from a simply connected domain into the unit disk as a stepping stone towards proving the general result for maps between any two proper simply connected domains.

Theorem 6.10 (Schwarz-Pick Lemma—Third Version) *Let D_1 and D_2 be proper simply connected domains. Let f be an analytic map from D_1 into D_2. Then,*

$$\rho_{D_2}(f(z))\,|f'(z)| \leq \rho_{D_1}(z), \quad z \in D_1, \tag{6.15}$$

with equality for one, and hence all, z in D_1 if and only if f is a conformal map of D_1 onto D_2.

For α and β in D_1,

$$d(f(\alpha), f(\beta); D_2) \leq d(\alpha, \beta; D_1), \tag{6.16}$$

with equality for some pair α and β of distinct points, and hence for all pairs α and β, if and only if f is a conformal map of D_1 onto D_2.

Moreover, if f is a conformal map of D_1 onto D_2 then γ in D_1 is a geodesic arc between its endpoints if and only if $f(\gamma)$ in D_2 is a geodesic arc between its endpoints.

Proof The method of proof is to bring everything back to the unit disk as much as is possible. Let $f : D_1 \to D_2$ be analytic and let z be in D_1. Choose any conformal mapping g_2 of D_2 onto $\mathbb{D}$. Then, $h = g_2 \circ f$ is analytic and maps D_1 into $\mathbb{D}$, as shown in Fig. 6.10. By (6.9),

$$\rho_{D_2}(f(z)) = \rho_{\mathbb{D}}(g_2(f(z)))\,|g_2'(f(z))|.$$

Fig. 6.10 The maps f, g_2 and h used in the proof of (6.15)

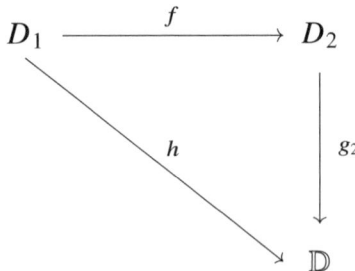

6.4 The Hyperbolic Metric in a Simply Connected Domain

By Exercise 6.6,

$$\begin{aligned}
\rho_{D_1}(z) &\geq \rho_{\mathbb{D}}(h(z))\,|h'(z)| \\
&= \rho_{\mathbb{D}}(g_2(f(z)))\,|(g_2 \circ f)'(z)| \\
&= \frac{\rho_{D_2}(f(z))}{|g_2'(f(z))|}\,|g_2'(f(z))|\,|f'(z)| \\
&= \rho_{D_2}(f(z))\,|f'(z)|,
\end{aligned}$$

with equality (by Exercise 6.6) for one z in D_1, and then for all z in D_1, if and only if h is a Riemann map of D_1 onto $\mathbb{D}$, which is in turn the case if and only if f is a conformal map of D_1 onto D_2. This establishes (6.15).

To establish (6.16), together with the equality statement, let α and β be distinct points in D_1. We apply (6.12) to the analytic map $h = g_2 \circ f$ of D_1 into $\mathbb{D}$ to deduce that

$$d(\alpha, \beta; D_1) \geq d\big(h(\alpha), h(\beta); \mathbb{D}\big). \tag{6.17}$$

We apply (6.12) again, but this time with the conformal mapping g_2 of D_2 onto $\mathbb{D}$, so that equality holds, and obtain

$$d\big(h(\alpha), h(\beta); \mathbb{D}\big) = d\big(g_2(f(\alpha)), g_2(f(\beta)); \mathbb{D}\big) = d\big(f(\alpha), f(\beta); D_2\big).$$

Together with (6.17), this proves (6.16). Again, equality holds in (6.17) if and only if h is a conformal map of D_1 onto $\mathbb{D}$, which is equivalent to f being a conformal map of D_1 onto D_2.

Finally, assume that f is, in fact, a conformal map of D_1 onto D_2. Then, $h = g_2 \circ f$ is a Riemann map of D_1 onto $\mathbb{D}$. A curve γ in D_1 is a geodesic arc if and only if $h(\gamma)$ is a geodesic arc in the unit disk $\mathbb{D}$ (this is part of the statement of Proposition 6.3). Again by Proposition 6.3, the curve $h(\gamma)$ is a geodesic arc in the unit disk $\mathbb{D}$ if and only if $g_2^{-1}(h(\gamma))$, which is simply $f(\gamma)$, is a geodesic arc in D_2. □

If f is a conformal map of the simply connected domain D_1 onto D_2 then, by the equality statement for (6.16),

$$d(f(\alpha), f(\beta); D_2) = d(\alpha, \beta; D_1), \quad \alpha, \beta \in D_1. \tag{6.18}$$

This is referred to as the *conformal invariance of the hyperbolic metric*.

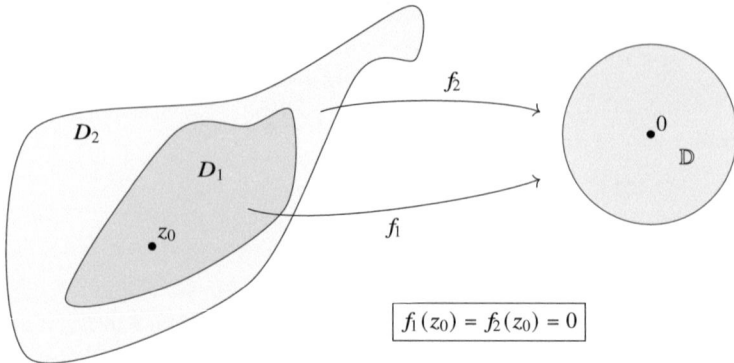

Fig. 6.11 Here $z_0 \in D_1 \subseteq D_2$ and so $\rho_{D_1}(z_0) \geq \rho_{D_2}(z_0)$

6.4.3 Monotonicity of the Hyperbolic Metric

Consider the setting of a point z_0 in a simply connected domain D_1 and a simply connected domain D_2 that contains D_1. Let f_1 be a conformal map of D_1 onto the unit disk $\mathbb{D}$ with $f_1(z_0) = 0$, and let f_2 be a conformal map of D_2 onto $\mathbb{D}$ with $f_2(z_0) = 0$, as shown in Fig. 6.11. Then, by (6.9),

$$\rho_{D_1}(z_0) = 2|f_1'(z_0)| \text{ and } \rho_{D_2}(z_0) = 2|f_2'(z_0)|.$$

Now $f = f_2 \circ f_1^{-1}$ is a self-map of the unit disk $\mathbb{D}$ with $f(0) = 0$ (we need $D_1 \subseteq D_2$ here so that f makes sense). By Schwarz's Lemma, $|f'(0)| \leq 1$ with equality if and only if f is a rotation, $f(w) = e^{i\theta} w$, some real θ. Since

$$f'(0) = f_2'(f_1^{-1}(0)) \times (f_1^{-1})'(0) = f_2'(z_0)/f_1'(f_1^{-1}(0)) = f_2'(z_0)/f_1'(z_0),$$

we deduce that $|f_2'(z_0)/f_1'(z_0)| \leq 1$, that is $\rho_{D_2}(z_0) \leq \rho_{D_1}(z_0)$ (this is effectively Exercise 6.4). The hyperbolic metric at z_0 relative to the larger domain is smaller than the hyperbolic metric at z_0 relative to the smaller domain.

Let us focus for a moment on the case of equality, whereby $f_2(f_1^{-1}(w)) = e^{i\theta} w$, $w \in \mathbb{D}$. Applying f_2^{-1} to both sides, we find that $f_1^{-1}(w) = f_2^{-1}(e^{i\theta} w)$, $w \in \mathbb{D}$, so that $f_2^{-1}(\mathbb{D}) = f_1^{-1}(\mathbb{D})$, that is $D_2 = D_1$. We summarise this work in the following theorem.

Theorem 6.11 *Let D_1 and D_2 be proper simply connected domains with $D_1 \subseteq D_2$. Then,*

$$\rho_{D_2}(z) \leq \rho_{D_1}(z), \quad z \in D_1. \tag{6.19}$$

If α and β are points in D_1 then

$$d(\alpha, \beta; D_2) \leq d(\alpha, \beta; D_1). \qquad (6.20)$$

If equality holds in (6.19) for some z in D_1, or in (6.20) for some pair of distinct points α and β in D_1, then $D_2 = D_1$.

Proof We already established (6.19) above using Schwarz's Lemma, including the case of equality.

For (6.20), let α and β be points in D_1. Let γ_0 be the geodesic arc joining α and β in D_1. Then, first by (6.19) and then by the definition of the hyperbolic distance as an infimum over hyperbolic lengths of curves,

$$d(\alpha, \beta; D_1) = \int_{\gamma_0} \rho_{D_1}(z)\,|dz| \geq \int_{\gamma_0} \rho_{D_2}(z)\,|dz|$$

$$\geq \inf\left\{\int_{\gamma} \rho_{D_2}(z)\,|dz| : \gamma \text{ is a curve in } D_2 \text{ joining } \alpha \text{ to } \beta\right\}$$

$$= d(\alpha, \beta; D_2).$$

If equality holds, then $\int_{\gamma_0} \rho_{D_2}(z)\,|dz| = \int_{\gamma_0} \rho_{D_1}(z)\,|dz|$ and so $\rho_{D_2}(z) = \rho_{D_1}(z)$ at some point z on γ (otherwise $\rho_{D_2}(z) < \rho_{D_1}(z)$ for every z on γ, which is impossible in this case). By the equality statement for (6.19), we again have $D_1 = D_2$. □

6.5 Some Special Domains and Their Hyperbolic Metrics

If we have an explicit conformal map f from a simply connected domain D to the disk $\mathbb{D}$ we can use (6.9), namely $\rho_D(z) = \rho_{\mathbb{D}}(f(z))\,|f'(z)|$, to explicitly compute the hyperbolic metric $\rho_D(z)$ at a point z in D. We have already discussed the hyperbolic metric in the upper half-plane $\mathbb{H}$ in Sect. 4.7. We briefly revisit this important example in the context of (6.9).

6.5.1 The Half-Plane

In Sect. 4.7, we saw that the Möbius map $w = M(z) = (z-i)/(z+i)$ maps the half-plane $\mathbb{H} = \{z : \Im z > 0\}$ conformally onto $\mathbb{D}$. Using $\rho_{\mathbb{H}}(z) = \rho_{\mathbb{D}}(M(z))\,|M'(z)|$, we computed the hyperbolic metric in the half-plane to be

$$\rho_{\mathbb{H}}(z) = \frac{1}{\Im z}, \quad z \in \mathbb{H}. \qquad (6.21)$$

6.5.2 The Strip

We don't always have to go back to the disk $\mathbb{D}$ as the next example shows. The case of equality in (6.15) tells us that, if f is a conformal map from the simply connected D_1 to the simply connected domain D_2 then their hyperbolic metrics are related by

$$\rho_{D_1}(z) = \rho_{D_2}(f(z))|f'(z)|, \quad z \in D_1. \tag{6.22}$$

In the case of the half-plane $D_1 = \mathbb{H}$, we chose D_2 to be the unit disk $\mathbb{D}$. We may choose the target domain D_2 to be any domain for which the hyperbolic metric is known. For example, let $\mathbb{S}$ be the strip

$$\mathbb{S} = \left\{ z : |\Im z| < \frac{\pi}{2} \right\}.$$

Then, the function $f(z) = \mathrm{i} e^z$ maps $\mathbb{S}$ conformally onto the upper half-plane $\mathbb{H}$, as shown in Fig. 6.12. In fact, since

$$e^z = e^{x+\mathrm{i}y} = e^x\, e^{\mathrm{i}y},$$

we see that the exponential function maps the vertical line segment at x in the strip $(-\infty < x < \infty)$ onto the semicircle with centre 0, radius e^x, in the right half-plane. Multiplying by i rotates this semicircle anticlockwise through an angle $\pi/2$. Thus, $f(z) = \mathrm{i} e^z$ is a conformal map of $\mathbb{S}$ onto $\mathbb{H}$. For $z = x + \mathrm{i}y$ in $\mathbb{S}$, (6.21) and (6.22) lead to

$$\rho_{\mathbb{S}}(z) = \rho_{\mathbb{H}}(\mathrm{i}e^z)|\mathrm{i}e^z| = \frac{1}{\Im(\mathrm{i}e^z)}\, e^x.$$

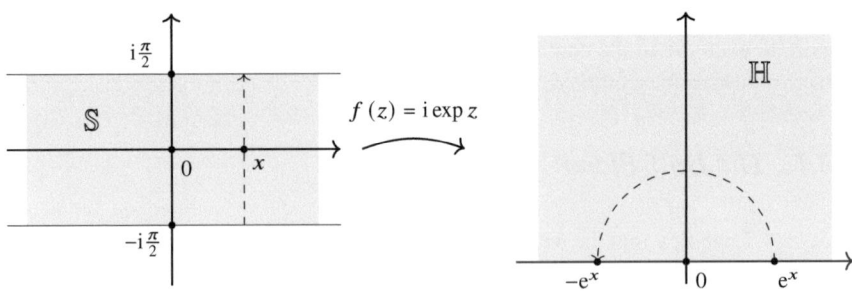

Fig. 6.12 Conformal map of the strip $\mathbb{S}$ onto the half-plane $\mathbb{H}$

6.5 Some Special Domains and Their Hyperbolic Metrics

Since $\Im(ie^{x+iy}) = e^x \cos y$, the hyperbolic metric in the strip is given explicitly as

$$\rho_{\mathbb{S}}(z) = \frac{1}{\cos y}, \quad z = x + iy \in \mathbb{S}.$$

6.5.3 Sectors, Including the Slit-Plane

Here we will compute the hyperbolic metric for a sector S_θ described by

$$S_\theta = \{z = re^{it} : r > 0, -\theta < t < \theta\}.$$

The opening of the sector is 2θ where $0 < \theta \leq \pi$. The case $\theta = \pi$ is that of a slit-plane, namely the complex plane slit along the negative real axis. The map $w = f(z) = z^{\pi/(2\theta)}$ maps S_θ conformally onto the right half-plane H, as shown in Fig. 6.13. To be precise,

$$f(z) = z^{\pi/(2\theta)} = \exp\left(\frac{\pi}{2\theta} \log z\right), \quad z \in S_\theta,$$

where the argument of z implicit in the logarithm function lies between $-\pi$ and π. Since $\rho_H(w) = 1/\Re w$, (6.22) leads to

$$\rho_{S_\theta}(z) = \rho_{S_\theta}(re^{it}) = \rho_H(f(z))|f'(z)|$$
$$= \frac{1}{\Re(z^{\pi/(2\theta)})} \left|\frac{\pi}{2\theta} z^{\pi/(2\theta)-1}\right|$$
$$= \frac{1}{r^{\pi/(2\theta)} \cos(\pi t/(2\theta))} \frac{\pi}{2\theta} r^{\pi/(2\theta)-1}.$$

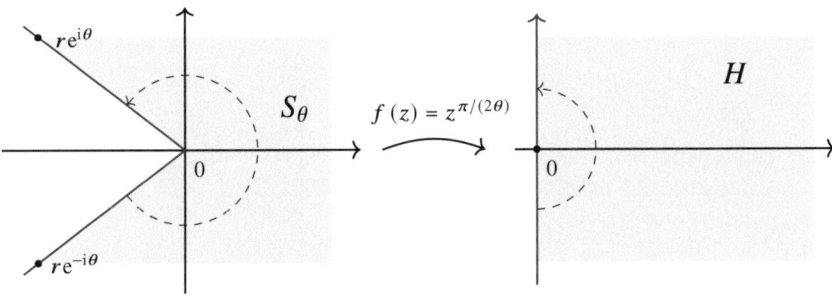

Fig. 6.13 Conformal map of the sector S_θ onto the right half-plane H

Thus,
$$\rho_{S_\theta}(re^{it}) = \frac{\pi}{2\theta r \cos(\pi t/(2\theta))}, \quad z = re^{it} \in S_\theta.$$

In particular, the hyperbolic metric in the slit-plane S_π is given by
$$\rho_{S_\pi}(re^{it}) = \frac{1}{2r \cos(t/2)}, \quad z = re^{it} \in S_\pi.$$

6.5.4 The Half-Disk and the Slit-Disk

Here we compute the hyperbolic metric in the half-disk $\mathbb{D}_h = \{z : |z| < 1, \Im z > 0\}$, the upper half of the unit disk. The Möbius transformation $M(z) = (1+z)/(1-z)$ maps the full disk $\mathbb{D}$ conformally onto the right half-plane H and, by symmetry, maps $\mathbb{D}_h$ onto the first quadrant $Q = \{z : \Re z > 0, \Im z > 0\}$ as shown in Fig. 6.14. Then, $f(z) = M(z)^2$ maps $\mathbb{D}_h$ onto the upper half-plane $\mathbb{H}$. For $z \in D$,

$$\rho_{\mathbb{D}_h}(z) = \rho_{\mathbb{H}}(f(z))|f'(z)| = \frac{|f'(z)|}{\Im(f(z))}.$$

Here we used the formula (6.21) for the hyperbolic metric in $\mathbb{H}$. Now,

$$\Im\left(\left(\frac{1+z}{1-z}\right)^2\right) = 2\Re\left(\frac{1+z}{1-z}\right)\Im\left(\frac{1+z}{1-z}\right)$$

$$= \frac{2}{|1-z|^4}\Re\big((1+z)(1-\bar{z})\big)\Im\big((1+z)(1-\bar{z})\big)$$

$$= \frac{4\Im z\,(1-|z|^2)}{|1-z|^4}. \tag{6.23}$$

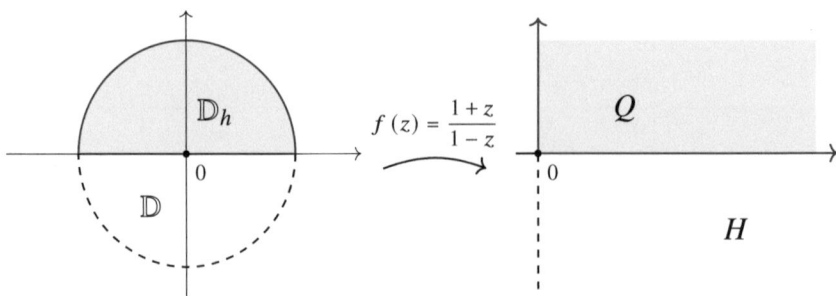

Fig. 6.14 Conformal map of the half-disk $\mathbb{D}_h$ onto the quadrant Q

6.5 Some Special Domains and Their Hyperbolic Metrics

Since $f'(z) = 4(1+z)/(1-z)^3$,

$$\rho_{\mathbb{D}_h}(z) = \frac{|1-z^2|}{\Im z\,(1-|z|^2)}. \tag{6.24}$$

Writing z in polar coordinates as $z = re^{i\theta}$, so that $0 < r < 1$ and $0 < \theta < \pi$,

$$\begin{aligned}
\rho_{\mathbb{D}_h}(re^{i\theta}) &= \frac{|1 - r^2 e^{2i\theta}|}{r\sin\theta\,(1-r^2)} \\
&= \frac{\sqrt{1 + r^4 - 2r^2\cos(2\theta)}}{r\sin\theta\,(1-r^2)} \\
&= \frac{\sqrt{(1-r^2)^2 + 4r^2\sin^2\theta}}{r\sin\theta\,(1-r^2)}.
\end{aligned} \tag{6.25}$$

Note that, for fixed θ with $0 < \theta < \pi$, we have $\rho_{\mathbb{D}_h}(re^{i\theta}) \approx 1/(1-r)$ as $r \to 1^-$. Similarly, for fixed r with $-1 < r < 1$, we have $\rho_{\mathbb{D}_h}(re^{i\theta}) \approx 1/(r\sin\theta)$ as $\sin\theta \to 0^+$. In particular, $\rho_{\mathbb{D}_h}(re^{i\theta})$ behaves like the reciprocal of the distance to the boundary in each case.

Now we turn to the slit-disk $\mathbb{D}_s = \mathbb{D} \setminus (-1, 0]$, that is the unit disk with the line segment from -1 to 0 removed. The conformal map $g(z) = \sqrt{z}$ maps $\mathbb{D}_s$ onto the right half-disk $\{z : |z| < 1, \Re z > 0\}$ that, in turn, is mapped to $\mathbb{D}_h$ by the rotation $z \to iz$. Hence, $f(z) = i\sqrt{z}$ maps $\mathbb{D}_s$ conformally onto $\mathbb{D}_h$, as shown in Fig. 6.15. Since we know the hyperbolic metric in the half-disk, we can compute the hyperbolic metric at a point z in the slit-disk by

$$\rho_{\mathbb{D}_s}(z) = \rho_{\mathbb{D}_h}(f(z))|f'(z)| = \frac{\rho_{\mathbb{D}_h}(i\sqrt{z})}{2|\sqrt{z}|} = \frac{|1+z|}{\Re(\sqrt{z})(1-|z|)2\sqrt{|z|}}. \tag{6.26}$$

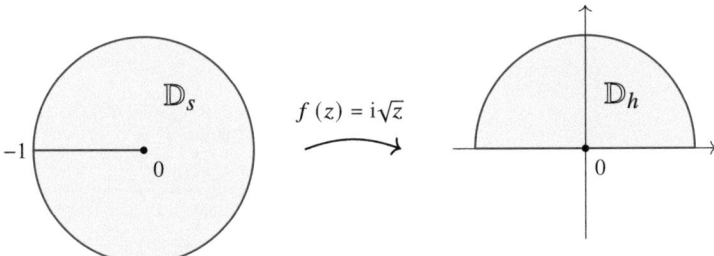

Fig. 6.15 Conformal map of the slit-disk $\mathbb{D}_s$ onto the half-disk $\mathbb{D}_h$

In a manner similar to (6.25), we can write this expression in polar form. Writing $z = re^{i\theta}$ where $0 < r < 1$ and $-\pi < \theta < \pi$, we find that

$$\rho_{\mathbb{D}_s}(re^{i\theta}) = \frac{\sqrt{(1-r)^2 + 4r\cos^2(\theta/2)}}{2r(1-r)\cos(\theta/2)}. \qquad (6.27)$$

As $r \to 1^-$ for fixed θ in $(-\pi, \pi)$, we see that $\rho_{\mathbb{D}_s}(re^{i\theta}) \approx 1/(1-r)$. As $\cos(\theta/2) \to 0^+$ for fixed r in $(0, 1)$, we see that $\rho_{\mathbb{D}_s}(re^{i\theta}) \approx 1/(2r\cos(\theta/2)) \approx 1/(r|\sin\theta|)$. As a consequence, $\rho_{\mathbb{D}_s}(re^{i\theta})$ behaves like the reciprocal of the distance to the boundary in both cases. On the other hand, setting $z = x, 0 < x < 1$, in (6.26) or (6.27), we see that $\rho_{\mathbb{D}_s}(x) \approx 1/(2x)$ as $x \to 0^+$, so that $\rho_{\mathbb{D}_s}(x)$ behaves like the reciprocal of twice the distance to the boundary as x approaches 0 along the positive real axis.

Exercise 6.8 Verify the identities (6.23), (6.24), (6.25), and (6.27).

6.6 Boundary Correspondence Under Conformal Mappings

For a conformal map f of a domain D_1 onto a domain D_2, to what extent can the map f be extended, in some sensible manner, to the boundary of D_1? This question is very general and has various answers depending on the meaning one attaches to 'sensible manner'. To begin to orientate ourselves, let's take a closer look at the conformal map of the complex plane slit along the negative real axis, $\mathbb{C} \setminus (-\infty, 0]$, onto the right half-plane $H = \{z : \Re z > 0\}$ given by $f(z) = \sqrt{z}$ (see Sect. 6.5.3 and Fig. 6.16). Here $\arg z$ is to lie between $-\pi$ and π and so, if $z = re^{i\theta}, -\pi < \theta < \pi$, then $f(z) = \sqrt{r}e^{i\theta/2}$. Under f, the sequence $\{z_n\}_{n=1}^{\infty} = \{e^{i(\pi-2/n)}\}_{n=1}^{\infty}$ is mapped to the sequence $\{f(z_n)\}_{n=1}^{\infty} = \{ie^{-i/n}\}_{n=1}^{\infty}$ while the sequence $\{w_n\}_{n=1}^{\infty} = \{e^{-i(\pi-2/n)}\}_{n=1}^{\infty}$ is mapped to $\{f(w_n)\}_{n=1}^{\infty} = \{-ie^{i/n}\}_{n=1}^{\infty}$. Thus both sequences $\{z_n\}_1^{\infty}$ and $\{w_n\}_1^{\infty}$ converge to the boundary point -1 of $\mathbb{C}\setminus(-\infty, 0]$ but their images

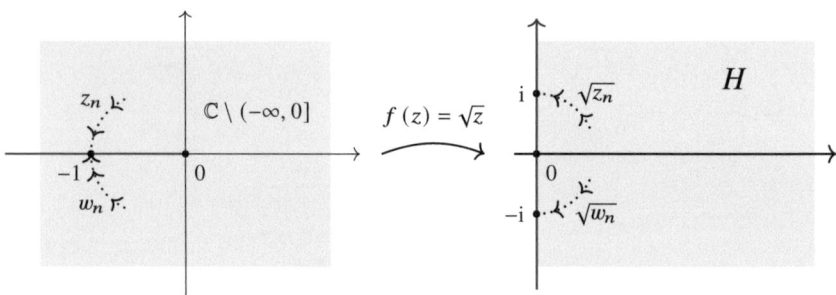

Fig. 6.16 An example of a conformal map that does not extend continuously to the boundary

6.6 Boundary Correspondence Under Conformal Mappings

under f converge to the distinct points i and $-$i, respectively, on the boundary of H. In particular, f does not extend continuously to the boundary of $\mathbb{C} \setminus (-\infty, 0]$, so we shouldn't necessarily interpret 'extend in some sensible manner' to mean 'extend continuously' in this case. For this conformal map, it is straightforward to see that $f(z)$ extends continuously to each point $x \in (-\infty, 0)$ on the negative real axis if either one restricts z to approach x from above the slit or else to approach x from below the slit—but with a different limit in each case. These approach regions correspond to the notion of a *prime end* so that, in this case, each boundary point on the slit (except for 0) corresponds to two prime ends, reflecting the interpretation that the slit has 'two sides'.

The theory of prime ends makes a distinction between the topological boundary of the domain D_1 and the 'boundary' of D_1 as seen by a conformal map f of D_1. That is, if we interpret the notion of 'boundary' correctly, then the conformal map f may extend continuously to this new boundary but not to the topological boundary.

We won't, however, follow these ideas here. The books [22] and [6] all have very nice accounts of boundary correspondence via prime ends, with Collingwood and Lohwater's book being a classic monograph on the topic. We will insist that the conformal map f extends continuously to the topological boundary of its domain and follow Kodaira [18, Sect. 5.2], who in turn credits Hurwitz. Kodaira phrases his main result [18, Thm. 5.2] in terms of piecewise-smooth Jordan curves. This has the potential of bringing the *Jordan curve theorem*[2] into play. The Jordan curve theorem is a non-trivial topological result—including a proof here would take us off course. Instead, we will introduce the notion of a *geometrically simple domain*—Kodaira effectively shows that a domain bounded by a Jordan curve is geometrically simple. For us it is sufficient to note that in any instance in which we make use of our result, it will be clear that the domains involved are geometrically simple.

Definition 6.4 A bounded domain D is said to be *geometrically simple* if, to each boundary point ζ of D, there corresponds a positive number ρ and continuous functions $\alpha(r)$ and $\beta(r)$ on $[0, \rho]$ with $\alpha(r) < \beta(r) \leq \alpha(r) + 2\pi$ such that

$$D \cap D(\zeta, \rho) = \{\zeta + re^{i\theta} : 0 < r < \rho,\ \alpha(r) < \theta < \beta(r)\}.$$

The definition says, roughly speaking, that each circle $C(\zeta, r), 0 < r < \rho$, meets D in a single arc and that these arcs combine to fill out the intersection of D with the disk $D(\zeta, \rho)$. This concept is illustrated in Fig. 6.17. Note that the slit-plane in Fig. 6.16 is not geometrically simple: circles centred at the point -1, for example, meet $\mathbb{C} \setminus (-\infty, 0]$ in two disjoint arcs.

---

[2] A *Jordan arc* is a continuous map $\gamma : [a, b] \to \mathbb{C}$ from a closed interval $[a, b]$ on the real line into the complex plane such that, for all s, t with $a \leq s < t < b$, we have $\gamma(s) \neq \gamma(t)$. This last condition says that the curve should not intersect itself. If, in addition, $\gamma(a) = \gamma(b)$ the curve is closed and is called a *Jordan curve*. The Jordan curve theorem states that a Jordan curve divides the complex plane into precisely two components that are identified as the *inside* and the *outside* of the Jordan curve. The difficulty in proving this result arises from the fact that Jordan curves can, to put it mildly, be quite complicated.

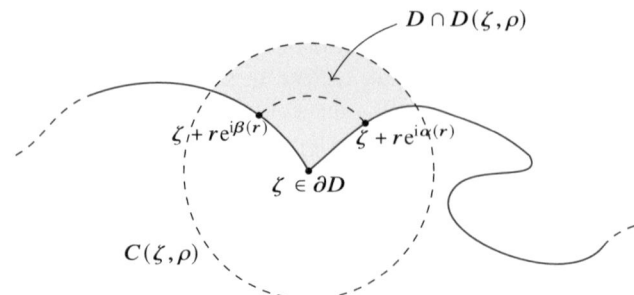

Fig. 6.17 The local boundary of a geometrically simple domain

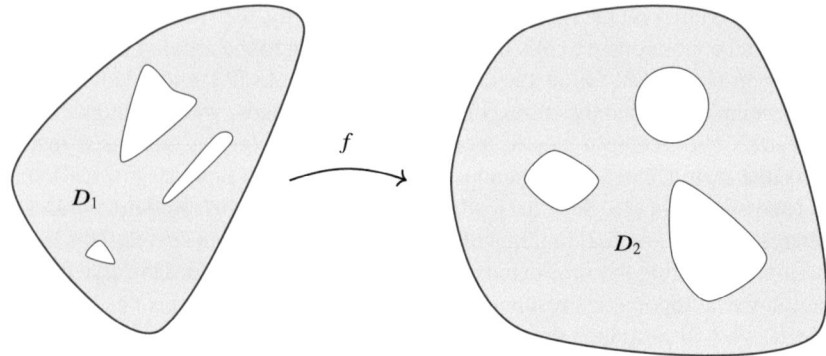

Fig. 6.18 A conformal map between geometrically simple domains

Our objective is to prove the following theorem. Recall that we write cl(E) for the closure of a subset E of the complex plane. The notation $\overline{E}$ is retained for complex conjugate, so that $\overline{E}$ denotes the set $\{\overline{z} : z \in E\}$, which is the reflection of E in the real axis.

Theorem 6.12 (Boundary Correspondence) *Let D_1 and D_2 be bounded domains, both geometrically simple. Suppose that f is a conformal map of D_1 onto D_2. Then f extends to a homeomorphism from* cl(D_1) *to* cl(D_2).

Let's clarify a few matters before we proceed. The assumption that a conformal map f of D_1 onto D_2 exists implies that D_1 and D_2 have the same topological connectivity. Of course, if D_1 and D_2 are simply connected the existence of a conformal map is guaranteed by the Riemann Mapping Theorem. This alone, however, is far from sufficient to guarantee the existence of such a conformal map in the multiply connected case, as we shall see even in the simplest case when D_1 and D_2 are both annuli. The content of Theorem 6.12 is, therefore, that *if* a conformal map exists between geometrically simple bounded domains D_1 and D_2, as shown in Fig. 6.18, then it extends to a homeomorphism of the closures.

6.6.1 Proof of Theorem 6.12

The conformal map $f(z) = \sqrt{z}$ from the slit-plane $\mathbb{C} \setminus (-\infty, 0]$ to the right half-plane (see Fig. 6.16) does not extend continuously to the boundary since there are points in the domain close to -1 (z_n and w_n in that example) that are mapped to points far apart in the image domain. Another way to express this is that the conformal map is not uniformly continuous on its domain. The core of the proof of Theorem 6.12 is to show that if f is a conformal map of D_1 onto D_2, where D_1 and D_2 are bounded geometrically simple domains, then f is automatically uniformly continuous on D_1: that is, given ε positive there exists δ positive such that

$$|f(z) - f(w)| \leq \varepsilon \text{ if } z, w \in D_1 \text{ with } |z - w| \leq \delta. \tag{6.28}$$

Interchanging the roles of D_1 and D_2, we see that f^{-1} too is uniformly continuous on D_2. Taking the uniform continuity of the conformal mapping f and its inverse for granted for the moment, let's see how this leads to a proof of Theorem 6.12.

Proof of Theorem 6.12, Assuming Uniform Continuity of the Mappings Let ζ be a boundary point of D_1 and let $\{z_n\}_{n=1}^{\infty}$ be any sequence of points in D_1 that converges to ζ. Since the conformal map f of D_1 onto D_2 is assumed to be uniformly continuous, given ε positive there exists δ positive such that $|f(z) - f(w)| \leq \varepsilon$ whenever z, w are in D_1 and $|z - w| \leq \delta$. Since $z_n \to \zeta$ as $n \to \infty$, there exists $N \in \mathbb{N}$ such that $|z_n - z_m| \leq \delta$ whenever n, m are both greater than N. It follows that $|f(z_n) - f(z_m)| \leq \varepsilon$ whenever $n, m \geq N$, that is $\{f(z_n)\}_{n=1}^{\infty}$ is a Cauchy sequence and hence convergent. We set

$$\tilde{f}(\zeta) := \lim_{n \to \infty} f(z_n). \tag{6.29}$$

We immediately need to verify that the value of $\tilde{f}(\zeta)$ so defined does not depend on the particular choice of sequence $\{z_n\}$. Let $\{w_n\}_{n=1}^{\infty}$ be another sequence in D_1 with limit ζ. With ε and δ as above, there exists N_1 such that $|z_n - w_n| \leq \delta$ for $n \geq N_1$, this because $\{z_n\}$ and $\{w_n\}$ have the same limit. But then $|f(z_n) - f(w_n)| \leq \varepsilon$ for $n \geq N_1$. Since ε was arbitrary, the convergent sequences $\{f(z_n)\}$ and $\{f(w_n)\}$ must have the same limit. Thus, $\tilde{f}(\zeta)$ is well-defined by (6.29).

What can we say about $\tilde{f}(\zeta)$? Since, in the definition (6.29) of $\tilde{f}(\zeta)$, the points z_n lie in D_1, the points $f(z_n)$ lie in D_2. Hence, $\tilde{f}(\zeta)$ is a limit point of D_2 and so lies in the closure $\mathrm{cl}(D_2)$ of D_2. It can't be that $\tilde{f}(\zeta)$ lies in D_2 itself, however. Suppose, to the contrary, that this was the case, say $\tilde{f}(\zeta) = w_\zeta \in D_2$. Choose a sequence $\{z_n\}_1^{\infty}$ in D_1 that converges to ζ. Then, by (6.29), $\{f(z_n)\}_1^{\infty} = \{w_n\}_1^{\infty}$, say, converges to w_ζ in D_2. Since f^{-1} is continuous at w_ζ, the sequence $\{z_n\}_1^{\infty} = \{f^{-1}(w_n)\}_1^{\infty}$ converges to $z_\zeta = f^{-1}(w_\zeta)$, a point in D_1, as $n \to \infty$. But $z_n \to \zeta$ as $n \to \infty$, so $\zeta = z_\zeta$. This doesn't make sense since $\zeta \in \partial D_1$ and $z_\zeta \in D_1$. We have shown, therefore, that $\tilde{f}(\zeta)$ has to be a boundary point of D_2.

We now define F on the closure $\text{cl}(D_1)$ of D_1 by

$$F(\zeta) = \begin{cases} f(\zeta) & \zeta \in D_1; \\ \tilde{f}(\zeta) & \zeta \in \partial D_1, \end{cases}$$

so that F is an extension of f on D_1 to its closure $\text{cl}(D_1)$.

Continuity of the Extension F We claim that F is uniformly continuous[3] on the closure $\text{cl}(D_1)$ of D_1. To see this, let ε positive be given and choose δ positive so that $|f(z) - f(w)| \leq \varepsilon/3$ whenever $z, w \in D_1$ and $|z - w| \leq \delta$ (here, again, we're making use of the uniform continuity of f on D_1). Now let ζ_1 and ζ_2 be points in $\text{cl}(D_1)$ with $|\zeta_1 - \zeta_2| \leq \delta/2$. We wish to show that $|F(\zeta_1) - F(\zeta_2)| \leq \varepsilon$. Choose a sequence $\{z_{n,1}\}_{n=1}^\infty$ in D_1 that converges to ζ_1 and a sequence $\{z_{n,2}\}_{n=1}^\infty$ in D_1 that converges to ζ_2. There is $N \in \mathbb{N}$ such that

$$|F(\zeta_1) - f(z_{n,1})| \leq \frac{\varepsilon}{3} \quad \text{and} \quad |F(\zeta_2) - f(z_{n,2})| \leq \frac{\varepsilon}{3}, \quad \text{for } n \geq N,$$

(if ζ_1 or ζ_2 lies in D_1 this is simply continuity of f on D_1, otherwise we use (6.29)). We can choose N larger if necessary to guarantee that $|z_{n,1} - \zeta_1| \leq \delta/4$ and $|z_{n,2} - \zeta_2| \leq \delta/4$, for $n \geq N$. Then, for $n \geq N$, and since $|\zeta_1 - \zeta_2| \leq \delta/2$,

$$|z_{n,1} - z_{n,2}| \leq |z_{n,1} - \zeta_1| + |\zeta_1 - \zeta_2| + |\zeta_2 - z_{n,2}| \leq \delta,$$

from which we deduce that $|f(z_{n,1}) - f(z_{n,2})| \leq \epsilon/3$. Finally, choosing any particular $n \geq N$, we find that

$$|F(\zeta_1) - F(\zeta_2)| \leq |F(\zeta_1) - f(z_{n,1})| + |f(z_{n,1}) - f(z_{n,2})| + |f(z_{n,2}) - F(\zeta_2)| \leq \varepsilon,$$

as required.

Interchanging the roles of D_1 and D_2, we see that the inverse conformal mapping f^{-1} of D_2 onto D_1 has a similar continuous extension $G : \text{cl}(D_2) \to \text{cl}(D_1)$.

F Is a Homeomorphism from $\text{cl}(D_1)$ to $\text{cl}(D_2)$ The extension F is uniformly continuous on $\text{cl}(D_1)$. It is also injective (that is, one-to-one). Suppose, to the contrary, that $\zeta_1, \zeta_2 \in \text{cl}(D_1)$ with $\zeta_1 \neq \zeta_2$ and $F(\zeta_1) = F(\zeta_2)$. This certainly can't happen if both ζ_1 and ζ_2 lie in D_1 as f is univalent in D_1. That we could have $\zeta_1 \in \partial D_1$ and $\zeta_2 \in D_1$ (or vice versa) is also ruled out as then $F(\zeta_1) \in \partial D_2$ and $F(\zeta_2) \in D_2$, so $F(\zeta_1)$ and $F(\zeta_2)$ can't be equal in this case either. The final possibility is that both ζ_1 and ζ_2 are on the boundary of D_1 with $\tilde{f}(\zeta_1) = \tilde{f}(\zeta_2) = \eta \in \partial D_2$. Set $\varepsilon = |\zeta_1 - \zeta_2|$ so that ε is positive. Now $g = f^{-1}$ is a conformal map of D_2 onto D_1, and it too is uniformly continuous on D_2 by assumption.

---

[3] Since $\text{cl}(D_1)$ is a closed and bounded subset of the complex plane, hence compact, once f is continuous on $\text{cl}(D_1)$ it is automatically uniformly continuous.

6.6 Boundary Correspondence Under Conformal Mappings

Thus, there is a positive δ such that $|g(w_1) - g(w_2)| \leq \varepsilon/2$ if $w_1, w_2 \in D_2$ and $|w_1 - w_2| \leq \delta$. Again, we choose a sequence $\{z_{n,1}\}_{n=1}^\infty$ in D_1 that converges to ζ_1 and a sequence $\{z_{n,2}\}_{n=1}^\infty$ in D_1 that converges to ζ_2. Then both sequences $\{f(z_{n,1})\}_1^\infty$ and $\{f(z_{n,2})\}_1^\infty$ converge to η. Thus, for sufficiently large n, we have $|f(z_{n,1}) - f(z_{n,2})| \leq \delta$. As a consequence of the uniform continuity of g, we deduce that, for all sufficiently large n,

$$|z_{n,1} - z_{n,2}| = |g(f(z_{n,1})) - g(f(z_{n,2}))| \leq \frac{\varepsilon}{2}.$$

In the limit as $n \to \infty$, we deduce that $|\zeta_1 - \zeta_2| \leq \varepsilon/2$, which is a contradiction showing that, in fact, F is injective on $\mathrm{cl}(D_1)$.

Finally, we show that F is onto $\mathrm{cl}(D_2)$. Recall that the corresponding extension to $\mathrm{cl}(D_2)$ of $g = f^{-1}$ on D_2 is denoted by G. Let $\eta \in \partial D_2$, and write ζ for $G(\eta)$. Then, $F(\zeta) = \eta$ since, by continuity of F on $\mathrm{cl}(D_1)$ and choosing a sequence $\{w_n\}_1^\infty$ in D_2 converging to η, we have

$$F(\zeta) = F(G(\eta)) = F\left(\lim_{n \to \infty} g(w_n)\right)$$

$$= \lim_{n \to \infty} F(g(w_n))$$

$$= \lim_{n \to \infty} f(g(w_n)) = \lim_{n \to \infty} w_n = \eta.$$

In summary, $F : \mathrm{cl}(D_1) \to \mathrm{cl}(D_2)$ is bijective and continuous with inverse function G that is similarly continuous. That is, F is a homeomorphism. □

Proof that f Is Uniformly Continuous on D_1 Suppose to the contrary that f is not uniformly continuous on D_1, so that (6.28) does *not* hold. That is, there is some positive ε_0 for which no corresponding δ can be found. In particular, for each natural number n there are points z_n and w_n in D_1 such that

$$|z_n - w_n| < \frac{1}{n} \text{ but } |f(z_n) - f(w_n)| \geq \varepsilon_0. \tag{6.30}$$

Since $\{z_n\}_{n=1}^\infty$ is a bounded sequence (it is confined to the bounded domain D_1), it has a convergent subsequence with limit p, say. Replacing $\{z_n\}_{n=1}^\infty$ by this subsequence, the sequence $\{f(z_n)\}_{n=1}^\infty$ is bounded (it is confined to the bounded domain D_2), and so it too has a convergent subsequence, with limit P, say. Replacing $\{z_n\}_{n=1}^\infty$ and $\{f(z_n)\}_{n=1}^\infty$ by these subsequences, both are now convergent. Replacing the sequences $\{z_n\}_{n=1}^\infty$ and $\{w_n\}_{n=1}^\infty$ by further subsequences, we may assume that the sequence $\{f(w_n)\}_{n=1}^\infty$ is also convergent, with limit Q, say. Since $|z_n - w_n| \to 0$ and $z_n \to p$ as $n \to \infty$, we automatically have that the sequence $\{w_n\}_{n=1}^\infty$ is convergent with the same limit p.

Thus, $z_n \to p$, $w_n \to p$, $f(z_n) \to P$, and $f(w_n) \to Q$ as $n \to \infty$, where $p \in \mathrm{cl}(D_1)$ and $P, Q \in \mathrm{cl}(D_2)$. Actually, we can say a little more about p, P and

Q. First, p cannot be a point in the domain D_1. If that was the case, the fact that $z_n \to p$ and $w_n \to p$ as $n \to \infty$, together with the continuity of f on D_1, would imply that $f(z_n) \to f(p)$ and $f(w_n) \to f(p)$ as $n \to \infty$, in turn implying that $|f(z_n) - f(w_n)| \to 0$ as $n \to \infty$ in contradiction of (6.30). Thus, p is a boundary point of D_1.

As a consequence, both P and Q are boundary points of D_2 (this is the same as the earlier argument showing that $\tilde{f}(\zeta)$, as defined by (6.29), had to be a boundary point of D_2). If P was a point in D_2 then the fact that $f(z_n) \to P$ as $n \to \infty$ together with the continuity of f^{-1} at P would imply that $z_n = f^{-1}(f(z_n))$ would converge to $f^{-1}(P)$, a point in D_1. But $z_n \to p$, a point that we have seen is on the boundary of D_1. The same argument with z_n replaced by w_n shows that Q is a boundary point of D_2.

Of course, by (6.30), $|P - Q| \geq \varepsilon_0$.

Since D_1 and D_2 are geometrically simple, there are positive numbers δ and ρ and functions α_i and β_i, $i = 1, 2, 3$, with $\alpha_i < \beta_i \leq \alpha_i + 2\pi$ such that

$$D_1 \cap D(p, \delta) = \{p + re^{i\theta} : 0 < r < \delta,\ \alpha_1(r) < \theta < \beta_1(r)\}$$

$$D_2 \cap D(P, \rho) = \{P + re^{i\theta} : 0 < r < \rho,\ \alpha_2(r) < \theta < \beta_2(r)\}$$

$$D_2 \cap D(Q, \rho) = \{Q + re^{i\theta} : 0 < r < \rho,\ \alpha_3(r) < \theta < \beta_3(r)\}.$$

We refer to these three regions as $\Delta_\delta(p)$, $\Delta_\rho(P)$ and $\Delta_\rho(Q)$, respectively. Moreover, we can choose ρ smaller if necessary so that $\rho < \varepsilon_0/3$. If $\zeta \in \Delta_\rho(P)$ and $\omega \in \Delta_\rho(Q)$ then, since $|P - Q| \geq \varepsilon_0$, it must be that ζ and ω are at least a distance ρ apart, that is $|\zeta - \omega| > \rho$. Let us write ζ_n for $f(z_n)$ and write ω_n for $f(w_n)$. Since $\zeta_n \to P$ and $\omega_n \to Q$ as $n \to \infty$, there is an integer N_1 such that $\zeta_n \in \Delta_\rho(P)$ and $\omega_n \in \Delta_\rho(Q)$ for every n greater than N_1. By choosing δ smaller if necessary, we can assume that neither z_{N_1} nor w_{N_1} lies in $\Delta_\delta(p)$—see Fig. 6.19.

Now we fix an arbitrary positive ε with $\varepsilon < \delta$. Since $z_n \to p$ and $w_n \to p$ as $n \to \infty$, there is $N_2 > N_1$ such that both z_n and w_n lie in $\Delta_\varepsilon(p)$ for $n \geq N_2$. Let

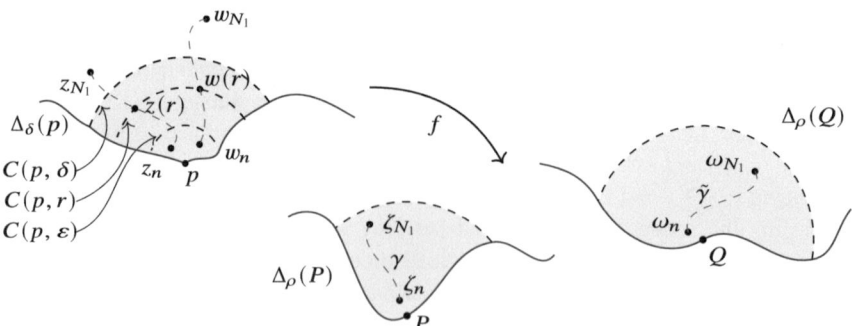

Fig. 6.19 Diagram for the proof that f is uniformly continuous

6.6 Boundary Correspondence Under Conformal Mappings

$n \geq N_2$. The points ζ_{N_1} and ζ_n lie in $\Delta_\rho(P)$, and so can be connected in $\Delta_\rho(P)$ by a curve γ. Then $f^{-1}(\gamma)$ is a curve in D_1 that connects z_{N_1} and z_n in D_1. Since $|z_{N_1} - p| > \delta$ and $|z_n - p| < \varepsilon$, this curve must cross the circle $C(p, r)$ for each r with $\varepsilon < r < \delta$ at some point (possibly more than one point), say $z(r)$. Similarly, the points ω_n and ω_{N_1} can be connected by a curve $\tilde{\gamma}$ in $\Delta_\rho(Q)$ whose image under f^{-1} joins w_n to w_{N_1} in D_1. Again, for each r with $\varepsilon < r < \delta$, this curve crosses the circle $C(p, r)$ at some point $w(r)$—see Fig. 6.19.

As observed earlier, knowing that $f(z(r)) \in \Delta_\rho(P)$ and $f(w(r)) \in \Delta_\rho(Q)$ we deduce that

$$|f(z(r)) - f(w(r))| \geq \rho.$$

Writing $z(r) = p + re^{i\phi}$ and $w(r) = p + re^{i\psi}$ (where $\phi = \phi(r)$ and $\psi = \psi(r)$ both lie between $\alpha_1(r)$ and $\beta_1(r)$),

$$|f(z(r)) - f(w(r))| = |f(p + re^{i\phi}) - f(p + re^{i\psi})|$$

$$= \left| \int_\psi^\phi \frac{d}{d\theta} f(p + re^{i\theta})\, d\theta \right|$$

$$= \left| \int_\psi^\phi f'(p + re^{i\theta}) ire^{i\theta}\, d\theta \right| \leq \int_{\alpha_1(r)}^{\beta_1(r)} |f'(p + re^{i\theta})| r\, d\theta.$$

Writing $|f'(p + re^{i\theta})| r$ as $\left(|f'(p + re^{i\theta})| \sqrt{r} \right) \sqrt{r}$, and applying the Cauchy-Schwarz inequality gives

$$\rho^2 \leq \left(\int_{\alpha_1(r)}^{\beta_1(r)} |f'(p + re^{i\theta})| r\, d\theta \right)^2$$

$$\leq \left(\int_{\alpha_1(r)}^{\beta_1(r)} |f'(p + re^{i\theta})|^2 r\, d\theta \right) \left(\int_{\alpha_1(r)}^{\beta_1(r)} r\, d\theta \right).$$

Consequently, and since $\beta_1(r) - \alpha_1(r) \leq 2\pi$,

$$\frac{\rho^2}{2\pi r} \leq \int_{\alpha_1(r)}^{\beta_1(r)} |f'(p + re^{i\theta})|^2 r\, d\theta.$$

This holds for each r between ε and δ, so we may integrate in r to obtain

$$\frac{\rho^2}{2\pi} \log\left(\frac{\delta}{\varepsilon}\right) \leq \int_\varepsilon^\delta \int_{\alpha_1(r)}^{\beta_1(r)} |f'(p + re^{i\theta})|^2 r\, dr\, d\theta \leq \int_{D_1} |f'(z)|^2\, dm(z).$$

Here $dm(z)$ stands for area measure; $dm(z) = r\, dr\, d\theta$. Now $\int_{D_1} |f'(z)|^2\, dm(z)$ represents the area of the image of D_1 under f, that is the area of D_2, assumed finite. Thus, no matter how small we take ε to be, we will have $\rho^2/(2\pi)\log(\delta/\varepsilon) \leq$ Area(D_2). But this is impossible since $\log(\delta/\varepsilon) \to +\infty$ as $\varepsilon \to 0^+$. This contradiction shows that f is uniformly continuous on D_1. $\square$

6.7 Conformal Mapping of Annuli

Recall that a domain D_2 is *conformally equivalent* to a domain D_1 if there is a conformal map of D_1 onto D_2. This defines an equivalence relation on the collection of domains in the complex plane, which is thereby divided into disjoint (conformal) equivalence classes. We learned from the Riemann Mapping Theorem that any two proper simply connected domains are conformally equivalent, that is to say there is just one equivalence class of proper simply connected domains. Also, the complex plane forms an equivalence class all on its own. The situation for multiply-connected domains, however, is more complicated as we shall see in the simplest case of annuli. By an *annulus* we mean a domain of the form

$$A(r, R) = \{z \in \mathbb{C} : r < |z| < R\}$$

where $0 \leq r < R \leq \infty$. It is clear that any two annuli are *topologically equivalent* in that any annulus can be mapped to any other annulus by a homeomorphism. The point is that they may not be conformally equivalent, that is there may not exist an analytic homeomorphism between the annuli.

Exercise 6.9 Write down an explicit homeomorphism between an annulus $A(r_1, R_1)$ and an annulus $A(r_2, R_2)$ where $0 \leq r_1 < R_1 \leq \infty$ and $0 \leq r_2 < R_2 \leq \infty$.

First, let's see if we can identify annuli that *are* conformally equivalent. Suppose that $0 < r_1 < R_1 < \infty$, that $0 < r_2 < R_2 < \infty$, and that

$$\frac{R_2}{r_2} = \frac{R_1}{r_1}.$$

Then $f(z) = (R_2/R_1)z$ is a simple, explicit conformal map of $A(r_1, R_1)$ onto $A(r_2, R_2)$, showing that $A(r_1, R_1)$ and $A(r_2, R_2)$ are conformally equivalent. In fact, $A(r_2, R_2)$ and $A(r_1, R_1)$ are simply scaled versions of each other in this case. This is illustrated in Fig. 6.20.

If $0 < R_1 < \infty$ and $0 < R_2 < \infty$, then the map $f(z) = (R_2/R_1)z$ also acts as a conformal map of $A(0, R_1)$ onto $A(0, R_2)$, showing that any two annuli $A(0, R)$, with $0 < R < \infty$, are conformally equivalent. Similarly, the dilation $f(z) = (r_2/r_1)z$ is a conformal map of the annulus $A(r_1, \infty)$, $(r_1 > 0)$, onto the annulus $A(r_2, \infty)$, $(r_2 > 0)$. The annulus $A(r, \infty)$ $(r > 0)$ is conformally equivalent, via the conformal map $f(z) = 1/z$, to the annulus $A(0, 1/r)$.

6.7 Conformal Mapping of Annuli

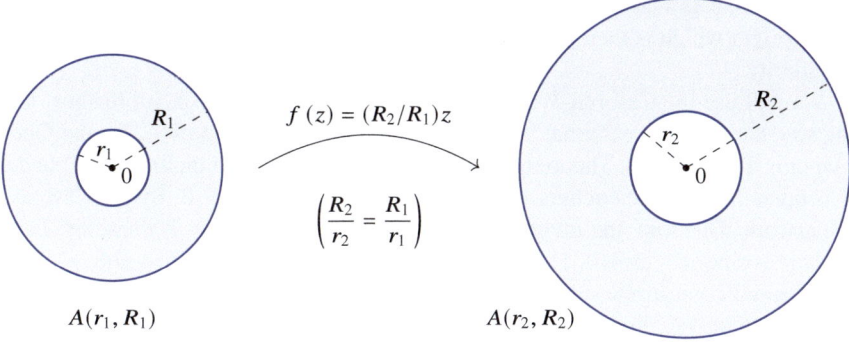

Fig. 6.20 Conformally equivalent annuli

This suggests partitioning the family of all annuli into disjoint collections $\mathscr{A}_\lambda$, $1 \leq \lambda \leq \infty$, as follows:

$$\mathscr{A}_1 = \{A(0, \infty)\}, \tag{6.31a}$$

$$\mathscr{A}_\lambda = \{A(r, R) : 0 < r < R < \infty \text{ and } R/r = \lambda\}, \quad 1 < \lambda < \infty, \tag{6.31b}$$

$$\mathscr{A}_\infty = \{A(0, R) : 0 < R < \infty\} \cup \{A(r, \infty) : 0 < r < \infty\}. \tag{6.31c}$$

The annuli in each collection $\mathscr{A}_\lambda$ are all conformally equivalent. While it may be surprising, we go on to show that this is actually an exact description of the conformally equivalent classes of annuli, that is if two annuli are conformally equivalent then they belong to the same collection $\mathscr{A}_\lambda$.

Theorem 6.13 *Let $0 \leq r_1 < R_1 \leq \infty$ and $0 \leq r_2 < R_2 \leq \infty$, and suppose that the annuli $A_1 = A(r_1, R_1)$ and $A_2 = A(r_2, R_2)$ are conformally equivalent. Then A_1 and A_2 belong to the same collection $\mathscr{A}_\lambda$ for some $\lambda \in [1, \infty]$.*

The proof we give is based on the Schwarz Reflection Principle (see Sect. 2.3).[4] Before this, we need a result on conformal mapping of the annulus $A(0, \infty)$, that is of the punctured plane $\mathbb{C} \setminus \{0\}$. It will be helpful to keep in mind that $f(z) = 1/z$ is a conformal self-map of $\mathbb{C} \setminus \{0\}$. The proof below has some overlap with the proof of Proposition 6.2.

Proposition 6.4 *Suppose that f is a conformal mapping of the punctured plane $\mathbb{C}\setminus\{0\}$. Then, either f has the form $f(z) = a + bz$ or f has the form $f(z) = a + b/z$, where $a, b \in \mathbb{C}$, $b \neq 0$. In either case, $f(\mathbb{C} \setminus \{0\}) = \mathbb{C} \setminus \{a\}$, the complex plane punctured at a.*

---

[4] Beliaev [3] gives three proofs of this result, two in Sect. 2.9.1 and another, using extremal length, in Sect. 4.2.1.

Proof Since f is analytic in the punctured plane $\mathbb{C}\setminus\{0\}$, f has an isolated singularity at the origin which is then either an essential singularity, or a pole, or a removable singularity.

We will use the Casorati-Weierstrass Theorem [16, Theorem 8.10] to show that the singularity being essential would contradict the univalence of f. By the Open Mapping Theorem [16, Theorem 10.7], the image of (say) the annulus $A(2,3)$ under f is open. It therefore contains some disk $D(a,r)$, $a \in \mathbb{C}$, $r > 0$. By the Casorati-Weierstrass Theorem, the image of the annulus $A(0,1)$ under f is dense in $\mathbb{C}$ and so there is a point z in $A(0,1)$ whose image w under f lies in the disk $D(a,r)$. This is a contradiction of the univalence of f since f now takes the value w (at least) twice, once in $A(2,3)$ and again in $A(0,1)$.

If f has a pole at the origin, then this must be a simple pole since meromorphic functions are n to 1 in a sufficiently small neighbourhood of a pole of order n—see Exercise 3.8 for a precise statement. If the order of the pole was greater than 1 this would again contradict the univalence of f. As f has a simple pole at 0, it has a Laurent series expansion about 0 of the form

$$f(z) = \frac{a_{-1}}{z} + a_0 + a_1 z + a_2 z^2 + a_3 z^3 + \ldots, \quad z \neq 0,$$

for some coefficients a_i, $i \geq -1$. If we allow a_{-1} to be equal to 0, this also covers the case that f has a removable singularity at 0.

Now, since f is a conformal mapping of the punctured disk $\mathbb{C}\setminus\{0\}$, so is $g(z) = f(1/z)$. The Laurent series expansion of g is, in terms of that of f,

$$g(z) = \ldots + \frac{a_3}{z^3} + \frac{a_2}{z^2} + \frac{a_1}{z} + a_0 + a_{-1}z, \quad z \neq 0.$$

Just as was argued for f, the conformal map g of $\mathbb{C}\setminus\{0\}$ can have at worst a simple pole at 0, and so the coefficients a_i must vanish for i greater than 1.

At this stage, we have shown that f must be of the form

$$f(z) = \frac{a_{-1}}{z} + a_0 + a_1 z, \quad z \neq 0,$$

for some coefficients a_{-1}, a_0 and a_1. The solution set of the equation $f(z) = 0$, $z \neq 0$, becomes the solution set of the equation

$$a_{-1} + a_0 z + a_1 z^2 = 0, \quad z \neq 0,$$

which is to have at most one solution since f is univalent. If $a_{-1} \neq 0$ then any solution of $a_{-1} + a_0 z + a_1 z^2 = 0$ is non-zero and so a_1 must vanish in order to avoid there being two non-zero solutions or a double zero. If $a_{-1} = 0$, the quadratic reduces to $0 = a_0 z + a_1 z^2 = z(a_0 + a_1 z)$, which has at most one non-zero solution.

In summary, f is given as either $f(z) = a_{-1}/z + a_0$ or $f(z) = a_0 + a_1 z$. To avoid f being constant (and so not univalent), we must have $a_{-1} \neq 0$ in the first

6.7 Conformal Mapping of Annuli

instance and $a_1 \neq 0$ in the second. If so, mappings of either form are univalent on the plane punctured at 0, and map $\mathbb{C} \setminus \{0\}$ onto the plane punctured at a_0. □

Now we turn to the proof of Theorem 6.13. We will shortly give a complete characterisation of the conformal self-maps of an annulus $A(r, R)$ ($0 < r < R < \infty$). In the meantime, note that

$$f(z) = \frac{rR}{z}, \quad z \in A(r, R),$$

is a conformal self-map of the annulus $A(r, R)$ that sends the boundary circle $C(0, R)$ onto $C(0, r)$, and vice versa.

Proof of Theorem 6.13 We consider an annulus $A_1 = A(r_1, R_1)$ where $0 \leq r_1 < R_1 \leq \infty$ and attempt to classify annuli $A_2 = A(r_2, R_2)$, $0 \leq r_2 < R_2 \leq \infty$, that are conformally equivalent to A_1. There are four cases (really three) depending on whether r_1 is zero or positive and whether R_1 is finite or infinite.

Case 1: $r_1 = 0$ *and* $R_1 = \infty$ In this case, $A_1 = A(0, \infty)$, the plane punctured at 0. By Proposition 6.4, the only annulus (in fact, up to translation, the only domain in the complex plane) to which the annulus $A(0, \infty)$ is conformally equivalent is itself. Thus $\mathscr{A}_1$, as given by (6.31a), is a conformal equivalence class of annuli consisting of the single annulus $A(0, \infty)$.

Case 2: $r_1 = 0$ *and* $R_1 < \infty$ In this case $A_1 = A(0, R_1)$, $R_1 < \infty$, so that $A_1 \in \mathscr{A}_\infty$. We consider a conformal mapping f from A_1 onto an annulus $A_2 = A(r_2, R_2)$ and wish to show that A_2 also belongs to $\mathscr{A}_\infty$.

As we have just seen in Case 1, the annulus A_2 cannot be the annulus $A(0, \infty)$. Thus, if $R_2 = \infty$ then $r_2 > 0$ and so A_2 has the form $A_2 = (r_2, \infty)$, which again belongs to $\mathscr{A}_\infty$ (see 6.31c). The other possibility is that $R_2 < \infty$ and that A_2 has the form $A(r_2, R_2)$. To show that $A_2 \in \mathscr{A}_\infty$ in this case as well, we need to show that $r_2 = 0$ which we now do.

The conformal map f of A_1 onto A_2 has an isolated singularity at 0. Since A_2 is a bounded domain and functions with an essential singularity or a pole are unbounded in the neighbourhood of the singularity, f must have a removable singularity at 0. With this singularity removed, f is analytic in the disk $D(0, R_1)$, univalent in $A_1 = D'(0, R_1)$, and $f(0) = a$, say. These properties guarantee that $a \notin A(r_2, R_2)$. Otherwise, $a \in A(r_2, R_2)$ and there is also some $z \in A_1$ for which $f(z) = a$. Choose δ positive so that the disks $D(0, \delta)$ and $D(z, \delta)$ don't intersect. By the Open Mapping Theorem, both $f(D(0, \delta))$ and $f(D(z, \delta))$ are then open subsets of $f(D(0, R_1))$ that contain a, and so these images have points in common other than a. Any such point will be taken on twice by f in $A(0, R_1)$, once in the punctured disk $D'(0, \delta)$ and once in the punctured disk $D'(z, \delta)$, contradicting the univalence of f on $A(0, R_1)$.

We have shown that f is analytic (in fact, univalent) in the disk $D(0, R_1)$ with range $A(r_2, R_2) \cup \{a\}$. In particular, $A(r_2, R_2) \cup \{a\}$ must be a domain. The only situation in which adding an extra point to an annulus $A(r_2, R_2)$ can result in a domain is if $r_2 = 0$ and the additional point is 0. We conclude, moreover, that f extends to a conformal map of the disk $D(0, R_1)$ to the disk $D(0, R_2)$.

Case 3: $r_1 > 0$ *and* $R_1 = \infty$ In this case $A_1 = A(r_1, \infty)$, $r_1 > 0$, so that $A_1 \in \mathscr{A}_\infty$. Suppose that an annulus $A_2 = A(r_2, R_2)$ is conformally equivalent to A_1. Again, we wish to show that A_2 also belongs to $\mathscr{A}_\infty$. The inversion $M(z) = 1/z$ maps A_1 conformally onto $\tilde{A}_1 = A(0, 1/r_1)$. Since A_2 is conformally equivalent to A_1, which is in turn conformally equivalent to $\tilde{A}_1$, we have A_2 conformally equivalent to $\tilde{A}_1$. But the annulus $\tilde{A}_1$ comes under Case 2, and we conclude from that case that A_2 must be an annulus in the class $\mathscr{A}_\infty$, as required.

Case 4: $r_1 > 0$ *and* $R_1 < \infty$ Set $\lambda = R_1/r_1$, so that $1 < \lambda < \infty$ and $A_1 = A(r_1, R_1)$ belongs to the class $\mathscr{A}_\lambda$. Now suppose that f is a conformal mapping of A_1 onto $A_2 = A(r_2, R_2)$ where $0 \leq r_2 < R_2 \leq \infty$. If r_2 was equal to 0, or if R_2 was equal to infinity, then A_2 would be an annulus covered under one of the previous three cases and so A_1, being conformally equivalent to A_2, would belong to either $\mathscr{A}_1$ or $\mathscr{A}_\infty$, which it doesn't. Thus, $r_2 > 0$ and $R_2 < \infty$. We want to show that, in fact, $A_2 \in \mathscr{A}_\lambda$, that is that $R_2/r_2 = \lambda$.

Both annuli A_1 and A_2 are bounded geometrically simple domains in the sense of Definition 6.4 and so f extends to a homeomorphism of the closure $\text{cl}(A_1)$ of A_1 onto the closure $\text{cl}(A_2)$ of A_2. In particular, f sends the boundary circle $C(0, r_1)$ of A_1 onto either the boundary circle $C(0, r_2)$ or $C(0, R_2)$ of A_2. If the latter, that is if $f(C(0, r_1)) = C(0, R_2)$, we replace f by the map $r_2 R_2/f(z)$ which is again a conformal map of A_1 onto A_2 but now $C(0, r_1)$ is mapped to $C(0, r_2)$, and $C(0, R_1)$ is mapped to $C(0, R_2)$.

In general, the reflection of the annulus $A(r, R)$ in the circle $C(0, R)$ is the annulus $A(R, R^2/r)$ (see (2.16)). The annulus $A(r, R)$ together with its reflection in the circle $C(0, R)$ (and the circle itself) is then the annulus $A(r, R^2/r)$. By the Schwarz Reflection Principle, Theorem 2.2, f can be extended by reflection in the circle $C(0, R_1)$ to a conformal map of $A(r_1, R_1^2/r_1)$ onto the annulus $A(r_2, R_2^2/r_2)$. The map extends continuously to the boundary of the larger annulus with inner and outer boundary circles corresponding as before.

Now we can reflect again, this time in the circle $C(0, R_1^2/r_1)$. We find that f can be extended to a conformal map of the annulus $A(r_1, R_1^4/r_1^3)$ onto the annulus $A(r_2, R_2^4/r_2^3)$. This process of reflection in the outer boundary circle of the annulus can be repeated indefinitely. After n reflections, f is extended as a conformal map of the annulus $A(r_1, R_1^{2^n}/r_1^{2^n-1})$ onto the annulus $A(r_2, R_2^{2^n}/r_2^{2^n-1})$. Since $r_i < R_i$, $i = 1, 2$, we have $R_i^{2^n}/r_i^{2^n-1} \to \infty$ as $n \to \infty$, so that f extends as a conformal map of the annulus $A(r_1, \infty)$ onto the annulus $A(r_2, \infty)$.

We now make one final reflection, but this time in the inner circle $C(0, r_1)$. In this way, f extends to a conformal map of the punctured plane $A(0, \infty)$ onto itself. By Proposition 6.4, the form of f is extremely restricted: there is a non-zero complex number b such that either $f(z) = bz$, $z \in \mathbb{C} \setminus \{0\}$, or $f(z) = b/z$, $z \in \mathbb{C} \setminus \{0\}$. In particular, the image of the annulus $A_1 = A(r_1, R_1)$ is the annulus $A(|b|r_1, |b|R_1)$

6.7 Conformal Mapping of Annuli

in the first case and the annulus $A(|b|/R_1, |b|/r_1)$ in the second case (we are in the first case, but no matter). Since $f(A_1) = A_2 = A(r_2, R_2)$, we have either

$$\frac{R_2}{r_2} = \frac{|b|R_1}{|b|r_1} \quad \text{or} \quad \frac{R_2}{r_2} = \frac{|b|/r_1}{|b|/R_1}$$

and so $R_2/r_2 = R_1/r_1$, showing that A_1 and A_2 belong to the same class $\mathscr{A}_\lambda$, as required. $\square$

Though there are shorter proofs of the above result, the proof presented makes nice use of the Schwarz Reflection Principle, which is an elementary result. The method also yields a little more, in that an examination of the proof shows that we have effectively found the automorphism group of a general annulus. We state this result separately, as it is important in its own right, even if the majority of the work needed to establish it has already been carried out.

Theorem 6.14 *Consider an annulus $A = A(r, R)$ where $0 \le r < R \le \infty$.*

If $A = A(0, \infty)$ then $\mathrm{Aut}(A)$ consists of all mappings of the form $f(z) = bz$, $b \ne 0$, or of the form $f(z) = b/z$, $b \ne 0$.
If $A \in \mathscr{A}_\infty$ then $\mathrm{Aut}(A)$ consists only of the rotations $f(z) = e^{i\theta}z$, $\theta \in [0, 2\pi)$.
If $A \in \mathscr{A}_\lambda$, $1 < \lambda < \infty$, with $R/r = \lambda$ then $\mathrm{Aut}(A)$ consists of the rotations $f(z) = e^{i\theta}z$, $\theta \in [0, 2\pi)$, together with all maps of the form

$$f(z) = \frac{rRe^{i\theta}}{z}, \quad \theta \in [0, 2\pi).$$

Proof The statement for $A = A(0, \infty) = \mathbb{C} \setminus \{0\}$ is effectively Proposition 6.4.

If $A \in \mathscr{A}_\infty$ with $A = A(0, R)$ and $R < \infty$, the argument for Case 2 in the proof of Theorem 6.13 shows that any automorphism of A extends to an automorphism f of $D(0, R)$ for which $f(0) = 0$. We have a complete description of the automorphisms of the disk (Theorem 4.2 and Exercise 4.7) from which it follows that any such automorphism is a rotation, as claimed.

If $A \in \mathscr{A}_\infty$ with $A = A(r, \infty)$ and $r > 0$, then $g(z) = 1/f(1/z)$ is an automorphism of the annulus $A(0, 1/r)$ if f is an automorphism of A. Since such an automorphism g is, as we have just seen, necessarily a rotation, we conclude that f itself is a rotation.

So far we have dealt with the conformal equivalence classes $\mathscr{A}_1$ and $\mathscr{A}_\infty$. Finally, suppose that $1 < \lambda < \infty$ and that $A \in \mathscr{A}_\lambda$, say $A = A(r, R)$ with $R/r = \lambda$. We saw from the argument for Case 4 in the proof of Theorem 6.13 that any automorphism f of A can be extended to an automorphism of the plane punctured at 0. Thus, f either has the form $f(z) = bz$, $b \ne 0$, or the form $f(z) = b/z$, $b \ne 0$. This alone is not sufficient, as f must map A onto itself. If $f(z) = bz$ this extra requirement forces $|b| = 1$, so that f is a rotation. In the case $f(z) = b/z$, the image of A under f is the annulus $A(|b|/R, |b|/r)$, so we need $|b|/R = r$ and $|b|/r = R$, that is $|b| = rR$. $\square$

We are far from dealing with general doubly connected domains in this brief discussion. We have, however, met all the main characters since it transpires that any doubly connected domain is conformally equivalent to some annulus (see [3], for example). Assuming this, we can associate a number λ to any doubly connected domain that is conformally equivalent to an annulus in $\mathscr{A}_\lambda$, $1 < \lambda < \infty$, this being 'R/r' for the particular annulus $A(r, R)$ to which the domain is conformally equivalent. This number λ is a conformally invariant quantity—that is, it is the same for any two conformally equivalent doubly connected domains. For reasons that we won't go into here, the conformal invariant is taken to be

$$\frac{1}{2\pi} \log\left(\frac{R}{r}\right),$$

or $1/(2\pi) \log \lambda$ rather than λ itself, and is called the *modulus* of the doubly connected domain.

Here is one example in which this result can be verified by finding an explicit conformal map onto an annulus.

Exercise 6.10 For $a \in (0, 1/2]$, let D be the domain $\mathbb{D} \setminus \overline{D(a, a)}$, the complement of the closed disk centre a and radius a relative to the unit disk.

▷ For $0 < a < 1/2$, find a conformal map of D onto a suitable annulus $A(r_0, 1)$. Determine the modulus of D in terms of a.

▷ For $a = 1/2$, the domain D is simply connected. Find a conformal map of D onto the upper half-plane in this case.

Chapter 7
Runge's Theorem and Further Characterisations of Simply Connected Domains

Our main objective in this chapter is to prove, as promised in Chap. 6, the characterisation of simply connected domains as those domains whose complement relative to the extended complex plane $\mathbb{C}_\infty$ is connected. Several other conditions each equivalent to simple connectivity are listed as Theorem 6.8. All but the last of these is analytic in nature. Even though the conditions 'D is simply connected' and '$\mathbb{C}_\infty \setminus D$ is connected' are both topological in nature, the route to proving their equivalence passes through the analytic in the form of Theorem 6.8. The proof of this equivalence also brings in some new ideas, first and foremost Runge's Theorem on approximation by rational functions and on approximation by polynomials. We also take this opportunity to firm up the concept of winding numbers. A simply connected domain is one for which no closed curve in the domain winds around any point in its complement, which is another way to express the idea that a simply connected domain 'has no holes'.

We begin, then, with Runge's Theorem.

7.1 Runge's Theorem

By a famous theorem of Weierstrass, any real-valued continuous function on a closed interval $[a, b]$ can be uniformly approximated by polynomials. That is, if f is continuous on $[a, b]$ and ε is any positive number there is a polynomial $p(x)$ such that $|f(x) - p(x)| \leq \varepsilon$ for all x in $[a, b]$. See, for example, [24, §35].

The corresponding result in the complex setting is Runge's Theorem which, like Weierstrass's Approximation Theorem, dates to 1885. In this setting, we are given a function f that is analytic on a neighbourhood[1] of a compact set K and want to approximate f uniformly on K by polynomials. At first sight, this is not always

---
[1] By a 'neighbourhood' of a compact set K we mean any domain D that contains K.

© The Author(s), under exclusive license to Springer Nature Switzerland AG 2024
T. Carroll, *Geometric Function Theory*, Springer Undergraduate Mathematics Series, https://doi.org/10.1007/978-3-031-73727-5_7

possible. For example, $f(z) = 1/z$ is analytic in a neighbourhood, say the annulus $D = \{z : \frac{1}{2} < |z| < 2\}$, of the circle $K = C(0, 1)$. Suppose that $1/z$ was a uniform limit of polynomials on K, in that P_n are polynomials for which $P_n \to f$ uniformly on $C(0, 1)$. The point is that $\int_{C(0,1)} \frac{1}{z}\, dz = 2\pi i$ and that $\int_{C(0,1)} P(z)\, dz = 0$ for any polynomial P (the latter by Cauchy's Theorem or direct computation). Then we would have, by the ML-inequality,

$$2\pi = \left| \int_{C(0,1)} \frac{1}{z}\, dz - \int_{C(0,1)} P_n(z)\, dz \right| \leq 2\pi \max \left\{ \left| \frac{1}{z} - P_n(z) \right| : z \in C(0, 1) \right\},$$

and a contradiction arises if, say, $\left| \frac{1}{z} - P_n(z) \right| \leq \frac{1}{2}$ on $C(0, 1)$. It transpires that this problem disappears if we are willing to expand the collection of approximating functions from polynomials to rational functions (in fact, $f(z) = 1/z$ in the above example is itself a rational function). This is our first version of Runge's Theorem.

Theorem 7.1 (Runge's Theorem—Rational Approximation) *Suppose that f is analytic on a neighbourhood D of a compact set K. Then, f can be uniformly approximated on K by rational functions all of whose poles lie in $D \setminus K$.*

The first step in the proof is the following proposition.

Proposition 7.1 *Suppose that K is a compact subset of a domain D. Then there exist finitely many horizontal or vertical directed line segments γ_i, $i = 1, \ldots N$, all of equal length and lying in $D \setminus K$, such that*

$$f(z) = \sum_{i=1}^{N} \frac{1}{2\pi i} \int_{\gamma_i} \frac{f(w)}{w - z}\, dw \tag{7.1}$$

for each function f analytic in D and for each $z \in K$.

Proof Set $d = \operatorname{dist}(K, D^c)$, the distance from K to the complement of D. Since K is compact, since D^c is closed and since K and D^c are disjoint, d is a positive number.

Divide the whole complex plane horizontally and vertically into a grid of closed, solid squares of side length l where $l < d/\sqrt{2}$ ($l = d/(2\sqrt{2})$ would be fine). To be clear, each square Q consists of its interior square $\operatorname{int}(Q)$ together with its four boundary sides ∂Q. Two adjacent squares share one side.

Let $\mathscr{Q} = \{Q_1, Q_2, \ldots, Q_M\}$ be the collection of those closed squares that have non-empty intersection with K—such a square may have only a single point on one of its sides in common with K, see Fig. 7.1. There are only finitely many such squares since K, being compact, is bounded. Each of these squares lies inside D. Otherwise a square would contain both a point of K and a point of D^c, and these would be at most a distance $\sqrt{2}l$ apart. But this is less than d by choice of the sidelength l of the squares in the grid.

7.1 Runge's Theorem

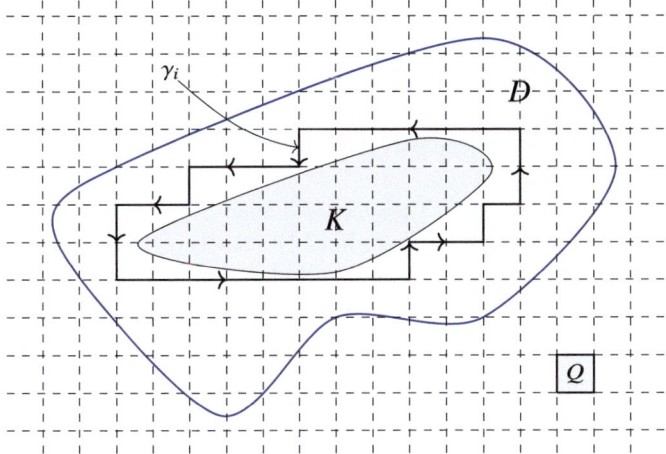

Fig. 7.1 Illustrative diagram for the proof of Proposition 7.1, showing domain D, compact subset K, grid with a typical square Q, and the line segments γ_i in (7.1)

Now suppose that f is analytic in D. Suppose that z is a point in K and in the interior of one of the squares Q_k in the collection $\mathscr{Q}$. Then, by Cauchy's Integral Formula,

$$f(z) = \frac{1}{2\pi i} \int_{\partial Q_k} \frac{f(w)}{w-z} \, dw.$$

None of the other squares $Q_i \in \mathscr{Q}, i \neq k$, contains z and so

$$\frac{1}{2\pi i} \int_{\partial Q_i} \frac{f(w)}{w-z} \, dw = 0, \quad i \neq k.$$

Thus, for each z in $K \setminus \bigcup_{i=1}^{M} \partial Q_i$,

$$f(z) = \sum_{i=1}^{M} \frac{1}{2\pi i} \int_{\partial Q_i} \frac{f(w)}{w-z} \, dw. \tag{7.2}$$

If γ is a common side of two of these squares, say Q_i and Q_j, then the contribution to the integral over ∂Q_i in (7.2) from the side γ cancels with the contribution to the integral over ∂Q_j from γ, as these contributions are taken in opposite directions. The right hand side of (7.2) therefore simplifies to a sum of integrals over those directed line segments that arise as the side of one and only one square in the collection $\mathscr{Q}$. Any such side cannot meet K as, if it did, both squares in the grid of which it is a common side would also meet K and hence both would belong to the collection $\mathscr{Q}$. Denoting those directed line segments that arise as the side of one and only one square in the collection $\mathscr{Q}$ by γ_i, $i = 1, \ldots N$, we obtain (7.1) for all points z in K that are interior points of a square in the collection $\mathscr{Q}$. By continuity, the identity (7.1) extends to all points z in K. □

It is perhaps believable, though by no means immediately obvious, that the line segments γ_i in (7.1) form closed polygonal curves. This requires proof, to which we return later as Theorem 7.3.

In the meantime, Runge's Theorem—Rational Approximation can now be proved by showing that a typical term in (7.1), namely $\int_\gamma f(w)/(w-z)\,\mathrm{d}w$, can be uniformly approximated by rational functions (since a finite sum of such terms can then be similarly approximated). That this can be achieved is the subject of the following proposition.

Proposition 7.2 *Suppose that D is a domain, that K is a compact subset of D and that γ is a line segment lying entirely in $D \setminus K$. Suppose that f is analytic on D. Then there is a sequence of rational functions $\{R_n(z)\}_{n=1}^\infty$ whose only poles are on γ and that converges uniformly to $\int_\gamma f(w)/(w-z)\,\mathrm{d}w$ on K. That is,*

$$\left| \int_\gamma \frac{f(w)}{w-z}\,\mathrm{d}w - R_n(z) \right| \to 0 \text{ as } n \to \infty \tag{7.3}$$

uniformly for z in K.

Proof Let $\gamma(t)$, $t \in [0, 1]$, be a parameterisation of γ. Then, for $z \in K$,

$$\int_\gamma \frac{f(w)}{w-z}\,\mathrm{d}w = \int_0^1 F(z,t)\,\mathrm{d}t \text{ where } F(z,t) = \frac{f(\gamma(t))}{\gamma(t)-z}\gamma'(t).$$

The expression $|\gamma(t) - z|$ is bounded away from 0 for z in K, as the trace of γ is a closed set that does not meet K. The function $F(z, t)$ is therefore continuous on $K \times [0, 1]$, in fact uniformly continuous since $K \times [0, 1]$ is a compact set. In particular, given ε positive, there is a positive δ such that

$$\left| F(z, t_1) - F(z, t_2) \right| \leq \varepsilon \text{ whenever } z \in K \text{ and } |t_1 - t_2| \leq \delta,\ t_1, t_2 \in [0, 1]. \tag{7.4}$$

Now approximate the integral $\int_0^1 F(z, t)\,\mathrm{d}t$ by Riemann sums as in, for a fixed n,

$$\int_0^1 F(z, t)\,\mathrm{d}t \approx \sum_{k=1}^n \frac{1}{n} F(z, k/n).$$

Set $t_k = k/n$ for $0 \leq k \leq n$. For each n, the function

$$R_n(z) = \sum_{k=1}^n \frac{1}{n} F(z, t_k) = \sum_{k=1}^n \frac{1}{n} \frac{f(\gamma(t_k))}{\gamma(t_k) - z} \gamma'(t_k)$$

7.1 Runge's Theorem

is a rational function of z (it is a sum of terms of the form $\alpha/(\beta - z)$) with its only poles at $\gamma(t_k)$, $k = 1, \ldots, n$, all of which lie on γ. Moreover, if $1/n < \delta$ and $z \in K$,

$$\left| \int_0^1 F(z,t)\,dt - R_n(z) \right| = \left| \sum_{k=1}^n \int_{t_{k-1}}^{t_k} \left(F(z,t) - F(z,t_k) \right) dt \right|$$

$$\leq \sum_{k=1}^n \int_{t_{k-1}}^{t_k} \left| F(z,t) - F(z,t_k) \right| dt \leq \sum_{k=1}^n \int_{t_{k-1}}^{t_k} \varepsilon\,dt = \varepsilon,$$

where (7.4) was used at the third step. This shows that (7.3) holds and the sequence $\{R_n(z)\}_{n=1}^\infty$ of rational functions, all of which only have poles on γ, converges uniformly to $\int_\gamma f(w)/(w-z)\,dw$ on K. □

Propositions 7.1 and 7.2 together prove Runge's Theorem—Theorem 7.1—on approximation of analytic functions by rational functions. In the current context, however, we are particularly interested in approximation of an analytic function by polynomials. There is one familiar setting in which this is always possible, namely that of an analytic function f in the unit disk $\mathbb{D}$. Such a function is given by a power series $f(z) = \sum_{k=0}^\infty a_k z^k$ convergent (at least) for $|z| < 1$ and, on $K = \{z : |z| \leq r\}$ ($0 < r < 1$), the sequence $\{P_n(z)\}$ of partial sums, where $P_n(z) = \sum_{k=0}^n a_k z^k$, is a sequence of polynomials that converges uniformly to f on the compact set K. Whereas uniform polynomial approximation of analytic functions on a compact subset of a disk is always possible, this is not always possible (as we have seen) in the case of analytic functions on a compact subset of an annulus. It transpires that the relevant difference between these two settings is that the complement of the compact set is connected in the case of the disk whereas it is not in the case of the annulus. This distinction is the key to Runge's Theorem—Polynomial Approximation.

Theorem 7.2 (Runge's Theorem—Polynomial Approximation) *Suppose that K is compact and that $K^c = \mathbb{C} \setminus K$ is connected. Suppose that D is a neighbourhood of K and that f is analytic on D. Then there is a sequence of polynomials that converges uniformly to f on K.*

Proof By Runge's Theorem—Rational Approximation, there is a sequence of rational functions $\{R_n(z)\}_1^\infty$ that converges uniformly to f on K. Specifically, by Proposition 7.1, f can first be written as a finite sum of terms of the form $\int_\gamma f(w)/(w-z)\,dw$ where γ is a directed line segment in $D \setminus K$. Then, from the proof of Proposition 7.2, each of these terms can, in turn, be approximated by finite sums of the form $\alpha/(\beta - z)$ where β lies on γ. Thus it suffices to show that, for each point z_0 in $D \setminus K$, the rational function $h(z) = 1/(z_0 - z)$ can be uniformly approximated on K by polynomials. This we now set out to do.

Fix z_0 in $D \setminus K$. To begin, suppose that $K \subset D(0, M)$ and choose a point z_* with, say, $|z_*| = 2M$ so that $z_* \in K^c$. Since K^c is connected and open it is path connected and so there is a smooth curve γ in K^c that joins z_0 to z_*. Since the trace of γ is closed and K is compact, the distance from γ to K is positive. Thus, $\rho = \frac{1}{2}\text{dist}(\gamma, K)$ is positive.

On the disk $D(0, M)$, and in particular on K, $g(z) = 1/(z_* - z)$ can be expanded as a power series[2]

$$g(z) = \frac{1}{z_* - z} = \frac{1}{z_*} \frac{1}{1 - z/z_*} = \sum_{n=0}^{\infty} \frac{z^n}{z_*^{n+1}},$$

this since $|z/z_*| \leq 1/2$ on $D(0, M)$. Moreover, $1/(z_* - z)$ is the uniform limit on K of the partial sums of this geometric series. That is, $1/(z_* - z)$ is the uniform limit of polynomials on K.

Let $z_0, z_1, z_2, \ldots, z_n = z_*$ be successive points on γ travelling from z_0 to z_*, and suppose that $|z_{i-1} - z_i| \leq \rho$ for each i, $1 \leq i \leq n$. We will show that if $1/(z_i - z)$ is the uniform limit of polynomials on K then so is $1/(z_{i-1} - z)$. This inductive step together with the knowledge that $1/(z_n - z)$ is the uniform limit of polynomials on K allows us to conclude that $1/(z_0 - z)$ is the uniform limit of polynomials on K, as required.

Suppose, then, that $1/(z_i - z)$ is the uniform limit of polynomials on K for some i between 1 and n. For $z \in K$,

$$\frac{1}{z_{i-1} - z} = \frac{1}{z_i - z} \times \frac{z - z_i}{z - z_{i-1}} = \frac{1}{z_i - z} \times \frac{z - z_i}{(z - z_i) - (z_{i-1} - z_i)}$$

and

$$\frac{z - z_i}{(z - z_i) - (z_{i-1} - z_i)} = \frac{1}{1 - \frac{z_{i-1} - z_i}{z - z_i}} = \sum_{n=0}^{\infty} \left(\frac{z_{i-1} - z_i}{z - z_i}\right)^n.$$

This geometric series is convergent since $|z_{i-1} - z_i| \leq \rho = \frac{1}{2}\text{dist}(\gamma, K) \leq \frac{1}{2}|z_i - z|$. The partial sums $\{P_n\}$ of this geometric series converge uniformly to $(z - z_i)/(z - z_{i-1})$ on K. Taken together with a sequence $\{Q_n\}$ of polynomials that converges, by assumption, uniformly to $1/(z_i - z)$ on K, we find that $\{P_n Q_n\}$ is a sequence of polynomials that converges uniformly to $1/(z_{i-1} - z)$ on K. □

Exercise 7.1 Verify the statement in the first paragraph of the proof of Theorem (7.2) that it is sufficient to approximate, for z_0 in $D \setminus K$, the rational function $h(z) = 1/(z_0 - z)$ by polynomials, uniformly on K.

Exercise 7.2 Verify the last statement in the proof of Theorem 7.2: if $\{P_n\}$ and $\{Q_n\}$ are sequences of polynomials that converge uniformly to analytic functions f and g respectively on the compact set K, then the sequence of polynomials $\{P_n Q_n\}$ converges uniformly to fg on K.

---

[2] On the unit disk $\mathbb{D}$, the analytic function $1/(1 - z)$ has the geometric series expansion $\sum_{n=0}^{\infty} z^n$.

7.2 A General Integral Formula

The integral formula (7.1) can be rewritten in a more familiar form by showing that the line segments occurring therein join together to form closed curves. This is not immediately obvious since the arrangement of the squares in the collection $\mathscr{Q}$ from the proof of Proposition 7.1 could potentially be quite complicated. We follow in outline the argument given by Remmert in [23, §12.4] who credits Burckel [5] with this form of Proposition 7.1.

By a *polygonal closed curve* $C = [p_1 p_2 \ldots p_n p_1]$ in this context we mean a curve C made up of a succession of horizontal or vertical line directed segments $[p_1, p_2], [p_2, p_3], \ldots, [p_{n-1}, p_n], [p_n, p_1]$ where each segment has the same length l, say. We call the curve a *step polygonal closed curve* if the points p_1 to p_n are distinct. A step polygonal closed curve is automatically simple. To see this, assume without loss of generality that $p_1 = 0$. Since all line segments have the same length l, the horizontal and the vertical coordinates of all points p_k, $1 \le k \le n$, are multiples of l and so lie on the lattice $\Lambda = \{(al, bl) : a, b \in \mathbb{Z}\}$. Moreover, these points p_k are the only points on the curve that lie on this lattice. The curve cannot self-intersect at one of the points p_k as these are distinct. Any point z of self-intersection is therefore an internal point of two line segments $[p_{i-1}, p_i]$ and $[p_{j-1}, p_j]$, $i \ne j$. If both line segments are horizontal or are both vertical then, as all endpoints p_k lie on the lattice Λ, they must coincide and this is not possible since the points p_1 to p_n are distinct. The line segments $[p_{i-1}, p_i]$ and $[p_{j-1}, p_j]$ therefore have opposite orientations. Take $[p_{i-1}, p_i]$ to be the horizontal line segment. Since z has the same vertical coordinate as p_i, its vertical coordinate is a multiple of l. Since z has the same horizontal coordinate as p_j, its horizontal coordinate is a multiple of l. That is, z is a point on the lattice Λ which is a contradiction as z is not one of the points p_k. The step polygonal closed curve C is therefore simple.

The following development of Proposition 7.1 can be viewed as an extension of the Cauchy Integral Formula, Theorem 6.2.

Theorem 7.3 *Suppose that K is a compact subset of a domain D. Then there exist finitely many step polygonal closed curves $C_1, \ldots, C_M$ lying in $D \setminus K$ such that*

$$f(z) = \sum_{n=1}^{M} \frac{1}{2\pi i} \int_{C_n} \frac{f(w)}{w - z} \, dw \qquad (7.5)$$

for each function f analytic in D and for each $z \in K$.

Proof We are in the setting of Proposition 7.1 so that (7.1) holds for any f analytic in D and any z in K, this in terms of the collection Γ of directed line segments γ_i, $i = 1, \ldots N$, constructed in the course of proving Proposition 7.1. The directed line segments all have equal length, lie in $D \setminus K$ and each is either horizontal or vertical. If γ is the directed line segment $[a, b]$, we refer to a as its initial point and to b as its terminal point. We now show that the number of times a point c occurs as an initial

Fig. 7.2 Diagrams showing that the number of directed line segments of which c is the initial point equals the number of directed line segments of which c is the terminal point

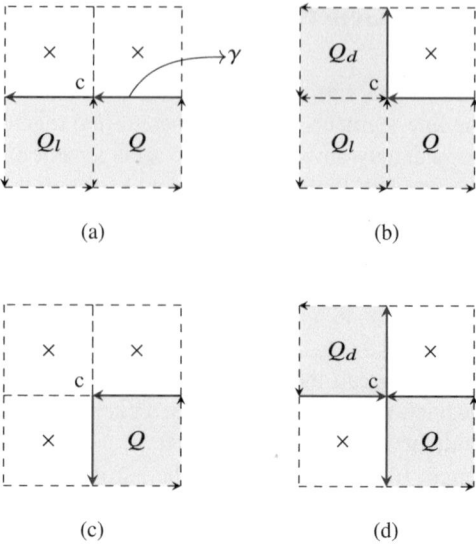

point of one of the directed line segments in Γ equals the number of times c occurs as a terminal point.

To do so, we return to the collection $\mathscr{Q}$ of squares Q_i from the proof of Proposition 7.1. The directed line segments γ_i are those sides of squares in the collection $\mathscr{Q}$ that belong to a single square in the collection. Let γ be any one of these directed line segments, and let Q be the square in the collection $\mathscr{Q}$ of which γ is a side. Up to rotation, the situation is as shown in Fig. 7.2: Q is a square in the collection $\mathscr{Q}$ and the side γ is the top side of Q directed from right to left. The square directly above Q does not belong the collection $\mathscr{Q}$ as γ is the side of only one square in this collection. This square is marked with an '×' in Fig. 7.2. We first focus on the terminal point c of γ. There are then four cases to consider depending on the status of the other two squares having c as a corner. In Fig. 7.2a, b and c, the point c is a terminal point only of γ and is the initial point of one other directed line segment. In Fig. 7.2d, the point c is the terminal point of two directed line segments and the initial point of two directed line segments as well. A similar analysis shows that if c is an initial point of one of the directed line segments γ then the number of directed line segments of which it is an initial point is again matched by the number of directed line segments of which it is a terminal point. It follows that, given any directed line segment γ_i in Γ, there is another directed line segment γ_j in Γ whose initial point is the same as the terminal point of γ_i, so that γ_j joins to γ_i.

Now choose any directed line segment $\gamma_1 \in \Gamma$, writing p_0 for its initial point and p_1 for its terminal point. Choose a directed line segment γ_2 from Γ that joins to γ_1 at p_1, and write p_2 for its terminal point. Again choose a directed line segment γ_3 from Γ that has initial point p_2 and terminal point p_3. Continuing in this manner, we produce a sequence of points $\{p_k\}_{k=0}^{\infty}$. We stop the first time that a point is repeated, say $p_n = p_i$, $0 \leq i < n$. This must happen since there are only finitely many

7.3 Further Characterisations of Simply Connected Domains

initial/terminal points of directed line segments in Γ. Then, $[p_i p_{i+1} \ldots p_{n-1} p_i]$ is a step polygonal closed curve.

Removing from Γ all directed line segments in this step polygonal closed curve, the remaining collection of directed line segments Γ' still has the property that the number of times a point c occurs as an initial point of one of the directed line segments in Γ' equals the number of times c occurs as a terminal point of a directed line segment in Γ'. Thus we may continue. Eventually we exhaust the collection Γ so that, in this way, the collection of directed line segments Γ constructed in the proof of (7.1) can be arranged to produce a finite number of step polygonal closed curves. The integral representation (7.5) now follows from (7.1). □

Exercise 7.3 Show that, if K is a compact subset of a domain D and if C_n, $n = 1, \ldots, M$, are closed curves in $D \setminus K$ for which (7.5) holds then

$$\sum_{n=1}^{M} \int_{C_n} f(w) \, dw = 0$$

for any function f analytic in D. [For f analytic in D and α in K, apply (7.5) to the function $f(z)(z - \alpha)$.]

7.3 Winding Numbers and Further Characterisations of Simply Connected Domains

Consider a closed curve γ parameterised by $\gamma(t)$, $t \in [0, 1]$, and a point a not on γ. The number of times that the curve winds around a can be measured by the net change in the argument of $\gamma(t) - a$, that is $\big[\arg(\gamma(1) - a) - \arg(\gamma(0) - a)\big]/(2\pi)$. This is known as the *winding number of the curve γ about a* and is denoted by $w(\gamma; a)$. The argument changes by 2π each time the curve winds around a anticlockwise and by -2π each time it winds around a in the clockwise direction. Since, formally, $\log z = \log|z| + i \arg z$, we can view the argument as the imaginary part of $\log z$ and we can write the winding number as

$$\frac{1}{2\pi i}\big[\log(\gamma(1) - a) - \log(\gamma(0) - a)\big] = \frac{1}{2\pi i} \int_0^1 \frac{\gamma'(t)}{\gamma(t) - a} \, dt.$$

Here, $\log|\gamma(1) - a| = \log|\gamma(0) - a|$ since the curve is closed and, formally, $\gamma'(t)/(\gamma(t) - a) = \frac{d}{dt}(\log(\gamma(t) - a))$.

These formal considerations motivate the definition of the winding number of the closed curve γ about a point a that is not on γ as

$$w(\gamma; a) = \frac{1}{2\pi i} \int_\gamma \frac{dz}{z - a}.$$

We first show that $w(\gamma; a)$ is, in fact, an integer. Again parametrising γ by $\gamma(t)$, $t \in [0, 1]$, set

$$\alpha(s) = \int_0^s \frac{\gamma'(t)}{\gamma(t) - a} \, dt,$$

so that $\alpha(0) = 0$ and $\alpha(1) = 2\pi i w(\gamma; a)$. Since (apart for possibly finitely many exceptions), $\alpha'(s) = \gamma'(s)/(\gamma(s) - a)$, setting

$$\beta(s) = (\gamma(s) - a) e^{-\alpha(s)},$$

we find that $\beta'(s) = 0$ (again with possibly finitely many exceptions). As a consequence, β is piecewise constant, and therefore constant since it is continuous. Since $\beta(0) = \gamma(0) - a$, $\beta(1) = (\gamma(1) - a) \exp(-\alpha(1))$ and $\gamma(0) = \gamma(1)$, we find that $\exp(-\alpha(1)) = 1$ so that $\alpha(1)$ is an integer multiple of $2\pi i$. Hence, $w(\gamma; a)$ is an integer.

Being integer-valued, the winding number $w(\gamma; a)$ is necessarily constant on each component of $\mathbb{C} \setminus \gamma$. Since the curve lies in a bounded region of the plane, there is only one unbounded component of $\mathbb{C} \setminus \gamma$. In this unbounded component, the winding number is identically zero since $w(\gamma; a) \to 0$ as $|a| \to \infty$.

Simply connected domains are often described as domains with 'no holes'. Intuitively, if a domain did have a 'hole' then there would be a curve in the domain that winds around points in this hole and that would give rise to a non-zero winding number. For example, if D is the annulus $\{z : 1/2 < |z| < 2\}$ then the circle $C(0, 1)$ lies in D and has non-zero winding number about the origin which lies in the complement of D. One way to capture the idea that a simply connected domain is one with no holes is that the winding number of any curve in the domain about any point in the complement of the domain should be zero.

We are now ready to complete the list of properties that we began in Theorem 6.8, each of which is equivalent to simple connectivity for a planar domain. In fact, we will use some of these in the course of proving Theorem 7.4. We first put in place the following lemma which is used in proving the implication (h) $\implies$ (i). Simmons [24, Chap. 6] is a good reference for connectedness in topological spaces. The result depends in an essential manner on the specific form of the compact sets K_n. For example, the result would be blatantly false in the case of a compact subset of the domain D consisting simply of a circle in D.

Lemma 7.1 *Suppose that D is a domain in the complex plane for which $\mathbb{C}_\infty \setminus D$ is connected and consider, for $n \in \mathbb{N}$, the compact subset K_n of D given by (5.13). Then, $\mathbb{C} \setminus K_n$ is also connected.*

Proof An alternative description of K_n is

$$K_n = \{z \in \mathbb{C} : \operatorname{dist}(z; \mathbb{C} \setminus D) \geq 1/n\} \cap \{z \in \mathbb{C} : |z| \leq n\}.$$

7.3 Further Characterisations of Simply Connected Domains

Since, in general, $(A \cap B)^c = A^c \cup B^c$,

$$\mathbb{C}_\infty \setminus K_n = \Big(\{\infty\} \cup \{z \in \mathbb{C} : \text{dist}(z; \mathbb{C} \setminus D) < 1/n\}\Big)$$
$$\cup \Big(\{\infty\} \cup \{z \in \mathbb{C} : |z| > n\}\Big)$$
$$= \Big(\{\infty\} \cup \bigcup_{w \in \mathbb{C} \setminus D} D(w, 1/n)\Big) \cup \Big(\{\infty\} \cup \{z \in \mathbb{C} : |z| > n\}\Big).$$

The set $E_2 = \{\infty\} \cup \{z \in \mathbb{C} : |z| > n\}$ is obviously connected—in fact, it is path connected and open. The set $E_1 = \{\infty\} \cup \bigcup_{w \in \mathbb{C} \setminus D} D(w, 1/n)$ is also connected. We may verify this as follows. First note that E_1 contains the set $\mathbb{C}_\infty \setminus D$ and this set is connected by assumption. Thus, $\mathbb{C}_\infty \setminus D$ lies in a single component C of E_1. Each disk $D(w, 1/n)$, where $w \in \mathbb{C} \setminus D$, is connected and so it too lies in a single component C_w of E_1. Since both components C and C_w contain the point w, they must coincide. In other words, $D(w, 1/n) \subseteq C$ whenever $w \in \mathbb{C} \setminus D$. That is, $E_1 \subseteq C$ so that $E_1 = C$ and E_1 is connected. Since the union of connected sets that have a point in common (the point at infinity in this case) is connected, we may conclude that $\mathbb{C}_\infty \setminus K_n = E_1 \cup E_2$ is connected. Since K_n is bounded, it follows that $\mathbb{C} \setminus K_n$ is connected. □

Theorem 7.4 *The following statements regarding a planar domain D are equivalent:*

(a) *D is simply connected;*
(g) *For all piecewise-smooth closed curves γ in D and for all points a in $\mathbb{C} \setminus D$, the winding number $w(\gamma; a)$ is 0;*
(h) *$\mathbb{C}_\infty \setminus D$ is connected;*
(i) *Any function f analytic in D is a locally uniform limit of polynomials.*

Note that (i) says that the polynomials are dense in $\mathscr{H}(D)$—see Theorem 5.3.

Proof Let D be a planar domain. Suppose that (a) holds, that the point a lies in the complement of D and that γ is a piecewise-smooth closed curve in D. Then the function $f(z) = 1/(z - a)$ is analytic in D. By Theorem 6.8, Property (b), $2\pi i\, w(\gamma; a) = \int_\gamma dz/(z - a) = 0$. This is Property (g).

Now suppose that (g) holds but that $\mathbb{C}_\infty \setminus D$ is *not* connected. Then $\mathbb{C}_\infty \setminus D$, being closed in $\mathbb{C}_\infty$, can be written as the union of two non-empty, disjoint, closed subsets F_1 and F_2 of $\mathbb{C}_\infty$. Since these sets are disjoint, only one of them can contain the point at infinity, say F_1. Then F_2 is a closed subset of the Riemann sphere that doesn't contain the point at infinity. As a consequence, F_2 is a bounded subset of the complex plane and therefore closed relative to the complex plane as well; see Lemma 3.2. In short, F_2 is compact. Now consider $D_1 = D \cup F_2 = \mathbb{C}_\infty \setminus F_1$. Then D_1 is an open set in the complex plane that contains the compact set F_2. By

Theorem 7.3 applied to the constant function $f(z) \equiv 1$ on D_1, there are finitely many step polygonal closed curves $C_1, \ldots, C_M$ in $D_1 \setminus F_2 = D$ such that

$$1 = \sum_{n=1}^{M} \frac{1}{2\pi i} \int_{C_n} \frac{dw}{w - z} = \sum_{n=1}^{M} w(C_n; z), \quad z \in F_2.$$

Choose z in the non-empty set F_2. Then, for some n, the winding number $w(C_n; z)$ of the curve C_n in D about this point z in the complement of D must be non-zero, contradicting the assumption that (g) holds. This shows that (g) implies (h).

Now suppose that (h) holds and that f is analytic in D. Consider the exhaustion of D by the compact sets $\{K_n\}_{n=1}^{\infty}$ in Lemma 5.2. From the assumption that $\mathbb{C}_\infty \setminus D$ is connected, it follows from Lemma 7.1 that $\mathbb{C} \setminus K_n$ is connected for each n. By Runge's Theorem—Polynomial Approximation (Theorem 7.2), there is a polynomial P_n such that

$$|f(z) - P_n(z)| \leq \frac{1}{n}, \quad z \in K_n. \tag{7.6}$$

Then, $\{P_n\}_{n=1}^{\infty}$ is a sequence of polynomials that converges locally uniformly to f—in fact, if K is a compact subset of D, we have $K \subseteq K_n$ for all sufficiently large n and so (7.6) holds for z in K for all sufficiently large n. That is, f is a locally uniform limit of polynomials in D, and so (i) holds.

Finally, suppose that (i) holds. Suppose that f is analytic in D and that γ is a piecewise-smooth closed curve in D. Then, the trace of γ is a compact subset of D and so there is a sequence of polynomials $\{P_n\}_{n=1}^{\infty}$ that converges uniformly to f on γ. But then $\int_\gamma P_n(z)\,dz = 0$ for each n and, moreover, by the ML-inequality,

$$\left|\int_\gamma f(z)\,dz\right| = \left|\int_\gamma f(z)\,dz - \int_\gamma P_n(z)\,dz\right| \leq \ell(\gamma) \times \max\{|f(z)-P_n(z)| : z \in \gamma\},$$

where $\ell(\gamma)$ is the length of the curve γ. From this we conclude that $\int_\gamma f(z)\,dz = 0$. The domain D therefore satisfies Theorem 6.8, Property (b), and so is simply connected.

In short, (a) $\Longrightarrow$ (g) $\Longrightarrow$ (h) $\Longrightarrow$ (i)[$\Longrightarrow$ (b)] $\Longrightarrow$ (a). $\square$

If D is a bounded domain then the condition that $\mathbb{C}_\infty \setminus D$ is connected coincides with the condition that $\mathbb{C} \setminus D$ is connected. In general, however, these two conditions are different. Consider the case that D is the strip $\mathbb{S} = \{z : |\Im z| < \frac{\pi}{2}\}$ discussed in Sect. 6.5.2. This domain is simply connected since it is, in fact, star-shaped about the origin. Its complement relative to the complex plane consists of two disjoint, closed half-planes, $H_1 = \{z : \Im z \geq \frac{\pi}{2}\}$ and $H_2 = \{z : \Im z \leq -\frac{\pi}{2}\}$ and so is not connected. Viewed as a subset of the Riemann sphere, $\mathbb{S}$ is shaped near the point at infinity somewhat like two regions each lying between two lines of longitude and each narrowing to a point at the North Pole. The half-planes H_1 and H_2 when viewed

7.3 Further Characterisations of Simply Connected Domains

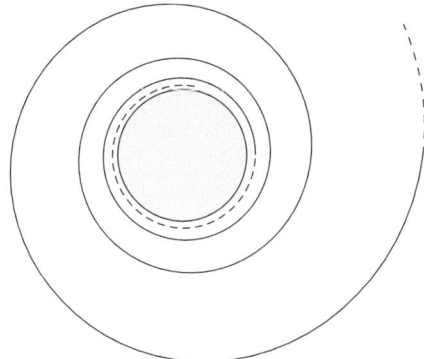

Fig. 7.3 The simply connected domain shown is the complement of the closed unit disk and of the spiral

on the Riemann Sphere are then connected through the North Pole. The complement of $\mathbb{S}$ relative to the Riemann sphere is indeed connected, as it should be.

Another example is the domain D that is the complement relative to the complex plane of the half-lines $(-\infty, -1]$ and $[1, \infty)$ on the real axis. The set $\mathbb{C} \setminus D$ is precisely these two half-lines and is disconnected. On the other hand, $\mathbb{C}_\infty \setminus D$ is the half great circle on the Riemann sphere through the points $1 = (1, 0, 0)$, the North Pole $(0, 0, 1)$, and $-1 = (-1, 0, 0)$, which is certainly connected and which shows that D is simply connected. On the other hand, let D be the domain formed by removing the line segment $I = [-1, 1]$ on the real axis from the complex plane. Then, $\mathbb{C} \setminus D = I$ is connected but $\mathbb{C}_\infty \setminus D = I \cup \{\infty\}$ is not and, in fact, D is not simply connected. Note also that the circle $C(0, 2)$ in D has winding number 1 about the origin in the complement of D and so (g) does not hold either.

An interesting example of a simply connected domain is shown in Fig. 7.3. The domain is the complement relative to the complex plane of the closed set

$$E = \overline{D(0, 1)} \cup \{(1 + e^t)e^{it}, \ t \in \mathbb{R}\}.$$

The curve $\gamma(t) = (1 + e^t)e^{it}$, $-\infty < t < \infty$, spirals infinitely often around the unit circle $\mathbb{T}$, being wound tighter and tighter around $\mathbb{T}$ for large negative t and spiralling out towards infinity for large positive t. The set $E \cup \{\infty\}$ is a closed, connected subset of $\mathbb{C}_\infty$ and so $D = \mathbb{C} \setminus E$ is a simply connected domain. Note that even though $\mathbb{C}_\infty \setminus D = E \cup \{\infty\}$ is connected, it is not path connected.

Chapter 8
Univalent Functions: The Basics

By the Riemann Mapping Theorem, any conformal mapping between proper simply connected domains can, in principle, be factored through the unit disk $\mathbb{D}$. That is, if $f : D_1 \to D_2$ is a conformal map between proper simply connected domains D_1 and D_2, we can write f as $h \circ g^{-1}$ where $g : \mathbb{D} \to D_1$ and $h : \mathbb{D} \to D_2$ are conformal maps. In this sense, it is sufficient to study conformal maps of the unit disk $\mathbb{D}$ and to choose the unit disk as the canonical simply connected domain.

Since any function analytic in a disk can be written as a power series, the question arises whether the coefficients of power series that correspond to conformal maps might possess special properties. The most famous question in this direction, the Bieberbach Conjecture, dominated and drove the field of univalent functions until it was finally resolved in the affirmative by Louis de Branges in 1984.

We will learn the rudiments of univalent function theory in this chapter and, in so doing, some basic estimates of the hyperbolic metric, in particular the Koebe 1/4-Theorem.

8.1 The Classes S and Σ of Univalent Functions

Suppose that f is a conformal mapping of the unit disk $\mathbb{D}$ (a *univalent function* in this context). A key *necessary* condition for the function to be univalent is that its derivative f' never vanishes[1] [16, Theorem 10.6]. Thus, in particular, $f'(0) \neq 0$ and may normalise f so that $f(0) = 0$ and also $f'(0) = 1$ by replacing f by

---

[1] An analytic function in a domain is said to be *locally univalent* if its derivative never vanishes. A non-vanishing derivative is far from sufficient to guarantee that a function is univalent. For example, the function $f(z) = \exp(2\pi z)$ is not univalent in the strip $S = \{z : |\Im z| < 1\}$ since $f(-i/2) = -1 = f(i/2)$, yet its derivative is never 0.

$(f - f(0))/f'(0)$—this transformation is a shift followed by a scaling, and so is univalent if and only if the original function f is.

The class **S** consists of those univalent functions f in the unit disk $\mathbb{D}$ that satisfy $f(0) = 0$ and $f'(0) = 1$. Each function in the class **S** has a power series expansion about the origin with radius of convergence at least 1, so that

$$f(z) = z + \sum_{n=2}^{\infty} a_n z^n, \quad |z| < 1. \tag{8.1}$$

Remember that the unit disk $\mathbb{D}$ corresponds to the southern hemisphere on the Riemann sphere under stereographic projection. The northern hemisphere corresponds to the exterior of the disk, $\mathbb{U} = \{z \in \mathbb{C}_\infty : |z| > 1\}$ (the equator corresponds to the unit circle). Also, 0 plays the same role in the southern hemisphere (or the unit disk $\mathbb{D}$) as ∞ plays in the northern hemisphere (or the exterior $\mathbb{U}$ of the unit disk). It is therefore natural to also consider functions that are univalent in $\mathbb{U}$, in which case these should mirror functions univalent in the unit disk. Since we normalise functions univalent in $\mathbb{D}$ so that they fix the origin, we should normalise functions univalent in $\mathbb{U}$ so that the fix the point at infinity; that is, these functions should have a pole at the point at infinity. By Exercise 3.8, this must be a simple pole as the function will not be univalent near the point at infinity if the order of the pole is greater than 1. By Definition 3.4 with $N = 1$, a function g analytic in $\mathbb{U}$ with a simple pole at infinity has a power series expansion

$$g(z) = \alpha z + b_0 + \frac{b_1}{z} + \frac{b_2}{z^2} + \dots$$

about the point at infinity. In the standard manner by which we interchange 0 and ∞ on the Riemann Sphere, we associate with g defined in $\mathbb{U}$ the function f defined in $\mathbb{D}$ by $f(z) = 1/g(1/z)$; and conversely the function f defined in $\mathbb{D}$ corresponds to $g(z) = 1/f(1/z)$ defined in $\mathbb{U}$. In order that f will be analytic in $\mathbb{D} \setminus \{0\}$, we need that g never vanishes in $\mathbb{U}$. Similarly, we need that f vanishes only at 0 as then g will be analytic in $\mathbb{U}$. In summary, f is analytic and univalent in $\mathbb{D}$ with $f(0) = 0$ if and only if g is analytic, non-zero and univalent in $\mathbb{U}$ with a simple pole at infinity. In terms of the above expansion for g, the derivative of f at 0 is $f'(0) = \lim_{h \to 0} f(h)/h = 1/\alpha$, so that $f'(0) = 1$ if $\alpha = 1$.

We then write Σ for the class of functions that are univalent and non-zero in $\mathbb{U}$ with the power series representation

$$g(z) = z + b_0 + \frac{b_1}{z} + \frac{b_2}{z^2} + \dots, \quad z \in \mathbb{U}. \tag{8.2}$$

There is then a direct, one-to-one correspondence between functions in the class **S** and functions in the class Σ. Specifically, if f is in the class **S** then $g(z) = 1/f(1/z)$, $z \in \mathbb{U}$, is in the class Σ. If g is in the class Σ then $f(z) = 1/g(1/z)$, $z \in \mathbb{D}$, is in the

8.1 The Classes **S** and Σ of Univalent Functions

class **S**. Notice that the image of $\mathbb{D}$ under f and the image of $\mathbb{U}$ under g correspond under the inversion $z \to 1/z$ of the Riemann sphere.

Exercise 8.1

▷ Suppose that f is a univalent function in the class **S** with power series (8.1). Show that the corresponding univalent function $g(z) = 1/f(1/z)$ in Σ has an expansion (8.2) that begins

$$g(z) = z - a_2 + \frac{a_2^2 - a_3}{z} + \frac{2a_2 a_3 - a_2^3 - a_4}{z^2} + \dots$$

▷ Suppose that g is a univalent function in the class Σ with power series (8.2). Show that the corresponding univalent function $f(z) = 1/g(1/z)$ in **S** has an expansion (8.1) that begins

$$f(z) = z - b_0 z^2 + (b_0^2 - b_1) z^3 + (2b_0 b_1 - b_0^3 - b_2) z^4 + \dots$$

It should be said that the identity function $f(z) = z$, $z \in \mathbb{D}$, is the simplest function in the class **S** while the identity function $g(z) = z$, $z \in \mathbb{U}$, is the simplest function in the class Σ.

The function

$$M(z) = \frac{1+z}{1-z}$$

is a conformal map of the disk $\mathbb{D}$ onto the right half-plane $\mathcal{H} = \{z : \Re z > 0\}$ (cf. Exercise 4.9 where the extra multiplicative factor i serves to rotate the right half-plane $\mathcal{H}$ to the upper half-plane $\mathbb{H}$.). The map M isn't normalised, so we work with

$$f(z) = \frac{M(z) - M(0)}{M'(0)} = \frac{1}{2}\left(\frac{1+z}{1-z} - 1\right) = \frac{z}{1-z}, \quad z \in \mathbb{D}.$$

This function is in the class **S** and maps $\mathbb{D}$ onto $(\mathcal{H} - M(0))/M'(0) = \frac{1}{2}(\mathcal{H} - 1)$, which is the half-plane $\{z : \Re z > -\frac{1}{2}\}$. The power series expansion for f, using the Geometric Series, is

$$f(z) = \sum_{n=1}^{\infty} z^n.$$

Returning to the mapping M above, $M^2(z)$ is also a conformal map of the unit disk. In fact, if $M^2(z_1) = M^2(z_2)$ then either $M(z_1) = M(z_2)$ or $M(z_1) = -M(z_2)$. The second case cannot happen since, if z_1 and z_2 are both in the unit disk then $M(z_1)$ and $M(z_2)$ are both in the right half-plane and so one cannot be the negative

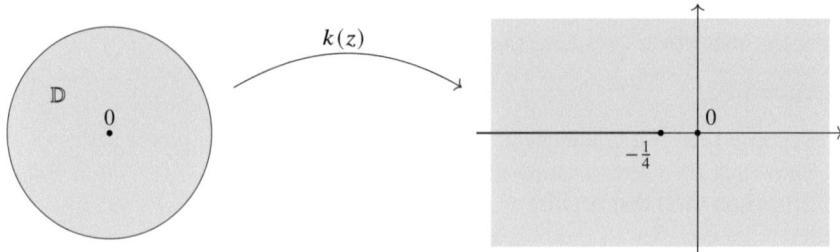

Fig. 8.1 The Koebe function $k(z) = 1/(1-z)^2$ and its image

of the other. Thus, $M(z_1) = M(z_2)$ hence, since M is univalent, $z_1 = z_2$.[2] The mapping M^2 is the composition of the square mapping with the mapping M of the unit disk onto the right half-plane. The square mapping, $z \to z^2$, maps the right half-plane conformally onto the complex plane slit along the negative real axis from $-\infty$ to 0 (remember that the square mapping doubles angles). Once again, M^2 is not in the class $\mathbf{S}$ as it is not normalised, so we translate and scale it to the mapping

$$k(z) = \frac{M^2(z) - M^2(0)}{(M^2)'(0)} = \frac{1}{4}\left(\left(\frac{1+z}{1-z}\right)^2 - 1\right) = \frac{z}{(1-z)^2}.$$

This mapping is called the *Koebe function*. The mapping $M^2(z) - 1$ maps the unit disk onto the complex plane slit along the negative real axis from $-\infty$ to -1. Dividing by 4, we see that the image of the unit disk $\mathbb{D}$ under the mapping k is the complex plane slit along the negative real axis from $-\infty$ to $-\frac{1}{4}$. This simply connected domain is known as the *Koebe region* and is shown in Fig. 8.1. If we view $1/(1-z)^2$ as the derivative of $1/(1-z)$, the latter having power series expansion $1 + z + z^2 + \ldots$, we find that the power series expansion of the Koebe function is

$$k(z) = \frac{z}{(1-z)^2} = \sum_{n=1}^{\infty} n z^n. \tag{8.3}$$

Exercise 8.2 Find the univalent function in Σ that corresponds (under the transformation $g(z) = 1/f(1/z)$ for $z \in \mathbb{U}$) to the function in $\mathbf{S}$ given by (i) $f(z) = z/(1-z)$ for $z \in \mathbb{D}$ and (ii) $k(z) = z/(1-z)^2$ for $z \in \mathbb{D}$.

In each case, determine the image of $\mathbb{U}$ under the corresponding mapping in Σ.

Example 8.1 Let us find all quadratics $f(z) = z + cz^2$ that belong to the class $\mathbf{S}$. We want to determine the range of the coefficient c for which this function is

---

[2] The typical method to show that a function f on a set X is one-to-one (or injective) is to suppose that $f(x_1) = f(x_2)$ and then conclude that $x_1 = x_2$. This implies that f cannot take the same value at different points.

8.1 The Classes **S** and Σ of Univalent Functions

univalent in the unit disk. The case $c = 0$ is the identity, which belongs to **S**, and so we assume that $c \neq 0$. Suppose, then, that z_1 and z_2 are in the disk and that $f(z_1) = f(z_2)$. This equation is equivalent to

$$z_1 - z_2 + c(z_1^2 - z_2^2) = 0$$

or

$$(z_1 - z_2)(1 + c(z_1 + z_2)) = 0.$$

If $|c| \leq \frac{1}{2}$ then

$$|1 + c(z_1 + z_2)| \geq 1 - |c|(|z_1| + |z_2|) \geq 1 - \frac{1}{2}(|z_1| + |z_2|) > 1 - \frac{1}{2}(1+1) = 0,$$

that is $1 + c(z_1 + z_2)$ is not zero. Consequently, $z_1 = z_2$ and f is univalent.

In the other direction, suppose that $|c| > 1/2$. Then $-1/(2c)$ lies in the unit disk. Set $\epsilon = 1 - 1/(2|c|)$, so that ϵ is positive. Put $z_1 = -1/(2c) + \epsilon/2$ and $z_2 = -1/(2c) - \epsilon/2$, so that both z_1 and z_2 lie in the unit disk $\mathbb{D}$, are distinct, and $z_1 + z_2 = -1/c$. Thus, $f(z_1) = f(z_2)$ and f is not univalent in this case. (Alternatively, $f'(z) = 1 + 2cz$ and, if $|c| > 1/2$, then f' has a zero in $\mathbb{D}$ at $-1/(2c)$, a fact that also rules out univalence.)

We have shown that $f(z) = z + cz^2$ is univalent in $\mathbb{D}$ if and only if $|c| \leq 1/2$.

Taking this example somewhat further, we begin to appreciate the subtlety of the univalence condition.

Exercise 8.3

▷ Let $Q(z) = 1 + b_1 z + b_2 z^2 + \ldots + b_n z^n$ where $|b_n| > 1$. Show that Q has at least one zero in $\mathbb{D}$.

▷ Consider the polynomial $P(z) = z + a_2 z^2 + a_3 z^3 + \ldots + a_n z^n$. Prove that P' has a zero in $\mathbb{D}$ if $|a_n| > \frac{1}{n}$. Deduce that if the polynomial P is univalent in $\mathbb{D}$ then $|a_n| \leq \frac{1}{n}$.

Show by means of an example of the form $P(z) = z + cz^2 + \frac{1}{3}z^3$ that this condition is not sufficient for univalence.

▷ Let $n \geq 2$. Prove that the polynomial $P(z) = z + cz^n$ is univalent in $\mathbb{D}$ if and only if $|c| \leq \frac{1}{n}$.

See the paper by Suffridge [26], for example, for much more on univalent polynomials.

Exercise 8.4 Prove that $f(z) = z + \sum_{n=2}^{\infty} a_n z^n$ is univalent in $\mathbb{D}$ if $\sum_{n=2}^{\infty} n|a_n| \leq 1$.

Another elementary *univalence criterion* is the following: if f is analytic in the unit disk and satisfies $\Re f' > 0$ there then f is univalent in $\mathbb{D}$. To see this, suppose that z_1 and z_2 are distinct points in the unit disk and parameterise the line segment joining them by $\gamma(t) = (1-t)z_1 + tz_2$, $0 \le t \le 1$. Since $f(z_2) - f(z_1) = \int_0^1 f'(\gamma(t))\gamma'(t)\,dt = (z_2 - z_1)\int_0^1 f'(\gamma(t))\,dt$,

$$\Re\left(\frac{f(z_2) - f(z_1)}{z_2 - z_1}\right) = \Re\left(\int_0^1 f'(\gamma(t))\,dt\right) = \int_0^1 \Re f'(\gamma(t))\,dt > 0,$$

so that $f(z_2) \ne f(z_1)$ and f is therefore univalent.

Even though univalence is an elusive concept, it is preserved under a number of transformations. Suppose that f is a univalent function in the class **S** with power series given by (8.1). It is straightforward to see that, for $\theta \in [0, 2\pi)$, the function f_1 given by

$$f_1(z) = e^{i\theta} f(e^{-i\theta} z), \quad z \in \mathbb{D}, \tag{8.4}$$

also belongs to the class **S**. We refer to f_1 as a *rotation* of the function f. The image of the unit disk under f_1 is its image under f rotated through an angle θ anticlockwise.

More generally, for any conformal self-map M of the disk $\mathbb{D}$ (as characterised in Theorem 4.2), $f \circ M$ is again univalent in the unit disk $\mathbb{D}$ as it is the composition of univalent maps. If $M(z) = (\alpha - z)/(1 - \bar{\alpha}z)$ where $|\alpha| < 1$, then

$$f_2(z) = \frac{(f \circ M)(z) - (f \circ M)(0)}{(f \circ M)'(0)} = \frac{f\left(\frac{\alpha-z}{1-\bar{\alpha}z}\right) - f(\alpha)}{f'(\alpha)(|\alpha|^2 - 1)}, \quad z \in \mathbb{D}, \tag{8.5}$$

belongs to the class **S**. The image of the unit disk under $f \circ M$ is the same as that under f; the difference is that the origin is mapped to $f(\alpha)$ rather than to 0. The result of implementing the normalisation for the class **S** is to shift, then scale and rotate the image of the unit disk under the original function f.

Exercise 8.5 In this exercise, we show by example that univalence is *not* necessarily preserved under addition of functions, that is the sum of two univalent functions need not be univalent. Consider $f(z) = z/(1-z)$ so that f is a univalent function in the class **S**. Then, $g(z) = -if(iz) = z/(1-iz)$ is also a univalent function in the class **S** (g corresponds to the case $\theta = -\pi/2$ in (8.4)). The function $F = (f+g)/2$ is analytic in $\mathbb{D}$, satisfies $F(0) = 0$ and $F'(0) = 1$. Show, however, that F is not univalent in $\mathbb{D}$.

Exercise 8.6 Let $k(z)$ be the Koebe function given by (8.3). For θ in $[0, 2\pi)$, set k_θ to be the rotation of the Koebe function $k_\theta(z) = e^{i\theta} k(e^{-i\theta} z)$, $z \in \mathbb{D}$. Find an explicit expression for k_θ and find its power series expansion. Describe the simply connected domain $k_\theta(\mathbb{D})$.

8.1 The Classes **S** and Σ of Univalent Functions

The function

$$f_3(z) = \overline{f(\bar{z})} = z + \sum_{n=1}^{\infty} \overline{a_n} z^n, \quad z \in \mathbb{D},$$

belongs to the class **S**. We refer to f_3 is the *conjugation* of f. The image of the unit disk under f_3 is the reflection in the real axis of its image under f.

The *omitted value transformation* works as follows. Let w be any complex number not in the range of f, so that the analytic function $f(z) - w$ never vanishes in $\mathbb{D}$. Then, $g(z) = 1/(f(z) - w)$ is also analytic in $\mathbb{D}$, and is actually univalent there (it takes different values at different points in $\mathbb{D}$ since f has this property). If we normalise this function g we obtain a univalent function in the class **S**. Computing, we find that $g(0) = -1/w$ and $g'(0) = -1/w^2$, so that

$$f_4(z) = \frac{g(z) - g(0)}{g'(0)} = \frac{wf(z)}{w - f(z)}, \quad z \in \mathbb{D}, \tag{8.6}$$

belongs to the class **S**.

Exercise 8.7 Suppose that $f \in \mathbf{S}$, has power series (8.1), and that w is an omitted value of f. Show that the power series for the function f_4 in **S**, given by (8.6), begins

$$f_4(z) = z + \left(a_2 + \frac{1}{w}\right)z^2 + \left(a_3 + \frac{2a_2}{w} + \frac{1}{w^2}\right)z^3 + \dots$$

A final important transformation is known as the *square root transformation*, by which each function in **S** can be written in terms of an odd univalent function. In general, an analytic function h in the disk is said to be *odd* if $h(-z) = -h(z), z \in \mathbb{D}$. Coefficients of all even powers of z in its power series expansion $h(z) = \sum_{n=0}^{\infty} c_n z^n$ must then vanish. This can be seen by writing

$$0 = h(z) + h(-z) = \sum_{n=0}^{\infty} c_n[1 + (-1)^n]z^n = \sum_{\substack{n=0 \\ n \text{ even}}}^{\infty} 2c_n z^n.$$

Since a power series is identically zero only if all coefficients are zero, it follows that $c_n = 0$ for n even. Similarly, in the case of an *even* analytic function h in the disk, that is one for which $h(-z) = h(z), z \in \mathbb{D}$, all coefficients of odd powers vanish. It is obvious that the converse holds in each case. For instance, a function h analytic in the disk is odd if and only if it has a power series expansion $h(z) = c_1 z + c_3 z^3 + c_5 z^5 + \dots, z \in \mathbb{D}$.

Proposition 8.1 (Square Root Transformation) *To each function f in* **S** *there corresponds a unique, odd univalent function h in* **S** *such that*

$$f(z^2) = h(z)^2, \quad z \in \mathbb{D}. \tag{8.7}$$

The converse to Proposition 8.1 also holds, namely to each odd univalent function h in **S** there corresponds a function f in **S** for which (8.7) holds, but this is not central to our story.

Proof Given f in **S** with power series (8.1),

$$g(z) = \frac{f(z)}{z} = 1 + \sum_{n=1}^{\infty} a_{n+1} z^n, \quad z \in \mathbb{D} \setminus \{0\},$$

has a removable singularity at 0. That is, g is analytic in the unit disk once we set $g(0) = 1$. It is also non-zero there since $f(z)/z \neq 0$ for $0 < |z| < 1$; being univalent, f only takes the value 0 at 0. Since it is a non-zero analytic function on a simply connected domain, the function g has an analytic square root, that is $g(z) = g_1^2(z)$ for a non-zero analytic function g_1 in $\mathbb{D}$ (and we may take $g_1(0) = 1$)—see the discussion after Theorem 6.6.

Now set

$$h(z) = z g_1(z^2) = z \sqrt{\frac{f(z^2)}{z^2}}, \quad z \in \mathbb{D}.$$

Then, $h(z)^2 = f(z^2)$ so that (8.7) holds. Moreover, h is an odd function since $h(-z) = (-z) g_1((-z)^2) = -z g_1(z^2) = -h(z)$. Also, $h(0) = 0$ and $h'(0) = g_1(0) = 1$, so that h has the correct normalisation. We next check that h is univalent. Suppose that $h(z_1) = h(z_2)$. Then, $f(z_1^2) = f(z_2^2)$ so that $z_1^2 = z_2^2$ and either $z_1 = z_2$ or $z_1 = -z_2$. If $z_1 \neq z_2$ then $0 \neq z_1 = -z_2$ and $0 \neq h(z_1) = h(-z_2) = -h(z_2) = -h(z_1)$ which is impossible. Thus, $z_1 = z_2$ and h is univalent.

Finally, we show that h is unique. Suppose that h_1 is a second odd function in **S** for which $h_1(z)^2 = f(z^2)$, $z \in \mathbb{D}$. Then, $h_1(z)^2 = h(z)^2$ so that, for each $z \in \mathbb{D}$, either $h_1(z) = h(z)$ or $h_1(z) = -h(z)$. It follows that the zeros of one of the analytic functions $h_1 - h$ or $h_1 + h$ has a limit point in $\mathbb{D}$ and hence is identically zero. This can't be $h_1 + h$ as the derivative of this function has the value 2 at 0, so $h_1 - h$ must be identically zero in $\mathbb{D}$. □

The image of the unit disk under the odd univalent function h in Proposition 8.1 is the square root of its image under f.

Exercise 8.8 Find a formula for the square root transform of the Koebe function. Describe the image of the unit disk under this conformal mapping.

Exercise 8.9 Suppose that $f \in \mathbf{S}$ with power series (8.1). Show that the power series for the square root transform h of f, given by (8.7), begins

$$h(z) = z + \frac{a_2}{2} z^3 + \left(\frac{a_3}{2} - \frac{a_2^2}{8} \right) z^5 + \ldots$$

Exercise 8.10 Show that the converse of Proposition 8.1 is true: that is, if h is an odd univalent function in the class **S** then there is a function f in the class **S** of which h is the square root transform.

8.2 Bieberbach's Coefficient Estimate and the Koebe 1/4-Theorem

The material in this next section is well-trodden ground indeed—it can be found, with little variation, in many of the references listed at the end of this book, for example any of [15, 22, 27] or [10]. Nevertheless, we repeat the arguments here. The story begins with what is known as the *Area Theorem* for functions in the class Σ.

8.2.1 The Area Theorem and Bieberbach's Coefficient Estimate

Theorem 8.1 *Let g be a univalent function in the class Σ whose expansion is given by (8.2). Then,*

$$\sum_{n=1}^{\infty} n|b_n|^2 \leq 1. \tag{8.8}$$

In particular, $|b_1| \leq 1$ with equality if and only if $g(z) = z - (\alpha + \beta) + (\alpha\beta)/z$ where $|\alpha| = |\beta| = 1$.

Proof Let g be a univalent function in the class Σ with expansion given by (8.2) and let $r > 1$. Since g is univalent in $\mathbb{U} = \{z : |z| > 1\}$, $g(C(0, r))$ is a simple closed smooth curve γ parameterised by $\gamma(t) = g(re^{it})$, $0 \leq t \leq 2\pi$. This curve γ encloses a bounded region Ω_r. We use the Divergence Theorem from multivariable calculus, (one version of) which states that

$$\iint_\Omega \Delta h \, dx \, dy = \int_{\partial \Omega} (\nabla h) \cdot n \, ds, \tag{8.9}$$

where h is any smooth function on the closure of the smooth bounded domain Ω, where Δh is the Laplacian of h, where ∇h is the gradient of h, and where n is the unit outward normal to Ω. We take $\Omega = \Omega_r$, and take $h(x, y) = \frac{1}{4}(x^2 + y^2)$ so that $\Delta h = h_{xx} + h_{yy} = 1$. Then,

$$\iint_{\Omega_r} \Delta h \, dx \, dy = \iint_{\Omega_r} dx \, dy = \text{Area}(\Omega_r). \tag{8.10}$$

In the case of Ω_r, 'ds' in (8.9) becomes $ds = |\gamma'(t)| dt$, and an outward normal vector to the boundary of Ω_r at $\gamma(t)$ is, in complex form, $-i\gamma'(t)$. The unit outward normal is then $n = -i\gamma'(t)/|\gamma'(t)|$. Here, $\nabla h = (h_x, h_y) = \frac{1}{2}(x,y)$, which is simply $\frac{1}{2}z$ in complex form. Noting that the dot product between two complex numbers z and w is $z.w = (x+iy).(u+iv) = xu + yv = \Re(\bar{z}w)$, the right hand side of (8.9) becomes, in complex form,

$$\int_{\partial\Omega_r} (\nabla h).n\, ds = \int_0^{2\pi} (\nabla h)(\gamma(t)).(-i\gamma'(t))\, dt$$

$$= \int_0^{2\pi} \frac{1}{2}\gamma(t).(-i\gamma'(t))\, dt$$

$$= \frac{1}{2}\int_0^{2\pi} \Re\left[\overline{\gamma(t)}(-i\gamma'(t))\right] dt$$

$$= \frac{1}{2}\int_0^{2\pi} \Re\left[\overline{g(re^{it})}re^{it}g'(re^{it})\right] dt$$

Now,

$$g(z) = z + b_0 + \sum_{n=1}^{\infty} \frac{b_n}{z^n} \quad \text{and} \quad zg'(z) = z - \sum_{n=1}^{\infty} \frac{nb_n}{z^n}.$$

Replacing z by re^{it} we find that

$$\int_{\partial\Omega_r} (\nabla h).n\, ds$$

$$= \frac{1}{2}\int_0^{2\pi} \Re\left[\overline{\left(re^{it} + b_0 + \sum_{n=1}^{\infty} \frac{b_n}{r^n e^{int}}\right)}\left(re^{it} - \sum_{m=1}^{\infty} \frac{mb_m}{r^m e^{imt}}\right)\right] dt$$

$$= \frac{1}{2}\Re\left[\int_0^{2\pi}\left(re^{-it} + \bar{b}_0 + \sum_{n=1}^{\infty} \frac{\bar{b}_n}{r^n e^{-int}}\right)\left(re^{it} - \sum_{m=1}^{\infty} \frac{mb_m}{r^m e^{imt}}\right) dt\right]$$

$$= \pi\left(r^2 - \sum_{n=1}^{\infty} n|b_n|^2 r^{-2n}\right) \tag{8.11}$$

The dramatic simplification at the last line of this computation might seem surprising at first. The series for both $g(z)$ and $zg'(z)$ converge uniformly on the circle $C(0,r)$ (since $r > 1$ and both series are convergent for $|z| > 1$). The series for $\overline{g(re^{it})}re^{it}g'(re^{it})$ is then given by the ordinary product of the two series as a sum of exponential terms of the form e^{ikt} as k ranges over the integers. Moreover, this series for the product $\overline{g(z)}(zg'(z))$, $z = re^{it}$, also converges uniformly on $[0, 2\pi]$

8.2 Bieberbach's Coefficient Estimate and the Koebe 1/4-Theorem

and so the order of integration and summation can be interchanged. But then we end up with integrals $\int_0^{2\pi} e^{ikt} dt$ which evaluate to either 2π or 0 depending on whether $k = 0$ or $k \neq 0$. The only terms that contribute to the integral above, therefore, are those corresponding to $re^{-it} \times re^{it} = r^2$, and $r^{-n} e^{int} \times r^{-m} e^{-imt}$ where $m = n$.

By (8.9), (8.10), (8.11), we deduce that

$$\text{Area}(\Omega_r) = \pi \left(r^2 - \sum_{n=1}^{\infty} n |b_n|^2 r^{-2n} \right)$$

so that, in particular,

$$r^2 \geq \sum_{n=1}^{\infty} n |b_n|^2 r^{-2n}.$$

Since this statement holds for every $r > 1$, it holds for $r = 1$, which is (8.8).

Geometrically, $1 - \sum_{n=1}^{\infty} n |b_n|^2$ represents the limit as $r \downarrow 1$ of the area of Ω_r. Thus, equality holds in the area theorem if and only of the area of the set omitted by g (that is, of $\mathbb{C}_\infty \setminus g(\mathbb{U})$) is zero.

Since $|b_1|^2 \leq \sum_{n=1}^{\infty} n |b_n|^2$, we have $|b_1| \leq 1$ and, if equality holds, then $b_n = 0$ for $n \geq 2$. That is, if $|b_1| = 1$ then $g(z) = z + b_0 + e^{i\theta}/z$, with $0 \leq \theta < 2\pi$. It is straightforward to check that any function of this form is univalent in $\mathbb{U}$. In order that g belongs to the class Σ, we also need g to be zero-free in $\mathbb{U}$. Solving $g(z) = 0$ leads to the quadratic equation $z^2 + b_0 z + e^{i\theta} = 0$ so that $g \in \Sigma$ only if the roots of this quadratic lie outside $\mathbb{U}$, that is in $\{z : |z| \leq 1\}$. Denoting the roots of this quadratic equation by α and β, let us assume this to be the case so that $|\alpha| \leq 1$ and $|\beta| \leq 1$. Since $\alpha\beta = e^{i\theta}$, we have $|\alpha| |\beta| = 1$, and so $|\alpha| = |\beta| = 1$. Then $\alpha + \beta = -b_0$, the sum of the roots.

Conversely, any g of the form $g(z) = z - (\alpha + \beta) + (\alpha\beta)/z$ where $|\alpha| = |\beta| = 1$ is univalent and zero-free in $\mathbb{U}$ and so belongs to Σ. □

Exercise 8.11 Consider $g(z) = z - (\alpha + \beta) + (\alpha\beta)/z$ where $|\alpha| = |\beta| = 1$.

▷ Find the function f in **S** that corresponds to g in Σ, where $f(z) = 1/g(1/z)$. Which choices of α and β give rise to the Koebe function and to its square root transform?

▷ Describe the image $g(\mathbb{U})$ of $\mathbb{U}$ under g.

Hint: If $g(z) = z + e^{i\theta}/z$, the image of a point $z = e^{it}$ on the unit circle under g is

$$g(e^{it}) = e^{it} + e^{i(\theta - t)} = e^{i\theta/2} \left(e^{i(t - \theta/2)} + e^{i(\theta/2 - t)} \right) = 2 e^{i\theta/2} \cos(t - \theta/2).$$

The Area Theorem leads to a sharp estimate for the second coefficient of a function in the class **S**.

Theorem 8.2 (Bieberbach's Coefficient Estimate) *Let $f(z) = z + \sum_{n=2}^{\infty} a_n z^n$ be a univalent function in the class* **S**. *Then, $|a_2| \leq 2$ with equality if and only if f is a rotation of the Koebe function k given by (8.3).*

Proof Suppose that f is a univalent function in the class **S** and has power series expansion $f(z) = z + \sum_{n=2}^{\infty} a_n z^n$. Then, by Proposition 8.1, there is an odd function h in **S** such that $h(z)^2 = f(z^2)$. By Exercise 8.9, h has a power series expansion that begins

$$h(z) = z + \frac{a_2}{2} z^3 + \left(\frac{a_3}{2} - \frac{a_2^2}{8}\right) z^5 + \cdots$$

Consider the function $g(z) = 1/h(1/z)$, $z \in \mathbb{U}$, which belongs to the class Σ of univalent functions and has, by Exercise 8.1, an expansion (8.2) that begins

$$g(z) = z - \frac{a_2}{2z} + \cdots$$

By the Area Theorem, Theorem 8.1, $|a_2/2| \leq 1$, that is $|a_2| \leq 2$. Moreover, by the equality statement in the Area Theorem, equality holds if and only if $g(z) = z - e^{i\theta}/z$ for some θ in $[0, 2\pi)$. In that case,

$$h(z) = 1/g(1/z) = \frac{z}{1 - e^{i\theta} z^2}, \quad z \in \mathbb{D}.$$

Since $h(z)^2 = f(z^2)$, we find that

$$f(z) = \frac{z}{(1 - e^{i\theta} z)^2}, \quad z \in \mathbb{D},$$

so that f is the rotation $k_{-\theta}$ of the Koebe function as in Exercise 8.6.

Conversely, the coefficient of z^2 in the power series expansion of the rotation k_θ of the Koebe function is of modulus 2. □

Had we applied the Area Theorem directly to the function $g(z) = 1/f(1/z)$, $z \in \mathbb{U}$, we would have concluded that

$$|a_2^2 - a_3| \leq 1$$

which, although correct, involves both a_2 and a_3. Passing first to the odd function h and then applying the Area Theorem led to an estimate involving only a_2.

Having proved that $|a_2| \leq 2$ for any function $f(z) = z + \sum_{n=2}^{\infty} a_n z^n$ in the class **S**, and noting that equality held precisely for the Koebe function and its rotations, Bieberbach conjectured (in 1916) the inequality $|a_n| \leq n$ for every n, not just $n = 2$. An enormous effort was expended in subsequent years in attempting to prove (or disprove) this conjecture, giving rise to much beautiful mathematics. Hayman's 1955 proof of the asymptotic Bieberbach conjecture [13] is one such beautiful piece

8.2 Bieberbach's Coefficient Estimate and the Koebe 1/4-Theorem

of mathematics. The conjecture was finally settled, and the inequality $|a_n| \leq n$ proved with equality only for the Koebe function and its rotations, in 1984 by Louis de Branges [9].

The odd univalent function h that corresponds to the Koebe function k in the sense of (8.7) is

$$h(z) = \frac{z}{1-z^2} = z + z^3 + z^5 + z^7 + \ldots.$$

Since the Koebe function was conjectured (by Bieberbach) to have the largest coefficients of all functions in **S**, it seems natural to think that $z/(1-z^2)$ might have the largest coefficients among all odd functions in **S**. That is, if $h(z) = z + \sum_{n=2}^{\infty} c_{2n-1} z^{2n-1}$ is in **S** then might it be the case that $|c_{2n-1}| \leq 1$ for each n? It is the case that $|c_3| \leq 1$ by considering $g(z) = 1/h(1/z)$ and applying the Area Theorem as we did in the proof of Theorem 8.2. But $|c_5| \leq 1$ fails in general as was shown by Fekete and Szegö. Setting $c_1 = 1$, Robertson made the weaker conjecture that

$$\sum_{k=1}^{n} |c_{2k-1}|^2 \leq n, \quad n \geq 1,$$

and this, in turn, implies the Bieberbach conjecture. To see this, let $f(z) = z + \sum_{n=1}^{\infty} a_n z^n$ be in the class **S** and h be the odd univalent function in **S** for which $h(z)^2 = f(z^2)$ with $h(z) = z + \sum_{n=2}^{\infty} c_{2n-1} z^{2n-1}$. The coefficient of z^{2n} in $h(z)^2$ is $\sum_{k=1}^{n} c_{2k-1} c_{2n-2k+1}$ which must then equal a_n. Thus, by the Cauchy-Schwarz inequality,

$$|a_n| = \left| \sum_{k=1}^{n} c_{2k-1} c_{2n-2k+1} \right|$$
$$\leq \left(\sum_{k=1}^{n} |c_{2k-1}|^2 \right)^{1/2} \left(\sum_{k=1}^{n} |c_{2n-2k+1}|^2 \right)^{1/2} = \sum_{k=1}^{n} |c_{2k-1}|^2 \leq n,$$

where at the last step we assumed Robertson's Conjecture. de Branges proved the Bieberbach Conjecture by proving a conjecture of Milin on the logarithmic coefficients of functions in **S**, that is on the coefficients of $\log(f(z)/z)$, $f \in \mathbf{S}$, that in turn implies the Robertson Conjecture. The logarithmic coefficients γ_n of a function f in **S** are given by the power series expansion

$$\log\left(\frac{f(z)}{z}\right) = 2 \sum_{n=1}^{\infty} \gamma_n z^n.$$

In the case of the Koebe function, the logarithmic coefficients are $\gamma_n = 1/n$ since

$$\log\left(\frac{k(z)}{z}\right) = \log\frac{1}{(1-z)^2} = 2\log\frac{1}{1-z} = 2\sum_{n=1}^{\infty}\frac{z^n}{n}.$$

Set, for each n, $\Delta_n = \sum_{k=1}^{n}(k|\gamma_k|^2 - 1/k)$. Milin's Conjecture is that, for each $n \geq 1$,

$$\sum_{j=1}^{n}\Delta_j = \sum_{k=1}^{n}(n-k+1)(k|\gamma_k|^2 - 1/k) \leq 0.$$

8.2.2 The Koebe 1/4-Theorem

By definition, any function f in the class **S** maps 0 to 0. Since analytic functions are open mappings [16, Theorem 10.7], the simply connected image $D = f(\mathbb{D})$ of the unit disk under f must contain *some* disk of positive radius centred at the origin. The Koebe 1/4-Theorem says that $f(\mathbb{D})$ always contains a disk of radius 1/4 centred at the origin, which is a uniform estimate over all functions in the class.

Theorem 8.3 (Koebe 1/4-Theorem) *Let f be any univalent function in the class **S**. Then the simply connected domain $D = f(\mathbb{D})$ contains the disk $D(0, 1/4)$. If $f(\mathbb{D})$ doesn't contain any larger disk centred at 0 then f is the Koebe function or a rotation of the Koebe function.*

Proof Let f be any univalent function in the class **S** and have power series expansion $f(z) = z + \sum_{n=1}^{\infty} a_n z^n$ as in (8.1). Let w be an omitted value of f (that is, w doesn't belong to the simply connected domain $D = f(\mathbb{D})$). Then, turning to (8.6) and Exercise 8.7, the univalent function f_4 given by

$$f_4(z) = \frac{wf(z)}{w - f(z)} = z + \left(a_2 + \frac{1}{w}\right)z^2 + \left(a_3 + \frac{2a_2}{w} + \frac{1}{w^2}\right)z^3 + \ldots, \quad z \in \mathbb{D},$$

belongs to the class **S**. By Bieberbach's Coefficient Theorem, Theorem 8.2,

$$\left|a_2 + \frac{1}{w}\right| \leq 2.$$

Then, using Bieberbach's Coefficient Theorem a second time,

$$\left|\frac{1}{w}\right| \leq \left|a_2 + \frac{1}{w}\right| + |a_2| \leq 2 + 2 = 4, \qquad (8.12)$$

8.2 Bieberbach's Coefficient Estimate and the Koebe 1/4-Theorem

so that $|w| \geq 1/4$. No values in the disk $D(0, 1/4)$ are omitted values of f and so, as claimed, $f(\mathbb{D})$ contains the disk $D(0, 1/4)$.

If $f(\mathbb{D})$ contains no larger disk centred at 0 then, for each natural number n, the disk $D(0, 1/4 + 1/n)$ meets the complement of $f(\mathbb{D})$. That is, there is some complex number w_n with $|w_n| < 1/4 + 1/n$ that is an omitted value of f. Since $\{w_n\}_{n=1}^\infty$ is a bounded sequence of complex numbers, it has a convergent subsequence, say $\{w_{n_k}\}_{k=1}^\infty$. Writing w for the limit of this subsequence, we see that w must be an omitted value for f (since the complement of D is closed) and $|w| = \lim_{k \to \infty} |w_{n_k}| = \lim_{k \to \infty} (1/4 + 1/n_k) = 1/4$. With this choice of omitted value w, we have equality in (8.12) and so $|a_2| = 2$. By the equality statement in Theorem 8.2, f must be the Koebe function or a rotation of the Koebe function. □

The Koebe 1/4-Theorem can be expressed in more generality by removing all normalisations, much as the basic Schwarz Lemma (Theorem 4.1) became the Schwarz-Pick Lemma—First Version.

For a domain D other than $\mathbb{C}$ and a point z in D, we write

$$\delta_D(z) = \inf\{|z - w| : w \in \mathbb{C} \setminus D\},$$

this being the distance from z to the complement of D or, equivalently, to the boundary of D.

Exercise 8.12 Show that, in fact, there is a point w on the boundary ∂D of D with $\delta_D(z) = |z - w|$. Thus, the infimum defining the distance from z to the complement of D, $\delta_D(z)$, is actually a minimum and there is always a (non-necessarily unique) point closest to z on the boundary of D.

Note that D contains a disk $D(z, r)$ if and only if $\delta_D(z) \geq r$. In other words, $\delta_D(z)$ is the radius of the largest disk centred at z and contained in D. The Koebe 1/4-Theorem can therefore be equivalently expressed as follows: if $f \in \mathbf{S}$ then

$$\delta_{f(\mathbb{D})}(0) \geq \frac{1}{4},$$

with equality if and only if f is the Koebe function or a rotation of the Koebe function.

Suppose, now, that D is any proper simply connected domain and let z_0 be a point in D. Choose a conformal map f of $\mathbb{D}$ onto D with $f(0) = z_0$. We can normalise f to be in the class $\mathbf{S}$: in fact,

$$\tilde{f}(z) = \frac{f(z) - z_0}{f'(0)}, \quad z \in \mathbb{D},$$

is a conformal map in **S**. Under $\tilde{f}$, the disk $\mathbb{D}$ is mapped onto a domain $\tilde{D}$ which is the domain D translated by $-z_0$ and scaled by $1/f'(0)$. Then,

$$\delta_{\tilde{D}}(0) = \frac{1}{|f'(0)|} \delta_D(z_0).$$

By the Koebe 1/4-Theorem,

$$\delta_D(z_0) = |f'(0)| \delta_{\tilde{D}}(0) \geq \frac{|f'(0)|}{4}.$$

The transformation formula (6.22) for the hyperbolic metric leads to $\rho_{\mathbb{D}}(0) = \rho_D(f(0))|f'(0)|$, so that $\rho_D(z_0) = 2/|f'(0)|$. Then, $\rho_D(z_0) \geq 1/(2\delta_D(z_0))$. Equality holds if and only if $\delta_{\tilde{D}}(0) = 1/4$, which is the case precisely when $\tilde{D}$ is the Koebe region or a rotation of same. In that case, the original domain D is a slit plane (that is, the complex plane with a half line removed) and z_0 lies on the continuation of the slit.

An analogous upper bound comes from monotonicity of the hyperbolic metric and comparison with a disk. Thus, up to absolute constants, the hyperbolic metric is comparable to the reciprocal of the distance to the boundary.

Theorem 8.4 *Let D be a proper simply connected domain and z a point in D. Then,*

$$\frac{1}{2\,\delta_D(z)} \leq \rho_D(z) \leq \frac{2}{\delta_D(z)} \tag{8.13}$$

Both inequalities are sharp.

Proof The lower bound, as we have seen, comes from the Koebe 1/4-Theorem. Equality holds if and only if D is a slit plane and z lies on the continuation of the slit.

For z in D, the domain D contains the disk $D(z, \delta_D(z))$ and so, by monotonicity of the hyperbolic metric (Theorem 6.11) and Exercise 6.7,

$$\rho_D(z) \leq \rho_{D(z,\delta_D(z))}(z) = \frac{2}{\delta_D(z)}.$$

By the equality statement for (6.19), equality holds if and only if D is a disk and z is the centre of the disk. □

In the case of a convex image domain, it is possible to improve the lower bound in (8.13). A domain D is *convex* if whenever z_1 and z_2 are in D then so is the line segment $[z_1, z_2]$ joining them. If f is in **S** and the image $D = f(\mathbb{D})$ of the unit disk under f is a convex domain we call f a *convex univalent function*. Consider a (proper) convex domain D and a point z_0 in D. As before, D contains the disk $D(z_0, \delta_D(z_0))$ where $\delta_D(z_0)$ is the distance from z_0 to the boundary of D. Choose a point w on the boundary of D on the circle $C(z_0, \delta_D(z_0))$ (cf. Exercise 8.12). Let

8.2 Bieberbach's Coefficient Estimate and the Koebe 1/4-Theorem

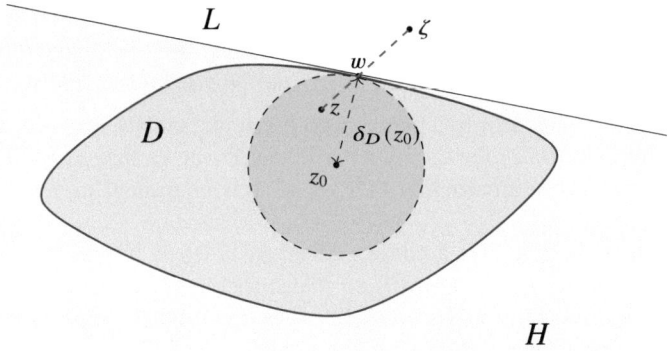

Fig. 8.2 Convex domain D, disk of largest radius $\delta_D(z_0)$ centred at z_0, and half-plane H with $D \subseteq H$

L be the line tangent to this circle at w and H the half-plane bounded by L and containing z_0. See Fig. 8.2.

The convexity of D implies that D lies in the half-plane H. In fact, the line through any point ζ lying outside the closure of H, and through w, will enter the disk $D(z_0, \delta_D(z_0))$. Take any point z on the line through ζ and w and in the disk $D(z_0, \delta_D(z_0))$. Since $z \in D$, it cannot be that $\zeta \in D$ as w lies on the line segment between them but is not in D. Since D is open, neither can any point on L belong to D. Hence, $D \subseteq H$.

By the monotonicity of the hyperbolic metric, Theorem 6.11, and the explicit formula for the hyperbolic metric in a half-plane (6.21),

$$\rho_D(z_0) \geq \rho_H(z_0) = \frac{1}{\delta_D(z_0)},$$

with equality only if $D = H$. In summary, we have the following result.

Theorem 8.5 *Let D be a proper convex domain and z a point in D. Then,*

$$\frac{1}{\delta_D(z)} \leq \rho_D(z) \leq \frac{2}{\delta_D(z)}.$$

Equality holds in the left hand inequality if and only if D is a half-plane and in the right hand inequality if and only if D is a disk and z is its centre.

When written in terms of the conformal mapping, this result takes the following form, similar to Theorem 8.3.

Theorem 8.6 *Let f be any convex univalent function in the class* **S**. *Then the convex domain $D = f(\mathbb{D})$ contains the disk $D(0, 1/2)$. If $f(\mathbb{D})$ doesn't contain any larger disk centred at 0 then $f(z) = z/(1-z)$ or a rotation of this mapping.*

Proof For any univalent function f with image domain $D = f(\mathbb{D})$, the transformation formula (6.22) for the hyperbolic metric shows that $\rho_D(f(0))|f'(0)| = \rho_{\mathbb{D}}(0) = 2$. The normalisation $|f'(0)| = 1$ is equivalent to the normalisation $\rho_D(f(0)) = 2$.

Now suppose that f is a convex univalent function in the class **S**. By Theorem 8.5, $2 = \rho_D(0) \geq 1/\delta_D(0)$ so that $\delta_D(0) \geq 1/2$. If equality holds then, by the case of equality in Theorem 8.5, D is a half-plane. Since 0 is in this half-plane, and the distance from 0 to the boundary is $1/2$, the half-plane must be a rotation of $H = \{z : \Re z > -1/2\}$. The normalised map to H is $f(z) = \big((1+z)/(1-z) - 1\big)/2 = z/(1-z)$ (see Exercise 4.9). The mapping of $\mathbb{D}$ onto the rotated half-plane $e^{i\theta} H$ is given by (8.4), specifically $f_1(z) = z/(1 - e^{-i\theta}z)$. $\square$

Here is another way of proving Theorem 8.6 (see [27, Theorem 4.2.3]).

Exercise 8.13 Let f be a convex univalent function in the class **S** and w an omitted value for f. Show that $h(z) = (f(z) - w)^2$ is univalent in the unit disk. By applying the Koebe 1/4-Theorem to the normalised form of h, deduce that $|w| \geq 1/2$.

8.3 Growth and Distortion Theorems for Functions in the Class S

The main reason that we work with normalised univalent functions is that the class **S** forms a normal family—it is a compact subset of the space $\mathscr{H}(\mathbb{D})$ of analytic functions on the unit disk $\mathbb{D}$ with the topology of uniform convergence on compact subsets. This is one consequence of the growth and distortion results in this section. By a 'growth theorem' we mean an estimate for the size $|f|$ of the function f and by a 'distortion theorem' we mean an estimate for the size of its derivative $|f'|$. These estimates stem, once again, from Bieberbach's bound for the second coefficient a_2.

Let f be a normalised univalent function in the class **S**, and let α be a point in the unit disk $\mathbb{D}$. As we saw in (8.5), the function

$$f_2(z) = \frac{(f \circ M)(z) - (f \circ M)(0)}{(f \circ M)'(0)} = \frac{(f \circ M)(z) - f(\alpha)}{f'(\alpha)(|\alpha|^2 - 1)}, \quad z \in \mathbb{D},$$

belongs to the class **S**, where $M(z) = (\alpha - z)/(1 - \bar{\alpha}z)$. Since

$$M'(z) = \frac{|\alpha|^2 - 1}{(1 - \bar{\alpha}z)^2}, \quad \text{and} \quad M''(z) = \frac{2\bar{\alpha}(|\alpha|^2 - 1)}{(1 - \bar{\alpha}z)^3},$$

8.3 Growth and Distortion Theorems for Functions in the Class S

we find that

$$M(0) = \alpha, \quad M'(0) = |\alpha|^2 - 1, \quad M''(0) = 2\overline{\alpha}(|\alpha|^2 - 1).$$

The power series expansion for f_2 begins

$$f_2(z) = z + a_2(\alpha)z^2 + a_3(\alpha)z^3 + \ldots, \quad z \in \mathbb{D},$$

where $a_2(\alpha)$ can be computed as $f_2''(0)/2$. Doing the computation,

$$f_2''(z) = \frac{1}{f'(\alpha)(|\alpha|^2 - 1)} \left[f''(M(z)) M'(z)^2 + f'(M(z)) M''(z) \right],$$

so that

$$a_2(\alpha) = \frac{1}{2f'(\alpha)(|\alpha|^2 - 1)} \left[f''(\alpha)(|\alpha|^2 - 1)^2 + 2\overline{\alpha}(|\alpha|^2 - 1) f'(\alpha) \right]$$

$$= \frac{|\alpha|^2 - 1}{2\alpha} \left[\alpha \frac{f''(\alpha)}{f'(\alpha)} - \frac{2|\alpha|^2}{1 - |\alpha|^2} \right].$$

By Bieberbach's Coefficient Estimate, Theorem 8.2, $|a_2(\alpha)| \leq 2$ with equality if and only if f_2, and hence f, is the Koebe function or a rotation of the Koebe function.

Replacing α by $z = re^{i\theta}$, we obtain the following result.

Theorem 8.7 *Let f be a normalised univalent function in the class S and let $z = re^{i\theta}$ be a point in the unit disk $\mathbb{D}$. Then,*

$$\left| \frac{zf''(z)}{f'(z)} - \frac{2r^2}{1 - r^2} \right| \leq \frac{4r}{1 - r^2}. \tag{8.14}$$

Equality holds if and only if f is the Koebe function or a rotation of the Koebe function.

From this result we can obtain a sharp distortion theorem. In the case of the Koebe function $k(z) = z/(1 - z)^2$, we have

$$k'(z) = \frac{1 + z}{(1 - z)^3}, \quad z \in \mathbb{D}.$$

On the circle $C(0, r)$, therefore, the maximum value of $|k'(z)|$ occurs at $z = r$ and is $(1+r)/(1-r)^3$ while the minimum value of $|k'(z)|$ is $(1-r)/(1+r)^3$ and occurs at $z = -r$. We see in the Distortion Theorem that these are actually the maximum and minimum values over the whole class **S**.

Theorem 8.8 (Distortion Theorem) *For a univalent function f in $\mathbf{S}$ and a point $z = re^{i\theta}$ in the unit disk $\mathbb{D}$,*

$$\frac{1-r}{(1+r)^3} \leq |f'(z)| \leq \frac{1+r}{(1-r)^3}. \tag{8.15}$$

Strict inequality holds for all f except for the Koebe function and its rotations.

Proof Since, in general, $|\Re w| \leq |w|$, we deduce from (8.14) that

$$-\frac{4r}{1-r^2} \leq \Re\left(\frac{zf''(z)}{f'(z)} - \frac{2r^2}{1-r^2}\right) \leq \frac{4r}{1-r^2},$$

so that

$$\frac{2r^2 - 4r}{1-r^2} \leq \Re\left(\frac{zf''(z)}{f'(z)}\right) \leq \frac{2r^2 + 4r}{1-r^2}. \tag{8.16}$$

Since f' is analytic and non-zero in $\mathbb{D}$, it has an analytic logarithm $g(z) = \log f'(z)$ such that $f'(z) = \exp(g(z))$. Since $f'(0) = 1$, we may choose $g(0) = 0$. Then, for fixed θ,

$$\frac{\partial}{\partial r} \log f'(re^{i\theta}) = \frac{1}{f'(re^{i\theta})} \frac{\partial}{\partial r} f'(re^{i\theta}) = \frac{1}{f'(re^{i\theta})} f''(re^{i\theta})e^{i\theta}, \tag{8.17}$$

so that

$$r \frac{\partial}{\partial r} \log f'(re^{i\theta}) = \frac{zf''(z)}{f'(z)}, \quad z = re^{i\theta}.$$

Taking the real part, we find that

$$\Re\left(\frac{zf''(z)}{f'(z)}\right) = r \frac{\partial}{\partial r} \log |f'(re^{i\theta})|. \tag{8.18}$$

Together, (8.16) and (8.18) lead to

$$\frac{2t-4}{1-t^2} \leq \frac{\partial}{\partial t} \log |f'(te^{i\theta})| \leq \frac{2t+4}{1-t^2}. \tag{8.19}$$

Integrating with respect to t from 0 to r, for fixed θ, and using $\log |f'(0)| = 0$, leads to

$$\int_0^r \frac{2t-4}{1-t^2} \, dt \leq \log |f'(re^{i\theta})| \leq \int_0^r \frac{2t+4}{1-t^2} \, dt. \tag{8.20}$$

8.3 Growth and Distortion Theorems for Functions in the Class S

The inequalities (8.15) follow by evaluating these integrals (by the method of partial fractions).

Suppose that equality holds in (8.15), say on the left, for some $z = re^{i\theta}$. Then equality holds in the left hand integrand in (8.20), and hence on the left in (8.19) for all t between 0 and r. Then,

$$\frac{2t-4}{1-t^2} = \frac{\partial}{\partial t}\log|f'(te^{i\theta})|, \quad 0 < t < r.$$

By continuity, equality holds at $t = 0$ so that, using also the identity (8.17),

$$-4 = \frac{\partial}{\partial t}\log|f'(te^{i\theta})|\Big|_{t=0} = \Re\left(\frac{f''(0)e^{i\theta}}{f'(0)}\right) = \Re(2a_2 e^{i\theta}) \geq -2|a_2|.$$

Here a_2 is the coefficient of z^2 in the power series expansion (8.1) of f. This forces $|a_2| \geq 2$ and so, by Bieberbach's Coefficient Theorem, $|a_2| = 2$. By the case of equality in the coefficient theorem, f is either the Koebe function or a rotation of the Koebe function. The same conclusion is reached if equality holds for some z in the right hand inequality of (8.15). □

The Growth Theorem now follows in a reasonably straightforward manner from the Distortion Theorem.

Theorem 8.9 (Growth Theorem) *For a univalent function f in $\mathbf{S}$ and a point $z = re^{i\theta}$ in the unit disk $\mathbb{D}$,*

$$\frac{r}{(1+r)^2} \leq |f(z)| \leq \frac{r}{(1-r)^2}. \tag{8.21}$$

Strict inequality holds for all f except for the Koebe function and its rotations.

Proof The right hand inequality follows directly from the right hand inequality in (8.15). In fact, for $z = re^{i\theta}$ in $\mathbb{D}$,

$$|f(z)| = \left|\int_0^r f'(te^{i\theta})\,dt\right| \leq \int_0^r |f'(te^{i\theta})|\,dt \leq \int_0^r \frac{1+t}{(1-t)^3}\,dt = \frac{r}{(1-r)^2}.$$

If equality holds then equality holds in the right hand inequality of (8.15) and hence f is either the Koebe function or a rotation of the Koebe function, this by the equality statement in the Distortion Theorem.

For the left hand inequality, fix r and set $D_r = f(D(0,r))$. The boundary of D_r is the image of the circle $C(0,r)$ under f. The domain D_r, and hence D itself, contains the disk $D(0, \delta_{D_r}(0))$, and $|f(z)| \geq \delta_{D_r}(0)$ if $|z| = r$. Choose a point w in $f(C(0,r))$ closest to 0. Let γ be the preimage of the line segment $[0, w]$ under

f, parameterised by $\gamma(t) = f^{-1}(tw)$, $0 \leq t \leq 1$. Then, γ lies in $D(0, r)$ and $f'(\gamma(t))\gamma'(t) = (f \circ \gamma)'(t) = (tw)'(t) = w$. Thus, for $|z| = r$ and using (8.15),

$$|f(z)| \geq \delta_{D_r}(0) = |w| = \int_0^1 |f'(\gamma(t))| \, |\gamma'(t)| \, dt$$
$$\geq \int_0^1 \frac{1 - |\gamma(t)|}{(1 + |\gamma(t)|)^3} |\gamma'(t)| \, dt$$
$$\geq \int_0^1 \frac{1 - |\gamma|(t)}{(1 + |\gamma|(t))^3} |\gamma|'(t) \, dt$$
$$= \int_0^r \frac{1 - u}{(1 + u)^3} \, du = \frac{r}{(1 + r)^2}.$$

Here we used the general fact, for a curve $\gamma(t)$, that $|\gamma'(t)| \geq |\gamma|'(t)$. One may verify this geometrically clear estimate by setting $\gamma(t) = \rho(t)e^{i\theta(t)}$, so that $|\gamma|(t) = \rho(t)$, and differentiating. Once again, if equality holds then equality holds in the left hand side of (8.15) and f is a rotation of the Koebe function. □

An immediate consequence of the Growth Theorem and Montel's Theorem is that the class **S** is compact.

Theorem 8.10 *The class* **S** *is a compact subset of* $\mathcal{H}(\mathbb{D})$.

Proof By the Growth Theorem, in particular the right hand inequality in (8.21), functions in the class **S** are uniformly bounded on each closed disk $\overline{D(0, r)}$ with $0 < r < 1$, in fact by $r/(1-r)^2$. Functions in the class **S** are therefore uniformly bounded on every compact subset of $\mathbb{D}$ since any such compact set is contained in the closed disk of radius r for some r between 0 and 1. By Montel's Theorem, Theorem 5.8, **S** is a relatively compact subset of $\mathcal{H}(\mathbb{D})$.

By Hurwitz's Theorem (Theorem 5.7), the locally uniform limit of a sequence of functions from **S** is either univalent or constant. But the latter possibility is ruled out because of the normalisation $f'(0) = 1$, which will continue to be satisfied by the limit of any sequence of functions from **S**. Hence, **S** is also closed and is therefore compact. □

Exercise 8.14 Suppose that $\mathscr{F}$ is a family of conformal maps of the unit disk. Suppose that there are finite constants C_1 and C_2 such that $|f(0)| \leq C_1$ and $|f'(0)| \leq C_2$ for each f in $\mathscr{F}$. Show that $\mathscr{F}$ is a normal family.

The compactness of **S** means that extremal problems arising from continuous functionals always have an extremiser in **S**. An important example is evaluation of the nth derivative. We saw in Remark 5.1 that, as a consequence of Theorem 5.5, the linear functional $\mathscr{D}_0^{(n)} : \mathcal{H}(\mathbb{D}) \to \mathbb{C}$ defined by $\mathscr{D}_0^{(n)}(f) = f^{(n)}(0)$ is continuous.

8.3 Growth and Distortion Theorems for Functions in the Class S

Since **S** is compact, $\mathscr{D}_0^{(n)}$ achieves its maximum modulus on **S**. That is, there is a function f_n in **S** for which

$$|f_n^{(n)}(0)| = \max\{|f^{(n)}(0)| : f \in \mathbf{S}\}.$$

As a consequence, there is a function f_n in **S** that maximises $|a_n|$ over all functions f in the class **S** with power series (8.1). The Bieberbach Conjecture is the assertion that, for each n, the extremisers f_n are precisely the Koebe function and its rotations.

We end this section with the following combined growth/distortion estimate.

Theorem 8.11 *For a univalent function f in **S** and $z = re^{i\theta}$ in the unit disk $\mathbb{D}$,*

$$\frac{1-r}{1+r} \leq \left|\frac{zf'(z)}{f(z)}\right| \leq \frac{1+r}{1-r}. \tag{8.22}$$

Strict inequality holds for all f except for the Koebe function and its rotations.

Proof For f in the class **S** and for z in the unit disk, set

$$g(w) = \frac{f\left(\frac{z-w}{1-\bar{z}w}\right) - f(z)}{f'(z)(|z|^2 - 1)}, \quad w \in \mathbb{D},$$

as in the transformation (8.5), so that g is also in the class **S**. Applying the Growth Theorem to g at $w = z = re^{i\theta}$ and using $f(0) = 0$, we find that

$$\frac{r}{(1+r)^2} \leq \frac{|f(z)|}{|f'(z)|(1-|z|^2)} \leq \frac{r}{(1-r)^2}.$$

Then,

$$\frac{(1-r)^2}{r(1-r^2)} \leq \frac{|f'(z)|}{|f(z)|} \leq \frac{(1+r)^2}{r(1-r^2)},$$

which is (8.22).

If equality holds on either the left or right of (8.22), then equality holds in the Growth Theorem when applied to g and so g, hence f itself, is the Koebe function or a rotation of the Koebe function. □

Chapter 9
Carathéodory Convergence of Domains and Hyperbolic Geodesics

Convergence of analytic functions was the focus of our attention for much of Chap. 5. In the case of univalent functions, we had the theorem of Hurwitz, Theorem 5.7, to the effect that a locally uniform limit of univalent functions will itself be either constant or univalent. The limit of a convergent sequence of conformal maps of the unit disk will therefore be either constant or itself a conformal mapping. It is natural to ask how, in the second case, the sequence of images of the unit disk might 'converge' to the image of the disk under the limit mapping. In essence, we wish to relate the analytic notion of convergence of the mapping functions to an appropriate geometric notion of convergence of the image domains. This idea of viewing the analytic and the geometric as two sides of the same coin is at the core of geometric function theory. Carathéodory has formulated the correct notion of convergence of a sequence of domains in this context—it is known as *kernel convergence*—and has shown that convergence of the conformal maps, suitably normalised, is equivalent to kernel convergence of the image domains. Kernel convergence and Carathéodory's theorem on the subject are discussed in Sect. 9.1.

Having established Carathéodory's result, we turn to a more in-depth study of hyperbolic geodesics in simply connected domains and, in particular, how the shape of a domain affects the shape of its geodesics. Since geodesics are length-minimising curves with respect to the hyperbolic metric and since, by Koebe's 1/4-Theorem (Theorem 8.4), the hyperbolic metric is comparable to the reciprocal of the distance to the boundary in a simply connected domain, one would expect hyperbolic geodesics to stay as far away from the boundary as possible. Of several results that give this heuristic statement concrete form, we discuss results of Jorgensen [17] in Sect. 9.2.

9.1 Carathéodory Convergence of Domains

By Hurwitz's Theorem, Theorem 5.7, the limit f of a sequence of conformal mappings $\{f_n\}_1^\infty$ of the unit disk is either constant or is itself a conformal mapping. We need to be careful when seeking a characterisation of the convergence of conformal maps in terms of the 'convergence' of the corresponding image domains since the image domain does not determine the conformal map: if h is a conformal map of $\mathbb{D}$ onto the simply connected domain D then so is $h \circ \phi$ for any automorphism ϕ of the disk $\mathbb{D}$ (these being described in full in Theorem 4.2). Whatever notion of 'convergence' of domains we eventually end up with, surely the constant sequence of domains $\{D_n\}_{n=1}^\infty$ with each D_n being the unit disk $\mathbb{D}$ *must* be 'convergent'? Any sequence $\{\phi_n\}_{n=1}^\infty$ of automorphisms of the disk, $\phi_n(z) = (\alpha_n - z)/(1 - \overline{\alpha_n} z)$, $z \in \mathbb{D}$, $\alpha_n \in \mathbb{D}$, maps to this constant sequence of image domains. However, this sequence of conformal maps is not convergent unless the sequence $\{\alpha_n\}$ is convergent. Even if we take each α_n to be zero, we could set $\phi_n(z) = \beta_n z$ where each β_n has modulus 1. Again, this sequence of mappings is only convergent if the sequence $\{\beta_n\}$ is convergent even though the sequence of image domains is 'convergent'. If we use the normalisation in the statement of the Riemann Mapping Theorem, Theorem 6.7, then the conformal map becomes unique so that not only does the conformal map determine the image domain, the image domain will determine the conformal map. To implement this normalisation, we assume that each image domain contains the same designated point, which may as well be the origin, and that the conformal maps in question fix the origin and have positive derivative at the origin.

Standing Assumption In this section, $\{f_n\}_{n=1}^\infty$ will stand for a sequence of conformal maps of the unit disk $\mathbb{D}$ each normalised so that $f_n(0) = 0$ and $f_n'(0)$ is positive. We denote the simply connected domain $f_n(\mathbb{D})$ by D_n. Also, f will be a conformal mapping of the unit disk, again normalised by $f(0) = 0$, $f'(0)$ positive, and we will denote the domain $f(\mathbb{D})$ by D.

The following two examples highlight the main issues involved in the question of how convergence of conformal maps is reflected in 'convergence' of the corresponding image domains.

Example 9.1 Let D_n be the domain formed by joining the disks $D(0, 1)$ and $D(3, 1)$ by a channel of width $2/n$. Formally, let R_n be the rectangle $(0, 3) \times (-1/n, 1/n)$ and set

$$D_n = D(0, 1) \cup D(3, 1) \cup R_n.$$

The domain D_n has a dumbbell shape as shown in Fig. 9.1. The disk $D(3, 1)$ is hyperbolically very far away from 0. To see this, take a point w in the disk $D(3, 1)$. Any curve γ joining 0 to w contains a subcurve $\tilde{\gamma}$ that travels along the channel and joins the circles $C(0, 1)$ and $C(3, 1)$. This subcurve has length at least 1 and

9.1 Carathéodory Convergence of Domains

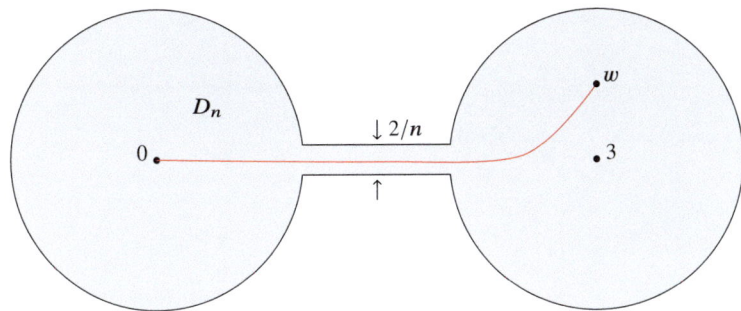

Fig. 9.1 The dumbbell-shaped domain D_n in Example 9.1. Also shown is a typical curve joining 0 to w in $D(3, 1) \subset D_n$

the distance to the boundary at points on the subcurve is at most $1/n$. Thus, by the Koebe 1/4-Theorem,

$$L_{\rho_{D_n}}(\gamma) \geq L_{\rho_{D_n}}(\tilde{\gamma}) \geq \int_{\tilde{\gamma}} \rho_{D_n}(\zeta) |d\zeta| \geq \int_{\tilde{\gamma}} \frac{1}{2\delta_{D_n}(\zeta)} |d\zeta| \geq \frac{n}{2} \int_{\tilde{\gamma}} |d\zeta| \geq \frac{n}{2}.$$

Since the hyperbolic distance $d(0, w; D_n)$ is the infimum of $L_{\rho_{D_n}}(\gamma)$ over all curves γ joining 0 to w in D_n, we see that $d(0, w; D_n) \geq n/2$ for $w \in D(3, 1) \subset D_n$.

With f_n being the normalised conformal map of $\mathbb{D}$ onto D_n and with z in the unit disk being the preimage under f_n of w in $D(3, 1)$ then, by the conformal invariance of the hyperbolic distance (6.18) and by the formula (4.22) for the hyperbolic distance in the disk, $d(0, w; D_n) = d(0, z; \mathbb{D}) = \log((1 + |z|)/(1 - |z|))$. Since $d(0, w; D_n) \to \infty$ as $n \to \infty$, we find that $|z| \to 1$, so that the preimage of the disk $D(3, 1)$ under f_n moves out to the boundary of the unit disk as n increases. In the limit, the preimage of the disk $D(3, 1)$ disappears completely so that f_n converges to the identity map $f(z) = z$ on $\mathbb{D}$ and the image domain D_n 'converges' to $\mathbb{D}$. From a geometrical point of view, as n increases and the channel narrows the domain D_n degenerates to two disjoint components $\mathbb{D}$ and $D(3, 1)$. The limit mapping f then picks out the component $\mathbb{D}$.

It should be noted that there is nothing special about the component $\mathbb{D}$ in this example. If we had chosen the conformal map f_n to map $\mathbb{D}$ onto D_n but with $f_n(0) = 3$ and $f_n'(0)$ positive then f_n would converge to $f(z) = z + 3$ in the limit with image domain $D(3, 1)$. Note also that if instead of the disk $D(3, 1)$ we had the half-plane $\{z : \Re z > 2\}$, say, the resulting limit function would still be the identity map $f(z) = z$, for $z \in \mathbb{D}$. This emphasises that how a domain looks from a Euclidean point of view may be totally unrelated to how it looks from a hyperbolic point of view.

Example 9.2 Set $f_n(z) = z$, $z \in \mathbb{D}$, for n even, and set $f_n(z) = 2z$, $z \in \mathbb{D}$, for n odd. Then, $D_n = \mathbb{D}$ for n even and $D_n = 2\mathbb{D}$ for n odd. See Fig. 9.2. Of course, this

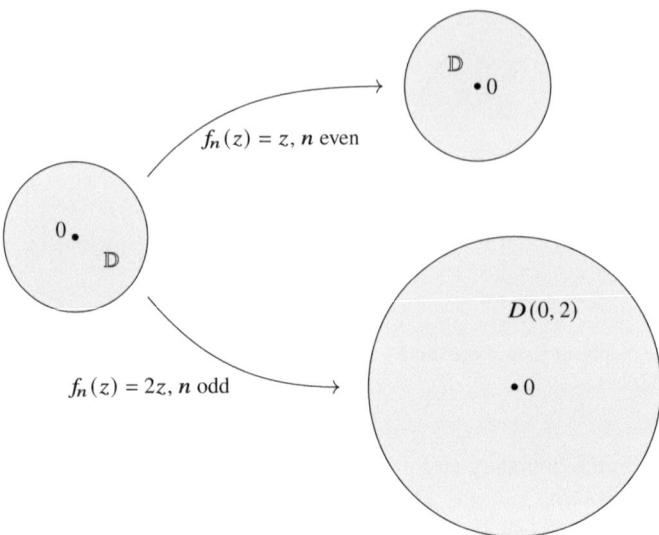

Fig. 9.2 A non-convergent sequence of conformal maps

sequence of conformal maps is not convergent. Yet both subsequences $\{f_{2n}\}_{n=1}^{\infty}$ and $\{f_{2n+1}\}_{n=0}^{\infty}$ are convergent to $f(z) = z$ and to $f(z) = 2z$ respectively.

With these examples in mind, suppose that the sequence of conformal maps $\{f_n\}_{n=1}^{\infty}$ of the unit disk $\mathbb{D}$ converges to the conformal map f. Using the notation in the earlier *standing assumption*, we now employ a variation on the proof of Hurwitz's Theorem to obtain a uniform version of that result. Suppose that w_0 lies in D and that z_0 is its preimage in $\mathbb{D}$ under f so that $f(z_0) = w_0$. Choose r positive so that the closed disk $\overline{D(z_0, r)}$ lies in the unit disk $\mathbb{D}$. Since f is univalent, f does not take the value w_0 on the circle $C(z_0, r)$ and so $\delta = \frac{1}{3}\min\{|f(z) - w_0| : z \in C(z_0, r)\}$ is positive. Not only does f take the value w_0 at z_0, it takes every value in the closed disk $\overline{D(w_0, \delta)}$ somewhere in the disk $D(z_0, r)$. This is a consequence of Rouché's Theorem. For $w \in \overline{D(w_0, \delta)}$ and for z on the circle $C(z_0, r)$, we have $|(f(z) - w_0) - (f(z) - w)| = |w - w_0| \leq \delta$, and $3\delta \leq |f(z) - w_0|$. Thus, for each w in $\overline{D(w_0, \delta)}$, we have

$$|(f(z) - w_0) - (f(z) - w)| < |f(z) - w_0| + |f(z) - w|, \quad z \in C(z_0, r).$$

By Rouché's Theorem, $f(z) - w_0$ and $f(z) - w$ have the same number of zeros in $D(z_0, r)$, namely exactly one. In particular, $\overline{D(w_0, \delta)}$ lies in the domain D.

Since f_n converges to f uniformly on the circle $C(z_0, r)$, there is an integer N such that $|f_n(z) - f(z)| \leq \delta$ for all z on $C(z_0, r)$ and all $n \geq N$. Fix w in $\overline{D(w_0, \delta)}$.

9.1 Carathéodory Convergence of Domains

For z on the circle $C(z_0, r)$, we have $|f(z) - w| \geq |f(z) - w_0| - |w_0 - w| \geq 3\delta - \delta = 2\delta$. Then, for $n \geq N$ and $z \in C(z_0, r)$,

$$|(f_n(z) - w) - (f(z) - w)| = |f_n(z) - f(z)| \leq \delta < 2\delta \leq |f_n(z) - w| + |f(z) - w|.$$

It follows from Rouché's Theorem that $f_n(z) - w$ and $f(z) - w$ have the same number of zeros in the disk $D(z_0, r)$, namely one. In particular, $\overline{D(w_0, \delta)}$ lies in each domain D_n for $n \geq N$.

We have shown the following: *if $w_0 \in D$ there is a positive integer N and a positive δ such that the closed disk $\overline{D(w_0, \delta)}$ lies in D and lies in D_n for all $n \geq N$.* This leads to the following definition.

Definition 9.1 The *pre-kernel* of a sequence of simply connected domains $\{D_n\}_{n=1}^{\infty}$ is the open set Ω consisting of all points w_0 for which there is a positive δ such that the closed disk $\overline{D(w_0, \delta)}$ lies in D_n for all sufficiently large n.

If w_0 lies in the pre-kernel Ω of the sequence of domains $\{D_n\}_{n=1}^{\infty}$, we refer to the component of Ω that contains w_0 as the *kernel of* $\{D_n\}_{n=1}^{\infty}$ *with respect to* w_0 and denote it by Ω_{w_0}. If w_0 does not lie in Ω we say that the kernel of $\{D_n\}_{n=1}^{\infty}$ with respect to w_0 does not exist.

Note that if K is a compact subset of the pre-kernel Ω then K is contained in D_n for all sufficiently large n.

In Example 9.1 the pre-kernel of the sequence of dumbbell domains is $\Omega = D(0, 1) \cup D(3, 1)$, showing that the pre-kernel may be disconnected and consist of a number of components. It is not difficult to imagine more elaborate versions of Example 9.1. In Example 9.2 the pre-kernel of the sequence of domains is $\Omega = \mathbb{D}$.

We're assuming that f_n converges to f and have shown that the domain D is then a subset of the pre-kernel Ω of the simply connected domains $\{D_n\}_{n=1}^{\infty}$. Being a domain containing 0, it must be that D is contained in the component of the kernel containing 0, that is in Ω_0, so that $D \subseteq \Omega_0$. We next show that $\Omega_0 \subseteq D$ so that, in fact, the domain D coincides with Ω_0. We also show that the inverse functions f_n^{-1} converge, in an appropriate sense, to the inverse of f.

Proposition 9.1 *Suppose that $f_n: \mathbb{D} \to D_n$, $n \geq 1$, is a sequence of conformal maps normalised by $f_n(0) = 0$ and $f_n'(0)$ positive. Suppose that $f_n \to f$ locally uniformly on $\mathbb{D}$ where $f: \mathbb{D} \to D$ and f is not constant. Then, 0 lies in the pre-kernel of the sequence of domains $\{D_n\}_{n=1}^{\infty}$, and D is the kernel of $\{D_n\}_{n=1}^{\infty}$ with respect to 0.*

Moreover, the sequence of inverse functions $\{f_n^{-1}\}_{n=1}^{\infty}$ converges to f^{-1} locally uniformly on D.

The final statement on the convergence of the inverse functions requires clarification since the inverse functions are defined on different domains! The statement that 'f_n^{-1} converges to f^{-1} locally uniformly on D' is to mean that f_n^{-1} converges to f^{-1} uniformly on each compact subset K of D. Since D lies in the pre-kernel of the sequence of domains $\{D_n\}_{n=1}^{\infty}$ on which the inverse functions are defined, the compact set K belongs to D_n for all sufficiently large n. Hence, f_n^{-1} is defined on K for all sufficiently large n and it makes sense to talk of f_n^{-1} converging to f^{-1} on K.

Proof We have already shown that 0, in fact all of D, is contained in the pre-kernel Ω of the sequence of image domains $\{D_n\}_{n=1}^{\infty}$, so that $D \subseteq \Omega_0$.

We wish to show the reverse inclusion, namely that $\Omega_0 \subseteq D$. First, a simple observation. Referring back to Remark 5.1 following Theorem 5.5, we see that $f_n'(0) \to f'(0)$ as $n \to \infty$. Since $f'(0)$ is finite, there is a number C for which $|f_n'(0)| \leq C < \infty$ for each n. Then, $|(f_n^{-1})'(0)| \geq 1/C > 0$ for each n.

Now let $\{K_j\}_{j=1}^{\infty}$ be an exhaustion of Ω_0 by compact sets (with the properties set out in Lemma 5.2), and set $V_j = \operatorname{int}(K_j)$. We will show that $V_j \subset D$ for each j, from which we conclude that $\Omega_0 \subseteq D$ since $\Omega_0 = \cup_{j=1}^{\infty} V_j$. Fix j large enough so that the origin lies in V_j. By the definition of the pre-kernel, K_{j+1} lies in D_n for all sufficiently large n, say $n \geq N$, and for each such n the inverse function f_n^{-1} maps V_{j+1} conformally into $\mathbb{D}$ with $f_n^{-1}(0) = 0$. By Montel's Theorem (Theorem 5.8), the bounded family of functions $\{f_n^{-1}\}_{n=N}^{\infty}$ forms a normal family on V_{j+1} and hence has a subsequence, say $\{f_{n_k}^{-1}\}_{k=1}^{\infty}$, that converges locally uniformly on V_{j+1}, say to g. Then, $f_{n_k}^{-1}$ converges uniformly to g on the compact subset K_j of V_{j+1}. By Hurwitz's Theorem, g is univalent on V_{j+1} (it cannot be constant since $|g'(0)| = \lim_{k\to\infty} |(f_{n_k}^{-1})'(0)| \geq 1/C > 0$). For each w in V_{j+1}, we have $g(w) = \lim_{k\to\infty} f_{n_k}^{-1}(w)$. This exhibits $g(w)$ as the limit of a sequence of points in $\mathbb{D}$ and so we must have $|g(w)| \leq 1$. By the univalence of g and the Open Mapping Theorem, $g(w) \in \mathbb{D}$. In short, $g(V_{j+1})$ lies in the unit disk and $f(g(w))$ is defined for all w in V_{j+1}.

Since K_j is a compact subset of V_{j+1}, the set $X = g(K_j)$ is compact in $g(V_{j+1})$. Fix w in V_j so that $g(w)$ lies in the open subset $g(V_j)$ of X. The sequence of points $\{f_{n_k}^{-1}(w)\}_k$ converges to $g(w)$ and therefore lies in X for sufficiently large k. By the assumption that f_n converges to f, the functions $\{f_{n_k}\}_k$ converge to f uniformly on X. It then follows from Exercise 5.7 that

$$w = \lim_{k\to\infty} f_{n_k}(f_{n_k}^{-1}(w)) = f(g(w)).$$

Being the image under f of a point $g(w)$ in $\mathbb{D}$, the point w lies in D. That is, $V_j \subseteq D$. Moreover, $g = f^{-1}$ on V_j since, for $w \in V_j$, we have $g(w) = (f^{-1} \circ f)(g(w)) = f^{-1}(f(g(w))) = f^{-1}(w)$.

To conclude, we show that the inverse functions f_n^{-1} converge locally uniformly to f^{-1} on D. It is sufficient to show that every subsequence $\{f_{n_k}^{-1}\}_{k=1}^{\infty}$ of $\{f_n^{-1}\}_{n=1}^{\infty}$ has a further subsequence that converges locally uniformly to f^{-1} on $\mathbb{D}$.[1] Then the original sequence $\{f_n^{-1}\}_{n=1}^{\infty}$ itself must converge to f^{-1} locally uniformly on $\mathbb{D}$. If

---

[1] This idea is sometimes known as the *Urysohn subsequence principle*.

9.1 Carathéodory Convergence of Domains

not, there would a compact subset K of D, a positive number c, and a subsequence $\{f_{n_k}^{-1}\}_{k=1}^{\infty}$ for which

$$\max\{|f_{n_k}^{-1}(w) - f^{-1}(w)| : w \in K\} \geq c$$

for each k. But then no subsequence of this subsequence can converge uniformly to f^{-1} on K, hence cannot converge locally uniformly to f^{-1} on $\mathbb{D}$.

To simplify notation, we denote a subsequence of $\{f_n^{-1}\}_{n=1}^{\infty}$ by $\{f_{0,n}^{-1}\}_{n=1}^{\infty}$. All we use about the each subsequence to follow is that the functions concerned converge locally uniformly to the same univalent function f. We again use the exhaustion $\{K_j\}_{j=1}^{\infty}$ of Ω_0 by compact sets, keeping in mind that we now know that Ω_0 coincides with D. In the earlier part of the proof, we effectively showed that there is a subsequence $\{f_{1,n}^{-1}\}_{n=1}^{\infty}$ of $\{f_{0,n}^{-1}\}_{n=1}^{\infty}$ that are each defined on V_2 and that converge uniformly on $V_1 \subset K_1 \subset V_2$ to f^{-1} (at the time, the limit function was denoted by g only to subsequently show that g was, in fact, f^{-1}). Repeating the process, we find a further subsequence $\{f_{2,n}^{-1}\}_{n=1}^{\infty}$ of $\{f_{1,n}^{-1}\}_{n=1}^{\infty}$ that are defined on V_3 and that converge uniformly on $V_2 \subset K_2 \subset V_3$ to f^{-1}. Continuing in this manner, we produce a nested sequence of subsequences of functions $\{f_{j,n}^{-1}\}_{n=1}^{\infty}$ that are defined on V_{j+1} and that converge uniformly on $V_j \subset V_{j+1}$ to f^{-1}. The diagonal sequence $\{f_{n,n}^{-1}\}_{n=1}^{\infty}$, being (eventually) a subsequence of $\{f_{j,n}^{-1}\}_{n=1}^{\infty}$ for each j, is eventually defined on V_{j+1} and converges uniformly on $V_j \subset V_{j+1}$ to f^{-1}. That is, $\{f_{n,n}^{-1}\}_{n=1}^{\infty}$ is a subsequence of $\{f_{0,n}^{-1}\}_{n=1}^{\infty}$ that converges uniformly to f^{-1} on each V_j, and hence on each K_j since $K_{j-1} \subseteq V_j$. □

Example 9.2 is instructive at this point. Here the sequence of conformal maps $\{f_n\}_{n=1}^{\infty}$ is not convergent: both the subsequence of even-indexed mappings $\{f_{2n}\}_{n=1}^{\infty}$ and the subsequence of odd-indexed mappings $\{f_{2n+1}\}_{n=0}^{\infty}$ converge but the limit mappings do not agree: in the first case the limit is $f(z) = z, z \in \mathbb{D}$, and in the second case it is $f(z) = 2z$. Notice that the kernel with respect to 0 of the subsequence of image domains for the even-indexed mappings is $\mathbb{D}$ while the kernel with respect to 0 of the subsequence of image domains for the odd-indexed mappings is $2\mathbb{D}$. That different subsequences of the original sequence of conformal maps have different limits is reflected in different kernels with respect to 0 of the corresponding subsequences of image domains.

Let's put this another way. Suppose that the sequence of normalised conformal maps $f_n : \mathbb{D} \to D_n$ converges to the conformal map f, say. Then the same can be said for any subsequence $\{f_{n_k}\}_{k=1}^{\infty}$ of the original sequence of maps and then, by Proposition 9.1, the kernel with respect to 0 of the corresponding subsequence of domains $\{D_{n_k}\}_{k=1}^{\infty}$ is $f(\mathbb{D})$. That is, every subsequence of the sequence of domains $\{D_n\}_{n=1}^{\infty}$ has the same kernel with respect to 0. This is sufficient to guarantee a converse to Proposition 9.1. We first formalise this concept in terms of a definition.

Definition 9.2 (Carathéodory Kernel Convergence) A sequence of simply connected domains $\{D_n\}_{n=1}^{\infty}$ is said to converge to the domain D in the sense of Carathéodory with respect to w_0 if

(i) w_0 lies in the pre-kernel Ω of the sequence of domains $\{D_n\}_{n=1}^{\infty}$;
(ii) D is the kernel with respect to w_0 of every subsequence $\{D_{n_k}\}_{k=1}^{\infty}$ of the sequence of domains $\{D_n\}_{n=1}^{\infty}$.

It is no loss of generality to assume that $w_0 = 0$, a normalisation that requires only a translation of each domain D_n.

Example 9.3 A simple, useful example of kernel convergence is that of an increasing sequence of simply connected domains $\{D_n\}_{n=1}^{\infty}$, so that $D_n \subseteq D_{n+1}$ for each n. The pre-kernel of this sequence of domains is their union $D = \bigcup_{n=1}^{\infty} D_n$. In fact, any point in the pre-kernel must lie in D and, if w lies in D, then it and a disk about it lies in some D_N, and then in D_n for all $n \geq N$. It is clear that every subsequence has the same pre-kernel, since the union of any subsequence of these domains is the same as the union of all of them. Since D is connected, the kernel with respect to w_0 is D for each w_0 in D. By definition, then, D_n converges to D in the sense of Carathéodory with respect to w_0 for each w_0 in D.

The goal of this section is now within reach: we show that convergence of normalised conformal maps is equivalent to kernel convergence of the image domains. Another example helps to clarify one of the assumptions in the statement of this result.

Example 9.4 The pre-kernel of a sequence of domains can be empty as would be the case for $D_n = D(0, 1/n)$, $n \geq 1$, for example. Here the point 0 lies in every domain D_n but there is no disk about 0 that lies in D_n for all sufficiently large n. The relevant mappings are $f_n(z) = z/n$ which converge to the function that is constantly zero.

For a diametrically opposite example, take $D_n = D(0, n)$, $n \geq 1$. Here both the pre-kernel and the kernel with respect to the origin are the whole complex plane $\mathbb{C}$. The relevant mappings are $f_n(z) = nz$ which don't converge to a limit function (at best, the function that is constantly infinity), reflecting the fact that there is no conformal mapping of the disk onto the complex plane.

In the statement of the Carathéodory kernel theorem, therefore, we need to rule out the extreme cases when 0 doesn't lie in the pre-kernel of the sequence of image domains and when the (pre-)kernel is the whole complex plane.

Theorem 9.1 (Carathéodory Kernel Theorem) *Suppose that $f_n : \mathbb{D} \to D_n$, $n \geq 1$, is a sequence of conformal maps normalised by $f_n(0) = 0$ and $f_n'(0)$ positive. Then f_n converges locally uniformly on the disk $\mathbb{D}$ to a conformal mapping f if and only if $\{D_n\}_{n=1}^{\infty}$ converges in the sense of Carathéodory with respect to 0 to a domain D other than the complex plane.*

In the case of convergence, the limit mapping f maps $\mathbb{D}$ onto D. Moreover, the sequence of inverse functions $\{f_n^{-1}\}_{n=1}^{\infty}$ converges to f^{-1} locally uniformly on D.

9.1 Carathéodory Convergence of Domains

Proof The 'only if' direction follows from Proposition 9.1 applied to each subsequence of the sequence of conformal maps $\{f_n\}_{n=1}^\infty$. The final statements concerning the case of convergence also come from Proposition 9.1.

In the other direction, suppose that 0 lies in the pre-kernel of the sequence of simply connected domains $\{D_n\}_{n=1}^\infty$ and that $\{D_n\}_{n=1}^\infty$ converges to a domain D in the sense of Carathéodory with respect to 0, where D is not the whole complex plane. Since 0 lies in the pre-kernel of $\{D_n\}_{n=1}^\infty$, there is a disk $D(0, \epsilon)$ that lies in D_n for all (sufficiently large) n. Since f_n^{-1} is a conformal map of D_n onto $\mathbb{D}$ with $f_n^{-1}(0) = 0$, (6.9) with $z = 0$ gives

$$\rho_{D_n}(0) = \rho_\mathbb{D}(0) \, |(f_n^{-1})'(0)| = 2/|f_n'(0)|.$$

Together with (8.13), this leads to

$$|f_n'(0)| = \frac{2}{\rho_{D_n}(0)} \geq \delta_{D_n}(0) \geq \epsilon. \tag{9.1}$$

Next we claim that there is a finite R with the property that, for each n, there is some point w_n of modulus less than R that lies in the complement of D_n. Were this not the case then, for each positive integer k, there would be a domain D_{n_k} such that $D(0, k) \subseteq D_{n_k}$. The kernel with respect to 0 of the subsequence of domains $\{D_{n_k}\}_{k=1}^\infty$ would then be the whole complex plane. This contradicts the assumption that D_n converges to D, with $D \neq \mathbb{C}$, in the sense of Carathéodory with respect to 0, since this assumption implies that *every* subsequence of $\{D_n\}_{n=1}^\infty$ has the same kernel with respect to 0 and that this isn't the whole complex plane. It now follows that $\delta_{D_n}(0) \leq R$ for each n and, in turn by (8.13), that

$$|f_n'(0)| = \frac{2}{\rho_{D_n}(0)} \leq 4\delta_{D_n}(0) \leq 4R. \tag{9.2}$$

Again by the Urysohn subsequence principle, it will be sufficient to show that every subsequence $\{f_{n_k}\}_{k=1}^\infty$ of $\{f_n\}_{n=1}^\infty$ has a further subsequence that converges locally uniformly to the same conformal mapping f of $\mathbb{D}$. Then the original sequence $\{f_n\}_{n=1}^\infty$ itself must converge to f locally uniformly on $\mathbb{D}$. If not, there would a subsequence $\{f_{n_k}\}_{k=1}^\infty$ and a positive number δ for which $\rho(f_{n_k}, f) \geq \delta$ for each k. Here, ρ is the metric on $\mathscr{C}(D, \mathbb{C})$ given by (5.15) and we recall that, by Theorem 5.3, locally uniform convergence of a sequence of functions is equivalent to convergence in the metric ρ. But then no subsequence of this subsequence of functions converges locally uniformly to f on $\mathbb{D}$, which is a contradiction.

Let $\{f_{n_k}\}_{k=1}^\infty$ be a subsequence of $\{f_n\}_{n=1}^\infty$. By (9.2), the derivatives of the univalent functions f_{n_k} at 0 are uniformly bounded. It follows from Exercise 8.14 that $\{f_{n_k}\}_{k=1}^\infty$ is a normal family. There is therefore a sub-subsequence $\{f_{n_{k(j)}}\}_{j=1}^\infty$ that converges locally uniformly on $\mathbb{D}$ to a limit function f. By (9.1), $f'(0) = \lim_{j \to \infty} f_{n_{k(j)}}'(0) \geq \epsilon > 0$ and so f is not constant. By Hurwitz's Theorem, f is a conformal mapping. At this point, we have a sub-subsequence $\{f_{n_{k(j)}}\}_j$ of conformal maps of the unit disk $\mathbb{D}$ that converges locally uniformly to the conformal mapping f. This puts us in the setting of Proposition 9.1, from which we may conclude that $f(\mathbb{D})$ is the kernel Ω_0 of the sequence of simply connected domains $\{D_{n_{k(j)}}\}_{j=1}^{\infty}$ with respect to 0. Since $\{D_n\}_{n=1}^{\infty}$ converges to D, the kernel of this (sub-)subsequence of domains must be the domain D, so that f is a conformal map of $\mathbb{D}$ onto D. □

It's now clear that the conformal maps discussed in Example 9.1 do indeed converge to the identity function on $\mathbb{D}$ as the domains $\{D_n\}_{n=1}^{\infty}$ converge to $\mathbb{D}$ with respect to 0.

A consequence of Carathéodory's kernel theorem, one that is perhaps not so obvious, is the following: if $\{D_n\}_{n=1}^{\infty}$ is a sequence of proper simply connected domains that converges in the sense of Carathéodory with respect to 0 to the domain D then D must be simply connected. This is clear if D is the complex plane. If D is not the complex plane then, by Carathéodory's kernel theorem, the sequence of normalised conformal maps $\{f_n\}$, where f_n maps the unit disk conformally onto D_n, converges to a conformal map of the unit disk onto D, hence D must be simply connected.

The Carathéodory's kernel theorem has implications for the convergence of the hyperbolic metric. Suppose that $\{D_n\}_{n=1}^{\infty}$ is a sequence of proper simply connected domains that converges in the sense of Carathéodory to the proper simply connected domain D with respect to 0. Let g_n be the conformal map of D_n onto $\mathbb{D}$ with $g_n(0) = 0$ and $g_n'(0)$ positive, and let g be the similarly normalised conformal map of D onto $\mathbb{D}$. By the Carathéodory kernel theorem, g_n converges locally uniformly to g on D. Just as in Theorem 5.5, the sequence of derivatives g_n' also converges locally uniformly to g' on D. To see this take, for example, the earlier exhaustion of D by compact sets $\{K_j\}_{j=1}^{\infty}$. For fixed j and for sufficiently large n, the function g_n is defined on K_j and g_n converges uniformly to g on K_j and hence on V_j. By Theorem 5.5, therefore, g_n' converges locally uniformly to g' on V_j. Since j was arbitrary, the local uniform convergence extends to all of D.

Now let K be a compact subset of D. From (6.9) we see that the hyperbolic metric for D_n at w in K is given in terms of the conformal map g_n by

$$\rho_{D_n}(w) = \rho_{\mathbb{D}}(g_n(w)) |g_n'(w)| = \frac{2|g_n'(w)|}{1 - |g_n(w)|^2}.$$

As $n \to \infty$, this expression converges uniformly on K to $2|g'(w)|/(1 - |g(w)|^2)$ which is $\rho_D(w)$. We have obtained the following result.

Theorem 9.2 *Suppose that $\{D_n\}_{n=1}^{\infty}$ is a sequence of proper simply connected domains that converges to the proper simply connected domain D in the sense of Carathéodory. Then, the hyperbolic metric for D_n converges to the hyperbolic metric for D uniformly on compact subsets of D.*

9.2 Comparison Results for the Hyperbolic Metric

The Koebe 1/4-Theorem is a comparison between the hyperbolic metric, which is an analytic quantity, and the distance to the boundary, which is a geometric quantity. Our objective in this section is to describe some further results, primarily due to Jørgensen [17], that relate the hyperbolic metric to the geometry of the domain. Most if not all results apply to general hyperbolic domains but we necessarily restrict ourselves to simply connected domains.

A set in Euclidean space is *convex* if the line segment joining any two points in the set also lies in the set. Thinking of this line segment as the Euclidean geodesic joining its endpoints, it is natural to define a set E in a simply connected domain D to be *hyperbolically convex* if, for each pair of points α and β in E, the hyperbolic geodesic in D joining α and β also lies in E. Jørgensen showed that any half-plane or disk in a simply connected domain is hyperbolically convex. That is, if the endpoints of a hyperbolic geodesic lie in a half-plane or disk then so does the geodesic itself. A result in the opposite direction, saying where a hyperbolic geodesic can't go, is the following: a hyperbolic geodesic will never enter any disk that is tangent to it. These are but a sample of the many results in the literature of a similar nature. An excellent reference in this regard is Chap. 4 in Pommerenke's book [21]. In fact, the proof of Theorem 9.7 below is taken from this reference.

To begin, we show that the hyperbolic metric, actually its logarithm, satisfies a simple partial differential equation.

9.2.1 The Laplacian and the Hyperbolic Metric

We know from a first course in complex analysis, see [16, Theorem 11.5] for example, that the real and imaginary parts of an analytic function are harmonic, this being an immediate consequence of the Cauchy-Riemann equations. The *Laplacian* of a twice continuously differentiable, real-valued function $u(x, y)$ of two real variables is $\Delta u = u_{xx} + u_{yy}$. The operator Δ is known as the *Laplace Operator*. A function is said to be *harmonic* in a domain D if $\Delta u = 0$ everywhere in D. If $f = u + iv$ (u and v being the real and imaginary parts of f, respectively) is analytic in D then, by the Cauchy-Riemann equations,

$$u_{xx} = (u_x)_x = (v_y)_x = v_{yx} \text{ and } u_{yy} = (u_y)_y = (-v_x)_y = -v_{xy}.$$

By the equality of mixed partial derivatives, we deduce that $\Delta u = 0$. In a similar manner, we see that the imaginary part v of f is also harmonic, (or regard v as the real part of $-if$). The converse holds in the sense that if u is harmonic in a disk then there is a function f analytic in that disk for which $u = \Re f$ [16, Theorem 11.6].

As an example, suppose that D is a simply connected domain and that g is analytic and non-zero in D. Then g has an analytic logarithm $h = \log g$ (or $g = e^h$)

in D by Theorem 6.6, and so

$$u(z) = \Re h(z) = \log|g(z)|$$

is harmonic in D. This is the case if D is simply connected and f is a conformal mapping of D so that f' is non-zero in D. Then $u(z) = \log|f'(z)|$ is harmonic in D.

It is helpful to have a formula for the Laplacian in polar coordinates. This involves using the Chain Rule for functions of two variables, the steps of which are set out in the following exercise.

Exercise 9.1 Show that Δu, in polar coordinates (r, θ), is given by

$$\Delta u = u_{rr} + \frac{1}{r} u_r + \frac{1}{r^2} u_{\theta\theta} = \frac{1}{r}\frac{\partial}{\partial r}(r u_r) + \frac{1}{r^2} u_{\theta\theta}. \quad (9.3)$$

[Write u as $u(x, y)$ where $x = r\cos\theta$ and $y = r\sin\theta$. Using the Chain Rule, show that

$$\begin{cases} u_r = u_x \cos\theta + u_y \sin\theta \\ u_\theta = -u_x r \sin\theta + u_y r \cos\theta. \end{cases}$$

Deduce that

$$\begin{cases} u_x = u_r \cos\theta - \frac{1}{r} u_\theta \sin\theta \\ u_y = u_r \sin\theta + \frac{1}{r} u_\theta \cos\theta. \end{cases}$$

Hence find expressions for u_{xx} and u_{yy} and, in turn, for Δu.]

Exercise 9.2 Find all radial harmonic functions, that is functions of the form $u(r, \theta) = f(r)$ that are harmonic in $\mathbb{C} \setminus \{0\}$.

The hyperbolic metric in the unit disk $\mathbb{D}$ is $\rho(z) = 2/(1 - |z|^2)$ (see (4.20)) or, in polar coordinates, $\rho(r, \theta) = 2/(1 - r^2)$. Then, with $u(r, \theta) = \log \rho(r, \theta) = \log 2 - \log(1 - r^2)$, we have $u_r = 2r/(1 - r^2)$ and

$$\Delta u = \frac{1}{r}\frac{\partial}{\partial r}(r u_r) = \frac{1}{r}\frac{\partial}{\partial r}\left(\frac{2r^2}{1 - r^2}\right) = \frac{4}{(1 - r^2)^2} = \rho^2 = e^{2u}.$$

That is, $\Delta \log \rho = \rho^2$ or, equivalently, the function $u = \log \rho$ satisfies the partial differential equation $\Delta u = e^{2u}$. We will shortly see that this applies not only to the hyperbolic metric of the unit disk but to the hyperbolic metric of any proper simply connected domain. This becomes easier if we have an expression for the Laplacian that works better in complex coordinates.

9.2 Comparison Results for the Hyperbolic Metric

Exercise 9.3 With $\rho(z) = 2/(1 + |z|^2)$, $z \in \mathbb{C}$, being the spherical metric on the complex plane, show that $u(z) = \log \rho(z)$ satisfies the partial differential equation $\Delta u = -e^{2u}$.

As will be familiar to those with some exposure to mathematical physics or applied mathematics, there are formulas for the Laplacian in three dimensions in spherical and in cylindrical coordinates analogous to that in polar coordinates. In dimension two, there is another complex-style expression that can be very useful. If $u = u(x, y)$ in Cartesian coordinates, write $x = (z + \bar{z})/2$ and $y = (z - \bar{z})/(2i)$. The function u can then be regarded, formally, as a function of z and $\bar{z}$. For example,

$$\rho_\mathbb{D}(z, \bar{z}) = \frac{2}{1 - |z|^2} = \frac{2}{1 - z\bar{z}}, \quad z \in \mathbb{D}.$$

By the Chain Rule applied to $u(z, \bar{z})$, with $z = x + iy$, $\bar{z} = x - iy$,

$$\begin{cases} u_x = u_z \dfrac{\partial z}{\partial x} + u_{\bar{z}} \dfrac{\partial \bar{z}}{\partial x} = u_z + u_{\bar{z}}, \\[6pt] u_y = u_z \dfrac{\partial z}{\partial y} + u_{\bar{z}} \dfrac{\partial \bar{z}}{\partial y} = iu_z - iu_{\bar{z}}. \end{cases}$$

Applying the first of these identities to u_x itself,

$$u_{xx} = (u_x)_x = (u_z + u_{\bar{z}})_x = (u_z + u_{\bar{z}})_z + (u_z + u_{\bar{z}})_{\bar{z}} = u_{zz} + 2u_{z\bar{z}} + u_{\bar{z}\bar{z}}.$$

Similarly, $u_{yy} = -u_{zz} + 2u_{z\bar{z}} - u_{\bar{z}\bar{z}}$, so that

$$\Delta u = u_{xx} + u_{yy} = 4u_{z\bar{z}}. \tag{9.4}$$

For example, with $u(z, \bar{z}) = \log \rho_\mathbb{D}(z) = \log(2/(1 - |z|^2))$,

$$\Delta u = 4 \frac{\partial}{\partial z} \frac{\partial}{\partial \bar{z}} \left(\log 2 - \log(1 - z\bar{z}) \right)$$

$$= 4 \frac{\partial}{\partial z} \left(\frac{z}{1 - z\bar{z}} \right)$$

$$= 4 \frac{(1 - z\bar{z}) - z(-\bar{z})}{(1 - z\bar{z})^2} = \frac{4}{(1 - |z|^2)^2} = \rho_\mathbb{D}(z)^2 = e^{2u},$$

as before.

Now let D be a proper simply connected domain and f a conformal mapping of D onto the unit disk $\mathbb{D}$. Then, by (6.9),

$$\rho_D(z) = \rho_\mathbb{D}(f(z)) |f'(z)| = \frac{2|f'(z)|}{1 - |f(z)|^2}, \quad z \in D,$$

so that $u_D(z) = \log \rho_D(z)$ is given by

$$u_D(z) = \log |f'(z)| + \log \rho_{\mathbb{D}}(f(z)). \tag{9.5}$$

Now, $\Delta(\log |f'(z)|) = 0$ since, as noted earlier, f' is analytic and non-zero in the simply connected domain D. Thus, $\Delta u_D(z) = \Delta(\log \rho_{\mathbb{D}}(f(z)))$.

In general, if f is analytic in a domain D_1 and takes values in a domain D_2, and if u is real-valued and twice continuously differentiable in D_2 then $v(z) = u(f(z))$ is twice continuously differentiable in D_1. With $w = f(z), z \in D_1$,

$$\begin{aligned}
\Delta v(z) = \Delta u(f(z), \overline{f(z)}) &= 4\frac{\partial}{\partial z}\frac{\partial}{\partial \bar{z}} u(f(z), \overline{f(z)}) \\
&= 4\frac{\partial}{\partial z}\left[u_w \frac{\partial f(z)}{\partial \bar{z}} + u_{\bar{w}} \frac{\partial \overline{f(z)}}{\partial \bar{z}}\right] \\
&= 4\frac{\partial}{\partial z}\left[\overline{f'(z)} u_{\bar{w}}(f(z), \overline{f(z)})\right] \\
&= 4\overline{f'(z)}\left[u_{\bar{w}w}\frac{\partial f(z)}{\partial z} + u_{\bar{w}\bar{w}}\frac{\partial \overline{f(z)}}{\partial z}\right] \\
&= 4\overline{f'(z)}u_{\bar{w}w}f'(z) = |f'(z)|^2 (\Delta u)(f(z)). \tag{9.6}
\end{aligned}$$

Here we used the identity (9.4) at the first and last steps, as well as the formulae

$$\frac{\partial f(z)}{\partial \bar{z}} = 0 = \frac{\partial \overline{f(z)}}{\partial z} \quad \text{and} \quad \frac{\partial f(z)}{\partial z} = f'(z), \quad \frac{\partial \overline{f(z)}}{\partial \bar{z}} = \overline{f'(z)},$$

these identities being clear from the local power series representations of f.

Returning to (9.5) and using (9.6),

$$\begin{aligned}
\Delta u_D(z) &= \Delta\big(\log \rho_{\mathbb{D}}(f(z))\big) = |f'(z)|^2 \big(\Delta \log \rho_{\mathbb{D}}\big)(f(z)) \\
&= |f'(z)|^2 \rho_{\mathbb{D}}^2(f(z)) = \rho_D(z)^2 = e^{2u_D(z)}.
\end{aligned}$$

In summary, we have the following proposition.

Proposition 9.2 *Let D be a proper simply connected domain and ρ_D the hyperbolic metric in D. Then,*

$$\Delta\big(\log \rho_D(z)\big) = \rho_D^2(z). \tag{9.7}$$

Equivalently, $u_D(z) = \log \rho_D(z)$ satisfies

$$\Delta u_D(z) = e^{2u_D(z)}. \tag{9.8}$$

9.2 Comparison Results for the Hyperbolic Metric

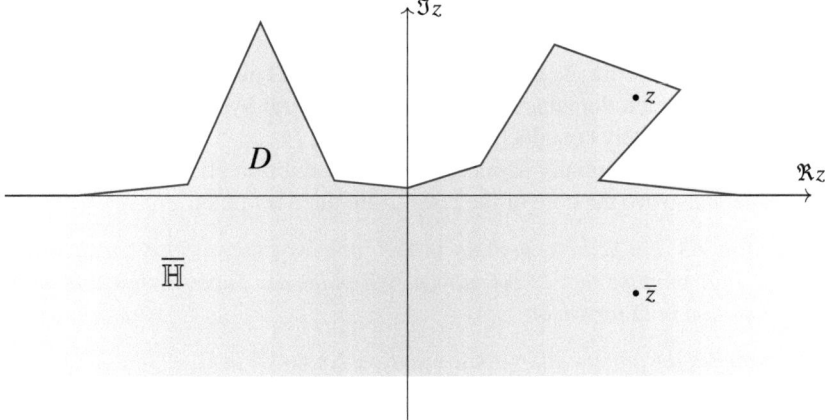

Fig. 9.3 A simply connected domain D containing the lower half-plane. With z and $\bar{z}$ as shown, $\rho_D(z) \geq \rho_D(\bar{z})$

9.2.2 Estimates for Solutions of $\Delta u = e^{2u}$

Imagine a simply connected domain D as shown in Fig. 9.3. It contains a half-plane which, by rotation and translation, we may take to be the lower half-plane $\overline{\mathbb{H}} = \{z : \Im z < 0\}$. We can interpret the Koebe 1/4-Theorem as telling us that the hyperbolic metric is large near the boundary. If z is a point in D in the upper half-plane then the boundary of D is closer to z than it is to $\bar{z}$, the reflection of z in the real axis. With this in mind, it is reasonable to infer that the hyperbolic metric at z should be larger than at $\bar{z}$. This is indeed the case as we will now see. This result is stated as Theorem 1 in Jørgensen [17]. Theorem 4 in that paper treats the case of a Euclidean disk. These results are naturally combined in the following theorem: we are, after all, dealing with generalised circles. We effectively follow Jørgensen's proof, which works for arbitrary hyperbolic domains. First, some notation. If H is a half-plane bounded by a line L, or if U is a Euclidean disk bounded by a circle C, we write z^* for the reflection of z in L or C, respectively (see Sect. 2.3.2).

Theorem 9.3 *Suppose that D is a proper simply connected domain. If D contains a disk $U = D(a, r)$ and if z lies in D with z^* in U then*

$$|z - a|\rho_D(z) > |z^* - a|\rho_D(z^*). \tag{9.9}$$

If D contains a half-plane H and if z lies in D with z^* in H then

$$\rho_D(z) \geq \rho_D(z^*), \tag{9.10}$$

with equality if and only if D is symmetric in the boundary line of H.

Since a half-plane can be thought of as a limit of expanding disks, (9.10) is effectively a limiting case of (9.9) and, in fact, we deduce the estimate in the half-plane case from the estimate in the disk case. Note that equality never holds in (9.9) for simply connected domains but it does for a general hyperbolic domain whose boundary is completely contained in the circle $C(a, r)$.

Theorem 9.3 is essentially about solutions to the differential equation (9.8) and so we state and prove the core of the result from this viewpoint.

Proposition 9.3 *Let Ω be a domain in the complex plane whose complement is bounded, and suppose that Ω contains a half-plane H. Suppose that v is a real-valued function in Ω for which*

(i) v satisfies the partial differential equation $\Delta v = e^{2v}$ in Ω,
(ii) $v(w) \to +\infty$ as w approaches any finite boundary point of Ω,
(iii) $v(w) + \log|w|^2$ tends to a finite limit as w tends to infinity through Ω.

Then, for $w \in \Omega$ with $w^* \in H$,

$$v(w) \geq v(w^*). \tag{9.11}$$

Equality holds in (9.11) if and only if the boundary of Ω lies on the boundary of H.

Proof Postponing the case of equality for the moment, we set out to prove (9.11). By translating and rotating the domain Ω, we may assume that the half-plane H is the lower half-plane $\overline{\mathbb{H}} = \{z : \Im z < 0\}$ as shown in Fig. 9.4. We then need to show, for $w \in \Omega$ with $\Im w > 0$, that $v(w) \geq v(\overline{w})$.

Since it may be that the boundary of Ω meets the real axis, we begin by working with reflection in a line $L_\epsilon = \{w : \Im w = -\epsilon\}$ for a fixed positive ϵ rather than directly with reflection in the real axis. Accordingly, we set $\overline{\mathbb{H}}_\epsilon = \{w : \Im w < -\epsilon\}$, set $\mathbb{H}_\epsilon = \{w : \Im w > -\epsilon\}$ and set $\Omega_\epsilon^+ = \Omega \cap \mathbb{H}_\epsilon$. Let R_ϵ denote reflection in the line L_ϵ, so that $R_\epsilon(x + iy) = x - i(y + 2\epsilon)$. Set

$$\phi(w) = v(w) - v(R_\epsilon(w)), \quad \text{for } w \text{ in } \Omega_\epsilon^+.$$

The function ϕ is well-defined since $R_\epsilon(w)$ lies in $\overline{\mathbb{H}}_\epsilon$ if w lies in Ω_ϵ^+, and $\overline{\mathbb{H}}_\epsilon$ itself lies in Ω. Now, noting that v takes finite values there, we see that ϕ vanishes on the boundary line L_ϵ of Ω_ϵ^+. As w approaches a finite boundary point ζ of Ω, we have that $v(R_\epsilon(w))$ approaches $v(R_\epsilon(\zeta))$, which is finite, and that $v(w)$ tends to $+\infty$ by Assumption (ii). Hence, $\phi(w) \to +\infty$ as $w \to \zeta$. Finally, as w tends to infinity through Ω_ϵ^+ then, by Assumption (iii), $v(w) + \log|w|^2 \to c$ for some finite c. Since $R_\epsilon(w)$ tends to infinity as w tends to infinity, the same holds for $v(R_\epsilon(w))$. We then see that, as w tends to infinity through Ω_ϵ^+,

$$\phi(w) = \left[v(w) + \log|w|^2\right] - \left[v(R_\epsilon(w)) + \log|R_\epsilon(w)|^2\right] + \log\frac{|R_\epsilon(w)|^2}{|w|^2} \longrightarrow 0.$$

9.2 Comparison Results for the Hyperbolic Metric

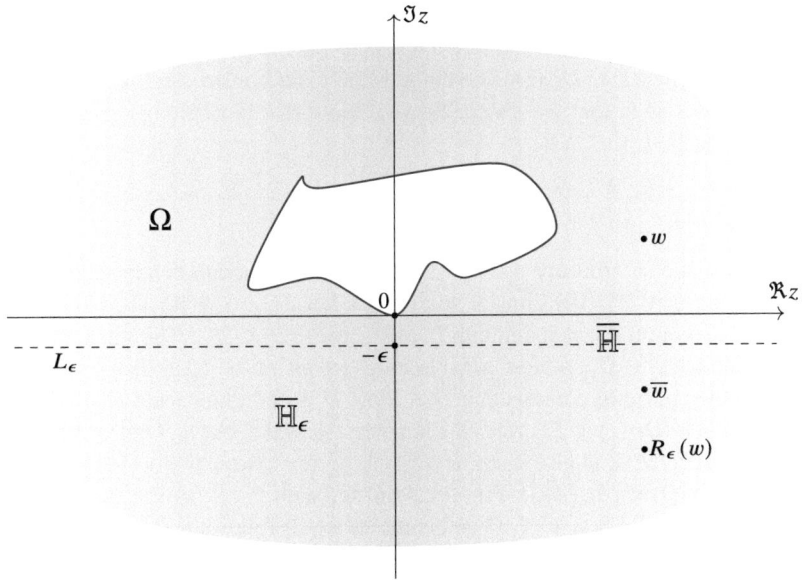

Fig. 9.4 A domain Ω as considered in Proposition 9.3, Ω being the unbounded exterior of the bounded region shown. Also shown is some of the notation used in the proof of Proposition 9.3

Here we used that $|R_\epsilon(w)|/|w| \to 1$ as w tends to infinity. It follows that if ϕ takes negative values in Ω_ϵ^+ it must have a local minimum somewhere in Ω_ϵ^+, say at w_0. The Laplacian of ϕ at w_0 must then be non-negative. However, by the assumption that $\Delta v = e^{2v}$ and since $\Delta v(R_\epsilon(w)) = (\Delta v)(R_\epsilon(w))$,

$$\Delta \phi(w_0) = e^{2v(w_0)} - e^{2v(R_\epsilon(w_0))} = e^{2v(w_0)}\left[1 - e^{-2\phi(w_0)}\right] < 0,$$

which is a contradiction. Hence, ϕ is non-negative everywhere in Ω_ϵ^+. That is, for each fixed w in $\Omega \cap \mathbb{H}$, we have $v(w) \geq v(R_\epsilon(w))$. As ϵ tends to 0, the point $R_\epsilon(w)$ tends to $\overline{w}$ and we conclude that $v(w) \geq v(\overline{w})$. □

Suppose that the boundary of Ω lies on the boundary of H, which we may take to be the lower half-plane $\overline{\mathbb{H}}$ so that $\partial\Omega \subseteq \mathbb{R}$. If $\Im w > 0$ then, by (9.11), $v(w) \geq v(\overline{w})$. On the other hand, Ω also contains the upper half-plane $\mathbb{H}$ and so, again by (9.11), $v(\overline{w}) \geq v(w)$. Thus, equality holds in (9.11) if $\partial\Omega \subseteq \partial H$.

The converse, that equality in (9.11) holds only in this case, seems to require more work. Following Jørgensen we give the details in the following lemmas. The first is a geometric lemma that may be useful in other contexts. We introduce the following notation, used only in this lemma. For a domain $\Omega \neq \mathbb{C}$ and z_0 in Ω, we write $D_\Omega(z)$ for the largest open disk centred at z and contained in Ω. That is, $D_\Omega(z) = D(z, \delta_\Omega(z))$ where $\delta_\Omega(z)$ is the distance from z to the boundary of Ω. Let z_0 be a point in Ω. For each point z_1 in the disk $D_\Omega(z_0)$, consider the disk

$D_\Omega(z_1)$. Repeat the process, forming the disk $D_\Omega(z_2)$ for each point z_2 in $D_\Omega(z_1)$. Continue in this manner, so that as each disk $D_\Omega(z)$ is produced, we then form all disks $D_\Omega(\tilde{z})$ where $\tilde{z} \in D_\Omega(z)$. The union of all disks constructed in this process is an open subset of Ω that we denote by Ω_{z_0}. Note that if z belongs to Ω_{z_0} then so does the disk $D_\Omega(z)$.

Lemma 9.1 *Let $\Omega \neq \mathbb{C}$ be a domain in the complex plane and z_0 be a point in Ω. Then, $\Omega_{z_0} = \Omega$.*

Proof Let z_1 be an arbitrary point of Ω. Since Ω is open and connected, there is a curve $\gamma(t)$, $t \in [0, 1]$, that joins z_0 to z_1 in Ω. Set $E = \{t \in [0, 1] : \gamma(t) \in \Omega_{z_0}\}$. The set E is non-empty as it contains 0, this because $z_0 \in \Omega_{z_0}$. The set E is open. If $t_0 \in E$ then $\gamma(t_0) \in \Omega_{z_0}$ and so $\gamma(t_0)$ belongs to one of the open disks, say $D_\Omega(z)$, arising in the course of constructing Ω_{z_0}. Since γ is continuous and $D_\Omega(z)$ is open, $\gamma(t)$ belongs to $D_\Omega(z) \subseteq \Omega_{z_0}$ for all t in some interval about t_0. On the other hand, the complement of E is also open. If $t_0 \notin E$, so that $w_0 = \gamma(t_0) \notin \Omega_{z_0}$, consider the disk $U = D(w_0, \delta_\Omega(w_0))$. For any point w_1 with $|w_1 - w_0| < \delta_\Omega(w_0)/4$, the disk $D(w_1, \delta_\Omega(w_0)/2) \subseteq U \subseteq \Omega$ and contains w_0. It follows that no such point w_1 belongs to Ω_{z_0} as if it did, we would have $w_0 \in D(w_1, \delta_\Omega(w_0)/2) \subseteq D_\Omega(w_1) \subseteq \Omega_{z_0}$, which is a contradiction. We have shown that $D(w_0, \delta_\Omega(w_0)/4)$ lies in the complement of Ω_{z_0} and hence so does $\gamma(t)$ for all t in an interval about t_0. Since $[0, 1]$ is connected and E is non-empty and both open and closed, we may conclude that $E = [0, 1]$ and that $z_1 \in \Omega_{z_0}$. □

The next technical lemma provides the key estimate that we need regarding solutions of the partial differential equation (9.8).

Lemma 9.2 *Suppose that v_1 and v_2 are bounded in the disk $D(0, R)$, that $v_1 > v_2$ there, and that v_1 and v_2 are both solutions of (9.8). Then there is a positive constant k such that*

$$(v_1 - v_2)(z) \geq k(R - |z|), \quad z \in D(0, R). \tag{9.12}$$

Proof Set $v = v_1 - v_2$ so that v is positive and bounded in $D(0, R)$. It is sufficient to establish (9.12) for z in the annulus $A = \{z : R/2 < |z| < R\}$. In fact, v is positive in the closed disk $\overline{D(0, R/2)}$ and therefore achieves its minimum there, say $v(z) \geq \delta > 0$ for $|z| \leq R/2$. Then, (9.12) holds for $|z| \leq R/2$ with $k = \delta/R$.

Now $\Delta v = e^{2v_1} - e^{2v_2} = e^{2v_2}(e^{2v} - 1)$ in $D(0, R)$. Since v_2 is bounded above, the term e^{2v_2} is bounded. Also, $e^{2v} - 1 = \int_0^{2v} e^t \, dt \leq e^{2v} 2v \leq Cv$, since v is bounded. Thus there is a finite constant K such that

$$(\Delta v)(z) \leq Kv(z), \quad z \in D(0, R). \tag{9.13}$$

9.2 Comparison Results for the Hyperbolic Metric

Now we produce a function ψ for which $(\Delta \psi)(z) \geq K \psi$. It will then follow that ψ is a lower bound for v in A (a *minorant* for v in A). We set, for c positive and $n \in \mathbb{N}$ yet to be chosen,

$$\psi(r) = c\left(\left(\frac{R}{r}\right)^n - 1\right), \quad 0 < r < R,$$

and $\psi(z) = \psi(|z|)$, $z \in D'(0, R)$. Then, $\psi'(r) = -ncR^n/r^{n+1}$. By the expression for the Laplacian in polar coordinates (9.3),

$$(\Delta \psi)(z) = \frac{1}{r}\frac{\partial}{\partial r}(r\psi'(r)) = \frac{cn^2 R^n}{r^{n+2}} = \psi(z)\frac{n^2}{r^2(1 - (r/R)^n)}.$$

Since $R^2 > r^2(1 - (r/R)^n)$, $0 < r < R$, we find that

$$(\Delta \psi)(z) \geq \frac{n^2}{R^2}\psi(z) \geq K\psi(z), \quad z \in D'(0, R), \tag{9.14}$$

if we choose n so that $n^2/R^2 \geq K$. Choose $c = \delta/(2^n - 1)$. Since $v(z) \geq \delta$ on $C(0, R/2)$,

$$\psi(z) = c(2^n - 1) = \delta \leq v(z), \quad |z| = R/2.$$

Now set $\phi(z) = v(z) - \psi(z)$. Since v is positive in $D(0, R)$ and $\psi(z) \to 0$ as $|z| \to R^-$, we have $\liminf \phi(z) \geq 0$ as $|z| \to R^-$. On the circle $C(0, R/2)$, $\phi(z) \geq 0$. If it was the case that the function ϕ was negative somewhere in the annulus A, it would therefore have to have a negative minimum somewhere in A, say at z_0. At z_0, then, we would have $\phi(z_0) < 0$ and $\Delta\phi(z_0) \geq 0$. But this is a contradiction since, by (9.13) and (9.14),

$$0 \leq \Delta\phi(z_0) = \Delta v(z_0) - \Delta\psi(z_0) \leq Kv(z_0) - K\psi(z_0) = K\phi(z_0) < 0.$$

It follows that ϕ is non-negative in A, that is $v(z) \geq \psi(z)$, $z \in A$. Thus, for $z \in A$ and $r = |z|$,

$$v(z) \geq c\left(\left(\frac{R}{r}\right)^n - 1\right) \geq c\left(\frac{R}{r} - 1\right) = \frac{c}{r}(R - r) \geq \frac{c}{R}(R - |z|).$$

The estimate (9.12) now follows with $k = \delta/((2^n - 1)R)$. □

Lemma 9.2 allows us to show that any two solutions of the partial differential equation (9.8) in a domain, one of which dominates the other, are either identical or never agree.

Lemma 9.3 *Suppose that v_1 and v_2 are both solutions of (9.8) in a domain Ω and that $v_1 \geq v_2$ in Ω. Then, either $v_1 = v_2$ everywhere in Ω or $v_1 > v_2$ everywhere in Ω.*

Proof Suppose that v_1 and v_2 are both solutions of the partial differential equation (9.8) in a domain Ω, and that $v_1 \geq v_2$ there. If v_1 and v_2 agree everywhere in Ω there is nothing to prove. Otherwise, there is some point z_0 at which $v(z) = v_1(z) - v_2(z)$ is positive. For any such point z_0, let $D(z_0, R)$ be the largest disk centred at z_0 in which v is positive (there is such a disk since $v(z_0)$ is positive and v is continuous at z_0). We claim that $D(z_0, R)$ touches the boundary of Ω. If not, the circle $C(z_0, R)$ lies entirely in Ω and there must be a point z_1 on $C(z_0, R)$ at which v vanishes (otherwise there would be a larger disk about z_0 on which v was positive). The conditions of Lemma 9.2 are therefore satisfied so that (9.12), with $z_t = (1-t)z_0 + tz_1$, $0 < t < 1$, leads to $v(z_t) \geq k(R - |z_t - z_0|) = k(1-t)R$. Since $v(z_1) = 0$, we find that the outward normal derivative to the disk $D(z_0, R)$ of v at z_1 satisfies

$$\frac{\partial v}{\partial n}(z_1) = \lim_{t \uparrow 1} \frac{v(z_1) - v(z_t)}{|z_1 - z_t|} = \lim_{t \uparrow 1} \frac{-v(z_t)}{(1-t)R} \leq -k < 0.$$

Since the outward normal derivative of v at z_1 is negative, there must be points in Ω just outside the disk $D(z_0, R)$ and on the ray from z_0 to z_1 at which v is negative. This contradicts the assumption that v is non-negative on Ω.

We have shown that if v is positive at z then v is positive on the disk $D_\Omega(z) = D(z, \delta_\Omega(z))$. Chose any point z_0 in Ω at which v is positive and form the open set Ω_{z_0}. Then, v is positive on Ω_{z_0}. In fact, each disk $D_\Omega(z)$ that arises in the construction of Ω_{z_0} is centred on a point z at which v is positive so that v is positive in the disk $D_\Omega(z)$ as well. By Lemma 9.1, $\Omega_{z_0} = \Omega$ and so v is positive everywhere in Ω if it is positive somewhere. □

With this result to hand, it is straightforward to show that if Ω in Proposition 9.3 has boundary points in the upper half-plane $\mathbb{H}$ (again taking the half-plane H in Proposition 9.3 to be the lower half-plane) then equality never holds in (9.11). In fact, both $v_1(w) = v(w)$ and $v_2(w) = v(\overline{w})$ satisfy (9.8) in the domain $D = \Omega \cap \mathbb{H}$ and, by Proposition 9.3, $v_1(w) \geq v_2(w)$ there. As w approaches a boundary point of Ω in the upper half-plane, $v_1(w) \to +\infty$ whereas $v_2(w) \to v(\overline{w})$ which is some finite value. There must then be points in D where v_1 is strictly bigger than v_2. Hence, by Lemma 9.3, v_1 is strictly greater than v_2 everywhere in D. Thus, equality never holds in (9.11) if Ω has boundary points in the upper half-plane.

This finally concludes the proof of Proposition 9.3, including the case of equality. We can now turn our attention to using Proposition 9.3 to prove Theorem 9.3.

Proof of Theorem 9.3 Suppose that the proper simply connected domain D contains the disk $U = D(a, r)$. Choose a point z_0 on $C = C(a, r)$ and lying in D (if this isn't possible then $D = U$ and there is nothing to prove). Set $M(z) = 1/(z_0 - z)$ and denote by Ω the image of the domain $D \setminus \{z_0\}$ under M. Since D contains a

9.2 Comparison Results for the Hyperbolic Metric

disk about z_0, the domain Ω contains the exterior of a disk about 0, so that the complement of Ω is bounded. Being a Möbius transformation, M maps the circle C to a straight line L and maps the disk U bounded by C to the half-plane $H = M(U)$ bounded by L. Using again that M is a Möbius transformation, if z and z^* are symmetric in the circle C, it follows from Proposition 2.4 that $w = M(z)$ and $w^* = M(z^*)$ are symmetric in L.

As usual, ρ_D is the hyperbolic metric in D and we set $u_D = \log \rho_D$ in D. For w in Ω, we set $f(w) = M^{-1}(w) = z_0 - 1/w$ and set[2]

$$v(w) = u_D(f(w)) + \log |f'(w)| = u_D(z_0 - 1/w) + \log \frac{1}{|w|^2}. \tag{9.15}$$

We show that v satisfies Conditions (i) to (iii) of Proposition 9.3. By (9.6) and (9.8), and since $\Delta \log |w| = 0$,

$$\Delta v(w) = |f'(w)|^2 (\Delta u_D)(f(w)) = e^{2u_D(f(w)) + 2\log|f'(w)|} = e^{2v(w)}.$$

This is Condition (i). Next, we show that v tends to infinity at the boundary of Ω. Let ζ be a finite boundary point of Ω. If $\zeta \neq 0$ then $\eta = f(\zeta)$ is a finite boundary point of D so that $u_D(z) = \log \rho_D(z) \to \infty$ as $z \to \eta$, $z \in D$. As $w \to \zeta$, $w \in \Omega$, we have $z = f(w) \to \eta$, and so $v(w) = u_D(f(w)) + \log(1/|w|^2) \to \infty$ since $\log(1/|w|^2) \to \log(1/|\zeta|^2)$ and $u_D(f(w)) \to \infty$. The remaining case is $\zeta = 0$ and $w \to 0$, $w \in \Omega$. Then, $z = z_0 - 1/w$ tends to the point at infinity through D. Choose any boundary point η of D. Then,

$$\delta_D(z) \leq |z - \eta| \leq 2|z| \text{ for } |z| \geq |\eta|.$$

By the Koebe 1/4-Theorem, $\rho_D(z) \geq 1/(2\delta_D(z)) \geq 1/(4|z|)$, for all sufficiently large $z \in D$. Then, as $w \to 0$ through Ω,

$$v(w) = \log \rho_D\left(z_0 - \frac{1}{w}\right) + \log \frac{1}{|w|^2} \geq \log \frac{|w|}{4|z_0 w - 1|} + \log \frac{1}{|w|^2}$$

$$= \log \frac{1}{4|w||z_0 w - 1|},$$

which tends to infinity as w approaches 0. In short, $v(w) \to +\infty$ as w approaches each boundary point ζ of Ω, which is Condition (ii) of Proposition 9.3.

As w approaches infinity through Ω, the point $z = z_0 - 1/w = f(w)$ approaches z_0. Hence, using (9.15),

$$v(w) + \log |w|^2 = u_D(z) + \log(1/|w|^2) + \log |w|^2 = u_D(z) \to u_D(z_0) \text{ as } w \to \infty.$$

---

[2] The function v is actually the hyperbolic metric for Ω, but we don't use this fact.

That is, v satisfies Condition (iii) of Proposition 9.3. By this proposition, then, $v(w) \geq v(w^*)$ for $w \in \Omega$ with $w^* \in H$.

Suppose that z lies in D with the point z^*, the reflection of z in the circle C, lying in U. Then $w = M(z)$ lies in Ω with $w^* = M(z^*)$ in H, so that

$$v(w) = u_D(f(w)) + \log \frac{1}{|w|^2} \geq u_D(f(w^*)) + \log \frac{1}{|w^*|^2} = v(w^*).$$

In terms of z and z^*, this becomes

$$u_D(z) \geq u_D(z^*) + \log \frac{|w|^2}{|w^*|^2} = u_D(z^*) + \log \frac{|z_0 - z^*|^2}{|z_0 - z|^2}.$$

Since $u_D = \log \rho_D$,

$$\rho_D(z) \geq \left| \frac{z_0 - z^*}{z_0 - z} \right|^2 \rho_D(z^*).$$

By Exercise 2.12 and by using the assumption that z and z^* are symmetric in the circle $C(a, r)$, that circle can be written as $C(z^*, z, k)$ where $k = |z^* - a|/r$. Since z_0 lies on the circle,

$$\left| \frac{z_0 - z^*}{z_0 - z} \right| = k.$$

Moreover, from (2.16), we have $|z^* - a| |z - a| = r^2$ and so

$$\rho_D(z) \geq \frac{|z^* - a|^2}{r^2} \rho_D(z^*) = \frac{|z^* - a|}{|z - a|} \rho_D(z^*)$$

which is (9.9). Equality can only hold if the boundary of Ω lies on L, which is equivalent to the boundary of D lying on the circle C. But then D would not be simply connected. The inequality (9.9) is therefore always strict.

Now we deduce (9.10), the half-plane estimate, from the estimate (9.9) for a disk. In order to simplify notation we may, by translating and rotating D, assume that D contains the lower half-plane $\overline{\mathbb{H}} = \{z : \Im z < 0\}$. Then, by translating horizontally, we may assume that the point z lies on the imaginary axis, say at $z = iy$ where $y > 0$, so that $\bar{z} = -iy$. Let C be the circle $C(a, r)$ where the centre a is chosen to be at $-ix$ on the negative imaginary axis (with $x > y$) and the radius r is chosen to be $\sqrt{x^2 - y^2}$. With these choices of a and r we can check that $(z - a)(\bar{z} - a) = r^2$ so that, by (2.16), z and $\bar{z}$ are symmetric in the circle $C(a, r)$. Since $D(a, r)$ is contained in D, we deduce from (9.9) that

$$\rho_D(z) \geq \frac{|\bar{z} - a|}{|z - a|} \rho_D(\bar{z}) = \frac{x - y}{x + y} \rho_D(\bar{z}).$$

9.2 Comparison Results for the Hyperbolic Metric   249

This holds for each choice of x and, letting $x \to \infty$, we deduce (9.10).

Equality holds if and only if the boundary of D lies entirely on the real axis. If the boundary of D does lie entirely on the real axis then the upper half-plane $\mathbb{H}$ also lies in D so that $\rho_D(\bar{z}) \geq \rho_D(z)$ and equality holds in (9.10). It follows from Lemma 9.3, just as it did in the case of Proposition 9.3, that this is the only case of equality. Both $u_1(z) = \log \rho_D(z)$ and $u_2(z) = \log \rho_D(\bar{z})$ satisfy the partial differential equation (9.8) in $D \cap \mathbb{H}$ and, as we have seen, $u_1(z) \geq u_2(z)$ there. If D has a boundary point in the upper half-plane $\mathbb{H}$, the function u_1 tends to $+\infty$ whereas u_2 remains bounded. Then, u_1 and u_2 do not agree everywhere in $D \cap \mathbb{H}$ and hence agree nowhere in $D \cap \mathbb{H}$ by Lemma 9.3. □

Exercise 9.4 Give a direct proof of the estimate (9.10) for a half-plane by the method used to prove the disk case (9.9).

9.2.3 Half-Planes and Disks Are Hyperbolically Convex

We show in this subsection how it follows from the inequality (9.9) that Euclidean disks are hyperbolically convex and from (9.10) that half-planes are hyperbolically convex. Recall what this means: if a proper simply connected domain contains a half-plane then the hyperbolic geodesic joining any two points in the half-plane never leaves the half-plane, with the analogous statement in the case of a disk.

Theorem 9.4 *Any Euclidean half-plane or any Euclidean disk is hyperbolically convex in any proper simply connected domain.*

Proof Let D be a proper simply connected domain. We first consider the case that D contains a half-plane. By rotation and translation, we may suppose that the half-plane in question is the lower half-plane $\overline{\mathbb{H}} = \{z : \Im z < 0\}$. Suppose that α and β are points in $\overline{\mathbb{H}}$, and write $\alpha = x_1 + iy_1$, $\beta = x_2 + iy_2$ where both y_1 and y_2 are negative. Again without loss of generality, we may assume that $y_1 \geq y_2$, and will then show that the hyperbolic geodesic (with respect to D) joining α and β lies in the closed half-plane $\{z : \Im z \leq y_1\}$. Let γ be any curve in D joining α and β. Suppose that γ contains a sub-arc C joining two points that both have imaginary part y_1 and that lies otherwise in $D \cap \{z : \Im z > y_1\}$, as shown in Fig. 9.5. Denoting reflection in the line $\Im z = y_1$ by R, it follows from (9.10) that $\rho_D(z) > \rho_D(R(z))$ for any z in $D \cap \{z : \Im z > y_1\}$ (we have strict inequality since D is not symmetric in the line $\Im z = y_1$). Thus, the hyperbolic metric at any point z in C is greater than the hyperbolic metric at the reflected point $R(z)$ on the reflected sub-arc $R(C)$. Hence, the hyperbolic length of $R(C)$ is less than that of C. Replacing the sub-arc C of γ by $R(C)$ we therefore produce a curve that is shorter than γ. Performing this reflection for each such sub-arc C of γ, we produce a curve joining α and β that is shorter than γ and that lies entirely in the closed half-plane $\{z : \Im z \leq y_1\}$. The hyperbolic geodesic joining α and β therefore lies in this closed half-plane and, in particular, in $\overline{\mathbb{H}}$ as was to be shown.

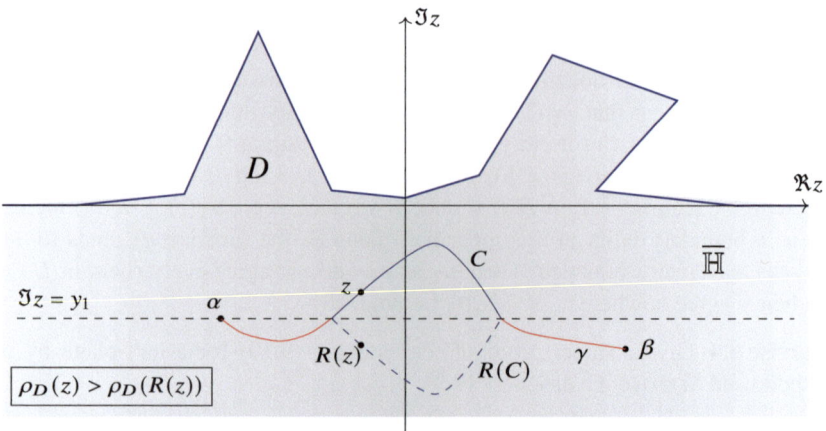

Fig. 9.5 The curve joining α to β in which the sub-arc C is replaced by $R(C)$ is shorter than the original curve γ

Next suppose that D contains a Euclidean disk $U = D(a, r)$ with centre a and radius r. Suppose that α and β lie in U. We may therefore choose r_1 such that both points lie in the smaller disk $D(a, r_1)$. Let γ be any curve in D joining α and β. Suppose that γ contains a sub-arc C joining two points that both lie on the circle $C(a, r_1)$ and that otherwise lies outside the disk $D(a, r_1)$. Denoting reflection in the circle $C(a, r_1)$ by R, if follows from (9.9) that, for each point z on C,

$$\rho_D(z) > \frac{|R(z) - a|}{|z - a|} \rho_D(R(z)). \tag{9.16}$$

Suppose that C is parameterised by $C(t)$ on $[0, 1]$. The reflection $R \circ C$ of the curve C then lies in the closed disk $\overline{D(a, r_1)}$ and, by (2.16), has the explicit parameterisation

$$(R \circ C)(t) = a + \frac{r_1^2}{\overline{C(t) - a}}$$

Differentiating, and then using (2.16) again, gives

$$|(R \circ C)'(t)| = \frac{r_1^2}{|C(t) - a|^2} |C'(t)| = \frac{|R(C(t)) - a|}{|C(t) - a|} |C'(t)|.$$

9.2 Comparison Results for the Hyperbolic Metric

By (3.14) and (9.16),

$$\begin{aligned} L_\rho(C) &= \int_0^1 \rho_D(C(t))|C'(t)|\,dt \\ &> \int_0^1 \rho_D(R(C(t))) \frac{|R(C(t)) - a|}{|C(t) - a|} |C'(t)|\,dt \\ &= \int_0^1 \rho_D(R(C(t))) |(R \circ C)'(t)|\,dt = L_\rho(R \circ C). \end{aligned}$$

If we therefore replace the sub-arc C in the curve γ by $R \circ C$, we obtain a curve shorter than γ. By replacing each sub-arc of γ that lies outside $D(a, r_1)$ by its reflection in the circle $C(a, r_1)$, we obtain a curve shorter than γ that joins α and β and lies entirely in the closed disk $\overline{D(a, r_1)}$ and so entirely in the disk $D(a, r)$. It follows that the hyperbolic geodesic joining α and β in D lies entirely in the disk $D(a, r)$, showing that this disk is hyperbolically convex. □

Brown Flinn [4, Theorem 2] has shown that half-planes and disks are the only domains that are always hyperbolically convex.

Another result that follows easily from Theorem 9.3 is the following.

Theorem 9.5 *Let D be a proper simply connected domain that contains the half-plane $\overline{\mathbb{H}} = \{z : \Im z < 0\}$ and that is not symmetric in the real axis. Let x be a point in D that lies on the real axis and let z be a point in D that lies in the upper half-plane $\mathbb{H}$. Then, $d(x, z; D) > d(x, \bar{z}; D)$.*

Proof Let γ be the hyperbolic geodesic joining x to z in D. On reflecting in the real axis those sub-arcs of γ that lie in the upper half-plane, we obtain a strictly shorter curve (by Theorem 9.3) that joins x to $\bar{z}$ in D. The hyperbolic geodesic joining x to $\bar{z}$ in D is, in turn, no longer than this second curve. Combining these two comparisons we conclude that $d(x, z; D) > d(x, \bar{z}; D)$. □

The estimate (9.9) from Theorem 9.3 is a comparison of the hyperbolic metric for two points symmetric with respect to a fixed circle. Jørgensen in [17, Theorem 4] fixes two points z_1 and z_2 in D and varies the circle. Suppose, then, that D is a proper simply connected domain and that z_1 and z_2 are two fixed points in D. With the notation of Sect. 2.1.1, suppose that the region $D(z_1, z_2, k)$ lies in D. It cannot be that k is greater than 1 as, in this case, $D(z_1, z_2, k)$ is the complement of a disk containing z_2 and, in turn, if such a region was contained in D then the complement of D relative to the complex plane would be bounded. This would then violate Condition (g) of Theorem 7.4, for example, as a circle with sufficiently large radius would lie in D and have non-zero winding number about z_2. Alternatively, $\mathbb{C}_\infty \setminus D$ would not be connected. Thus, $k \leq 1$ and the region $D(z_1, z_2, k)$ is either a disk if $0 < k < 1$ or a half-plane if $k = 1$. Suppose that $k < 1$. Since z_1 lies inside the circle $C(z_1, z_2, k)$ and z_2 lies outside this circle, and since z_1 and z_2 are symmetric in the circle, it follows from (9.9) that

$$\rho_D(z_2) \geq \frac{|z_1 - a|}{|z_2 - a|}\rho_D(z_1),$$

where a is the centre of $C(z_1, z_2, k)$. By Exercise 2.1, $|z_1 - a|/|z_2 - a| = k^2$ and so $\rho_D(z_2) \geq k^2 \rho_D(z_1)$. The best such estimate is the one with the largest k, so we set k_0 to be the largest k for which $D(z_1, z_2, k)$ is contained in D. Since the circles $C(z_1, z_2, k)$ expand with k, we have

$$k_0 = \min\left\{\frac{|\zeta - z_1|}{|\zeta - z_2|} : \zeta \in \partial D\right\}.$$

Interchanging the role of z_1 and z_2, we find that $\rho_D(z_1) \geq k_1^2 \rho_D(z_2)$ where

$$k_1 = \min\left\{\frac{|\zeta - z_2|}{|\zeta - z_1|} : \zeta \in \partial D\right\} = 1 \bigg/ \max\left\{\frac{|\zeta - z_1|}{|\zeta - z_2|} : \zeta \in \partial D\right\}.$$

Then, $\rho_D(z_2)/\rho_D(z_1) \leq 1/k_1^2$. Putting all this together we have the following result.

Theorem 9.6 *Let D be a proper simply connected domain. Let z_1 and z_2 be points in D. Then,*

$$\min\left\{\left|\frac{\zeta - z_1}{\zeta - z_2}\right|^2 : \zeta \in \partial D\right\} \leq \frac{\rho_D(z_2)}{\rho_D(z_1)} \leq \max\left\{\left|\frac{\zeta - z_1}{\zeta - z_2}\right|^2 : \zeta \in \partial D\right\}.$$

Exercise 9.5

▷ Suppose that D is a proper simply connected domain that is symmetric in a Euclidean line L. Suppose that the line segment $[\alpha, \beta]$ lies on $L \cap D$. Show that this Euclidean line segment $[\alpha, \beta]$ is the geodesic arc in D joining α and β.
▷ Show that if a hyperbolic geodesic arc for a proper simply connected domain D coincides with a line segment on a Euclidean line L then the domain D is symmetric in L.

We end this chapter with one last result about hyperbolic geodesics to the effect that hyperbolic geodesics tend to keep travelling straight ahead. We follow the proof given by Pommerenke [22, Proposition 4.22].

Theorem 9.7 *Let γ be a geodesic in the hyperbolic metric of a proper simply connected domain D. Let U be a disk in D that is tangent to γ. Then γ does not enter the disk U.*

Before we embark on the proof of Theorem 9.7, we need a version of the maximum principle for harmonic functions that is sufficient for many applications.

9.2 Comparison Results for the Hyperbolic Metric

Proposition 9.4 (A Maximum Principle for Harmonic Functions) *Suppose that u is harmonic in a bounded domain D and that u is continuous on the closure of D. Then u achieves its maximum value on the boundary of D, and only on the boundary of D unless u is constant.*

The key to this maximum principle is the mean-value property. Let u be a function harmonic in a domain D and let z_0 be in D. By [16, Theorem 11.6], there is a function f analytic in the disk $U = D(z_0, \delta_D(z_0))$ for which $u = \Re f$. Here, $\delta_D(z_0)$ is once again the distance from z_0 to the boundary of D and U is the disk of largest radius centred at z_0 and contained in D. The Cauchy Integral Formula [16, Theorem 7.1] leads to the integral representation, for any ρ less than $\delta_D(z_0)$,

$$f(z_0) = \frac{1}{2\pi i} \int_{C(z_0, \rho)} \frac{f(z)}{z - z_0} \, dz.$$

Parameterising $C(z_0, \rho)$ by $\gamma(t) = z_0 + \rho e^{it}, 0 \leq t \leq 2\pi$,

$$\begin{aligned} f(z_0) &= \frac{1}{2\pi i} \int_0^{2\pi} \frac{f(\gamma(t))}{\gamma(t) - z_0} \gamma'(t) \, dt \\ &= \frac{1}{2\pi i} \int_0^{2\pi} \frac{f(z_0 + \rho e^{it})}{\rho e^{it}} i\rho e^{it} \, dt \\ &= \frac{1}{2\pi} \int_0^{2\pi} f(z_0 + \rho e^{it}) \, dt. \end{aligned}$$

This is the mean-value property for the analytic function f. Taking real parts, we obtain the mean-value property for u, namely

$$u(z_0) = \frac{1}{2\pi} \int_0^{2\pi} u(z_0 + \rho e^{it}) \, dt \qquad (9.17)$$

valid for $\rho < \delta_D(z_0)$. With the mean-value property at our disposal, we can prove the maximum principle.

Proof of Proposition 9.4 As D is assumed to be bounded, its closure $\mathrm{cl}(D)$ is a compact set. Since u is continuous on the closure of D, it achieves its maximum value there. Set $M = \max\{u(z) : z \in \mathrm{cl}(D)\}$ and let us assume that there is a point z_0 in D at which $u(z_0) = M$. We claim that u is constant in this case. Thus, if u is not constant then u is less than M on D and only assumes its maximum value on the boundary of D.

Suppose, then, that z_0 is in D and that $u(z_0) = M$. We deduce from the mean-value property that u has the constant value M on the disk $U = D(z_0, \delta_D(z_0))$. If not, there is a point z_1 in U for which $u(z_1) = u(z_0) - \epsilon$ for some positive ϵ. With

$z_1 = z_0 + \rho e^{it_1}$, there is an interval I of t about t_1 such that $u(z_0 + \rho e^{it}) \le u(z_0) - \epsilon/2$ for $t \in I$. Then, by the mean-value property and with $|I|$ being the length of I,

$$\begin{aligned} M = u(z_0) &= \frac{1}{2\pi} \int_{[0,2\pi]\setminus I} u(z_0 + \rho e^{it})\,dt + \frac{1}{2\pi} \int_I u(z_0 + \rho e^{it})\,dt \\ &\le \frac{1}{2\pi} \int_{[0,2\pi]\setminus I} M\,dt + \frac{1}{2\pi} \int_I (M - \tfrac{\epsilon}{2})\,dt \\ &= \frac{1}{2\pi} M(2\pi - |I|) + \frac{1}{2\pi}(M - \tfrac{\epsilon}{2})|I| \\ &= M - \frac{|I|}{4\pi}\epsilon, \end{aligned}$$

which is a contradiction. Thus, if $z_0 \in D$ with $u(z_0) = M$ then u takes the constant value M on the disk $D(z_0, \delta_D(z_0))$. Starting with one such z_0 (if it exists), we form the open subset D_{z_0} of D as in Lemma 9.1. The function u is then constant on this open set and so, by Lemma 9.1, is constant on D since D_{z_0} coincides with D. □

The maximum principle is a key tool in the proof of Theorem 9.7 as given by Pommerenke [22].

Proof of Theorem 9.7 We begin by arranging for a convenient normalisation of the setting in the theorem. By translation, we may assume that U is tangent to γ at the origin. Then, by a rotation, we may assume that γ, hence also U, is tangent to the real axis at 0. By rotation through π if necessary, we may further assume that U lies in the lower half-plane $\overline{\mathbb{H}}$. Now set f to be a conformal mapping of the unit disk $\mathbb{D}$ onto D with $f(0) = 0$. Then, $f^{-1}(\gamma)$ is a geodesic in $\mathbb{D}$ that passes through 0 and hence, by Theorem 4.6, is a diameter of $\mathbb{D}$. By precomposing f with a rotation of the disk, we may assume that this diameter is the interval $(-1, 1)$. Then, γ is parameterised by $\gamma(t) = f(t)$, $-1 < t < 1$, and $\gamma'(0) = f'(0)$ is real. By replacing $f(z)$ by $f(-z)$ if necessary, we may suppose that $f'(0)$ is positive. Finally, replacing f by $f/f'(0)$, which simply scales the domain D by $1/f'(0)$, we may suppose that $f'(0) = 1$. This completes the various normalisations as illustrated in Fig. 9.6.

We prove the result for any disk U tangent to γ at 0 that is compactly contained in D. It then follows that γ doesn't enter the largest disk in D tangent to γ at 0. For $0 < r < 1$, set $g(z) = f(rz)/r$. Then, $g(0) = 0$, $g'(0) = 1$, and $g(-1, 1)$ is the sub-arc $\gamma_r = f(-r, r)$ of γ scaled by $1/r$ which remains tangent to U at 0. Taking r sufficiently large (depending on U), we may assume that $U \subset g(\mathbb{D}) = \tfrac{1}{r} f(r\mathbb{D})$, the latter being a bounded domain. The point is that g is univalent in $\mathbb{D}$ and continuous on the closure $\mathrm{cl}(\mathbb{D})$ of $\mathbb{D}$. Set

$$h(z) = \frac{1}{g(z)} - \frac{1}{z} - z, \quad z \in \mathbb{D}.$$

9.2 Comparison Results for the Hyperbolic Metric

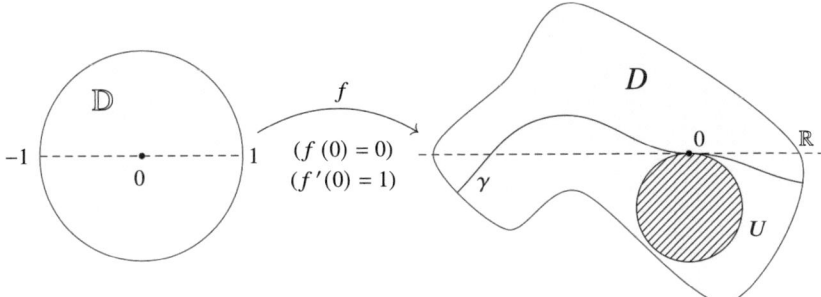

Fig. 9.6 An illustration of Theorem 9.7 including the normalised conformal map f: the hyperbolic geodesic γ never enters the tangential disk U

Since g has a power series expansion of the form $g(z) = z + a_2 z^2 + a_3 z^3 + \ldots$, its reciprocal $1/g$ has the expansion $1/g(z) = 1/z - a_2 + (a_2^2 - a_3)z + \ldots$. This shows that h has a removable singularity at the origin (with $h(0) = -a_2$) and so may be considered to be analytic in the unit disk.

Under the inversion $w \to 1/w$, the disk U is mapped to a half-plane $H = \{w : \Im w > a\}$ ($a > 0$, in fact $a = 1/(2y)$ if U is the disk $D(-iy, y)$). The rest of the domain $g(\mathbb{D})$, and in particular its boundary, is mapped into the closed half-plane $\{w : \Im w \leq a\}$ under inversion so that, in particular, $\Im(1/g(\zeta)) \leq a$ for ζ on the unit circle $\mathbb{T}$. On the unit circle, with $z = e^{i\theta}$, we have $z + 1/z = 2\cos\theta$ and so $\Im(z + 1/z)$ vanishes on the unit circle. Hence,

$$\Im(h(\zeta)) = \Im\left(\frac{1}{g(\zeta)}\right) \leq a, \quad \zeta \in \mathbb{T}.$$

By the Maximum Principle, Proposition 9.4, the harmonic function $\Im(h(z))$ is bounded above by a on the whole unit disk $\mathbb{D}$. In particular,

$$\Im(h(x)) = \Im\left(\frac{1}{g(x)} - \frac{1}{x} - x\right) = \Im\left(\frac{1}{g(x)}\right) \leq a, \quad -1 < x < 1.$$

It follows that $g(x)$, $-1 < x < 1$, does not lie in U (as otherwise $1/g(x)$ would lie in the half-plane H), that is γ_r does not enter U. Since this is the case for every r, we conclude that γ does not enter U. $\square$

Chapter 10
Uniformisation of Planar Domains

The aim in this final chapter is to show that the disk can act as a covering space for the vast majority of planar domains, in fact for any planar domain whose complement contains at least two points. Such domains are called *hyperbolic*. The only domains homeomorphic to a disk are those that are simply connected. For a domain that may not be simply connected, a covering map is the next best thing: its inverse, as a multi-valued function, is locally a homeomorphism. This idea is made precise in the definition of a covering map to follow later.

A classical example is the Modular Function, which is a covering of the twice punctured plane by the half plane $\mathbb{H}$. In Sect. 10.1 we set out the bare rudiments of covering spaces and lifts of curves, in particular proving (a version of) the Monodromy Theorem. In Sect. 10.2 we construct the Modular Function and describe its automorphism group. In Sect. 10.3 we show how the modular function can be used to prove the Little Picard Theorem and Montel's Second Theorem, the latter having been a key ingredient in the proof of the Great Picard Theorem in Chap. 5. Finally, in Sect. 10.4, we work through a proof of the Uniformisation Theorem for planar domains, to the effect that every hyperbolic planar domain can be covered by the disk, the key point being that the disk is simply connected. In this sense, the Uniformisation Theorem is a generalisation of Theorem 6.7, the Riemann Mapping Theorem.

10.1 Covering Spaces and the Monodromy Theorem

The notion of *covering space* is topological and is properly treated from a topological point of view. Nevertheless, we will take a more restricted view and consider only planar domains rather than more general topological spaces. The development and proofs are essentially the same in the general topological setting—we give ourselves the luxury of avoiding the finer topological points and, in any case, planar domains are our sole concern. Both Conway [8] and Ahlfors and Sario [2], for example, contain excellent accounts of covering spaces in this greater generality.

Definition 10.1 (Covering Space) Let D be a planar domain. An open set Ω in the complex plane is said to be a *covering space* of D if there is a continuous mapping f of Ω onto D with the following property:

> to each point a in D there corresponds a neighbourhood Δ_a of a such that each component of $f^{-1}(\Delta_a)$ is mapped homeomorphically onto Δ_a by f.

The mapping f in this definition is called a *covering map*. We vary between referring to Ω as a covering space or, more properly, referring to (Ω, f) as a covering space. The neighbourhood Δ_a is said to be *evenly covered* by f. If Δ in D is evenly covered by f and if $\tilde{\Delta}$ is open and connected in Δ then $\tilde{\Delta}$ is also evenly covered by f. If we wish, therefore, we can take the evenly covered neighbourhood Δ_a of a to be simply a disk about a. If the map f is not just continuous but is analytic, we refer to f as an *analytic covering map* and we refer to Ω (or (Ω, f)) as an *analytic covering space*. In this case, each component in Ω of the preimage of an evenly covered neighbourhood Δ in D is mapped by f conformally onto Δ.

For the remainder of this section, (Ω, f) will be a covering space of a domain D. If $a \in D$ and if $a^* \in \Omega$ with $f(a^*) = a$, we say that a^* *lies over* a. The collection of all points $f^{-1}(a)$ lying over a is called the *fibre* of a. Let us also, for convenience, denote the plane punctured at 0, that is $\mathbb{C} \setminus \{0\}$, by $\mathbb{C}_0$.

Example 10.1 The pair $(\Omega, f) = (\mathbb{C}, \exp)$ is an analytic covering space of the punctured plane $D = \mathbb{C}_0$. For a non-zero a with $a = re^{i\theta}$, $(r > 0, \theta \in [0, 2\pi))$, we have $f^{-1}(a) = \{a_n^* = \log r + i(\theta + 2n\pi) : n \in \mathbb{Z}\}$. This is a sequence of points on the vertical line through $\log r$ and spaced 2π apart. The preimage of a small open polar rectangle Δ centred at a comprises identical open rectangles Δ_n^* about each a_n^*. Each of these rectangles is mapped conformally onto Δ as shown in Fig. 10.1.

Example 10.2 For each natural number n, the pair $(\Omega, f) = (\mathbb{C}_0, z^n)$ is also an analytic covering space of the punctured plane $D = \mathbb{C}_0$. The preimage of a non-zero point $a = re^{i\theta}$, $(r > 0, \theta \in [0, 2\pi))$, comprises

$$f^{-1}(a) = \left\{a_k^* = r^{1/n} e^{i(\theta/n + 2k\pi/n)} : k = 0, 1, \ldots, n-1\right\}.$$

The preimage of a small disk about a then has n components, one for each point a_k^*, and each component is mapped conformally onto Δ by f. Note that $f(z) = z^n$, $n > 1$, does not extend to a covering $f : \mathbb{C} \to \mathbb{C}$ of the complex plane $\mathbb{C}$ by $\mathbb{C}$ itself since the preimage of an open disk $\Delta = D(0, r)$ is the open disk $\Delta = D(0, r^{1/n})$ but f is not univalent on this disk.

Example 10.3 Fix $R > 1$ and let $\mathbb{S}_R$ be the strip $\{z : 0 < \Re z < \log R\}$. Then, $(\Omega, f) = (\mathbb{S}_R, \exp)$ is a covering space of the annulus $A(1, R) = \{z : 1 < |z| < R\}$, as shown in Fig. 10.2. In fact, each vertical line $\Re z = x$ is wrapped infinitely

10.1 Covering Spaces and the Monodromy Theorem

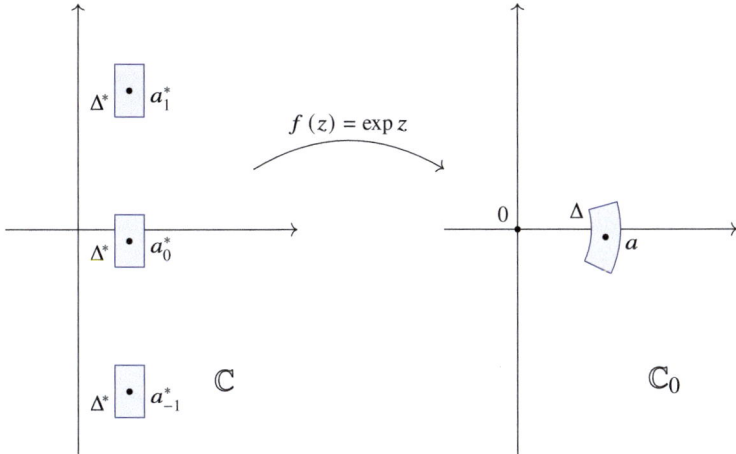

Fig. 10.1 $(\mathbb{C}, \exp)$ as an analytic covering space of the punctured plane $D = \mathbb{C}_0$. Here, $a = re^{i\theta}$, ($r > 0, \theta \in [0, 2\pi)$) and $a_n^* = \log r + i(\theta + 2n\pi)$

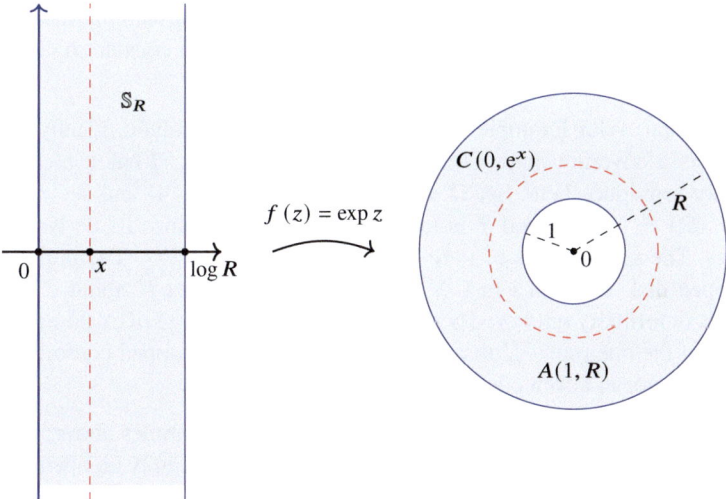

Fig. 10.2 $(\mathbb{S}_R, \exp)$ as a covering map of the annulus $A(1, R)$. The line $\Re z = x$ in the strip is wrapped infinitely often around the circle $C(0, e^x)$ in the annulus

often around the circle of radius e^x. Note that, up to conformal equivalence, $A(1, R)$ is a typical annulus in the family $\mathscr{A}_\lambda$, $1 < \lambda < \infty$, set out in (6.31b), where $\lambda = R$.

Example 10.4 If Ω and D are (simply connected) domains and $f : \Omega \to D$ is a conformal map, then (Ω, f) is trivially an analytic covering space of D.

If (Ω, f) is an analytic covering space of D and if g is a conformal map of Ω_1 onto Ω then $(\Omega_1, f \circ g)$ is again an analytic covering space of D. In particular, if $(\mathbb{D}, f)$ is an analytic covering space of a planar domain D then (Ω, h) is also an analytic covering space of D if Ω is a proper simply connected domain and we take $h = f \circ g$ where g is any conformal map of Ω onto $\mathbb{D}$.

If (Ω, f) is an analytic covering space of D and if g is a conformal map of D onto D_1 then $(\Omega, g \circ f)$ is an analytic covering space of D_1.

Example 10.5 $(\mathbb{C}, \phi)$ is a covering space of the unit disk $\mathbb{D}$ if we take, for example,

$$\phi(z) = \frac{z}{1 + |z|}, \quad z \in \mathbb{C}.$$

Note that ϕ shrinks each ray $\{re^{i\theta} : 0 \leq r < \infty\}$ in $\mathbb{C}$ to the radius $\{te^{i\theta} : 0 \leq t < 1\}$ of the unit disk. In fact, ϕ is a homeomorphism of $\mathbb{C}$ onto $\mathbb{D}$.

However, $(\mathbb{C}, f)$ cannot be an analytic covering space of the unit disk since any such f would be a bounded analytic function and therefore constant by Liouville's Theorem.

Example 10.6 With Example 10.1 in mind, not every analytic, locally univalent function is a covering map. Conway [7, Chap. IX, Sect. 7] has a nice example which we reproduce here. Set $\Omega = \{z : 0 < \arg z < 5\pi/4\}$ and set $f(z) = z^2$. Then, $f(\Omega) = \mathbb{C} \setminus \{0\}$ and f is locally univalent on Ω since its derivative never vanishes. The square roots of i are $e^{\pi i/4}$ and $e^{5\pi i/4} = -e^{\pi i/4}$. The preimage of a small open disk Δ about i in $\mathbb{C} \setminus \{0\}$ includes an open set U about $e^{\pi i/4}$ that is mapped conformally onto Δ. The remainder of the preimage of Δ comprises only that half of the open set $-U$ that lies in Ω and this is not mapped conformally onto Δ. In fact, it doesn't even contain the point $e^{5\pi i/4}$.

The covering space Ω was connected in each of the examples above, and this is the most natural setting. The reason for allowing Ω to be simply an open planar set in Definition 10.1 is to be able to satisfactorily treat Example 10.7 on the restriction of a covering map. In order to do so, we need the following results. The first is that covering maps are open (noting that this is automatic in the case of analytic covering maps).

Lemma 10.1 *Covering mappings are open mappings.*

Proof Suppose that (Ω, f) is a covering map of D and that U is open in Ω. We aim to show that $f(U)$ is open in D. Suppose that $a \in f(U)$ with $a = f(u)$, $u \in U$, say. Choose a neighbourhood Δ of a that is evenly covered by f. Then, u belongs to some component, say Δ^*, of $f^{-1}(\Delta)$ and $f : \Delta^* \to \Delta$ is a homeomorphism. Since

10.1 Covering Spaces and the Monodromy Theorem

$\Delta^* \cap U$ is open and contains u, $f(\Delta^* \cap U)$ is open, contains a and is contained in $f(U)$ thereby verifying that $f(U)$ is open. □

We now show that we can effectively restrict attention to connected covering spaces.

Lemma 10.2 *Suppose that (Ω, f) is a covering space of D in the sense of Definition 10.1. Suppose that Ω_1 is a connected component of Ω. Then, $(\Omega_1, f|_{\Omega_1})$ is also a covering space of D.*

Proof We begin by showing that $f : \Omega_1 \to D$ is surjective. We set $D_1 = f(\Omega_1)$. Since Ω_1 is non-empty, D_1 is non-empty. Since Ω_1 is connected and f is continuous, D_1 is connected. We aim to show that D_1 is both open and closed in D as we can then conclude that D_1 must coincide with D.

That D_1 is open follows, by Lemma 10.1, from the fact that f is an open mapping and that Ω_1 is open. On the other hand, $D \setminus D_1$ is also open. To see this, suppose that $z \notin f(\Omega_1)$. Since z is in D, there is a neighbourhood Δ_z of z that is evenly covered by f. If any component, say Δ_z^*, of $f^{-1}(\Delta_z)$ had non-empty intersection with Ω_1 then that component would lie completely in Ω_1. This is because Ω_1 is a maximal connected subset of Ω. Since $f : \Delta_z^* \to \Delta_z$ is a homeomorphism, this would force $z \in \Delta_z$ to be in the image of Ω_1 under f after all. Thus, no component of $f^{-1}(\Delta_z)$ can meet Ω_1 or, equivalently, $f^{-1}(\Delta_z)$ doesn't meet Ω_1 or, equivalently, Δ_z lies in $D \setminus D_1$, showing that $D \setminus D_1$ is open.

To complete the proof, we verify the covering property of $f : \Omega_1 \to D$. Take any point a in D and a neighbourhood of a that is evenly covered by f. With $\{\Delta_i^*\}_i$ being the connected components of $f^{-1}(\Delta)$ in Ω, the map $f : \Delta_i^* \to \Delta$ is a homeomorphism in each case. Each component Δ_i^* of the preimage either lies completely in Ω_1 or else doesn't meet Ω_1 at all. Again, this is because Δ_i^* is connected and Ω_1 is a maximal connected subset of Ω. The components $\{\Delta_i^*\}_i$ of Δ divide into those that lie completely in Ω_1, the ones that we are interested in, and those that lie completely in $\Omega \setminus \Omega_1$. Since, as already shown, $f : \Omega_1 \to D$ is surjective, there is at least one non-empty component Δ_i^* in Ω_1. It follows that $f|_{\Omega_1}$ maps each component of $(f|_{\Omega_1})^{-1}(\Delta)$ homeomorphically onto Δ, this being the key property of a covering map. □

Example 10.7 (Restriction of a Covering Map) Suppose that (Ω, f) is a covering space of D and that U is a domain lying in D. Let V be a connected component of $f^{-1}(U)$. Then, $(V, f|_V)$ is a covering space of U. We can verify this as follows. Given a in U and an evenly covered neighbourhood Δ_a of a for the covering space (Ω, f), replacing Δ_a by a sufficiently small disk about a, we may assume that $\Delta_a \subseteq U$. Since each connected component of the preimage of Δ_a under f lies in $f^{-1}(U)$, Δ_a is also an evenly covered neighbourhood of a for the function f restricted to $f^{-1}(U)$. That is, $(f^{-1}(U), f|_{f^{-1}(U)})$ is a covering space of U. By Lemma 10.2, a further restriction of the mapping function to the component V of $f^{-1}(U)$ still results in a covering map of U.

A key concept in the context of covering maps is the *lift of a curve* γ in D to Ω, this being a curve γ^* in Ω for which $f \circ \gamma^* = \gamma$. First we show that each curve in D has a unique lift once the initial point of the lift is specified.

Lemma 10.3 *Let (Ω, f) be a covering space of a domain D. Let γ be a continuous curve in D with initial point a in D. Let a^* in Ω lie over a. Then there is a unique lift γ^* of γ with initial point a^*.*

Proof Let $\gamma(t)$, $0 \leq t \leq 1$, be a continuous parametrisation of the curve γ. We first verify the uniqueness of any lift of γ with initial point a^*. Suppose that $\gamma_1^*(t)$ and $\gamma_2^*(t)$, $0 \leq t \leq 1$, are both lifts of γ with the same initial point a^* lying over a. Let E be the set of all t in $[0, 1]$ such that $\gamma_1^*(t) = \gamma_2^*(t)$. Since γ_1^* and γ_2^* are continuous, the set E is closed. To see that E is open, let t_0 be in E, choose an evenly covered neighbourhood Δ of $\gamma(t_0)$ and then the neighbourhood Δ^* of the point $\gamma_1^*(t_0)$ (which coincides with $\gamma_2^*(t_0)$) lying above $\gamma(t_0)$ that is homeomorphic to Δ under f. By continuity of γ_1^* and γ_2^*, there is some δ such that $\gamma_1^*(t)$ and $\gamma_2^*(t)$ both lie in Δ^* for $t \in (t_0 - \delta, t_0 + \delta)$ [here $t_0 = 0$ and $t_0 = 1$ are special cases]. Since f is a homeomorphism on Δ^* and $f(\gamma_1^*(t)) = f(\gamma_2^*(t)) = \gamma(t)$, it must be that $\gamma_1^*(t) = \gamma_2^*(t)$, $t \in (t_0 - \delta, t_0 + \delta)$. In short, E is open. Since the interval $[0, 1]$ is connected and $0 \in E$, it must be that $E = [0, 1]$.

Now we prove existence. Choose an evenly covered neighbourhood Δ of a. There is then a neighbourhood Δ^* of a^* that is mapped homeomorphically by f onto Δ. An initial portion of the curve γ lies in Δ, say $\gamma(t)$, $0 \leq t \leq t_0$. Since $f^{-1} : \Delta \to \Delta^*$ is continuous, the preimage under f in Δ^* of this initial section of the curve γ is a curve $\gamma^*(t)$, $0 \leq t \leq t_0$, for which $f(\gamma^*(t)) = \gamma(t)$. Clearly, γ^* has initial point a^*. In this way, we have a lift of an initial section of the curve γ, that is of $\gamma(t)$, $0 \leq t \leq t_0$.

Now let E be the set of all t_0 in $(0, 1]$ such that $\gamma(t)$, $0 \leq t \leq t_0$, has a lift $\gamma^*(t)$, $0 \leq t \leq t_0$, to Ω with initial point a^*. The lift of this portion of the curve will be unique, as argued earlier. If E isn't all of $(0, 1]$ then $E = (0, t_0)$ for some t_0 in $(0, 1)$. In fact, if t_0 is in E then, arguing as above, $\gamma(t)$ can be lifted with initial point a^* for $0 \leq t \leq t_0 + \epsilon$ for some positive ϵ. Suppose, then, that $E = (0, t_0)$ for $t_0 < 1$ and let $\gamma^*(t)$ be the lift of the open subarc $\gamma(t)$, $0 \leq t < t_0$, with initial point a^*. Let Δ be an evenly covered neighbourhood of $\gamma(t_0)$. There is a t' such that $\gamma(t)$, $t' \leq t < t_0$, lies in Δ and so $\gamma^*(t)$, $t' \leq t < t_0$, lies in some particular component of $f^{-1}(\Delta)$, say Δ^*. This is the case since, for any $t_1 \in (t', t_0)$, the curve $\gamma^*(t)$, $t' \leq t \leq t_1$, is continuous and therefore lies in a single component of $f^{-1}(\Delta)$. Set $\gamma^*(t_0)$ to be the point in Δ^* lying above $\gamma(t_0)$. Then, $\gamma^*(t) \to \gamma^*(t_0)$ as $t \uparrow t_0$ since $\gamma^*(t) = f^{-1}(\gamma(t))$, $t' \leq t < t_0$, and since $f^{-1} : \Delta \to \Delta^*$ is continuous. We have shown that $\gamma^*(t)$ extends to t_0 and hence, by previous arguments, beyond t_0. □

We can now state the Monodromy Theorem, to the effect that two curves that are fixed endpoint (FEP) homotopic in D (see Definition 6.2) will lift to fixed endpoint homotopic curves in the covering space Ω assuming, of course, that the lifts have a common initial point.

10.1 Covering Spaces and the Monodromy Theorem

Theorem 10.1 (Monodromy Theorem) *Suppose that (Ω, f) is a covering space of D, that a and b are in D, and that a^* lies over a. Let γ_0 and γ_1 be curves in D, both of which join a to b, and suppose that these curves are fixed endpoint homotopic in D. Let γ_0^* and γ_1^* be the lifts of γ_0 and γ_1, respectively, to Ω both with initial point a^*. Then, γ_0^* and γ_1^* have the same endpoint b^* in Ω lying over b and, moreover, γ_0^* and γ_1^* are fixed endpoint homotopic in Ω.*

Proof Let $\Gamma(t, s)$, with $(t, s) \in I_2 = [0, 1] \times [0, 1]$, be a fixed endpoint homotopy of γ_0 and γ_1 in the sense of Definition 6.2. Thus, Γ is continuous on I_2, $\Gamma(t, 0) = \gamma_0(t)$, $\Gamma(t, 1) = \gamma_1(t)$, $0 \leq t \leq 1$, and $\Gamma(0, s) = a$, $\Gamma(1, s) = b$, $0 \leq s \leq 1$.

Let us write $\gamma_s(t)$ for the curve $\gamma_s(t) = \Gamma(t, s)$, $0 \leq t \leq 1$. By Lemma 10.3, each curve γ_s has a unique lift $\gamma_s^*(t)$, $0 \leq t \leq 1$, to Ω with initial point a^*. We set

$$\Gamma^*(t, s) = \gamma_s^*(t), \quad (t, s) \in I_2,$$

with the aim of showing that Γ^* is continuous on I_2. Let's assume this for now. Since $f(\gamma_s^*(t)) = \gamma_s(t)$, $0 \leq t \leq 1$, we have $f(\gamma_s^*(1)) = \gamma_s(1) = b$ for each s. That is, the endpoint $\gamma_s^*(1)$ of γ_s^* lies over b for each s. But since $\gamma_s^*(1) = \Gamma^*(1, s)$ is continuous in s and the points in Ω lying over b are isolated, $\gamma_s^*(1)$ doesn't vary with s. Calling this common endpoint b^*, we see that each curve γ_s^* has endpoints a^* (of course) and b^*. In particular, γ_0^* and γ_1^* have the same endpoints, as claimed. On the assumption that Γ^* is continuous on I_2, we see that Γ^* is a fixed endpoint homotopy of γ_0^* and γ_1^*, in the sense of Definition 6.2.

The argument to show that Γ^* is continuous on I_2 has similarities with the proof of Lemma 10.3. Note that, for each fixed s in $[0, 1]$, we have that $\Gamma^*(t, s)$, $0 \leq t \leq 1$, is continuous in t since it is a lift of the continuous curve γ_s. It is continuity in s that is the issue. Note also that, because of how Γ^* is defined,

$$f(\Gamma^*(t, s)) = f(\gamma_s^*(t)) = \gamma_s(t) = \Gamma(t, s), \quad (t, s) \in I_2. \tag{10.1}$$

Fix s_0 in $[0, 1]$. Let E be the set of all τ in $[0, 1]$ for which $\Gamma^*(t, s)$, restricted to the rectangle $[0, \tau] \times [0, 1]$, is continuous on the line segment $0 \leq t \leq \tau$, $s = s_0$. We need to show that $1 \in E$. The set E is either an open interval $[0, t_0)$ or a closed interval $[0, t_0]$ (in common with the proof of Lemma 10.3, the argument requires minor modifications in the cases $E = [0, 1)$ and $E = \{0\}$). Choose an evenly covered neighbourhood Δ of $\Gamma(t_0, s_0)$ and the component Δ^* of the preimage of Δ under f that contains the point $\Gamma^*(t_0, s_0)$. We write Q_δ for the square $[t_0 - \delta, t_0 + \delta] \times [s_0 - \delta, s_0 + \delta]$ in I_2. By the continuity of Γ, there exists a positive δ such that $\Gamma(t, s)$ lies in Δ for $(t, s) \in Q_\delta$ and, by continuity of $\gamma_{s_0}^*(t)$ in t, such that $\Gamma^*(t, s_0)$ lies in Δ^* for $t_0 - \delta \leq t \leq t_0 + \delta$. If E is closed, we set $t_1 = t_0$ and, if E is open, we set $t_1 = t_0 - \delta/2$. Then, $\Gamma^*(t_1, s_0)$ lies in Δ^* and, by the continuity in s of $\Gamma^*(t_1, s)$ at (t_1, s_0), there is a $\delta_1 \leq \delta$ such that $\Gamma^*(t_1, s) \in \Delta^*$ for $s \in [s_0 - \delta_1, s_0 + \delta_1]$.

Now define a function $\widetilde{\Gamma}(t, s)$ on the rectangle $R = [t_0 - \delta, t_0 + \delta] \times [s_0 - \delta_1, s_0 + \delta_1]$, contained in Q_δ, by

$$\widetilde{\Gamma}(t, s) = \begin{cases} \Gamma^*(t, s) & t_0 - \delta \leq t \leq t_1, \; s_0 - \delta_1 \leq s \leq s_0 + \delta_1; \\ f^{-1}(\Gamma(t, s)) & t_1 \leq t \leq t_0 + \delta, \; s_0 - \delta_1 \leq s \leq s_0 + \delta_1. \end{cases}$$

If $(t, s) \in [t_1, t_0 + \delta] \times [s_0 - \delta_1, s_0 + \delta_1]$ then $(t, s) \in Q_\delta$ and so $\Gamma(t, s)$ lies in Δ. Thus, $f^{-1}(\Gamma(t, s))$ makes sense in terms of the inverse map $f^{-1} : \Delta \to \Delta^*$. By (10.1), $\Gamma^*(t, s) = f^{-1}(\Gamma(t, s))$ on the rectangle $t_0 - \delta \leq t \leq t_1$, $s_0 - \delta_1 \leq s \leq s_0 + \delta_1$, this in terms of the same branch of the inverse function f^{-1}. It follows that $\widetilde{\Gamma}$ is continuous on the rectangle R since Γ is continuous on I_2 and f^{-1} is continuous on Δ. Now, for each s in the interval $[s_0 - \delta_1, s_0 + \delta_1]$, $f^{-1}(\Gamma(t, s))$, $t_1 \leq t \leq t_0 + \delta$, is a lift of the curve γ_s, $t_1 \leq t \leq t_0 + \delta$, with initial point $\Gamma^*(t_1, s)$. But so is $\Gamma^*(t, s)$, $t_1 \leq t \leq t_0 + \delta$. By uniqueness of lifts, Lemma 10.3, $\widetilde{\Gamma}(t, s)$ must coincide with $\Gamma^*(t, s)$ on R, showing that E must be all of $[0, 1]$. □

Simply connected covering spaces play a special role in the theory. To describe why this is the case, we need the concept of a homomorphism between covering spaces.

Definition 10.2 (Covering Space Homomorphism) Let (Ω_1, f_1) and (Ω_2, f_2) be covering spaces of the domain D. A *homomorphism* from (Ω_1, f_1) to (Ω_2, f_2) is a continuous function $g : \Omega_1 \to \Omega_2$ such that $f_1 = f_2 \circ g$ (see Fig. 10.3).

If g is a homeomorphism (that is, continuous with continuous inverse) then g^{-1} is a homomorphism from (Ω_2, f_2) to (Ω_1, f_1), in which case the covering spaces (Ω_1, f_1) and (Ω_2, f_2) are said to be *isomorphic* and g is referred to as an *isomorphism* between covering spaces.

One way to interpret Definition 10.2 of a homomorphism between covering spaces is that the covering map f_1 can be recovered from f_2 via the continuous map g. Expressed in a different way, the covering map f_1 'factors through' (Ω_2, f_2) via g. For a in D, the homomorphism g maps the fibre of a in Ω_1 to the fibre of a in Ω_2. Consequently, the fibre of a in Ω_2 is 'smaller' than that of a in Ω_1 (a function cannot increase the cardinality of a set). Consider, for instance, the case $(\Omega_1, f_1) = (\mathbb{C}, \exp)$ and $(\Omega_2, f_2) = (\mathbb{C}_0, z^n)$, $n > 1$, from Examples 10.1 and 10.2.

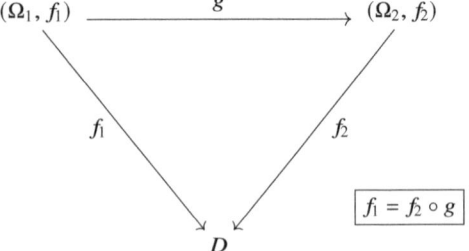

Fig. 10.3 The continuous function g is a homomorphism from the covering space (Ω_1, f_1) of D to the covering space (Ω_2, f_2) of D

10.1 Covering Spaces and the Monodromy Theorem

Both are covering spaces of $D = \mathbb{C}_0$. Then, $g(z) = \exp(z/n)$ is a homomorphism of (Ω_1, f_1) to (Ω_2, f_2). The fibre of $a \in \mathbb{C}_0$ in Ω_1 consists of infinitely many points separated vertically by $2\pi i$ while the fibre of a in Ω_2 consists of only n points. In this sense, (Ω_1, f_1) is a 'higher level' covering space than (Ω_2, f_2). Note that Ω_1 is simply connected in this example.

If (Ω_1, f_1), (Ω_2, f_2) and (Ω_3, f_3) are all covering spaces of D, if g is a homomorphism from (Ω_1, f_1) to (Ω_2, f_2) and h is a homomorphism from (Ω_2, f_2) to (Ω_3, f_3) then $h \circ g$ is a homomorphism from (Ω_1, f_1) to (Ω_3, f_3). In fact, on Ω_1, we have

$$f_3 \circ (h \circ g) = (f_3 \circ h) \circ g = f_2 \circ g = f_1.$$

If (Ω_1, f_1) and (Ω_2, f_2) are isomorphic then they are effectively equivalent covering spaces of D—one may pass from one to the other via the homeomorphism g.

Exercise 10.1 Show that isomorphism is an equivalence relation on covering spaces (Ω, f) of D.

Since we are working in the special setting of covering spaces, the homomorphism g in Definition 10.2 automatically has stronger properties over and above being a continuous function. In fact, g is itself a covering map!

Proposition 10.1 *Let g be a homomorphism of the covering space (Ω_1, f_1) of D to the covering space (Ω_2, f_2) of D. Then, (Ω_1, g) is a covering space of Ω_2, and g is uniquely determined by its value at a single point of Ω_1. Moreover, if (Ω_1, f_1) and (Ω_2, f_2) are both analytic covering spaces, then $g: \Omega_1 \to \Omega_2$ is analytic.*

Proof We first show that $g: \Omega_1 \to \Omega_2$ is surjective. Choose a point a_1^* in Ω_1 and set $a_2^* = g(a_1^*)$. Then, $f_1(a_1^*) = f_2(g(a_1^*)) = f_2(a_2^*)$. Thus, both a_1^* in Ω_1 and a_2^* in Ω_2 lie above the same point, a_0 say, in D. Let z_2^* be any point in Ω_2. Let $\gamma_2^*(t)$, $t \in [0, 1]$, be a path in Ω_2 joining a_2^* to z_2^*. Then, $\gamma = f_2(\gamma_2^*)$ is a path in D joining a_0 to $f_2(z_2^*)$, that is γ_2^* is the lift of γ to Ω_2 with initial point a_2^*. Now lift γ to Ω_1 with initial point a_1^*, and call the lifted path γ_1^*. Finally, $g(\gamma_1^*)$ is also the lift of γ to Ω_2 with initial point $g(a_1^*) = a_2^*$, since

$$f_2(g(\gamma_1^*(t))) = f_1(\gamma_1^*(t)) = \gamma(t), \quad t \in [0, 1].$$

By uniqueness of lifts, Lemma 10.3, $g(\gamma_1^*(1)) = \gamma_2^*(1) = z_2^*$, showing that g is onto.

In order to show that $g: \Omega_1 \to \Omega_2$ is a covering map we need to produce, for arbitrary a_2^* in Ω_2, an evenly covered neighbourhood of a_2^*. To this end, we fix a_2^* in Ω_2 and, as above, set $a_0 = f_2(a_2^*)$. Choose an evenly covered neighbourhood Δ of a_0 for the covering map (Ω_2, f_2) of D. Let Δ_2^* be the corresponding component of $f_2^{-1}(\Delta)$ that contains a_2^* so that $f_2: \Delta_2^* \to \Delta$ is a homeomorphism. We claim that Δ_2^* is an evenly covered neighbourhood of a_2^* for the map $g: \Omega_1 \to \Omega_2$.

To establish this claim, let Δ_1^* be any component of $g^{-1}(\Delta_2^*)$. We aim to show that $g: \Delta_1^* \to \Delta_2^*$ is a homeomorphism. Now $f_1(\Delta_1^*) = f_2(g(\Delta_1^*)) \subseteq f_2(\Delta_2^*) = \Delta$. Hence, Δ_1^* is contained in some component, say Δ_1^{**}, of $f_1^{-1}(\Delta)$. Moreover, since f_1 is a covering map, $f_1: \Delta_1^{**} \to \Delta$ is a homeomorphism. Consider $g(\Delta_1^{**}) = \Delta_2^{**}$, which is connected since Δ_1^{**} is connected and g is continuous. Again using $f_2 \circ g = f_1$, we see that $\Delta = f_1(\Delta_1^{**}) = f_2(g(\Delta_1^{**})) = f_2(\Delta_2^{**})$. That is, Δ_2^{**} is connected and its image under f_2 is all of Δ. Since f_2 is a covering map, this forces Δ_2^{**} to be a connected component of $f_2^{-1}(\Delta)$ (otherwise, Δ^{**} would be strictly contained in a connected component of $f_2^{-1}(\Delta)$ and f_2 would then not be injective on this component) in which case $f_2: \Delta_2^{**} \to \Delta$ is a homeomorphism. We claim that this forces $\Delta_2^{**} = \Delta_2^*$. To see this, note that both Δ_2^* and Δ_2^{**} are connected components of $f_2^{-1}(\Delta)$ that are homeomorphic to Δ. Therefore, if they intersect then they must coincide. Otherwise their union is a connected component of $f_2^{-1}(\Delta)$ that is not homeomorphic to Δ. But they do, in fact, intersect since $\Delta_2^{**} = g(\Delta_1^{**}) \supseteq g(\Delta_1^*)$ and $g(\Delta_1^*)$ has non-empty intersection with Δ_2^*.

This, in turn, leads to $\Delta_1^* \subseteq \Delta_1^{**} \subseteq g^{-1}(\Delta_2^*)$. But Δ_1^* is a connected component of $g^{-1}(\Delta_2^*)$, hence Δ_1^{**}, being connected, cannot be larger than Δ_1^*. That is, $\Delta_1^{**} = \Delta_1^*$.

Now, let us write f_2^{-1} for the branch of the inverse function of f_2 that maps Δ homeomorphically onto Δ_2^*. Let z_1^* be an arbitrary point in Δ_1^* and set $z = f_1(z_1^*)$. Then, $f_2^{-1}(z)$ lies in Δ_2^* and over z. On the other hand, $g(z_1^*)$ lies in Δ_2^* and also lies over z as $f_2(g(z_1^*)) = f_1(z_1^*) = z$. Since there is only a single point in Δ_2^* that lies over z, we conclude that $g(z_1^*) = (f_2^{-1} \circ f_1)(z_1^*)$, $z_1^* \in \Delta_1^*$. That is,

$$g = f_2^{-1} \circ f_1 \text{ on } \Delta_1^*. \tag{10.2}$$

Since $f_1: \Delta_1^* \to \Delta$ and $f_2^{-1}: \Delta \to \Delta_2^*$ are both homeomorphisms, we conclude that $g: \Delta_1^* \to \Delta_2^*$ is a homeomorphism as required.

Note also that, if (Ω_1, f_1) and (Ω_2, f_2) are analytic covering maps, then f_1 is analytic on Δ_1^* and f_2^{-1} is analytic on Δ, so that g is analytic on Δ_1^*.

To show that the homomorphism g is uniquely determined by its value at a single point, suppose that h is a second homomorphism of (Ω_1, f_1) to (Ω_2, f_2) and that there is a point a^* in Ω_1 at which $g(a^*) = h(a^*)$. We aim to show that $g = h$. To establish this, set $E = \{z_1^* \in \Omega_1 : g(z_1^*) = h(z_1^*)\}$. Since g and h are both continuous, E is closed. To see that E is open, let a_1^* be in E and set $a_2^* = g(a_1^*) = h(a_1^*)$. By the earlier part of this proof, there is a neighbourhood Δ_2^* of a_2^* in Ω_2 and neighbourhoods $\Delta_{1,g}^*$ and $\Delta_{1,h}^*$ of a_1^* in Ω_1 such that, by (10.2), $g = f_2^{-1} \circ f_1 = h$ on $\Delta_1^* = \Delta_{1,g}^* \cap \Delta_{1,h}^*$. That is, $g = h$ on Δ_1^* and we see that E is open. Since E is non-empty (by assumption), and is both open and closed, $E = \Omega_1$. □

Exercise 10.2 Show that covering space $(\mathbb{C}_0, z^n)$ of $\mathbb{C}_0$ from Example 10.2 is not homomorphic to the covering space $(\mathbb{C}, \exp)$ from Example 10.1.

10.1 Covering Spaces and the Monodromy Theorem

Exercise 10.3 Let $(\mathbb{C}_0, f_n)$ be the covering space of $\mathbb{C}_0$ in Example 10.2 where $f_n(z) = z^n$. When is it the case that $(\mathbb{C}_0, f_n)$ is homomorphic to $(\mathbb{C}_0, f_m)$ where $n \in \mathbb{N}$ and $m \in \mathbb{N}$?

As was said at the beginning of this section, simply connected covering spaces are special. This is made explicit in the following two results. In the first, we show that any continuous map from a simply connected domain to D can be factored through any covering space of D. As a consequence of this, we deduce that if Ω_1 is simply connected and (Ω_1, f_1) and (Ω_2, f_2) both cover D there then exists a homomorphism from (Ω_1, f_1) to (Ω_2, f_2). In the context of the discussion after Definition 10.2, we can interpret this as saying that simply connected covers are the 'highest level' of covering map.

Theorem 10.2 *Suppose that Ω_1 is a simply connected domain and that $f_1 : \Omega_1 \to D$ is a continuous map of Ω_1 into a domain D. Suppose that (Ω_2, f_2) is a covering space of D. Given a_1^* in Ω_1, and given a_2^* in Ω_2 lying over $f_1(a_1^*)$, there exists a unique continuous function $g : \Omega_1 \to \Omega_2$ for which $f_1 = f_2 \circ g$ and $g(a_1^*) = a_2^*$.*

If f_1 is analytic and if (Ω_2, f_2) is an analytic covering space then g is analytic.

Proof There is a certain resemblance between the proof of existence of the map g and the proof of Theorem 6.5. We define $g : \Omega_1 \to \Omega_2$ as follows. Set $a = f_1(a_1^*)$. For z_1^* in Ω_1, let γ_1^* be a curve in Ω_1 joining a_1^* to z_1^*, set $\gamma = f_1(\gamma_1^*)$ so that γ is a curve in D with initial point a and, finally, let γ_2^* be the lift of γ to Ω_2 with initial point a_2^*. Set $g(z_1^*) = z_2^*$ where z_2^* is the final point of γ_2^*. Writing z_0 for the final point of γ, we have $f_1(z_1^*) = z_0 = f_2(z_2^*)$, so that $f_1(z_1^*) = (f_2 \circ g)(z_1^*)$ as shown in Fig. 10.4.

It is crucial that this construction does not depend on the choice of curve γ_1^*. If $\tilde{\gamma}_1^*$ is a second curve in Ω_1 joining a_1^* to z_1^* then, since Ω_1 is simply connected, γ_1^* and $\tilde{\gamma}_1^*$ are FEP-homotopic in Ω_1. Since f_1 is continuous, the image curves $f_1(\gamma_1^*)$ and $f_1(\tilde{\gamma}_1^*)$ are FEP-homotopic in D—if $\Gamma(t, s)$ is a FEP-homotopy of γ_1^* and $\tilde{\gamma}_1^*$ in Ω_1 in the sense of Definition 6.2, then $f_1(\Gamma(t, s))$ is a FEP-homotopy of $f_1(\gamma_1^*)$ and $f_1(\tilde{\gamma}_1^*)$ in D. By the Monodromy Theorem, the lifts γ_2^* and $\tilde{\gamma}_2^*$ of $f_1(\gamma_1^*)$ and $f_1(\tilde{\gamma}_1^*)$ respectively to Ω_2 with initial point a_2^* are also FEP-homotopic and have the same final point. The function g is therefore well-defined.

Next we show that g is continuous at an arbitrary point z_1^* in Ω_1. We set $z_0 = f_1(z_1^*)$ and set $z_2^* = g(z_1^*)$. Let Δ be an evenly covered neighbourhood of z_0, and

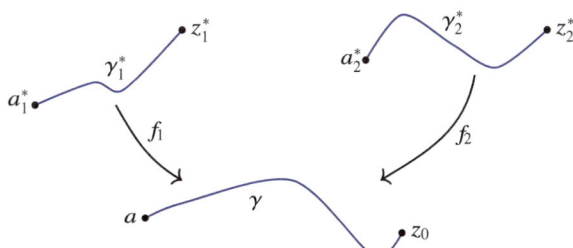

Fig. 10.4 Illustration of the construction of the function g in Theorem 10.2. Here $f_1(z_1^*) = z_0 = f_2(z_2^*)$ and $g(z_1^*) = z_2^*$

let Δ_2^* be a neighbourhood of z_2^* in Ω_2 that is homeomorphic to Δ under f_2. Since f_1 is continuous at z_1^*, there is a neighbourhood Δ_1^* of z_1^* such that $f(\Delta_1^*) \subseteq \Delta$. Now let w_1^* be any point in Δ_1^* and let $\gamma_1^*(t)$, $t \in [0, 1]$, be a curve in Δ_1^* from z_1^* to w_1^*. Once again, let γ be the curve $f_1(\gamma_1^*)$ in Δ so that γ has initial point z_0, and let γ_2^* be the lift of γ to Ω_2 with initial point z_2^*. By uniqueness of lifts (Lemma 10.3), $\gamma_2^* = f_2^{-1}(\gamma)$ where $f_2^{-1} : \Delta \to \Delta_2^*$. Moreover, $g(w_1^*) = w_2^*$ where w_2^* is the final point of the lifted curve γ_2^*. [This is because we obtain a curve joining the base point a_1^* to w_1^* by taking a curve joining a_1^* to z_1^* and following that with the short curve γ_1^* from z_1^* to w_1^*. Following the progress of this curve as it is first projected to D and then lifted back up to Ω_2 with initial point a_2^* we see, by uniqueness of lifts, that the final point of this lifted curve, which is $g(w_1^*)$ by definition, coincides with the final point w_2^* of γ_2^*.] Thus,

$$g(w_1^*) = \gamma_2^*(1) = f_2^{-1}(\gamma(1)) = f_2^{-1}(f_1(\gamma_1^*(1))) = f_2^{-1}(f_1(w_1^*)),$$

so that $g = f_2^{-1} \circ f_1$ on Δ_1^*. Being a composition of continuous functions, g is therefore continuous on Δ_1^*, in particular at z_1^*. Moreover, if f_1 is analytic and f_2 is an analytic covering map, then g is analytic at z_1^*.

To complete the proof of Theorem 10.2, we verify the uniqueness of g. Suppose that $h: \Omega_1 \to \Omega_2$ is continuous, that $f_1 = f_2 \circ h$ on Ω_1 and that $h(a_1^*) = a_2^*$. Set $E = \{z_1^* \in \Omega_1 : g(z_1^*) = h(z_1^*)\}$. Since $a_1^* \in E$, the set E is non-empty. It is immediate that E is closed since both g and h are continuous. In analogy with the argument in the proof of Proposition 10.1, we show that E is open as follows. Suppose that $z_1^* \in E$, set $z_0 = f_1(z_1^*)$ and set $z_2^* = g(z_1^*) = h(z_1^*)$. Then, z_2^* in Ω_2 lies over z_0. Choose an evenly covered neighbourhood Δ of z_0 and a neighbourhood Δ_2^* of z_2^* that is homeomorphic to Δ under f_2. Since g and h are both continuous and both send z_1^* to z_2^*, there are neighbourhoods Δ_1^a and Δ_1^b of z_1^* such that $g(\Delta_1^a) \subseteq \Delta_2^*$ and $h(\Delta_1^b) \subseteq \Delta_2^*$. Set $\Delta_1^* = \Delta_1^a \cap \Delta_1^b$. For w_1^* in Δ_1^*, both $g(w_1^*)$ and $h(w_1^*)$ are points in Δ_2^* that lie over the same point $f_1(w_1^*)$. Since f_2 is one-to-one on Δ_2^*, it must be that $g(w_1^*) = h(w_1^*)$ for w_1^* in Δ_1^*. Thus, E is open and, being non-empty, must be all of Ω_1. $\square$

We now use Theorem 10.2 to show how simply connected covering spaces play a special role in the theory.

Theorem 10.3 *Suppose that Ω_1 is a simply connected domain and that (Ω_1, f_1) and (Ω_2, f_2) are both covering spaces of a domain D. Suppose that a is a point in D, and that a_1^* in Ω_1 and a_2^* in Ω_2 both lie over a. Then there exists a unique homomorphism g from (Ω_1, f_1) to (Ω_2, f_2) for which $g(a_1^*) = a_2^*$.*

In particular, any two simply connected covering spaces of D are isomorphic.

Proof Theorem 10.2 applies since the domain Ω_1 is simply connected and the covering map f_1 is continuous. The resulting map $g: \Omega_1 \to \Omega_2$ from Theorem 10.2 then acts as the homomorphism from (Ω_1, f_1) to (Ω_2, f_2) for which $g(a_1^*) = a_2^*$.

10.1 Covering Spaces and the Monodromy Theorem

If Ω_2 is also a simply connected domain, there is therefore a unique homomorphism h from (Ω_2, f_2) to (Ω_1, f_1) for which $h(a_2^*) = a_1^*$. Then, $h \circ g$ is a homomorphism from (Ω_1, f_1) to itself that fixes the point a_1^*. The identity function on Ω_1 has the same properties and so, by the uniqueness statement in Proposition 10.1, $(h \circ g)(z_1^*) = z_1^*$ for each z_1^* in Ω_1. This implies that g is injective and, being a homomorphism, it is surjective by Proposition 10.1. That is, $g: \Omega_1 \to \Omega_2$ is a bijection. It then follows that h is g^{-1} since $h(g(z_1^*)) = z_1^*$ on Ω_1. Since h is a homomorphism, it is continuous and so g is a homeomorphism as required. □

If (Ω, f) is a covering space of D and Ω is simply connected, then (Ω, f) is called *the universal covering space* of D. The space is *universal* in the sense that there exists a homomorphism from (Ω, f) to any other covering space of D. In fact, by Proposition 10.1, (Ω, f) covers every other covering space of D. It is *the* universal covering space in that any other simply connected covering space of D is isomorphic to it. We need to keep in mind that, for now, the existence or otherwise of the universal covering space is far from being decided.

An *automorphism* of a covering space (Ω, f) of D is a covering space isomorphism from (Ω, f) to itself. The term *deck transformation* is also commonly used. The collection of all such automorphisms is denoted by $\mathrm{Aut}(\Omega, f)$. In part by Exercise 10.1, $\mathrm{Aut}(\Omega, f)$ forms a group under composition. It certainly contains the identity function and, if g and h are both isomorphisms from (Ω, f) to itself, so are $h \circ g$ and g^{-1}. Any such isomorphism g satisfies $f \circ g = f$ on Ω. In fact, $\mathrm{Aut}(\Omega, f)$ consists precisely of those homeomorphisms $g: \Omega \to \Omega$ for which $f \circ g = f$.

From now on, we will focus almost exclusively on analytic covering spaces. In that case, an isomorphism of the covering space is an analytic homeomorphism, that is a conformal map. In particular, by the last statement in Theorem 10.3, *any two simply connected analytic covering spaces of a domain D are conformally equivalent*. If (Ω, f) is an analytic covering space then any covering automorphism of (Ω, f) will therefore be a conformal self-map of Ω, that is an automorphism of Ω in the sense of Definition 4.2. In the case of an analytic covering space (Ω, f), therefore,

$$\mathrm{Aut}(\Omega, f) = \{g \in \mathrm{Aut}(\Omega) : f \circ g = f\}. \tag{10.3}$$

Example 10.8 (Universal Analytic Covering Space of the Punctured Plane) As we saw in Example 10.1, $(\mathbb{C}, \exp)$ is an analytic covering space of the punctured plane $\mathbb{C}_0$. Since $\mathbb{C}$ is simply connected, $(\mathbb{C}, \exp)$ is the universal covering space of $\mathbb{C}_0$.

Suppose that (Ω, f) is an isomorphic analytic covering space of $\mathbb{C}_0$, with isomorphism $g: \mathbb{C} \to \Omega$. Then, g is an entire univalent function so that, by Proposition 6.2, $g(z) = az + b$ for some non-zero a and some b. As a consequence, $\Omega = \mathbb{C}$ and $e^z = f(g(z))$, $z \in \mathbb{C}$. Applying the last relation to $g^{-1}(z) = (z-b)/a$, we see that $f(z) = e^{cz+d}$ where $c \neq 0$.

Thus, all universal analytic covering spaces of $\mathbb{C}_0$ are of the form $(\mathbb{C}, e^{cz+d})$ where $c \neq 0$. Imposing the additional condition that $e^{az+b} = e^{g(z)} = e^z, z \in \mathbb{C}$, we see that

$$\text{Aut}(\mathbb{C}, \exp) = \{g : \mathbb{C} \to \mathbb{C} : g(z) = z + 2n\pi i, \ n \in \mathbb{Z}\}.$$

Example 10.9 Let us compute $\text{Aut}(\mathbb{S}_R, \exp)$ from Example 10.3, which is the universal covering space of the annulus $A(1, R) = \{z : 1 < |z| < R\}$. Let $g \in \text{Aut}(\mathbb{S}_R)$ be a conformal mapping of $\mathbb{S}_R$ onto itself and suppose that $\exp(g(z)) = \exp z, z \in \mathbb{S}_R$ (this is the covering isomorphism condition, as in (10.3)). It follows, first individually for each z and then from continuity of g, that there is an integer n such that $g(z) = z + 2n\pi i$. Thus,

$$\text{Aut}(\mathbb{S}_R, \exp) = \{g : \mathbb{S}_R \to \mathbb{S}_R : g(z) = z + 2n\pi i, \ n \in \mathbb{Z}\}.$$

Referring back to Example 10.4, conformal mappings are trivially covering maps. We complete this section with some results on coverings of a simply connected domain. We first show that any analytic covering of a simply connected domain is a conformal mapping. This is a consequence of the Monodromy Theorem and of Theorem 10.3.

Proposition 10.2 *Suppose that Ω is connected and that (Ω, f) is a covering space of a simply connected domain D. Then, Ω is also simply connected and f is a homeomorphism.*

Proof Let γ_1^* and γ_2^* be two curves in Ω that share the same initial point a^* and the same final point b^*. Let γ_1 and γ_2 be the curves $\gamma_1 = f(\gamma_1^*)$ and $\gamma_2 = f(\gamma_2^*)$ in D, which have the same endpoints $a = f(a^*)$ and $b = f(b^*)$. Since D is simply connected, γ_1 and γ_2 are FEP-homotopic by Lemma 6.1. By the Monodromy Theorem, their lifts γ_1^* and γ_2^* are also FEP-homotopic in Ω. Again, by Lemma 6.1, Ω is simply connected.

To show that $f : \Omega \to D$ is a homeomorphism, we apply Theorem 10.3 with $(\Omega_1, f_1) = (\Omega, f)$ and $(\Omega_2, f_2) = (D, \text{Id})$ where $\text{Id} : D \to D$ is the identity function. By Theorem 10.3, there is a homeomorphism $g : \Omega \to D$ such that $f = \text{Id} \circ g$, so that $f = g$ is a homeomorphism. $\square$

Using the machinery in this section, we are able to show that simply connected neighbourhoods are always evenly covered.

Theorem 10.4 *Suppose that (Ω, f) is a covering space of a domain D and that U is a simply connected domain contained in D. Then, U is evenly covered by f.*

Proof Let U^* in Ω be a connected component of $f^{-1}(U)$. By Example 10.7, $(U^*, f|_{U^*})$ is a covering map of U. Since U is simply connected, it follows from Proposition 10.2 that U^* is simply connected and that $f|_{U^*}$ is a homeomorphism. That is, $f : U^* \to U$ is a homeomorphism. Since this is the case for each component of $f^{-1}(U)$, we see that U is evenly covered by f. $\square$

10.2 The Modular Function

By Example 10.8, no proper simply connected domain can be an analytic covering space of the punctured plane $\mathbb{C}_0$. Alternatively, recall that $(\mathbb{C}, \exp)$ is an analytic covering space of $\mathbb{C}_0$; this is Example 10.1. If (Ω, f) was an analytic covering space of $\mathbb{C}_0$ with Ω a proper simply connected domain, it would follow from Theorem 10.3 that Ω and $\mathbb{C}$ would be conformally equivalent, which they aren't. If they were, since Ω is conformally equivalent to the unit disk $\mathbb{D}$ by the Riemann Mapping Theorem, the complex plane $\mathbb{C}$ would then be conformally equivalent to $\mathbb{D}$, which in turn would violate Liouville's Theorem.

We will prove in Sect. 10.4 that the punctured plane is the only exceptional case, in that any other domain in the plane, that is any domain whose complement contains at least two points, has the unit disk as its universal analytic covering space. A major step towards obtaining this fundamental result is to show that the unit disk is an analytic covering space of the plane punctured in two points. The covering map is called the/a *modular function*. Since all proper simply connected domains are equivalent from this point of view (see Example 10.4), it is convenient (and conventional) to choose the upper half-plane $\mathbb{H}$ as the covering space. Let us write $\mathbb{C}_{a,b}$ for the plane punctured at the distinct points a and b, so that $\mathbb{C}_{a,b} = \mathbb{C} \setminus \{a, b\}$. The Möbius map $M(z) = a + (b - a)z$ is a conformal map of $\mathbb{C}_{0,1}$ onto $\mathbb{C}_{a,b}$. By Example 10.4, therefore, once we have a covering space $(\mathbb{H}, \lambda)$ of $\mathbb{C}_{0,1}$ then $(\mathbb{H}, M \circ \lambda)$ is a covering space of $\mathbb{C}_{a,b}$. In short, we may assume that $a = 0$ and that $b = 1$.

Theorem 10.5 (The Modular Function) *There is an analytic function $\lambda \colon \mathbb{H} \to \mathbb{C}_{0,1}$ for which $(\mathbb{H}, \lambda)$ is an analytic covering space of the twice punctured plane $\mathbb{C}_{0,1}$.*

In keeping with the tone of this book, we adopt a direct, computational approach. We break the construction of the Modular Function into several steps. The main ingredient of this construction is a series of reflections together with the Schwarz Reflection Principle for circles.

The region we start with is the domain T_0 in the upper half-plane $\mathbb{H}$ bounded by the lines $\Re z = 0$, $\Re z = 1$ and by the upper semicircle of the circle $C(1/2, 1/2)$. Since all three sides are geodesics in the hyperbolic metric of the half-plane $\mathbb{H}$ (see Sect. 4.7), T_0 is a *hyperbolic triangle* with vertices 0, 1 and ∞ in the hyperbolic geometry of the half-plane $\mathbb{H}$. The domain T_0 is simply connected and so, by the Riemann Mapping Theorem, there is a conformal map $\widetilde{f}$ from T_0 onto $\mathbb{H}$. By Theorem 6.12, $\widetilde{f}$ extends to a homeomorphism of the closure of T_0 to the closure of the half-plane $\mathbb{H}$. With $\widetilde{f}(0) = x$, $\widetilde{f}(1) = y$, $x < y$, and $M(z) = (z - x)/(y - x)$ (the latter being an automorphism of the half-plane $\mathbb{H}$), the map $f_0 = M \circ \widetilde{f}$ maps T_0 conformally onto $\mathbb{H}$ and extends to a homeomorphism of the closure of T_0 to the closure of $\mathbb{H}$ for which $f_0(0) = 0$ and $f_0(1) = 1$. This mapping is shown in Fig. 10.5 where the three sides of T_0 are labelled a, b and c. Under f_0, the open 'a' side of T_0 maps to the interval $I_1 = (-\infty, 0)$ on the real line, the

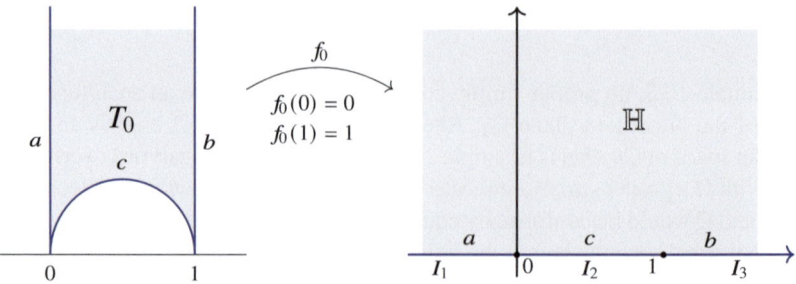

Fig. 10.5 The hyperbolic triangle T_0 and the mapping f from T_0 to the half-plane $\mathbb{H}$

open semicircular 'c' side maps to the interval $I_2 = (0, 1)$, and the open 'b' side maps to the interval $I_3 = (1, \infty)$. The covering map λ is produced by reflecting the function f_0 in each of the sides of T_0, and then repeating this reflection over and over again in the sides of the subsequent hyperbolic triangles. Our next task, therefore, will be to study these reflections; we postpone the study of the reflected functions for the moment. Reflection in the vertical sides is straightforward—we obtain a succession of horizontal translates of the half-strip—so we focus first on reflection in the semicircular 'c' side of T_0.

10.2.1 A Series of Reflections

For a circle $C(a, r)$ with a real, we write $C^+(a, r)$ for the semicircular arc of $C(a, r)$ lying in the upper half-plane. We first reflect T_0 in the semicircle $C^+(1/2, 1/2)$. Recall that, by Exercise 2.11, the reflection of a generalised circle in a generalised circle is again a generalised circle. The reflection of the line $\Re z = 0$ in the circle $C(1/2, 1/2)$ is therefore again a circle. This circle is symmetric in the real axis and passes through 0, which is fixed under this reflection, and through $1/2$, which is the reflection of the point at infinity. See Fig. 10.6. The reflection of the 'a' side of T_0 as shown in this figure is therefore the semicircle $C^+(1/4, 1/4)$. Similarly, the reflection of the 'b' side of T_0 is the semicircle $C^+(3/4, 1/4)$. The semicircular 'c' side of T_0 stays fixed. The reflection of T_0 in its 'c' side is therefore the hyperbolic triangle T_1 shown in Fig. 10.6 with vertices 0, $1/2$ and 1.

The next step is to reflect T_1 in each of its three sides. From now on, we denote reflection in an 'a', 'b' or 'c' side of a hyperbolic triangle by R_a, R_b and R_c, respectively. Reflection in the 'c' side simply brings us back to T_0. Consider reflection in the 'a' side (as shown in Fig. 10.7). The reflection of the 'c' side of T_1 in the 'a' side of T_1 is a semicircle with endpoints 0, which is fixed under

10.2 The Modular Function

Fig. 10.6 The hyperbolic triangle T_1 is the reflection of T_0 in the semicircle $C^+(\frac{1}{2}, \frac{1}{2})$. The 'a', 'b' and 'c' sides correspond, respectively

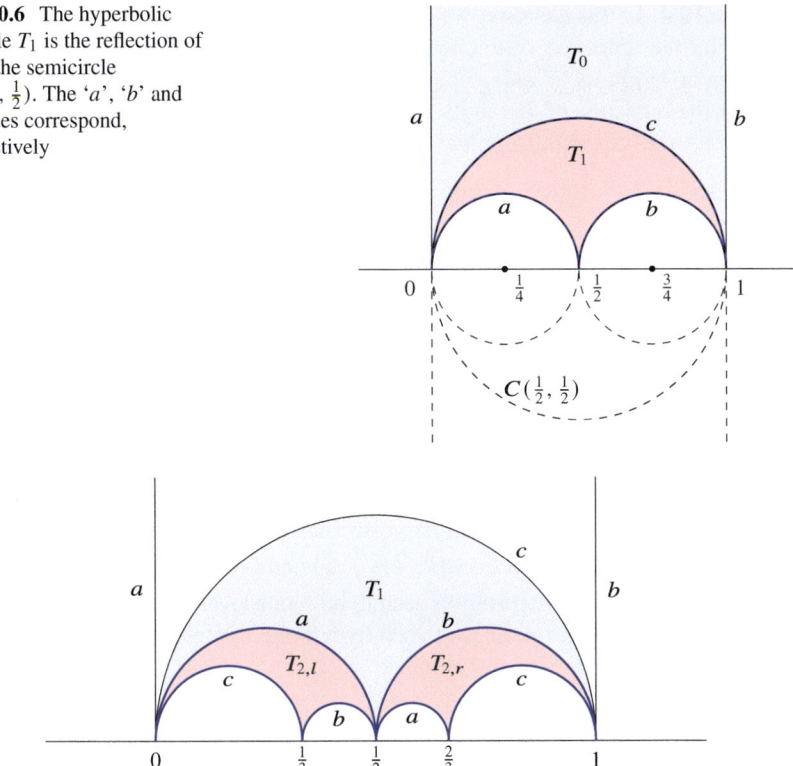

Fig. 10.7 The hyperbolic triangle $T_{2,l}$ is the hyperbolic triangle T_1 reflected in its 'a' side. The hyperbolic triangle $T_{2,r}$ is the reflection of T_1 in its 'b' side. The 'a', 'b' and 'c' sides correspond, respectively

this reflection, and the reflection of the point 1. The 'a' side is the semicircle $C^+(1/4, 1/4)$. Using the formula (2.15) for reflection,

$$R_a(1) = \frac{1}{4} + \frac{(1/4)^2}{1 - 1/4} = \frac{1}{3}.$$

The reflection of the 'b' side of T_1 in the 'a' side of T_1 is the semicircle with endpoints 1/2, which is fixed under this reflection, and 1/3 which, as we have just seen, is the reflection of the point 1. The reflection of the hyperbolic triangle T_1 in its 'a' side is therefore the hyperbolic triangle $T_{2,l}$ with vertices 0, 1/3 and 1/2 shown in Fig. 10.7. Exactly the same considerations hold for the reflection of T_1 in its 'b' side. Since $R_b(0)$ is seen to be 2/3, the reflection of T_1 in its 'b' side is the hyperbolic triangle $T_{2,r}$ with vertices 1/2, 2/3 and 1 shown in Fig. 10.7.

Exercise 10.4 In this exercise, we compute the reflection of $T_{2,l}$ in its 'c' and 'b' sides. Find the reflection of the point $1/2$ in the 'c' side of $T_{2,l}$. Hence, draw the reflection of $T_{2,l}$ in its 'c' side.

Draw the reflection of $T_{2,l}$ in its 'b' side.

Do the same for the reflection of $T_{2,r}$ in its 'a' and 'c' sides.

It is possible to notice, at this stage, a curious pattern in the vertices of the successive hyperbolic triangles. Let's write the two vertices of T_0 on the real axis, namely 0 and 1, as $\frac{0}{1}$ and as $\frac{1}{1}$ respectively. In Fig. 10.6, the new vertex of T_1 is $\frac{0}{1} \oplus \frac{1}{1} = \frac{1}{2}$ where $\oplus$ is the Farey sum of the (simple) fractions $\frac{p}{q}$ and $\frac{r}{s}$ defined by[1]

$$\frac{p}{q} \oplus \frac{r}{s} = \frac{p+r}{q+s}.$$

The vertices of T_1 in Fig. 10.6 are $\{\frac{0}{1}, \frac{1}{2}, \frac{1}{1}\}$. Moving on to Fig. 10.7, we see that the vertices of $T_{2,l}$ are $\{\frac{0}{1}, \frac{1}{3}, \frac{1}{2}\}$ and that the vertices of $T_{2,r}$ are $\{\frac{1}{2}, \frac{2}{3}, \frac{1}{1}\}$: notice that $\frac{1}{3} = \frac{0}{1} \oplus \frac{1}{2}$ and $\frac{2}{3} = \frac{1}{2} \oplus \frac{1}{1}$. From Exercise 10.4, we see that the two triangles one level down from $T_{2,l}$ have vertices $\{\frac{0}{1}, \frac{0}{1} \oplus \frac{1}{3}, \frac{1}{3}\}$ and $\{\frac{1}{3}, \frac{1}{3} \oplus \frac{1}{2}, \frac{1}{2}\}$, since $\frac{0}{1} \oplus \frac{1}{3} = \frac{1}{4}$ and $\frac{1}{3} \oplus \frac{1}{2} = \frac{2}{5}$. The same pattern is seen to hold one level down from $T_{2,r}$. We can also notice that adjacent vertices of each hyperbolic triangle are *Farey neighbours* in the sense of the following definition.

Definition 10.3 Two simple fractions $\frac{p}{q}$ and $\frac{r}{s}$, with $\frac{0}{1} \leq \frac{p}{q} < \frac{r}{s}$, are said to be *Farey neighbours* if they satisfy $qr - ps = 1$.

For example, $\frac{1}{3}$ and $\frac{1}{2}$ are Farey neighbours, as are $\frac{1}{2}$ and $\frac{2}{3}$, as well as $\frac{2}{5}$ and $\frac{1}{2}$. Another example is $\frac{0}{1}$ and $\frac{1}{n}$ for any natural number n. Note that the Farey neighbours $\frac{p}{q}$ and $\frac{r}{s}$ satisfy $\frac{r}{s} - \frac{p}{q} = \frac{1}{sq}$.

Exercise 10.5 Suppose that $\frac{p}{q}$ and $\frac{r}{s}$ are Farey neighbours as in the above definition.

▷ Show that $\frac{p}{q} \oplus \frac{r}{s}$ is a simple fraction.
▷ Show that $\frac{p}{q} \oplus \frac{r}{s}$ lies between $\frac{p}{q}$ and $\frac{r}{s}$. Show that $\frac{p}{q}$ and $\frac{p}{q} \oplus \frac{r}{s}$ are Farey neighbours, as are $\frac{p}{q} \oplus \frac{r}{s}$ and $\frac{r}{s}$.
▷ Show that any fraction that lies between $\frac{p}{q}$ and $\frac{r}{s}$ has denominator at least $q + s$.

Now we show that this pattern in the vertices of successive hyperbolic triangles holds in general.

Lemma 10.4 *Consider a hyperbolic triangle T in the upper half-plane $\mathbb{H}$ with vertices $\frac{p}{q}, \frac{p}{q} \oplus \frac{r}{s}$ and $\frac{r}{s}$, where $\frac{p}{q}$ and $\frac{r}{s}$ are Farey neighbours. Label the sides of*

---

[1] Farey addition is the completely wrong way of adding fractions, although favoured by some students! A fraction $\frac{m}{n}$ is *simple* if m and n have no proper common factor.

10.2 The Modular Function

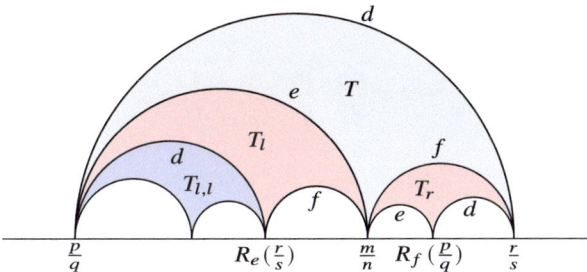

Fig. 10.8 The hyperbolic triangle T_l is the reflection of the hyperbolic triangle T in its 'e' side. The hyperbolic triangle T_r is the reflection of the hyperbolic triangle T in its 'f' side. The 'd', 'e' and 'f' sides correspond, respectively

T as 'd', 'e' and 'f' as shown in Fig. 10.8 and set $\frac{m}{n} = \frac{p}{q} \oplus \frac{r}{s}$. Then, the reflection of T in its 'e' side is the hyperbolic triangle T_l with vertices $\frac{p}{q}$, $\frac{p}{q} \oplus \frac{m}{n}$ and $\frac{m}{n}$. The reflection of T in its 'f' side is the hyperbolic triangle T_r with vertices $\frac{m}{n}$, $\frac{m}{n} \oplus \frac{r}{s}$ and $\frac{r}{s}$.

Proof The 'e' side of T is the semicircle $C^+(x, \rho)$ with centre $x = (\frac{m}{n} + \frac{p}{q})/2$ and radius $\rho = (\frac{m}{n} - \frac{p}{q})/2$. We need to find the reflection $R_e(\frac{r}{s})$. The reflection of T in its 'e' side will then be the hyperbolic triangle with vertices $\frac{p}{q}$, $R_e(\frac{r}{s})$ and $\frac{m}{n}$. We will be done if we can show that

$$R_e\left(\frac{r}{s}\right) = \frac{p}{q} \oplus \frac{m}{n} = \frac{p}{q} \oplus \left(\frac{p}{q} \oplus \frac{r}{s}\right) = \frac{p}{q} \oplus \frac{p+r}{q+s} = \frac{2p+r}{2q+s}. \tag{10.4}$$

This is simply a computation. The centre x of the 'e' semicircle is given by

$$2x = \frac{m}{n} + \frac{p}{q} = \frac{p+r}{q+s} + \frac{p}{q} = \frac{ps + rq + 2pq}{q(q+s)}.$$

Using $qr = 1 + ps$, we find that

$$x = \frac{p}{q} + \frac{1}{2q(q+s)}. \tag{10.5}$$

The radius ρ of the 'e' semicircle is given by

$$2\rho = \frac{m}{n} - \frac{p}{q} = \frac{p+r}{q+s} - \frac{p}{q} = \frac{qr - ps}{q(q+s)} = \frac{1}{q(q+s)}, \tag{10.6}$$

(or this is clear from (10.5)). Next,

$$\frac{r}{s} - x = \frac{r}{s} - \frac{p}{q} - \rho = \frac{1}{qs} - \rho,$$

so that

$$\frac{\rho^2}{\frac{r}{s} - x} = \frac{\frac{1}{(2q(q+s))^2}}{\frac{1}{qs} - \frac{1}{2q(q+s)}} = \frac{1}{2q(q+s)} \frac{s}{2q+s}.$$

Putting all this together, we see by (2.15) that

$$R_e\left(\frac{r}{s}\right) = x + \frac{\rho^2}{\frac{r}{s} - x} = \frac{p}{q} + \frac{1}{2q(q+s)}\left(1 + \frac{s}{2q+s}\right)$$

$$= \frac{p}{q} + \frac{1}{q(2q+s)}$$

$$= \frac{2pq + (ps+1)}{q(2q+s)} = \frac{2pq + rq}{q(2q+s)} = \frac{2p+r}{2q+s}.$$

This is (10.4). By a similar computation, one may show that

$$R_f\left(\frac{p}{q}\right) = \frac{p + 2r}{q + 2s}. \tag{10.7}$$

Alternatively, one may reflect in the line $\Re z = \frac{m}{n}$, use (10.4), and then reflect back across this line again. □

Exercise 10.6 Now that we see that the direct calculation proof of Lemma 10.4 leads to an interesting conclusion, we can attempt to find a more systematic approach to the same result. It is natural to normalise the calculation, for example so that the outer vertices are 0 and 1.

▷ With p/q and r/s as in the statement of Lemma 10.4, find an automorphism M of the upper half-plane $\mathbb{H}$ that sends $\frac{p}{q}$ to 0 and sends $\frac{r}{s}$ to 1 (and fixes the point at infinity).
▷ Find the image, call it x, of $\frac{p}{q} \oplus \frac{r}{s}$ under M.
▷ Find the image of 1 under reflection in the circle with diameter $[0, x]$. Find the image of 0 under reflection in the circle with diameter $[x, 1]$.
▷ By applying the inverse of the Möbius transformation M, verify (10.4) and (10.7).

Let us refer to the hyperbolic triangles T_l and T_r in Fig. 10.8 as the *children* of the hyperbolic triangle T. Starting with T_0, we reflect in its 'c' side, as in Fig. 10.6, to produce its child T_1. Reflecting T_1 in its 'a' and 'b' sides, we produce its two children $T_{2,l}$ and $T_{2,r}$ as shown in Fig. 10.7. These hyperbolic triangles are the

10.2 The Modular Function

two *grandchildren* of T_0. Each of $T_{2,l}$ and $T_{2,r}$, in turn, has two more children in accordance with Lemma 10.4. These four hyperbolic triangles are the grandchildren of T_1. As we continue to reflect, we produce successive generations of hyperbolic triangles. With T_0 being the zero-th generation and T_1 being the first generation, the nth-generation, $n \geq 1$, consists of 2^{n-1} hyperbolic triangles and, thanks to Lemma 10.4, their vertices can be computed explicitly as Farey fractions. Each hyperbolic triangle in the nth-generation has four grandchildren in the $(n+2)$nd-generation (the zero-th generation T_0 being the only exception).

Taken together, the union of the children from all generations fills the half-strip $S_0 = \{z : 0 < \Re z < 1, \Im z > 0\}$. In fact, the heights of the triangles tend to 0 as we progress through the generations; this is clear from the formula (10.6) for the radius of the bounding semicircles. We have effectively constructed a triangulation of S_0, a decomposition of S_0 into disjoint hyperbolic triangles. By reflecting S_0 in its two vertical sides and repeatedly reflecting in the sides of the resulting half-strips, we have a series of reflections of T_0 that triangulates the upper half-plane $\mathbb{H}$.

10.2.2 Proof of Theorem 10.5: The Reflected Mappings

Step 1: Construction of the Function λ Recall that f_0 is a conformal map of T_0 onto the upper half-plane $\mathbb{H}$, that f_0 extends as a homeomorphism to the closure of T_0, sending the boundary of T_0 onto the real axis with $f_0(0) = 0$ and $f_0(1) = 1$ as shown in Fig. 10.5. When T_0 is reflected in its 'c' side, the mapping f_0 is correspondingly reflected to a mapping λ of T_1 to the lower half-plane $\overline{\mathbb{H}}$. To see this, recall Exercise 2.13 and the formula (2.17) for the reflected function, namely $f(z) = R_{C_2}(f(R_{C_1}(z)))$. In the present case, C_1 is the semicircular side 'c' of T_0 (see Fig. 10.6) and $C_2 = \mathbb{R}_\infty$. Since reflection in the real axis is simply the complex conjugate, the reflected function λ, which is defined on $T_0 \cup c \cup T_1$, satisfies

$$\lambda(z) = \overline{\lambda(R_c(z))}, \quad z \in T_0 \cup c \cup T_1,$$

where $\lambda|_{T_0} = f_0$. In fact, λ is univalent in $T_0 \cup c \cup T_1$ by Theorem 2.2. Taking note of how f_0 maps the boundary of T_0 to the real axis (again, see Fig. 10.5) and that, in particular, f_0 maps the open semicircle c to the interval $I_2 = (0, 1)$, we find that λ maps $T_0 \cup c \cup T_1$ conformally onto the complex plane minus the slits $(-\infty, 0]$ and $[1, \infty)$ on the real axis. Since f_0 is a homeomorphism on the closure of T_0, the reflected function λ maps the open semicircular 'a' and 'b' sides of T_1 homeomorphically to the intervals $I_1 = (-\infty, 0)$ and $I_3 = (1, \infty)$ respectively. For example, $\lambda(1/2)$ is the point at infinity.

Now we move to Fig. 10.7 and reflect the function λ defined on the closure of T_1 in the 'a' side of T_1. Since the 'a' side of T_1 is sent, by λ, to the interval $(-\infty, 0)$ on the real axis, another application of Theorem 2.2 shows that λ extends to a conformal mapping of $T_1 \cup a \cup T_{2,l}$ onto the complex plane minus the slit $[0, \infty)$ on the real axis. The reflected function maps $T_{2,l}$ onto the upper half-plane, since $T_{2,l}$ is mapped to the complex conjugate of the image of T_1 which itself is the lower half-plane. Since λ is a homeomorphism on the closure of T_1, the reflected function λ maps the open semicircular 'c' and 'b' sides of $T_{2,l}$ homeomorphically to the intervals $I_2 = (0, 1)$ and $I_3 = (1, \infty)$ respectively. Similarly, reflecting λ in the 'b' side of T_1, we obtain a conformal map of $T_1 \cup b \cup T_{2,r}$ onto the complex plane minus the slit $(-\infty, 1]$ on the real axis, and under which $T_{2,r}$ is mapped to the upper half-plane and the 'a' and 'c' sides of $T_{2,r}$ are sent to $(-\infty, 0)$ and $(0, 1)$ respectively.

It is clear that we can continue to reflect in this way through the successive generations of hyperbolic triangles from Sect. 10.2.1. Having reflected the function λ multiple times, we are in the setting of Fig. 10.8 and Lemma 10.4. With the notation of that lemma, λ will be a conformal mapping of the hyperbolic triangle T onto either the upper half-plane $\mathbb{H}$ or the lower half-plane $\overline{\mathbb{H}}$. Moreover, λ will extend to the closure of T as a homeomorphism mapping the open semicircular sides 'd', 'e' and 'f' of T to the intervals I_1, I_2 and I_3 on the real axis in some order. Reflecting λ in the 'e' side of T extends λ to a conformal mapping of $T \cup e \cup T_l$ onto the complex plane less the closure of the image under λ of the 'd' and the 'f' sides. Similarly, reflecting λ in the 'f' side of T extends λ to a conformal mapping of $T \cup f \cup T_r$ onto the complex plane less the closure of the image under λ of the 'e' and the 'd' sides. In the case of T_l (and, similarly, of T_r), the mapping λ extends to a homeomorphism of the closure of T_l sending the sides of T_l to I_1, I_2 and I_3 in the relevant order. A key point is that λ always takes real values on the three sides of each hyperbolic triangle.

Proceeding by successive reflection, we extend λ to a function analytic in the half-strip S_0. At no point does λ take the value 0 or 1. With λ defined in the half-strip S_0, it can be reflected in the lines $\Re z = 0$ and $\Re z = 1$, then in the lines $\Re z = -1$ and $\Re z = 2$. Continuing to reflect in this manner, we obtain a function λ analytic on the upper half-plane $\mathbb{H}$ that never takes the values 0 or 1. This concludes the construction of λ by repeated reflection.

Step 2: Mapping Properties of λ Starting with a hyperbolic triangle T, let us call the collection of hyperbolic triangles including T, the grandchildren of T, then their grandchildren, then these grandchildren's grandchildren, and so forth, the *grand-descendants* of T. Let us use the terminology *double reflection* for the combined action of the two successive reflections in semicircles that sends a hyperbolic triangle T to one of its grandchildren. With this notation, we can give concise descriptions of $\lambda^{-1}(\mathbb{H})$ and $\lambda^{-1}(\overline{\mathbb{H}})$.

Lemma 10.5 *The components of the preimage under λ of the upper half-plane $\mathbb{H}$ comprise (a) T_0 and its grand-descendants together with (b) the reflection in the line $\Re z = 1$ of T_1 and of each of its grand-descendants, together with (c) horizontal translations of any hyperbolic triangle from (a) or (b) by any multiple of 2. Each of these hyperbolic triangles is mapped by λ conformally onto $\mathbb{H}$.*

The components of the preimage under λ of the lower half-plane $\overline{\mathbb{H}}$ comprise (d) T_1 and its grand-descendants together with (e) the reflection in the line $\Re z = 1$ of T_0 and of each of its grand-descendants, together with (f) horizontal translations of

10.2 The Modular Function

any hyperbolic triangle from (d) or (e) by any multiple of 2. Each of these hyperbolic triangles is mapped by λ conformally onto $\overline{\mathbb{H}}$.

Proof The hyperbolic triangle T_0 is mapped by λ (which agrees with f_0) conformally onto the upper half-plane $\mathbb{H}$. Reflecting in the 'c' side of T_0, we find that λ maps T_1 conformally onto the lower half-plane $\overline{\mathbb{H}}$. Reflecting again in the 'a' side of T_1, see Fig. 10.7, the hyperbolic triangle $T_{2,l}$ is mapped conformally by λ onto the upper half-plane $\mathbb{H}$ again. The same applies if we perform a double reflection from T_0 to $T_{2,r}$ in that λ is a conformal mapping of $T_{2,r}$ onto $\mathbb{H}$.

This is the case in general. If a hyperbolic triangle T is mapped conformally by λ onto the upper half-plane $\mathbb{H}$ then any grandchild of T is also mapped conformally onto $\mathbb{H}$. That is, T_0 and each of its grand-descendants are mapped conformally onto $\mathbb{H}$. Similarly, since T_1 is mapped conformally to the lower half-plane $\overline{\mathbb{H}}$, it and each of its grand-descendants are mapped conformally onto $\overline{\mathbb{H}}$, hence (a) and (d).

If T_1 or any of its grand-descendants is reflected in the line $\Re z = 1$, which is the 'b' side of T_0, we arrive at a hyperbolic triangle that is mapped conformally by λ onto $\mathbb{H}$. This is because T_1 and each of its grand-descendants are mapped conformally by λ onto the lower half-plane. The reflection of T_0, or of any of its grand-descendants, in the line $\Re z = 1$ is mapped by λ to the lower half-plane. Both (b) and (e) follow.

That portion of the preimage of $\mathbb{H}$ under λ that lies in the strip $S = \{z \colon 0 < \Re z < 2\}$ consists of the hyperbolic triangles described by (a) and (b), and the portion of the preimage of $\overline{\mathbb{H}}$ consists of the hyperbolic triangles described by (d) and (e). In general, the reflection of the point $z = x + iy$ in the line $\Re z = a$ is the point $2a - x + iy$ or $2a - \overline{z}$. The reflection in the line $\Re z = 2$ of a point z in the strip S is the point $4 - \overline{z}$. Reflecting again in the line $\Re z = 3$ results in the point $6 - \overline{(4 - \overline{z})}$ or $z + 2$. The value of λ is unchanged under this pair of reflections, that is under translation by 2, hence under translation by any multiple of 2, which leads to the hyperbolic triangles in (c) and (f). □

Step 3: λ Is a Covering Map of $\mathbb{C}_{0,1}$ To see that $(\mathbb{H}, \lambda)$ is a covering space of the twice punctured plane $\mathbb{C}_{0,1}$, we consider three cases: w is a point in $\mathbb{H}$; w is a point in $\overline{\mathbb{H}}$; $w \in \mathbb{R} \setminus \{0, 1\}$; and aim to show the existence of an evenly covered neighbourhood in each case.

In the first case, with $w \in \mathbb{H}$, we can choose the half-plane $\mathbb{H}$ itself as an evenly covered neighbourhood of w since the components of $\lambda^{-1}(\mathbb{H})$ are precisely the hyperbolic triangles listed under (a), (b) and (c) in Lemma 10.5 and each is mapped by λ conformally onto $\mathbb{H}$. Similarly, if w is a point in the lower half-plane $\overline{\mathbb{H}}$, we can take $\overline{\mathbb{H}}$ itself as an evenly covered neighbourhood of w. From Lemma 10.5, the components of $\lambda^{-1}(\overline{\mathbb{H}})$ are precisely the hyperbolic triangles listed under (d), (e) and (f) and each is mapped by λ conformally onto $\overline{\mathbb{H}}$.

The final case is that of a point x on the real line other than 0 and 1. If $x \in (0, 1)$ then, again referring to Fig. 10.5, there is a point x^* on the semicircle c such that $f_0(x^*) = x$. Choose a small disk Δ about x. Let Δ^* be the component of $\lambda^{-1}(\Delta)$ containing x^*. The components of the preimage of Δ under λ are all reflections of Δ^* lying in the half-strip S, including reflection in the line $\Re z = 1$, and all translates of these by a multiple of 2. Each meets a 'c' side of a hyperbolic triangle and is mapped conformally onto Δ. Similar considerations apply to points x in the intervals $(-\infty, 0)$ and $(1, \infty)$. This completes the proof of Theorem 10.5.

10.2.3 The Automorphism Group of the Modular Function

We now identify the automorphism group of the covering space $(\mathbb{H}, \lambda)$ of the twice punctured plane $\mathbb{C}_{0,1}$, which was constructed in the previous section. Each such automorphism is, in particular, an automorphism of the upper half-plane and so, by Theorem 4.3, each is a Möbius transformation of the form $\tau(z) = (az+b)/(cz+d)$ where a, b, c and d are real, and $ad - bc > 0$.

We again repeatedly use (2.15) in the case where the centre a of the circle lies on the real axis. The reflection $R(z)$ of z in the circle $C(a, r)$, with a real, is

$$R(z) = a + \frac{r^2}{\bar{z} - a}. \tag{10.8}$$

If $r = a$, the formula (10.8) for the reflected point simplifies to $R(z) = a\bar{z}/(\bar{z} - a)$. We begin by identifying two key automorphisms of the covering map.

Lemma 10.6 *Both*

$$\tau_1(z) = \frac{z}{2z+1} \quad \text{and} \quad \tau_2(z) = z + 2, \quad z \in \mathbb{H},$$

are automorphisms of the covering space $(\mathbb{H}, \lambda)$.

Proof Consider first a point z in T_0 (see Fig. 10.6) and its double reflection $\tau_1(z)$ first in the 'c' side of T_0, then in the 'a' side of T_1. The 'c' side of T_0 is the semicircle $C^+(\frac{1}{2}, \frac{1}{2})$. Reflection in the circle $C(\frac{1}{2}, \frac{1}{2})$ is given by

$$R(z) = \frac{\bar{z}}{2\bar{z} - 1}.$$

The reflection of this point $R(z)$ in the 'a' side of T_1, which is the semicircle $C^+(\frac{1}{4}, \frac{1}{4})$, is then

$$\tau_1(z) = \frac{\overline{R(z)}}{4\overline{R(z)} - 1} = \frac{z}{2z+1}.$$

By Theorem 4.3, this τ_1 is an automorphism of $\mathbb{H}$. Moreover, as we saw in the previous section, $\lambda(\tau_1(z)) = \lambda(z)$ for $z \in T_0$ (on the image side, $\lambda(\tau_1(z))$ is $\lambda(z)$ reflected twice in the real axis). The function $\lambda(\tau_1(z)) - \lambda(z)$ is analytic on $\mathbb{H}$ and

10.2 The Modular Function

vanishes on T_0, hence it vanishes on all of $\mathbb{H}$. That is, τ_1 is an automorphism of the covering space $(\mathbb{H}, \lambda)$.

In exactly the same manner, $\tau_2(z) = z + 2$ is an automorphism of $(\mathbb{H}, \lambda)$. As we saw in the proof of Lemma 10.5, reflection of a point z in T_0 in the 'b' side of T_0, that is in the line $\Re z = 1$, results in $R_b(z) = 2 - \bar{z}$. Reflecting again in the line $\Re z = 2$ results in the point

$$\tau_2(z) = 4 - \overline{(2-\bar{z})} = z + 2.$$

Again, in the image domain $\mathbb{C}_{0,1}$, $\lambda(\tau_2(z))$ is the point $\lambda(z)$ reflected twice in the real axis and so $\lambda(\tau_2(z)) = \lambda(z)$ for z in T_0, and hence for all z in $\mathbb{H}$. □

Since both τ_1 and τ_2 both belong to Aut$(\mathbb{H}, \lambda)$, so do all elements of the subgroup of Aut$(\mathbb{H})$ generated by τ_1 and τ_2 (see (10.3)). We will show that this is actually a complete description of Aut$(\mathbb{H}, \lambda)$.

Theorem 10.6 *The automorphism group* Aut$(\mathbb{H}, \lambda)$ *of the covering space* $(\mathbb{H}, \lambda)$ *of the twice punctured plane* $\mathbb{C}_{0,1}$ *is the subgroup $\mathcal{G}$ of* Aut$(\mathbb{H})$ *generated by the automorphisms* $\tau_1 = z/(2z+1)$ *and* $\tau_2(z) = z + 2$.

The proof we give of Theorem 10.6 involves explicit computations. Conway, for example, provides a much simpler proof as he shows in [8, Theorem 16.1.8] that the automorphism group of the universal cover of D is isomorphic (as a group) to the fundamental group of D. The computations to follow have the advantage that, at least in principle, they show how this group of transformations acts on the triangulation of the half-plane $\mathbb{H}$. We divide the proof of Theorem 10.6 into a number of steps. The first is to compute the automorphism τ_3 that corresponds to the reflection of T_0 first in its 'b' side, the line $\Re z = 1$, and then in the semicircle $C^+(\frac{3}{2}, \frac{1}{2})$. Refection in the line $\Re z = 1$ is $z \to 2 - \bar{z}$ while reflection in the semicircle $C^+(\frac{3}{2}, \frac{1}{2})$ is, by (10.8),

$$R(z) = \frac{3}{2} + \frac{(1/2)^2}{\bar{z} - 3/2} = \frac{3\bar{z} - 4}{2\bar{z} - 3}.$$

Applying R to $2 - \bar{z}$ gives

$$\tau_3(z) = \frac{3(2-z) - 4}{2(2-z) - 3} = \frac{2 - 3z}{1 - 2z} = 2 + \frac{z}{1 - 2z} = (\tau_2 \circ \tau_1^{-1})(z). \quad (10.9)$$

Here we used $\tau_1^{-1}(z) = z/(1 - 2z)$. The double reflection τ_3 maps T_0 onto the reflection of T_1 in the line $\Re z = 1$. In the image domain $\mathbb{C}_{0,1}$ this corresponds to a double reflection in the real axis and so, once again, $\lambda(\tau_3(z)) = \lambda(z)$, first for $z \in T_0$ and then for all $z \in \mathbb{H}$. Alternatively, the simple fact that $\tau_3 = \tau_2 \circ \tau_1^{-1}$ shows that τ_3 is an automorphism of $(\mathbb{H}, \lambda)$.

In order to prove Theorem 10.6, it will be enough to show that each double reflection can be written in terms of τ_1 and τ_2. Note that each double reflection will map the upper half-plane onto itself and, by (10.8), will be a Möbius transformation. Hence, it will be an automorphism of $\mathbb{H}$.

Proposition 10.3 *Each double reflection, which maps a hyperbolic triangle to one of its grandchildren, can be expressed as an element of the group $\mathscr{G}$ generated by τ_1 and τ_2.*

Exercise 10.7 Find a formula for the Möbius map τ corresponding to reflection in the 'c' side of T_0 followed by reflection in the 'b' side of T_1 (as in Fig. 10.6).

Write τ as an element of the group $\mathscr{G}$ of automorphisms of $\mathbb{H}$.

Exercise 10.8 Find a formula for the Möbius map τ corresponding to reflection in the 'a' side of T_1 followed by reflection in the 'b' side of $T_{2,l}$ (as in Fig. 10.7).

Write τ as an element of the group $\mathscr{G}$ of automorphisms of $\mathbb{H}$.

Exercise 10.9 Find a formula for the Möbius map τ corresponding to reflection in the 'a' side of $T_{2,r}$ (see Fig. 10.7) followed by reflection in the 'c' side of $T_{2,r,l}$. This takes z in $T_{2,r}$ with vertices $\{\frac{1}{2}, \frac{2}{3}, 1\}$ to $\tau(z)$ in $T_{2,r,l,r}$ with vertices $\{\frac{3}{5}, \frac{5}{8}, \frac{2}{3}\}$. Write τ as an element of the group $\mathscr{G}$ of automorphisms of $\mathbb{H}$. (Hint: it simplifies matters to first apply τ to the automorphism from Exercise 10.7 that sends T_0 to $T_{2,r}$.)

Proof of Theorem 10.6 Assuming Proposition 10.3 Suppose that τ is an automorphism of the covering space $(\mathbb{H}, \lambda)$ of $\mathbb{C}_{0,1}$. In particular, τ is a conformal mapping of $\mathbb{H}$ onto $\mathbb{H}$, and is a Möbius mapping. Since $\lambda(\tau(z)) = \lambda(z)$ for $z \in \mathbb{H}$ and $\lambda|_{T_0}$ is a conformal mapping of T_0 onto $\mathbb{H}$, we find that $\lambda(\tau(T_0)) = \lambda(T_0) = \mathbb{H}$. Since $\tau(T_0)$ is also connected and λ is a covering map, we deduce that $\tau(T_0)$ is a connected component of $\lambda^{-1}(\mathbb{H})$. Since, by Lemma 10.5, we have a complete description of the connected components of $\lambda^{-1}(\mathbb{H})$, we see that τ maps T_0 onto a hyperbolic triangle T that is either (a) T_0 or one of its grand-descendants, or (b) the reflection in the line $\Re z = 1$ of T_1 or of one of its grand-descendants, or (c) a horizontal translate by a multiple of 2 of one of the hyperbolic triangles from (a) or (b). Fix a point z_0 in T_0 and set $w_0 = \lambda(z_0)$. Then, $\tau(z_0)$ in T also lies over w_0 in that $\lambda(z_0) = \lambda(\tau(z_0))$.

In Case (c), we can replace τ by $\tau_2^n \circ \tau$ for some integer power n and thereby assume that we are in Case (a) or (b). In Case (a), there is a sequence of double reflections, say $z \to R(z)$, that maps T_0 to T. Since the value of λ is invariant under a double reflection, we have $\lambda(R(z)) = \lambda(z)$, $z \in T_0$. Hence, $\lambda(R(z)) - \lambda(z) = 0$, first for $z \in T_0$ and then for all $z \in \mathbb{H}$, so that R is an automorphism of $(\mathbb{H}, \lambda)$. Since $\lambda(R(z_0)) = \lambda(z_0) = w_0 = \lambda(\tau(z_0))$, we must have $R(z_0) = \tau(z_0)$ as λ is a conformal map of T onto $\mathbb{H}$ and both $R(z_0)$ and $\tau(z_0)$ lie in T. Since an automorphism is uniquely determined by its value at a single point (see Proposition 10.1), we may conclude that $\tau = R$. Hence, the automorphism τ can be realised as a sequence of double reflections each of which is an element of $\mathscr{G}$ (assuming Proposition 10.3), so that τ itself is an element of $\mathscr{G}$.

The argument in Case (b) is similar. First map T_0 to the reflection of T_1 in the line $\Re z = 1$ by the automorphism $\tau_3 = \tau_2 \circ \tau_1^{-1}$ (see (10.9)). By a sequence of double reflections, which we again call R, we can map the reflection of T_1 in the line $\Re z = 1$ to its grand-descendant T. We again find that $R \circ \tau_3$ is an automorphism of $(\mathbb{H}, \lambda)$ and that both $(R \circ \tau_3)(z_0)$ and $\tau(z_0)$ lie in T with λ taking the same value

10.2 The Modular Function

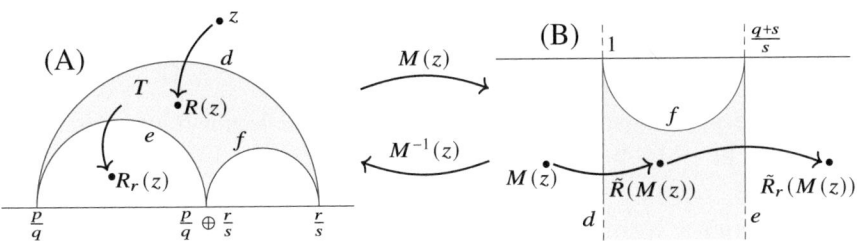

Fig. 10.9 Under $M(z) = (r/s - p/q)/(z - p/q)$, the 'd' and 'e' sides of the hyperbolic triangle T with vertices $\frac{p}{q}$, $\frac{p}{q} \oplus \frac{r}{s}$ and $\frac{r}{s}$ are mapped to the vertical lines $\Re z = 1$ and $\Re z = (q+s)/s$, respectively

w_0 at both points. Hence, $(R \circ \tau_3)(z_0) = \tau(z_0)$ and so $\tau = R \circ \tau_3$. Assuming Proposition 10.3, we conclude once more that τ is an element of $\mathscr{G}$. □

In order to prove Proposition 10.3, we find explicit formulas for generic double reflections in Euclidean circles.

Lemma 10.7 *Consider the hyperbolic triangle T with vertices $\frac{p}{q}$, $\frac{p}{q} \oplus \frac{r}{s}$ and $\frac{r}{s}$, where $\frac{p}{q} < \frac{r}{s}$ and where $\frac{p}{q}$ and $\frac{r}{s}$ are Farey neighbours. With d, e, and f being the sides of T as shown in Fig. 10.9, let R_r be the double reflection first in the 'd' side of T then in the 'e' side of T. Then,*

$$R_r(z) = \frac{(1+2pq)z - 2p^2}{2q^2 z + (1-2pq)} \qquad (10.10)$$

Proof We first normalise this picture by mapping it under the Möbius transformation

$$M(z) = \frac{r/s - p/q}{z - p/q} = \frac{1}{s(qz-p)}.$$

Here we used $qr - ps = 1$. Since $M(p/q) = \infty$, the images of both the 'd' and 'e' sides of T lie on straight lines. Since M satisfies $M(\bar{z}) = \overline{M(z)}$, both straight lines are symmetric in the real axis, hence vertical. The image of the 'd' side of T is half of the line $\Re w = 1$ since $M(r/s) = 1$. The image of the 'e' side of T is half of the line $\Re w = (q+s)/s$ since

$$M\left(\tfrac{p}{q} \oplus \tfrac{r}{s}\right) = \frac{1}{s\left(q\frac{p+r}{q+s} - p\right)} = \frac{q+s}{s}.$$

By Proposition 2.4, Möbius transformations preserve symmetric points. It follows that any of the reflections on the (A) side of Fig. 10.9 can be realised by first mapping with M, carrying out the corresponding reflections on the (B) side of Fig. 10.9, then mapping back with M^{-1}. For example, with R being reflection in the semicircle d and $\tilde{R}$ being reflection in the line $\Re w = 1$ then $M(R(z)) = \tilde{R}(M(z))$ so that $R(z) = (M^{-1} \circ \tilde{R} \circ M)(z)$.

The reflection of w in the line $\Re w = 1$ is $2 - \overline{w}$. Reflection of this point in the line $\Re w = (q+s)/s$ is

$$\tilde{R}_r(w) = 2\left(\frac{q+s}{s}\right) - \overline{(2-\overline{w})} = 2\left(\frac{q+s}{s} - 1\right) + w = w + \frac{2q}{s}.$$

Then, since $M^{-1}(w) = 1/(sqw) + p/q$,

$$R_r(z) = (M^{-1} \circ \tilde{R}_r \circ M)(z) = M^{-1}\left(M(z) + \frac{2q}{s}\right)$$

$$= M^{-1}\left(\frac{1}{s(qz-p)} + \frac{2q}{s}\right)$$

$$= M^{-1}\left(\frac{2q^2z + (1-2pq)}{s(qz-p)}\right)$$

$$= \frac{qz - p}{q[2q^2z + (1-2pq)]} + \frac{p}{q}$$

which simplifies to (10.10). □

In analogy with Lemma 10.7 and referring back to Fig. 10.8, we have a similar formula for reflection first in the 'd' side of T then in the 'f' side of T. For consistency of notation, we interchange the role of $\frac{p}{q}$ and $\frac{r}{s}$.

Lemma 10.8 *Consider the hyperbolic triangle T with vertices $\frac{r}{s}, \frac{r}{s} \oplus \frac{p}{q}$ and $\frac{p}{q}$, where $\frac{r}{s} < \frac{p}{q}$ and where $\frac{r}{s}$ and $\frac{p}{q}$ are Farey neighbours. With d and f being the sides of T as shown in Fig. 10.10, let R_l be reflection first in the 'd' side of T then in the 'f' side of T. Then,*

$$R_l(z) = \frac{(1-2pq)z + 2p^2}{-2q^2z + (1+2pq)}. \tag{10.11}$$

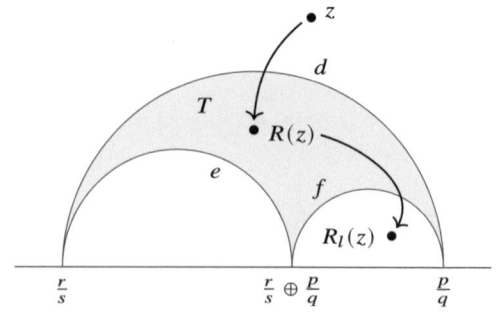

Fig. 10.10 Diagram for Lemma 10.8. R_l is reflection first in the 'd' side of the hyperbolic triangle T with vertices $\frac{r}{s}, \frac{r}{s} \oplus \frac{p}{q}$ and $\frac{p}{q}$, then in the 'f' side of T

10.2 The Modular Function

Exercise 10.10 Establish the formula (10.11), for example by adapting the method used to establish (10.10).

Notice that both R_r and R_l are Möbius transformations of the form $M(z) = (az + b)/(cz + d)$ where a, b, c and d are integers and $ad - bc = 1$.

Also, the formulas (10.10) for R_r and (10.11) for R_l depend only on $\frac{p}{q}$ and not on $\frac{r}{s}$. For example, the formula for R_r in the case $\frac{p}{q} = \frac{1}{2}$ is the same irrespective of whether $\frac{r}{s}$ equals $\frac{2}{3}$ or $\frac{3}{5}$ or $\frac{4}{7}$. Furthermore, bearing in mind Exercise 2.3, it is straightforward to verify that

$$R_r^{-1}(z) = R_l(z). \tag{10.12}$$

Proof of Proposition 10.3 The construction of the modular function in Sect. 10.2.2 proceeds by a series of reflections that produce hyperbolic triangles in the half-plane $\mathbb{H}$ with vertices on the real axis. At each stage, a new vertex $\frac{p}{q} \oplus \frac{r}{s}$ is produced between the Farey neighbour vertices $\frac{p}{q}$ and $\frac{r}{s}$, see Lemma 10.4. Let us set $\alpha = \frac{p}{q}$, $\beta = \frac{r}{s}$ and write $R_{r,\alpha}$ and $R_{l,\alpha}$ for the double reflections whose formulas are given by (10.10) and (10.11).

Since each double reflection is of the form R_r or R_l, it will be sufficient to prove by induction that if $R_{r,\alpha}$, $R_{l,\alpha}$, $R_{r,\beta}$ and $R_{l,\beta}$ belong to the group $\mathscr{G}$ generated by τ_1 and τ_2 then so do $R_{r,\alpha\oplus\beta}$ and $R_{l,\alpha\oplus\beta}$. We begin the induction by computing, using (10.10) and (10.11),

$$R_{r,0}(z) = \frac{z}{2z+1} = \tau_1(z)$$

$$R_{l,1}(z) = \frac{z-2}{2z-3} = (\tau_1 \circ \tau_2^{-1})(z)$$

$$R_{l,2}(z) = \frac{8-3z}{5-2z} = (\tau_2 \circ \tau_1^{-1} \circ \tau_2^{-1})(z).$$

By (10.12), each of $R_{l,0}$, $R_{r,1}$ and $R_{r,2}$ also belongs to $\mathscr{G}$.

Suppose that the double reflections R_r and R_l from Lemmas 10.7 and 10.8 and for both vertices α and β belong to the group $\mathscr{G}$ generated by τ_1 and τ_2. We will show that the double reflections R_r and R_l for the new vertex $\alpha \oplus \beta$ also belong to $\mathscr{G}$ by showing that they can be written in terms of the double reflections at α and β. In fact, it transpires that

$$R_{l,\alpha\oplus\beta} = R_{r,\alpha} \circ R_{r,\beta} \tag{10.13}$$

and that

$$R_{r,\alpha\oplus\beta} = R_{l,\beta} \circ R_{l,\alpha}. \tag{10.14}$$

Since the identity (10.14) follows from (10.13) by taking inverses and using (10.12), it is enough to verify (10.13). In terms of matrices representing these Möbius transformations, (10.13) becomes

$$\begin{pmatrix} 2(p+r)(q+s)-1 & -2(p+r)^2 \\ 2(q+s)^2 & -1-2(p+r)(q+s) \end{pmatrix}$$
$$\simeq \begin{pmatrix} 1+2pq & -2p^2 \\ 2q^2 & 1-2pq \end{pmatrix} \begin{pmatrix} 1+2rs & -2r^2 \\ 2s^2 & 1-2rs \end{pmatrix}. \quad (10.15)$$

Since the matrix representing a Möbius transformation is only determined up to a constant—see Exercise 2.3 and the discussion immediately preceding it—we took the liberty of multiplying all entries in the matrix for $R_{l,\alpha\oplus\beta}$ by -1. Here, $\simeq$ means equal up to multiplication by a scalar. The identity (10.15) can be checked by multiplying the matrices on the right and making judicious use of the relationship $qr - ps = 1$. For example, for the $(1, 1)$ entry,

$$(1+2pq)(1+2rs) - (2p^2)(2s^2) = 1 + 4pqrs + 2pq + 2rs - 4p^2s^2$$
$$= 2pq + 2rs + 1 + 4ps(qr - ps)$$
$$= 2pq + 2rs + 1 + 2ps + 2(qr - 1)$$
$$= 2(p+r)(q+s) - 1.$$

For the $(1, 2)$ entry,

$$(1+2pq)(-2r^2) - 2p^2(1-2rs) = -2r^2 - 2p^2 - 4r^2pq + 4p^2rs$$
$$= -2r^2 - 2p^2 - 4pr(qr - ps)$$
$$= -2(r^2 + p^2 + 2pr)$$
$$= -2(p+r)^2.$$

Similar calculations show that the remaining two entries in the matrix on the left of (10.15) match those in the matrix product on the right.

Proposition 10.3 now follows by induction from (10.13) and (10.14). □

Exercise 10.11 Complete the verification of the matrix identity (10.15), thereby establishing (10.13).

Exercise 10.12 In each case express, in terms of τ_1 and τ_2, the automorphism τ of $(\mathbb{H}, \lambda)$ that sends T_0 to the hyperbolic triangle T with the given vertices: (i) T: $\{\frac{0}{1}, \frac{1}{5}, \frac{1}{4}\}$; (ii) T: $\{\frac{1}{3}, \frac{3}{8}, \frac{2}{5}\}$; (iii) T: $\{\frac{3}{5}, \frac{5}{8}, \frac{2}{3}\}$.

Exercise 10.13 Use the identities (10.12), (10.13), and (10.14) to redo Exercises 10.7, 10.8, and 10.9.

10.3 The Picard Theorems: Reprise

The existence of the modular function opens up a different route to proving the Picard theorems. We first give a short proof of the Little Picard Theorem and then a second proof of Montel's Second Theorem which was the main step in the proof of the Great Picard Theorem.

Theorem 10.7 (Little Picard Theorem) *An entire function that omits two values must be constant.*

Proof Suppose that f is an entire function that omits two values, which we may take to be 0 and 1, so that $f\colon \mathbb{C} \to \mathbb{C}_{0,1}$ is analytic. Let $(\mathbb{D}, \tau)$ be an analytic covering space of $\mathbb{C}_{0,1}$ (one may take $\tau = \lambda \circ M$ where $M(z) = \mathrm{i}(1+z)/(1-z)$ maps $\mathbb{D}$ conformally onto the half-plane $\mathbb{H}$ and λ is the modular function). Since $\mathbb{C}$ is simply connected, Theorem 10.2 applies and we deduce that there is an analytic function $g\colon \mathbb{C} \to \mathbb{D}$ such that $f = \tau \circ g$. Since g is a bounded entire function it is constant by Liouville's Theorem, hence f is constant as well. □

We now give an alternative proof of Montel's Second Theorem, Theorem 5.11, to the effect that the family $\mathscr{F}$ of analytic functions on a domain D all of which omit the same two values is a normal family. In essence, Montel's First Theorem and the modular function combine to yield Montel's Second Theorem.

Alternative Proof of Montel's Second Theorem As in the proof of Montel's Second Theorem in Chap. 5, we may assume that D is the unit disk $\mathbb{D}$ and that all functions in the family $\mathscr{F}$ omit the values 0 and 1.

Let $\{f_n\}_{n=1}^{\infty}$ be a sequence of functions from $\mathscr{F}$. We will show that $\{f_n\}_{n=1}^{\infty}$ either has a subsequence that converges uniformly on compact subsets of $\mathbb{D}$ to an analytic function f, so that the sequence converges in $\mathscr{H}(\mathbb{D})$, or has a subsequence that converges to the constant function $\infty_{\mathbb{D}}$ uniformly on compact subsets of $\mathbb{D}$.

First, some subsequence of the sequence of points $\{f_n(0)\}_{n=1}^{\infty}$ converges either to a point α in the complex plane or to the point at infinity. Replace $\{f_n\}_{n=1}^{\infty}$ by this subsequence. There are a number of cases depending on the value of α.

The main case is that of $\alpha \in \mathbb{C}_{0,1}$. Let $(\mathbb{D}, \tau)$ be an analytic covering space of $\mathbb{C}_{0,1}$ and let Δ be an evenly covered neighbourhood of α. Let $\Delta^* \subset \mathbb{D}$ be homeomorphic to Δ under τ. By choosing Δ to be smaller if necessary, we may assume that $\mathrm{cl}(\Delta^*)$ does not meet the unit circle. Theorem 10.2 can now be used to replace the sequence of functions $\{f_n\}_{n=1}^{\infty}$ by a sequence $\{g_n\}_{n=1}^{\infty}$ of *bounded* functions by factoring each function f_n through the cover $\mathbb{D}$. Since $f_n(0) \to \alpha$ as $n \to \infty$, we may assume that $f_n(0) \in \Delta$ for each n. Then, by Theorem 10.2, there is an analytic function $g_n\colon \mathbb{D} \to \mathbb{D}$ for which $f_n = \tau \circ g_n$ and $g_n(0) \in \Delta^*$. Since the functions $\{g_n\}_{n=1}^{\infty}$ are uniformly bounded, Montel's First Theorem (Theorem 5.8) applies. Hence, some subsequence, say $\{g_{n_k}\}_{k=1}^\infty$, converges uniformly on compact subsets of $\mathbb{D}$ to an analytic function g. Since $g(0) = \lim_{k\to\infty} g_{n_k}(0)$ and each $g_{n_k}(0)$ lies in Δ^*, it follows that $g(0) \in \text{cl}(\Delta^*) \subset \mathbb{D}$. It follows that $g: \mathbb{D} \to \mathbb{D}$ even if g happens to be constant, so that the function $f = \tau \circ g$ is defined and analytic in $\mathbb{D}$. We claim that $f_{n_k} \to f$ locally uniformly on $\mathbb{D}$. To verify this, fix $r < 1$, set $\rho = \max\{|g(z)|: |z| = r\}$ and set $\rho' = (\rho + 1)/2$ so that ρ' is also less than 1. Since g_{n_k} converges to g uniformly on $\overline{D(0, r)}$, we have $|g_{n_k}(z)| \leq \rho'$ if $|z| \leq r$ for all sufficiently large k. Since τ is uniformly continuous on $\overline{D(0, \rho')}$, it follows that $f_{n_k} = \tau(g_{n_k})$ converges to $f = \tau(g)$ uniformly on $\overline{D(0, r)}$.

The next case we consider is $\alpha = 1$ so that $f_n(0) \to 1$ as $n \to \infty$. Since each f_n is zero-free in $\mathbb{D}$, Theorem 6.8 applies and there is an analytic function h_n in $\mathbb{D}$ for which $h_n^2 = f_n$. By replacing h_n by $-h_n$ if necessary, we can arrange that $h_n(0) \to -1$ as $n \to \infty$. Thus, $\{h_n\}_{n=1}^\infty$ is a sequence of functions analytic in $\mathbb{D}$ all of which omit the values 0 and 1 and to which the previous case applies with $\alpha = -1$. It follows that some subsequence $\{h_{n_k}\}_k$ converges locally uniformly to a limit function h, in which case $\{f_{n_k}\}_k$ converges to $f = h^2$.

To deal with the case $\alpha = 0$, set $h_n = 1 - f_n$ for each n. Then, $\{h_n\}_{n=1}^\infty$ is a sequence of functions analytic in $\mathbb{D}$ all of which omit the values 0 and 1 and to which the previous case $\alpha = 1$ applies. Thus, $\{h_n\}_n$ has a convergent subsequence which immediately implies that $\{f_n\}_n$ has a convergent subsequence.

The final case to consider is $\alpha = \infty$. In this case we set $h_n = 1/f_n$, which is analytic since f_n never takes the value 0. Also, h_n omits both 0 and 1 and $h_n(0) = 1/f_n(0) \to 0$ as $n \to \infty$. The previous case in which $\alpha = 0$ applies, hence there is a subsequence $\{h_{n_k}\}_k$ that converges locally uniformly to an analytic function h on $\mathbb{D}$. Since $h_{n_k}(0) \to 0$, it must be that $h(0) = 0$. Thus, each h_{n_k} omits the value 0 whereas the limit function h does not. By Hurwitz's Theorem, this can only happen if the limit function h is identically zero. Thus, $1/f_{n_k} \to 0$ locally uniformly on $\mathbb{D}$, that is $f_{n_k} \to \infty$ locally uniformly on $\mathbb{D}$, as $k \to \infty$.

In each case, whether $\alpha \in \mathbb{C}_{0,1}$, $\alpha = 1$, $\alpha = 0$ or $\alpha = \infty$, we have produced a subsequence of $\{f_n\}_{n=1}^\infty$ that converges locally uniformly whether this be to an analytic function f in the first three cases or to the constant function $\infty_\mathbb{D}$ in the final case. Hence, $\mathscr{F}$ satisfies Definition 5.4 and is a normal family. $\square$

10.4 Uniformisation of Planar Domains

Our goal in this section is to prove the Uniformisation Theorem for planar domains.

Theorem 10.8 *Any planar domain D whose complement consists of two or more points has the disk $\mathbb{D}$ as its universal analytic covering space.*

The modular function λ constructed in Sect. 10.2 is the special case when the complement of the domain D consists of exactly two points. This special case will be used in the proof for a general hyperbolic domain—a hyperbolic domain is one

10.4 Uniformisation of Planar Domains

whose complement consists of two or more points. We follow quite closely the proof set out in Fisher et al. [11], and that of Zakeri [30, Chap. 13] that, in turn, is based on an outline proof in *Advanced Complex Analysis* course notes by C. McMullen (2017). In fact, Zakeri's book is yet another excellent treatment of the main topics in complex analysis, including many of those treated here. Theorem 10.8 is a significant special case of the uniformisation theorem for Riemann Surfaces.

Following Zakeri, we break the proof of Theorem 10.8 into a number of steps. At Step 3, in which the compactness of the family of functions $\mathscr{F}$ is established, we depart somewhat from Zakeri's proof and provide an essentially complex analysis argument that does not, in particular, require a priori knowledge of the existence of a (topological) universal covering map. In doing so, we need the following result.

Lemma 10.9 *Let D be a planar domain and α be a point in D. In the case of each point z in D, there is a simply connected subdomain U of D that contains both α and z.*

Proof Let γ_1 be a curve in D joining α to z, parameterised on $[0, 1]$, say, with $\gamma_1(0) = \alpha$ and $\gamma_1(1) = z$. Since the trace of γ_1 is a compact subset of D, the distance ϵ from γ_1 to the complement of D is positive. Since γ_1 is uniformly continuous on $[0, 1]$, there is an integer m such that $|\gamma_1(k/m) - \gamma_1((k+1)/m)| \leq \epsilon/4$, $0 \leq k \leq m-1$. Set $z_k = \gamma_1(k/m)$. Let γ_2 be a polygonal path joining the points $z_0, z_1, \ldots z_m$ in succession by line segments that are parallel to the coordinate axes. Since these points are no more than $\epsilon/4$ apart, the path γ_2 also lies in D and is a distance at least $3\epsilon/4$ from the boundary of D. This path can have only finitely many overlaps or self-intersections. Removing these, we arrive at a simple polygonal path γ joining α to z that consists of line segments parallel to the coordinate axes with successive endpoints $w_0, w_1, \ldots w_n$, say.

Any two successive line segments meet at a single common point. If two line segments are not successive then they do not meet. Set δ to be the minimum distance between any two non-successive line segments. Set $l = \min\{\delta/4, \epsilon/4\}$. For each k, let Q_k be the open square centre w_k and sides parallel to the coordinate axes of length l. For $1 \leq k \leq n$, let R_k be the rectangle that is the convex hull of Q_{k-1} and Q_k. Finally, set $U_k = \cup_{j=1}^{k} R_j$ and $U = U_n$. Since $l \leq \epsilon/4$, each rectangle R_k lies in D. Since $l \leq \delta/4$, no two rectangles intersect except when consecutive rectangles R_k and R_{k+1} intersect in the square Q_k. Clearly, $U_1 = R_1$ is simply connected as it is a rectangle. Since $U_{k+1} = U_k \cup R_{k+1}$ and $U_k \cap R_{k+1} = Q_k$ is connected, it follows by induction from Exercise 6.5 that U_{k+1} is simply connected. Then, U is a simply connected subdomain of D that contains both α and z. □

Proof of Theorem 10.8 Without loss of generality, we may assume that D is contained in $\mathbb{C}_{0,1}$, the plane punctured at 0 and at 1. Fix α in D. Let $\mathscr{F}$ be the family of all analytic functions $f : \mathbb{D} \to \mathbb{C}_{0,1}$ with $f(0) = \alpha$ for which there is a subdomain Ω_f of $\mathbb{D}$ containing 0 such that $(\Omega_f, f|_{\Omega_f})$ is a covering map of D.

Step 1: $\mathscr{F}$ Is Non-empty Using the modular function λ, we can construct the universal covering map $\mu : \mathbb{D} \to \mathbb{C}_{0,1}$ of $\mathbb{C}_{0,1}$ by $\mathbb{D}$ for which $\mu(0) = \alpha$ (we can take $\mu = \lambda \circ M$ where M is a conformal map of $\mathbb{D}$ onto $\mathbb{H}$ with $\lambda(M(0)) = \alpha$).

Then, $\mu \in \mathscr{F}$ since we may take Ω_μ to be the connected component of $\mu^{-1}(D)$ that contains 0. By Example 10.7, $(\Omega_\mu, \mu|_{\Omega_\mu})$ is a covering space of D.

Step 2: Bound for $|f'(0)|$ when $f \in \mathscr{F}$ Set $r = \delta_D(\alpha)$, the distance from α to the boundary of D, so that r is positive and $U = D(\alpha, r)$ is contained in D. For f in $\mathscr{F}$, it follows from Theorem 10.4 that U is an evenly covered neighbourhood for the covering map $(\Omega_f, f|_{\Omega_f})$ of D. Let U_f^* be the component of $f^{-1}(U)$ that contains 0. Then, $U_f^* \subseteq \Omega_f \subseteq \mathbb{D}$ and $f: U_f^* \to U$ is a conformal mapping. The inverse mapping f^{-1} maps U to $U_f^* \subseteq \mathbb{D}$. By Schwarz's Lemma applied to $g(z) = f^{-1}(\alpha + rz)$, $z \in \mathbb{D}$, we find that $|(f^{-1})'(\alpha)| \leq 1/r$. As a consequence, $|f'(0)| \geq r$ for every f in $\mathscr{F}$.

Step 3: $\mathscr{F}$ Is a Compact Subset of $\mathscr{H}(\mathbb{D})$ Let $\{f_n\}_n$ be a sequence in $\mathscr{F}$. We need to show that $\{f_n\}_n$ has a subsequence that converges locally uniformly to a function f in $\mathscr{F}$. Since all functions in $\mathscr{F}$ omit the same two values 0 and 1, it follows from Montel's Second Theorem (Theorem 5.11) that $\{f_n\}_n$ has a subsequence that converges in $\mathscr{H}(\mathbb{D})$ to some function f. Replacing the original sequence by this subsequence, we have that $f_n \to f$ locally uniformly in $\mathbb{D}$. Since $|f_n'(0)| \geq r$ for each n, we have that $|f'(0)| \geq r$ and so f is non-constant. Hence, by Hurwitz's Theorem, f also omits the values 0 and 1.

While this limit function f is certainly analytic in the unit disk, maps $\mathbb{D}$ into $\mathbb{C}_{0,1}$ and maps 0 to α, the challenge is to show that f belongs to the family $\mathscr{F}$ by showing that there is a subdomain Ω_f of $\mathbb{D}$ containing 0 such that $(\Omega_f, f|_{\Omega_f})$ is a covering map of D. In order to establish the existence of the subdomain Ω_f we will show that, in the case of each simply connected subdomain U of D that contains α, each connected component of $f^{-1}(U)$ is mapped conformally onto U. If we set Ω_f to be the union of all preimages of all such simply connected subdomains of D, $(\Omega_f, f|_{\Omega_f})$ will then be a covering space of D. In fact, since each such subdomain U contains α, and $f(0) = \alpha$, at least one component of $f^{-1}(U)$ contains 0 and is therefore non-empty: this component is then mapped *onto* U. Since, by Lemma 10.9, the union of these simply connected subdomains is all of D, we see that $f|_{\Omega_f}$ is onto D. Given z in D, any simply connected subdomain U of D that contains both α and z will act as an evenly covered neighbourhood of z. If Ω_f is not connected, we use Lemma 10.2 to replace it by the component of Ω_f containing 0.

At this point, therefore, we fix a simply connected subdomain U of D that contains α, and a non-empty connected component U^* of $f^{-1}(U)$. Choose a^* in U^* and set $a = f(a^*)$ so that $a \in U$ (as stated in the previous paragraph, $a^* = 0$ and $a = \alpha$ is one concrete instance of this). Since f is non-constant, there is a positive δ such that the disk $\overline{D(a^*, \delta)}$ is contained in U^* and f never takes the value a in this closed disk except at a^*. With $r \leq \delta$, an application of Hurwitz's Theorem shows that f_n and f take the value a the same number of times in the disk $D(a, r)$, namely once, for all sufficiently large n depending on r. Thus, by omitting some initial terms in the sequence of functions $\{f_n\}$, we may assume that there is a sequence of points $\{a_n^*\}$ in $D(a^*, \delta) \subseteq U^*$ for which $f_n(a_n^*) = a$ and $a_n^* \to a^*$ as $n \to \infty$.

10.4 Uniformisation of Planar Domains

In the case of each function f_n, there is a subdomain Ω_n of $\mathbb{D}$ such that $(\Omega_n, f_n|_{\Omega_n})$ is a covering space of D. By Theorem 10.4, U is evenly covered by this covering map as U is simply connected. For clarity, let us write h_n for $f_n|_{\Omega_n}$. Since $f_n(a_n^*) \in U$, there is therefore a unique component, U_n^* say, of $h_n^{-1}(U)$ that contains a_n^* and that is mapped conformally by h_n onto U. Let us write $g_n: U \to U_n^*$ for the inverse conformal map. Since each map g_n is bounded by 1, Montel's First Theorem, Theorem 5.8, applies and shows that a subsequence of $\{g_n\}_n$ converges locally uniformly on U to a function g, say. For simplicity, let us replace the sequence $\{g_n\}$ by this subsequence so that $g_n \to g$ as $n \to \infty$ locally uniformly on U. Being a limit of conformal maps, the function g is either conformal or constant. The sequence $\{g_n'\}$ converges locally uniformly to g' on U, which is zero if g is constant and, in particular, we would have $g_n'(a) \to 0$ as $n \to \infty$ in this case. Then, $f_n'(a_n^*) = 1/g_n'(a)$ tends to infinity. Finally, since $f_n' \to f'$ locally uniformly on $\mathbb{D}$ and a_n^* tends to a^*, we deduce from Exercise 5.7 that $f'(a^*) = \lim_{n \to \infty} f_n'(a_n^*)$ is infinite, which is impossible. We conclude that the limit function g is a conformal map and we set $V^* = g(U)$. Note that $g(a) = \lim_{n \to \infty} g_n(a) = \lim_{n \to \infty} a_n^* = a^*$, so that $a^* \in V^*$. For z in U, we have $g_n(z)$ in U_n^* and so $f_n(g_n(z)) = h_n(g_n(z)) = z$. Since $g_n(z) \to g(z)$ and since f_n converges locally uniformly to f on $\mathbb{D}$, another application of Exercise 5.7 shows that $f(g(z)) = z$, $z \in U$. That is, $f = g^{-1}$ on V^*, so that $f: V^* \to U$ is a conformal map.

Finally, we show that V^* and U^* coincide. First, $V^* \subseteq U^*$ since V^* is connected (it is the image of U under g), lies in $f^{-1}(U)$, contains a^*, and U^* is, by definition, the connected component of $f^{-1}(U)$ containing a^*. Suppose that U^* contained a point z^* not in V^*. A curve in U^* joining a^* in V^* to z^* in the complement of V^* must meet the boundary of V^*, say at w^*, before reaching z^*. Then, $f(w^*)$ should lie on the boundary of U since $f: V^* \to U$ is conformal. However, $f(w^*)$ should also lie in U since $w^* \in U^* \subseteq f^{-1}(U)$. This contradiction shows that $U^* \subseteq V^*$. We have shown that $f: U^* \to U$ is conformal.

Step 4: Selection of an Extremal Function Each function f in $\mathscr{F}$ maps a subdomain Ω_f of the unit disk onto the domain D with $f(0) = \alpha$. The larger the domain Ω_f, the *smaller* we would expect the derivative of f at 0 to be. Using the established fact that $\mathscr{F}$ is compact and that evaluation of the derivative at 0 is a continuous function, there is a function μ in $\mathscr{F}$ that minimises $|f'(0)|$. That is,

$$\mu \in \mathscr{F} \text{ with } |\mu'(0)| = \min\{|f'(0)|: f \in \mathscr{F}\}.$$

This is our candidate for the covering map of D by $\mathbb{D}$.

Step 5: $\Omega_\mu = \mathbb{D}$ The uniformisation theorem for planar domains is proved once we show that $\Omega_\mu = \mathbb{D}$ as $(\mathbb{D}, \mu)$ is then a covering map of D (with $\mu(0) = \alpha$). Suppose, to the contrary, that $\Omega_\mu \subsetneq \mathbb{D}$ and that $a \in \mathbb{D} \setminus \Omega_\mu$. We have a contradiction if we can produce $g \in \mathscr{F}$ with $|g'(0)| < |\mu'(0)|$. Here we employ a 'square' trick analogous to the 'square root' trick from the proof of the Riemann mapping theorem in Sect. 6.3. We use the square function $s(z) = z^2$, $z \in \mathbb{D}$, instead of the square root, as our functions take the disk to D rather than the other way around and we want a function with a smaller derivative. Choose one of the two non-zero square roots of a, call it b. Recall the notation M_a for the automorphism of the disk $\mathbb{D}$ given by (4.7). Set $\phi = M_a \circ s \circ M_b$ and set

$$g = \mu \circ \phi, \quad z \in \mathbb{D}.$$

Note that $\phi(0) = 0$ and that, by Schwarz's Lemma (as ϕ is not a rotation), $|\phi'(0)| < 1$. In fact, one may compute $|\phi'(0)| = 2\sqrt{|a|}/(1 + |a|)$.

Since Ω_μ is a subdomain of $\mathbb{D}$ that contains 0 but not a, the subdomain $M_a(\Omega_\mu)$ of $\mathbb{D}$ contains $M_a(0) = a$ but not $M_a(a) = 0$. Just as was the case in Example 10.2, $s \colon \mathbb{D} \setminus \{0\} \to \mathbb{D} \setminus \{0\}$ is a covering map. By Example 10.7, the restriction of s to the component W of $s^{-1}(M_a(\Omega_\mu))$ that contains b is again a covering map. Then, $\phi \colon M_b(W) \to \Omega_\mu$ is a covering map and so $g = \mu \circ \phi \colon M_b(W) \to D$ is a covering map (as any point has at most two preimages under ϕ—that is, ϕ is a covering map of degree at most 2). Hence, g belongs to $\mathscr{F}$ and $|g'(0)| = |\mu'(0)| |\phi'(0)| < |\mu'(0)|$ which is the contradiction we were looking for. $\square$

Glossary

$\mathbb{C}$	Field of complex numbers
$\mathbb{C}_\infty$	extended complex plane
$\Re z$	Real part of a complex number z
$\Im z$	Imaginary part of a complex number z
$\bar{z}$	complex conjugate of a complex number z
$\|z\|$	modulus of a complex number z
$\mathbb{R}$	Field of real numbers
$\mathbb{R}_\infty$	extended real line
$C(a,r)$	circle $\{z : \|z-a\| = r\}$, centre a and radius r
$D(a,r)$	disk $\{z : \|z-a\| < r\}$, centre a and radius r
$\overline{D(a,r)}$	closed disk $\{z : \|z-a\| \leq r\}$, centre a and radius r
$D'(a,r)$	punctured disk $\{z : 0 < \|z-a\| < r\}$, centre a and radius r
$\mathbb{D}$	unit disk $\{z : \|z\| < 1\}$
$\mathbb{T}$	unit circle $\{z : \|z\| = 1\}$
$\mathbb{H}$	upper half-plane $\{z : \Im z > 0\}$
int (E)	interior of a set E
cl(E)	closure of a set E
$\overline{E}$	reflection of E in the real axis, that is $\{\bar{z} : z \in E\}$
$\mathscr{S}$	unit sphere in $\mathbb{R}^3$
$\mathscr{S}'$	unit sphere in $\mathbb{R}^3$ punctured at the North Pole $\mathbf{N} = (0, 0, 1)$
$\mathscr{P}$	Stereographic projection
$d(z, w; \Omega)$	Riemannian distance between points z and w in a domain Ω
$d^\#(z, w)$	spherical distance between points z and w in $\mathbb{C}_\infty$
$\delta^\#(\alpha, \beta)$	pseudo-spherical distance between α and β in $\mathbb{C}_\infty$
Aut(D)	group of all automorphisms of a domain D
$d(z, w; \mathbb{D})$	distance between z and w in the hyperbolic metric of the disk $\mathbb{D}$
$\mathscr{C}(X,Y)$	the space of continuous functions $f : X \to Y$, X and Y being metric spaces
$\mathscr{H}(D)$	the space of analytic functions on a domain D

$\mathscr{M}(D)$	the space of meromorphic functions on a domain D
$\mathscr{M}^*(D)$	$\mathscr{M}(D) \cup \{\infty_D\}$ where ∞_D is the constant function ∞ on D
S	class of normalised univalent functions in the unit disk $\mathbb{D}$
$\mathbb{U}$	the exterior $\{z : |z| > 1\}$ of the unit disk $\mathbb{D}$
Σ	the class of functions that are univalent and non-zero in $\mathbb{U}$

Solutions to the Exercises

Exercises from Chap. 2

2.1

▷ Using the formula (2.3) for the centre a of $C(z_1, z_2, k)$,

$$z_1 - a = z_1 - \frac{z_1 - k^2 z_2}{1 - k^2} = \frac{z_1(1 - k^2) - z_1 + k^2 z_2}{1 - k^2} = \frac{k^2}{1 - k^2}(z_2 - z_1),$$

as required. By (2.4), the radius of $C(z_1, z_2, k)$ is given by $r = k|z_2 - z_1|/|1 - k^2|$, and so $|z_1 - a| = kr$. The given formulae for $z_2 - a$ and $|z_2 - a|$ follow by a completely similar calculation.

▷ In general, three distinct points z_1, z_2, z_3 are collinear if and only if $(z_3 - z_1)/(z_2 - z_1)$ is real. In our setting, with $k \neq 1$ and using the previous part of this exercise, $(z_2 - a)/(z_1 - a) = 1/k^2$, which is real.

2.2 Since $C(z_1, z_2, k) = C(z_2, z_1, 1/k)$ and $1/k$ decreases from 1 to 0 as k increases from 1 to ∞, the regions $D(z_1, z_2, k)$ increase continuously from the half-plane containing z_1 and bounded by the perpendicular bisector of $[z_1, z_2]$ to the complement of the single point z_2 as k increases from 1 to ∞.

2.3 With the notation from the exercise itself,

$$(M_2 \circ M_1)(z) = \frac{a_2 M_1(z) + b_2}{c_2 M_1(z) + d_2}$$

$$= \frac{a_2(a_1 z + b_1) + b_2(c_1 z + d_1)}{c_2(a_1 z + b_1) + d_2(c_1 z + d_1)}$$

$$= \frac{(a_2 a_1 + b_2 c_1)z + (a_2 b_1 + b_2 d_1)}{(c_2 a_1 + d_2 c_1)z + (c_2 b_1 + d_2 d_1)}$$

$$\sim \begin{pmatrix} a_2 & b_2 \\ c_2 & d_2 \end{pmatrix} \times \begin{pmatrix} a_1 & b_1 \\ c_1 & d_1 \end{pmatrix}$$

$$= A_{M_2} \times A_{M_1}.$$

2.4 Let $M(z) = (az+b)/(cz+d)$, $z \in \mathbb{C}_\infty$, with $ad - bc \neq 0$. If $c = 0$ then $ad \neq 0$ and $M(z)$ reduces to a linear polynomial of the form $M(z) = \tilde{a}z + \tilde{b}$ where $\tilde{a} \neq 0$. This is clearly a bijection from $\mathbb{C}_\infty$ to $\mathbb{C}_\infty$ (with $M(\infty) = \infty$).

Now suppose that $c \neq 0$. Then M only takes the value ∞ at $z = -d/c$, and M only takes the value a/c at ∞. Next suppose that z_1 and z_2 are in $\mathbb{C}$ with $M(z_1) = M(z_2) \neq \infty$. It is then straightforward to check, making use of $ad - bc \neq 0$, that $z_1 = z_2$.

A calculation shows that, if $z \in \mathbb{C}_\infty$, then $M(w) = z$ where $w = (-dz + b)/(cz - a)$. In summary, this shows that M is onto and its inverse is again a Möbius transformation, so that $M : \mathbb{C}_\infty \to \mathbb{C}_\infty$ is a bijection. One can also see that $M^{-1}(z) = (-dz + b)/(cz - a)$ since the matrix

$$\begin{pmatrix} -d & b \\ c & -a \end{pmatrix}$$

is, up to a constant, the inverse of the matrix A_M.

2.5

▷ $M(z) = \dfrac{1+z}{1-z}$;
▷ $M(z) = z - 1$;
▷ $M(z) = \dfrac{z}{z - 1 - i}$;
▷ $M(z) = z_1 + z_2 - z$.

2.6

▷ When $x_0 = 0$, the image of the imaginary axis under $z \to 1/z$ is again the imaginary axis.

For $x_0 \neq 0$, the hint suggests that the image should be a circle passing through $1/\infty = 0$ and $1/x_0$, and should be symmetric in the real axis. This points to the circle centre $1/(2x_0)$ and radius $1/(2x_0)$. A point on the line $\Re z = x_0$ has the general form $z = x_0 + iy$, $-\infty < y < \infty$. Then, $1/z = (x_0 - iy)/(x_0^2 + y^2)$. The squared distance between this point and $1/(2x_0)$ is

$$\left(\frac{x_0}{x_0^2 + y^2} - \frac{1}{2x_0}\right)^2 + \frac{y^2}{(x_0 + y^2)^2}$$

which equals $1/(4x_0^2)$, as expected. The image of the line $\Re z = x_0$ is indeed the circle $C(\frac{1}{2x_0}, \frac{1}{2x_0})$.

Solutions to the Exercises

▷ Let $M(z) = (az + b)/(cz + d)$ with $ad - bc \neq 0$. Suppose that M is not the identity. Suppose first that $c = 0$ and normalise so that $d = 1$ and $M(z) = az+b$. Then, ∞ is a fixed point and any fixed point in the complex plane satisfies $(a - 1)z + b = 0$. If $b \neq 0$ this equation has at most one solution. If $b = 0$ then $a \neq 1$, as otherwise M would be the identity, and $z = 0$ is the only solution. In total, there are at most 2 fixed points.

Suppose that $c \neq 0$. Then ∞ is not a fixed point since $M(\infty) = a/c \neq \infty$. Neither is $z = -d/c$ a fixed point since its image under M is ∞. Thus any fixed points lie in the complex plane and are solutions of $(az + b) = z(cz + d)$. Since $c \neq 0$, this is a quadratic in z and therefore has at most two solutions.

▷ $M = M_1^{-1} \circ M_2$ is a Möbius transformation that fixes each of α, β and γ. Indeed, $M(\alpha) = M_1^{-1}(M_2(\alpha)) = M_1^{-1}(M_1(\alpha)) = \alpha$, and similarly for β and γ. By the previous part of this exercise, $M(z) = z$ for $z \in \mathbb{C}_\infty$. Thus, for arbitrary $z \in \mathbb{C}_\infty$, $M_1(z) = M_1(M(z)) = M_2(z)$, as required.

2.7 Let M be a Möbius transformation that fixes the extended real line $\mathbb{R}_\infty$, with M given by (2.6). If $c = 0$, normalise so that $d = 1$ in which case $M(z) = az + b$. Then $M(0) = b$ must be real. Since $M(1)$ is also real, $a = (a + b) - b$ is real as well.

If $c \neq 0$, normalise so that $c = 1$. Then $M(-d) = \infty$, so d is real (otherwise the image of $\mathbb{R}_\infty$ under M is a Euclidean circle). If $d \neq 0$ then $M(0) = b/d$ is real and so b is real. In this case, $0 = M(-b/a)$ so either $a = 0$ or $-b/a$ is real, hence a is real. If, however, $d = 0$ then $M(z) = (az + b)/z$. Then, $M(1) = a + b$ and $M(2) = a + b/2$ are both real from which we conclude that a and b are both real.

In summary, if M fixes the extended real line then it can be written as (2.6) with each of a, b, c, d real. The converse is clear. Any such M satisfies $M(z) = \overline{M(\bar{z})}$.

2.8 Here $f(z) = z^2$ satisfies the conditions of the Schwarz Reflection Principle with $D^+ = \{z : |z| < 1, \Im z > 0\}$, J the interval $(-1, 1)$, and $D^- = \{z : |z| < 1, \Im z < 0\}$ (f is continuous on $D^+ \cup J$, is analytic on D^+ and takes real values on J). Its extension from $D^+ \cup J$ to all of the unit disk $\mathbb{D}$ given by (2.12) is simply $\tilde{f}(z) = z^2$.

Even though f is univalent in D^+, it is not univalent in $D^+ \cup J$ (since $f(x) = f(-x)$, $0 < x < 1$). Moreover, $f(D^+)$ doesn't lie entirely in either $\mathbb{H}$ or $\overline{\mathbb{H}}$. The extension is not univalent in $\mathbb{D}$ as is clear.

Moreover, if we take D^+ to be the sector $\{z : |z| < 1, 0 < \arg z < 3\pi/4\}$ then $f(z) = z^2$ extends continuously to J which is the line segment $(0, 1)$ and is one-to-one on $D^+ \cup J$. The extension $\tilde{f}$ of f to $D = \{z : |z| < 1, -3\pi/4 < \arg z < 3\pi/4\}$ is not univalent, however. What goes wrong here is that $f(D^+)$ intersects both $\mathbb{H}$ and $\overline{\mathbb{H}}$.

On the other hand, if we take $D^+ = \{z : |z| < 1, 0 < \arg z < \pi/2\}$ then all conditions in the Schwarz Reflection Principle are satisfied.

2.9 If $z = x + iy$ then its reflection in the imaginary axis is $-x + iy$. This last equates to $-\bar{z} = -(x - iy)$.

2.10 By (2.15) with $a = i\lambda$ and $r = \lambda$,

$$R_{C_\lambda}(z) = i\lambda + \frac{\lambda^2}{(z - i\lambda)} = i\lambda + \frac{\lambda^2}{\bar{z} + i\lambda} = \left(\frac{i\lambda}{\bar{z} + i\lambda}\right)\bar{z},$$

which tends to $\bar{z}$ as $\lambda \to \infty$.

2.11

▷ First,

$$R_C(z) = a + \frac{r^2}{\bar{z} - \bar{a}} = \overline{M(z)} \text{ where } M(z) = \frac{\bar{a}z + (r^2 - |a|^2)}{z - a}.$$

Here '$ad - bc$' equals $-r^2$ which is non-zero. Since generalised circles are preserved by Möbius transformations and by taking complex conjugates, the reflection of any generalised circle in C will again be a generalised circle.

▷ The circle $C(a + r/2, r/2)$ passes through both a, whose reflection is ∞, and through $a + r$ which says fixed under reflection as it lies on C. The reflection of this circle is therefore a straight line through infinity and $a + r$ and is, by symmetry, the tangent line to C at $a + r$.

▷ With $C = C(0, 1)$, the reflection of 2 is $R_C(2) = 1/2$ and $R_C(\infty) = 0$. The image of the line $\Re z = 2$ in the unit circle is then the circle $C(1/4, 1/4)$ and the image of H is $D(1/4, 1/4)$.

2.12

▷ With a being the centre of the circle $C(z_1, z_2; k)$, $k \neq 1$, using Exercise 2.1 and (2.4)

$$(z_1 - a)\overline{(z_2 - a)} = \frac{k^2}{1 - k^2}(z_2 - z_1) \times \frac{1}{1 - k^2}\overline{(z_2 - z_1)}$$

$$= \frac{k^2}{(1 - k^2)^2}|z_2 - z_1|^2 = r^2.$$

Thus, the condition (2.16) for points z_1 and z_2 to be symmetric in a circle $C(a, r)$ is satisfied in this context. If $k = 1$ then $C(z_1, z_2; 1)$ is the perpendicular bisector of $[z_1, z_2]$ and so z_1 and z_2 are again symmetric.

▷ Let C be the circle $C(a, r)$ and let z_1 lie inside C but not at the centre of C. Set $k = |z_1 - a|/r$ so that $0 < k < 1$, and $z_2 = a + (z_1 - a)/k^2$. Then, $C(z_1, z_2, k)$ is the circle $C(a', r')$ where, by (2.3) and (2.4),

$$a' = \frac{z_1 - k^2 z_2}{1 - k^2} = \frac{z_1 - k^2 a - (z_1 - a)}{1 - k^2} = a$$

Solutions to the Exercises   299

and

$$r' = \frac{k}{1-k^2}|z_2 - z_1| = \frac{1}{k(1-k^2)}|k^2 z_2 - k^2 z_1|$$

$$= \frac{1}{k(1-k^2)}|(1-k^2)(z_1 - a)| = \frac{|z_1 - a|}{k} = r.$$

This shows that, fixing a point z_1 inside the circle $C(a, r)$, this circle can then be described as $C(z_1, z_2, k)$ in terms of the point z_2 and the parameter k. But

$$z_2 = a + \frac{z_1 - a}{k^2} = a + \frac{(z_1 - a)r^2}{|z_1 - a|^2} = R_C(z_1),$$

the reflection of z_1 in $C(a, r)$, by (2.15).

2.13 Suppose that z and z^* are symmetric in C_1 so that $z^* = R_{C_1}(z)$. Then, $M_1^{-1}(z)$ and $M_1^{-1}(z^*)$ are symmetric in $M_1^{-1}(C_1) = \mathbb{R}_\infty$, that is $M_1^{-1}(z^*) = \overline{M_1^{-1}(z)}$. Since $\tilde{F} = M_2 \circ \tilde{f} \circ M_1$, we have $\tilde{F} \circ M_1^{-1} = M_2 \circ \tilde{f}$.

By (2.13), the extension $\tilde{F}$ of F to all of $\tilde{D}$ satisfies $\tilde{F}(w) = \overline{\tilde{F}(\overline{w})}$, $w \in \tilde{D}$. Then,

$$(M_2 \circ \tilde{f})(z) = \tilde{F}(M_1^{-1}(z)) = \overline{\tilde{F}\left(\overline{M_1^{-1}(z)}\right)}$$

$$= \overline{\tilde{F}(M_1^{-1}(z^*))} = \overline{(\tilde{F} \circ M_1^{-1})(z^*)} = \overline{(M_2 \circ \tilde{f})(R_{C_1}(z))}.$$

That is, $M_2(\tilde{f}(z)) = \overline{M_2(\tilde{f}(R_{C_1}(z)))}$, so that $M_2(\tilde{f}(z))$ and $M_2(\tilde{f}(R_{C_1}(z)))$ are symmetric in $\mathbb{R}_\infty$, namely in $M_2(C_2)$. It follows that $\tilde{f}(z)$ and $\tilde{f}(R_{C_1}(z))$ are symmetric in C_2. In other words, $\tilde{f}(z) = R_{C_2}(\tilde{f}(R_{C_1}(z)))$ and (2.17) holds as required.

2.14 There is no such function. Theorem 2.2 applies here with $C_1 = C_2 = C(0, 1)$, with $D = \mathbb{C} \setminus \{0\}$ and with $J = C(0, 1)$. Taking D^+ to be $\mathbb{D} \setminus \{0\}$, the function f is continuous on $D^+ \cup J$, is analytic on D^+ and maps J into J. By the Schwarz Reflection Principle, f extends to an analytic function on $\mathbb{C} \setminus \{0\}$, in fact on $\mathbb{C}$ since f was assumed to be analytic on $\mathbb{D}$. But then, by Cauchy's Theorem, we would have $\int_C f(z)\,dz = 0$ whereas $\int_C \overline{z}\,dz = \int_C dz/z = 2\pi i$.

Exercises from Chap. 3

3.1 Let ϕ be the common latitude of points $\hat{z}$ on the intersection of the sphere and the plane $x_3 = k(r)$. By Proposition 3.1, $\phi = 2 \arctan r - \pi/2$. Then,

$$k(r) = \sin \phi = -\cos(\phi + \pi/2) = \frac{\sin^2(\phi/2 + \pi/4) - \cos^2(\phi/2 + \pi/4)}{\sin^2(\phi/2 + \pi/4) + \cos^2(\phi/2 + \pi/4)}$$

$$= \frac{\tan^2(\phi/2 + \pi/4) - 1}{\tan^2(\phi/2 + \pi/4) + 1}$$

$$= \frac{r^2 - 1}{r^2 + 1}.$$

As $r \to +\infty$, $k(r) \to 1$, so that the preimage of $C(0, r)$ under stereographic projection is a small circle close to the North Pole.

3.2

▷ If $P = (\theta, \phi)$ in spherical coordinates on the sphere then $-P = (\hat{\theta}, -\phi)$ where

$$\hat{\theta} = \begin{cases} \theta + \pi & \text{if } 0 \le \theta < \pi, \\ \theta - \pi & \text{if } \pi \le \theta < 2\pi. \end{cases}$$

▷ Let $\alpha = r_1 e^{it_1}$ and $\beta = r_2 e^{it_2}$ be antipodal, where $0 < r_1, r_2 < \infty$ and $0 \le t_1, t_2 < 2\pi$. Then, by Proposition 3.1 and in spherical coordinates,

$$\mathscr{P}^{-1}(\alpha) = (t_1, 2\arctan r_1 - \tfrac{\pi}{2})$$

$$\mathscr{P}^{-1}(\beta) = (t_2, 2\arctan r_2 - \tfrac{\pi}{2}).$$

By (i), since $\mathscr{P}^{-1}(\alpha)$ and $\mathscr{P}^{-1}(\beta)$ are antipodal on the sphere, comparing the longitudinal coordinates θ of each shows that $t_2 = t_1 + \pi$ if $0 \le t_1 < \pi$ or $t_2 = t_1 - \pi$ if $\pi \le t_1 < 2\pi$. In either case, $e^{it_2} = -e^{it_1}$ since $e^{i\pi} = e^{-i\pi} = -1$.

Comparing the latitude coordinates ϕ, we see that r_1 and r_2 need to satisfy

$$2\arctan r_2 = \pi - 2\arctan r_1$$

which becomes $r_2 = \tan[\pi/2 - \arctan r_1]$. In general,

$$\tan\left(\frac{\pi}{2} - \phi\right) = \frac{\sin(\pi/2 - \phi)}{\cos(\pi/2 - \phi)} = \frac{\cos \phi}{\sin \phi} = \frac{1}{\tan \phi},$$

and so $r_2 = 1/\tan[\arctan r_1] = 1/r_1$.

The conditions on the longitudinal coordinates and the latitudinal coordinates together show that

$$\beta = r_2 e^{it_2} = \frac{1}{r_1}(-e^{it_1}) = -\frac{1}{r_1 e^{-it_1}} = -\frac{1}{\bar{\alpha}}.$$

Solutions to the Exercises

It is straightforward to reverse this argument to check that if $\beta = -1/\bar{\alpha}$ then α and β are antipodal.

Summary: points α and β in the complex plane are antipodal if and only if $\bar{\alpha}\beta = -1$.

3.3 We want to minimise $f(x_1, x_2, x_3) = x_1^2 + x_2^2 + x_3^2$ subject to $ax_1 + bx_2 + cx_3 = d$. By the method of Lagrange multipliers, we look for critical points of

$$g(x_1, x_2, x_3) = f(x_1, x_2, x_3) - \lambda\big[ax_1 + bx_2 + cx_3 - d\big]$$
$$= x_1^2 + x_2^2 + x_3^2 - \lambda ax_1 - \lambda bx_2 - \lambda cx_3 + \lambda d.$$

The partial derivatives are

$$\frac{\partial g}{\partial x_1} = 2x_1 - \lambda a, \quad \frac{\partial g}{\partial x_2} = 2x_2 - \lambda b, \quad \frac{\partial g}{\partial x_3} = 2x_3 - \lambda c.$$

Setting these derivatives to 0, we find that $x_1 = \lambda a/2$, $x_2 = \lambda b/2$, $x_3 = \lambda c/2$. Then, $d = ax_1 + bx_2 + cx_3 = \lambda(a^2 + b^2 + c^2)/2$, so that $\lambda = 2d/(a^2 + b^2 + c^2)$. We find that

$$(x_1, x_2, x_3) = \left(\frac{ad}{a^2 + b^2 + c^2}, \frac{bd}{a^2 + b^2 + c^2}, \frac{cd}{a^2 + b^2 + c^2}\right)$$

is the point on the plane $ax_1 + bx_2 + cx_3 = d$ that is closest to the origin. The squared distance from this point to the origin is

$$x_1^2 + x_2^2 + x_3^2 = \frac{1}{(a^2 + b^2 + c^2)^2}\big[a^2d^2 + b^2d^2 + c^2d^2\big] = \frac{d^2}{a^2 + b^2 + c^2}.$$

The condition for this distance to be less than 1, so that the plane meets the sphere, is $a^2 + b^2 + c^2 > d^2$.

3.4 Let $\hat{\mathscr{C}}$ be the intersection of the plane $x_1 + x_2 + x_3 = 0$ with the sphere $\mathscr{S}$. If $z \in \mathscr{P}(\hat{\mathscr{C}})$, $z = x + iy$, then by (3.3),

$$\mathscr{P}^{-1}(z) = \left(\frac{2x}{|z|^2 + 1}, \frac{2y}{|z|^2 + 1}, \frac{|z|^2 - 1}{|z|^2 + 1}\right) = (x_1, x_2, x_3).$$

Since $x_1 + x_2 + x_3 = 0$, we find that

$$2x + 2y + (|z|^2 - 1) = 0,$$

that is $x^2 + y^2 + 2x + 2y = 1$ or $(x + 1)^2 + (y + 1)^2 = 3$. Thus, $\mathscr{P}(\hat{\mathscr{C}})$ is the circle in the plane with centre $a = -1 - i$ and radius $r = \sqrt{3}$, that is the circle $|z + 1 + i| = \sqrt{3}$.

3.5 Let $\hat{C}_t$ be the circle given by the intersection of the plane $x_1 = t$ with the sphere $\mathscr{S}$. If $z \in \mathscr{P}(\hat{C}_t)$, $z = x + iy$, then (again referring to (3.3))

$$x_1 = \frac{2x}{|z|^2 + 1} = t,$$

that is $2x = t(x^2 + y^2 + 1)$.
  When $t = 0$, then $\mathscr{P}(C_t)$ is the imaginary axis.
  When $t \neq 0$, $-1 < t < 1$, then

$$x^2 + y^2 - \frac{2x}{t} + 1 = 0.$$

Completing the square in x, this becomes the equation

$$\left(x - \frac{1}{t}\right)^2 + y^2 = \frac{1}{t^2} - 1,$$

which is a circle with centre $1/t$ and radius $\sqrt{1/t^2 - 1}$.

3.6 Revisiting the proof of Proposition 3.3, we see from (3.7) with $d = 0$ that if $z = x + iy$ lies on the image of a great circle under stereographic projection then $2ax + 2by + c(|z|^2 - 1) = 0$. Putting $2x = z + \bar{z}$, $2y = i(\bar{z} - z)$ this equation becomes

$$c|z|^2 + (a - ib)z + (a + ib)\bar{z} - c = 0, \quad (\dagger)$$

which has the form $A|z|^2 + \bar{B}z + B\bar{z} + C = 0$ with $A = c$ and $C = -c$ real, $B = a + ib$, and $|B|^2 > AC$ since $a^2 + b^2 > -c^2$). This is the locus (2.1) of a generalised circle with $C = -A$. Conversely, any such generalised circle is the image of a great circle formed as the intersection of the plane $ax_1 + bx_2 + cx_3 = 0$ with the sphere where $a = \Re B$, $b = \Im B$, and $c = A$.
  If $c = 0$ then ($\dagger$) represents a straight line through the origin in the complex plane.
  If $c \neq 0$ then ($\dagger$) represents a circle in the complex plane with, by Proposition 2.1, centre $z_0 = -B/A = -(a + ib)/c$ and radius r where

$$r^2 = \frac{|B|^2 - AC}{A^2} = \frac{a^2 + b^2 + c^2}{c^2} = 1 + |z_0|^2.$$

That is, ($\dagger$) represents the circle $|z - z_0| = \sqrt{1 + |z_0|^2}$ where $z_0 = -a/c - ib/c$. Conversely, any circle of this form arises as the image of the great circle corresponding to the plane $(\Re z_0)x_1 + (\Im z_0)x_2 - x_3 = 0$.

3.7 The North and South poles are interchanged under g. Now consider a point $x = (\theta, \phi)$ in spherical coordinates on the sphere, not the North or South Pole. By

Proposition 3.1, in fact (3.1),

$$\mathscr{P}(x) = \tan\left(\frac{\phi}{2} + \frac{\pi}{4}\right) e^{i\theta}.$$

Then,

$$(f \circ \mathscr{P})(x) = 1/\mathscr{P}(x) = \frac{1}{\tan\left(\frac{\phi}{2} + \frac{\pi}{4}\right)} e^{-i\theta}.$$

Since, in general, $\tan(\pi/2 - t) = 1/\tan t$,

$$(f \circ \mathscr{P})(x) = \tan\left(-\frac{\phi}{2} + \frac{\pi}{4}\right) e^{i(2\pi - \theta)}$$

By (3.2), $(\mathscr{P}^{-1} \circ f \circ \mathscr{P})(x) = (2\pi - \theta, -\phi)$ in spherical coordinates. This corresponds to reflection in the equatorial $x_1 x_2$-plane combined with reflection in the vertical plane $x_1 x_3$-plane. In Cartesian coordinates on the sphere, the composition of these two reflections is $g(x_1, x_2, x_3) = (x_1, -x_2, -x_3)$. Thus, inversion in the extended complex plane corresponds to a rotation on the sphere through an angle π about the x_1-axis.

The continuity of $f(z) = 1/z$ on $\mathbb{C}_\infty$ can be checked by examining the image of the three basic open sets and using the fact that this inversion maps generalised circles to generalised circles. For example, if D is an open disk then $f(D)$ is another open disk if 0 doesn't lie inside D or on its boundary, is an open half plane if 0 lies on the boundary of the disk, and is the exterior of a closed disk together with the point at infinity if 0 lies inside D. Alternatively, writing $f = \mathscr{P} \circ g \circ \mathscr{P}^{-1}$ and noting that g is continuous on the sphere shows that f is continuous.

3.8 Since f has a pole at a, $f(z) \to \infty$ as $z \to a$, so we may choose ε positive such that f is not only analytic but also non-zero in $D'(a, \varepsilon)$. Let N be the order of the pole of f at a. Then, $f(z) = (z-a)^{-N} h(z)$ where h is analytic and non-zero in $D(a, \varepsilon)$. Thus, $g(z) = 1/f(z) = (z-a)^N (1/h(z))$ is analytic in $D(a, \varepsilon)$ and has a zero of order N at a.

By [16, Theorem 10.6], choosing ε smaller if necessary, there is a positive δ such that every ζ in $D'(0, \delta)$ has N distinct preimages under g in the punctured disk $D'(a, \epsilon)$. Set $r = 1/\delta$ and suppose that $|w| > r$. Since $0 < |1/w| < \delta$, there are N distinct points z_i, $i = 1 \ldots N$, with $z_i \in D'(a, \epsilon)$ and $g(z_i) = 1/w$. Then, $f(z_i) = w$ for each i, and we see that w has N preimages under f in the punctured disk $D'(a, \varepsilon)$.

In case f has a pole of order N at ∞, then $g(z) = f(1/z)$ has a pole of order N at 0. Thus, there is a positive ϵ and a finite r such that every w with $|w| > r$ has N preimages in $D'(0, \varepsilon)$ under g. Any such w then has exactly N preimages in $\{z : |z| > 1/\epsilon\}$ under f.

3.9 In each case, the behaviour of $f(z)$ at ∞ corresponds to the behaviour of $g(z) = f(1/z)$ at 0.

▷ With $f(z) = z$, $g(z) = f(1/z) = 1/z$ has a simple pole at 0, so that f has a simple pole at ∞.

▷ With $f(z) = 1/z$, $g(z) = f(1/z) = z$ has a simple zero at 0, so that f has a simple zero at ∞.

▷ With $f(z) = z + 1/z$, $g(z) = f(1/z) = z + 1/z$ has a simple pole at 0, so that f has a simple pole at ∞.

▷ With $f(z) = e^z$, $g(z) = f(1/z) = e^{1/z}$ which has an essential singularity at 0. Thus, f has an essential singularity at ∞.

▷ In this case, $P(z) = a_0 + a_1 z + \cdots + a_n z^n$ with $a_n \neq 0$. Then,

$$g(z) = P(1/z) = a_0 + \frac{a_1}{z} + \cdots + \frac{a_n}{z^n},$$

so that g has a pole of order n at 0, showing that P has a pole of order n at ∞.

▷ Suppose that f has a pole of order n at $a \in \mathbb{C}$. In this case, f can be expanded as $f(z) = (z-a)^{-n} h(z)$ in some punctured disk about a where h is analytic and non-zero in this disk. Then, $1/f(z) = (z-a)^n (1/h(z))$ has a zero of order n at a. Conversely, if $1/f$ has a zero of order n at a then $1/f(z) = (z-a)^n h(z)$ in some disk about a where h is analytic and non-zero in the disk. Taking reciprocals shows that f has a pole of order n at a.

In case $a = \infty$, $f(z)$ has a pole of order n at ∞ if and only if $g(z) = f(1/z)$ has a pole of order n at 0, that is if and only if $1/g(z)$ has a zero of order n at 0, which is in turn if and only if $1/f(z)$ has a zero of order n at ∞, since $g(1/z) = f(z)$.

▷ If f and g both have a pole at $a \in \mathbb{C}$ then, for z in some punctured disk about a, we can expand f and g as $f(z) = (z-a)^{-n_1} h_1(z)$ and $g(z) = (z-a)^{-n_2} h_2(z)$. Here n_1 and n_2 are the orders of the poles of f and g at a respectively, and h_1 and h_2 are analytic near a and non-zero at a. Then, $f + g$ has a pole of order at most the maximum of n_1 and n_2 or a removable singularity, while fg has a pole of order $n_1 + n_2$ at a. The same conclusion applies if $a = \infty$ as this corresponds to the case $a = 0$ but for the functions $f(1/z)$ and $g(1/z)$.

3.10

▷ Writing n and m for the degrees of P and Q respectively, we need to classify the singularity of

$$g(z) = f(1/z) = \frac{P(1/z)}{Q(1/z)} \quad \text{at } z = 0.$$

Now, for coefficients a_0 to a_n, with $a_n \neq 0$,

$$P(1/z) = a_0 + \frac{a_1}{z} + \frac{a_2}{z^2} + \cdots + \frac{a_n}{z^n} = z^{-n} \left[a_0 z^n + a_1 z^{n-1} + a_2 z^{n-2} + \cdots + a_n \right].$$

Solutions to the Exercises 305

Thus $P(1/z) = z^{-n}\tilde{P}(z)$ near 0 where $\tilde{P}$ is a polynomial with $\tilde{P}(0) \neq 0$. Similarly, we can write $Q(1/z) = z^{-m}\tilde{Q}(z)$ where $\tilde{Q}$ is a polynomial with $\tilde{Q}(0) \neq 0$. Then, $g(z) = z^{m-n}h(z)$ where $h = \tilde{P}/\tilde{Q}$ is analytic and non-zero at 0. In particular, g is analytic at 0 (and equivalently f is analytic at ∞) if $n = m$; g has a zero of order $m - n$ at 0 (and equivalently f has a zero of order $m - n$ at ∞) if $\deg(P) < \deg(Q)$; g has a pole of order $n - m$ at 0 (and equivalently f has a pole of order $n - m$ at ∞) if $\deg(P) > \deg(Q)$.

▷ The rational function f has n zeros in the complex plane, counting multiplicity, and m poles there. If $n = m$ then, by the previous part, the numbers of zeros and poles in the complex plane are equal and f has neither a zero nor a pole at ∞. If $n < m$ then f has a zero of order $m - n$ at ∞ and so a total of $n + (m - n) = m$ zeros which is the same as the number of poles, m, in the complex plane. If $m < n$ then f has a pole of order $n - m$ at ∞ and so a total of $m + (n - m) = n$ poles; this is the same as the number n of zeros of f in the complex plane. In all cases, the total number of zeros and poles of f in $\mathbb{C}_\infty$ are the same, and have common value $\max\{\deg(P), \deg(Q)\}$.

3.11 Since f is analytic at ∞, we can expand f around ∞ as

$$f(z) = a_0 + \frac{a_1}{z} + \frac{a_2}{z^2} + \cdots, \quad |z| > R,$$

for some finite R. Then, $f(z) \to a_0$ as $|z| \to \infty$. Thus the entire function f is bounded on $\mathbb{C}$ and is therefore constant by Liouville's Theorem.

Alternatively, by Theorem 3.3, f is a rational function $f = P/Q$ for some polynomials P and Q with no common factors. It must be that Q is constant as any zero of Q would result in a pole of f. Thus, f is a polynomial. But then f has a pole at infinity of order $\deg(P)$ (see Exercise 3.9), and we conclude that the degree of P is zero, again forcing f to be constant.

3.12 Let f be meromorphic on $\mathbb{C}_\infty$. Then, by Theorem 3.3, there are polynomials P and Q with no common factors such that $f = P/Q$. Set $M = \max\{n, m\}$ where $n = \deg(P)$ and $m = \deg(Q)$. We claim that f assumes every value in the extended complex plane exactly M times. Note that every zero of Q is a pole of f, and every zero of P is a zero of f, since the zeros of P and Q do not cancel.

Suppose first that $\deg(P) \neq \deg(Q)$. By Exercise 3.10, f has either a pole or a zero at ∞. Consider first the case of a non-zero complex number a. Then $f(z) = a$ with $z \in \mathbb{C}$ if and only if $P(z) = aQ(z)$, that is $(P - aQ)(z) = 0$. Since $P - aQ$ is a polynomial of degree M, it has exactly M solutions in the complex plane, counting multiplicities, and f does not take the value a at ∞. In the case $a = 0$, the only zeros of f in $\mathbb{C}$ are the n zeros of P. If $n > m$ then f has a pole at ∞, hence a total of $n = M$ zeros in $\mathbb{C}_\infty$. If $n < m$ then f has a zero of order $m - n$ at ∞, hence a total of $n + (m - n) = m = M$ zeros in $\mathbb{C}_\infty$. The last case is $a = \infty$. We have m solutions of $f(z) = \infty$ in the complex plane, namely the m roots of Q. If $n < m$, there is no pole at ∞, so a total of $m = M$ poles. If $n > m$ there is a pole of order $n - m$ at ∞ and so a total of $(n - m) + m = n = M$ poles.

Now suppose that $\deg(P) = \deg(Q) = M$. In this case f is analytic at ∞ and takes the value p_M/q_M where p_M and q_M are the (non-zero) coefficients of z^M in P and Q respectively. Then, f has M poles in $\mathbb{C}_\infty$. For any a in $\mathbb{C}$ except $a = p_M/q_M$, the polynomial $P - aQ$ has degree M and so $f(z) = a$ has exactly M roots in $\mathbb{C}_\infty$. For $a = p_M/q_M$, the polynomial $P - aQ$ has degree $M - k$, say. Working out the expansion of $P(1/z)/Q(1/z) - p_M/q_M$ about 0, one may show that $f(z) - p_M/q_M$ has a zero of order k at ∞. Thus, this value too is taken on exactly M times by f.

Alternatively, we can use Exercise 3.10 to obtain this result. With $f = P/Q$ and n, m and M as above, by Exercise 3.10 f has M poles in $\mathbb{C}_\infty$ counting multiplicity. Let $a \in \mathbb{C}$. We claim that $f(z) = a$ also has M solutions in $\mathbb{C}_\infty$, counting multiplicity. Let $M(z) = 1/(z - a)$ which is, by Exercise 3.7, a homeomorphism of $\mathbb{C}_\infty$ with $M(a) = \infty$. Thus, $f(z) = a$ if and only if $M(f(z)) = M(a) = \infty$, that is if and only if z is a pole of $M \circ f$. Now

$$M(f(z)) = \frac{1}{f(z) - a} = \frac{1}{P(z)/Q(z) - a} = \frac{Q(z)}{P(z) - aQ(z)}.$$

The number of poles of the rational function $M \circ f$ in $\mathbb{C}_\infty$ therefore equals $N = \max\{\deg(Q), \deg(P - aQ)\}$. Considering the cases $n = m$, $n < m$ and $m < n$ separately (for example, if $m < n$ then $\deg(P - aQ) = \deg(P)$) we find that, in all cases, $N = M$. Therefore, $M \circ f$ has exactly M poles in $\mathbb{C}_\infty$, and so f takes the value a exactly M times in $\mathbb{C}_\infty$.

Finally, if f is a bijection then we need $M = \max\{\deg(P), \deg(Q)\} = 1$ so that f is a Möbius transformation.

3.13 Let $\tilde{\gamma}(s)$, $s \in [c, d]$, be an equivalent parameterisation of the curve γ so that $\tilde{\gamma}(s) = \gamma(\phi(s))$ for $s \in [c, d]$ where $\phi \colon [c, d] \to [a, b]$ with $\phi(c) = a$, $\phi(d) = b$, and with $\phi'(s) > 0$ on $[c, d]$. Computed with reference to the parameterisation $\tilde{\gamma}$, the ρ-length of the curve is

$$\int_c^d \rho(\tilde{\gamma}(s)) |\tilde{\gamma}'(s)| \, ds = \int_c^d \rho(\gamma(\phi(s))) |\gamma'(\phi(s))| |\phi'(s)| \, ds$$

$$= \int_c^d \rho(\gamma(\phi(s))) |\gamma'(\phi(s))| \phi'(s) \, ds$$

$$= \int_a^b \rho(\gamma(u)) |\gamma'(u)| \, du,$$

having first used $\phi'(s) > 0$ (so that the absolute value sign was redundant) and then made the substitution $u = \phi(s)$. This last agrees with the ρ-length of the curve computed with respect to the parameterisation γ.

3.14 A parameterisation of $-\gamma$ is

$$\tilde{\gamma}(t) = \gamma(a + b - t), \quad t \in [a, b].$$

Solutions to the Exercises

The ρ-length of the curve $-\gamma$ is then

$$\begin{aligned} L_\rho(-\gamma) &= \int_a^b \rho(\tilde\gamma(t)) \, |\tilde\gamma'(t)| \, dt \\ &= \int_a^b \rho(\gamma(a+b-t)) \, |\gamma'(a+b-t)| \, dt \\ &= -\int_a^b \rho(\gamma(u(t))) \, |\gamma'(u(t))| \, u'(t) \, dt \\ &= \int_a^b \rho(\gamma(u)) \, |\gamma'(u)| \, du = L_\rho(\gamma). \end{aligned}$$

Here we substituted $u(t) = a + b - t$.

3.15 From Definition 3.5 and (3.14),

$$L_\rho(\gamma) = \int_a^b \rho(\gamma(t)) \, |\gamma'(t)| \, dt \leq M \int_a^b |\gamma'(t)| \, dt = M \, L(\gamma).$$

$$L_\rho(\gamma) = \int_a^b \rho(\gamma(t)) \, |\gamma'(t)| \, dt \geq m \int_a^b |\gamma'(t)| \, dt = m \, L(\gamma).$$

3.16 <u>East-West displacement</u> (θ, ϕ) to $(\theta+\Delta\theta, \phi)$. On the sphere, we move through an angle $\Delta\theta$ on a circle of radius $\cos\phi$, so the displacement is $|\Delta\theta|\cos\phi$. In the plane, we move through an angle $\Delta\theta$ on a circle of radius $\tan(\phi/2 + \pi/4)$, so the displacement is $|\Delta\theta|\tan(\phi/2 + \pi/4)$. Now

$$\begin{aligned} \cos\phi &= \cos^2(\phi/2) - \sin^2(\phi/2) \\ &= \big[\cos(\phi/2) + \sin(\phi/2)\big]\big[\cos(\phi/2) - \sin(\phi/2)\big] \\ &= 2\sin(\phi/2 + \pi/4)\cos(\phi/2 + \pi/4). \end{aligned}$$

The magnification factor is therefore

$$M_\theta(\hat z) = \frac{|\Delta\theta|\tan(\phi/2 + \pi/4)}{|\Delta\theta|\cos\phi} = \frac{1}{2\cos^2(\phi/2 + \pi/4)}.$$

<u>North-South displacement</u> (θ, ϕ) to $(\theta, \phi + \Delta\phi)$ on the sphere. On the sphere, we move through an angle $\Delta\phi$ on a circle of radius 1, so the displacement is $|\Delta\phi|$. In the plane, the displacement is radial from $\tan(\phi/2+\pi/4)$ to $\tan((\phi+\Delta\phi)/2+\pi/4)$ which, as in the text, is approximately $(|\Delta\phi|/2)\sec^2(\phi/2+\pi/4)$. The magnification factor is

$$M_\phi(\hat z) = \frac{|\Delta\phi|/2 \, \sec^2(\phi/2 + \pi/4)}{|\Delta\phi|} = \frac{1}{2\cos^2(\phi/2 + \pi/4)}.$$

We see that the magnification factors $M_\theta(\hat{z})$ and $M_\phi(\hat{z})$ are the same.

3.17 $\gamma(t) = t/(1-t)$, $t \in [0, 1]$, is one such parameterisation. Then, $\gamma'(t) = 1/(1-t)^2$ and $\rho(t/(1-t)) = 2(1-t)^2/(2t^2 - 2t + 1)$. The ρ-length of the positive real axis is then given by the integral

$$\int_0^1 \rho(\gamma(t))\, |\gamma'(t)|\, dt = \int_0^1 \rho\left(\frac{t}{1-t}\right) \frac{1}{(1-t)^2}\, dt = \int_0^1 \frac{2}{2t^2 - 2t + 1}\, dt.$$

We complete the square and use $d/dt(\arctan t) = 1/(1+t^2)$ to compute this integral as

$$\int_0^1 \frac{2}{2t^2 - 2t + 1}\, dt = \int_0^1 \frac{4}{1 + (2t-1)^2}\, dt = 2\arctan(2t-1)\Big|_0^1 = \pi.$$

The preimage of the positive real axis $[0, \infty]$ is a semicircle joining the south and north poles on the sphere, which has length π in agreement with the previous computation.

3.18

▷ Consider $g(x) = f(x+r)/f(x) = (1+x^2)/(1+(x+r)^2)$, $x \in [0, \infty)$. Then g is positive and continuous on $[0, \infty)$. Also, $g(x) \to 1$ as $x \to \infty$. Hence there is a positive R such that $g(x) \geq 1/2$ on $[R, \infty)$. Set $c_1 = \min\{g(x) : 0 \leq x \leq R\}$, noting that c_1 is positive. We may then take $c = \min\{c_1, 1/2\}$.

▷ We begin by showing that $\rho(z+a) \leq C\rho(z)$ for all complex z with C depending only on $|a|$. First suppose that $|z| \leq |a|$. Then $\rho(z+a) \leq \rho(0) = 2$ while $\rho(z) \geq \rho(a) = 2/(1+|a|^2)$, and so

$$\rho(z+a) \leq 2 = (1+|a|^2)\rho(a) \leq (1+|a|^2)\rho(z).$$

Next suppose that $|z| > |a|$. Then $|z+a| \geq |z| - |a| > 0$, and

$$\rho(z+a) = \frac{2}{1+|z+a|^2} \leq \frac{2}{1+(|z|-|a|)^2} \leq \frac{1}{c}\frac{2}{1+|z|^2} = \frac{1}{c}\rho(z).$$

Here we used the first part of this exercise with $x = |z| - |a|$ and $r = |a|$ so that c is positive and depends only on $r = |a|$. We may take $C = \max\{1/c, 1+|a|^2\}$.

Replacing a by $-a$ and then z by $z+a$ gives $\rho(z) \leq C\rho(z+a)$.

Let γ be a curve parameterised by $\gamma(t)$ on $[0, 1]$ say. Then,

$$L_\rho(\gamma + a) = \int_0^1 \rho(\gamma(t) + a)\, |\gamma'(t)|\, dt \leq \int_0^1 C\rho(\gamma(t))\, |\gamma'(t)|\, dt = C\, L_\rho(\gamma).$$

The inequality $L_\rho(\gamma) \leq C\, L_\rho(\gamma + a)$ follows as above.

Solutions to the Exercises

Now, since the distance between points is the infimum of the lengths of curves joining them, and $\gamma + a$ joins $z + a$ and $w + a$ if and only if γ joins z and w, the inequalities for spherical distance for points in $\mathbb{C}$ follow. Letting z approach the point at infinity the inequality for all z and w in $\mathbb{C}_\infty$ follows.

▷ First we show that, for $r > 0$, there is a finite constant C depending only on r such that $f(rx) \geq Cf(x)$ for all x in $[0, \infty)$. Set

$$g(x) = \frac{f(rx)}{f(x)} = \frac{1 + x^2}{1 + r^2 x^2}, \quad x \in [0, \infty).$$

If $0 < r \leq 1$ then $g(x) \geq 1$ on $[0, \infty)$. If $r > 1$ then g is decreasing on $[0, \infty)$ since $g'(x) = 2x(1 - r^2)/(1 + r^2 x^2)^2$. Since $\lim_{x \to \infty} g(x) = 1/r^2$ we find that $g(x) \geq \min\{1, 1/r^2\}$, $x \in [0, \infty)$, in all cases.

Now suppose that $a \neq 0$ and that z is complex. Then

$$\frac{\rho(az)}{\rho(z)} = \frac{f(|a||z|)}{f(|z|)} \geq c_1 \text{ where } c_1 = \min\{1, 1/|a|^2\}.$$

Thus, $\rho(az) \geq c_1 \rho(z)$, $z \in \mathbb{C}$. Replacing a by $1/a$ and z by az, we find that $\rho(z)/\rho(az) \geq c_2$ where $c_2 = \min\{1, |a|^2\}$, so that $\rho(az) \leq (1/c_2)\rho(z)$. That is,

$$\min\{1, 1/|a|^2\}\rho(z) \leq \rho(az) \leq \max\{1, 1/|a|^2\}\rho(z), \quad z \in \mathbb{C}.$$

The corresponding inequalities for the lengths of curves follow, namely

$$\min\{|a|, 1/|a|\} L_\rho(\gamma) \leq L_\rho(a\gamma) \leq \max\{|a|, 1/|a|\} L_\rho(\gamma),$$

and then, in turn, the inequalities for spherical distance.

3.19

▷ For points x, y in X, $d(x, y) = 0$ if and only if $x = y$. Suppose that x and y are in X with $x \neq y$. To show that f is injective, we need to show that $f(x) \neq f(y)$. Since $x \neq y$, we have $d(x, y) > 0$. By the isometry property, $d(f(x), f(y)) > 0$ as well, and hence $f(x) \neq f(y)$.

▷ Take $\delta = \varepsilon$ in the ε-δ definition of continuity.

▷ Suppose, in addition, that X is compact. Then $f(X)$ is a compact subset of X (the continuous image of a compact set is compact) and hence is closed. Thus, $X \setminus f(X)$ is open. Suppose, if possible, that f is *not* onto and that $x_0 \in X \setminus f(X)$. Then there is a positive r such that the ball $B(x_0, r) \subseteq X \setminus f(X)$, that is $d(x_0, x) \geq r$ for every x in $f(X)$.

Now consider the orbit of x_0 under f, that is the sequence $\{x_n\}_{n=1}^\infty$ where $x_n = f^n(x_0)$ and f^n is the composition $f \circ f \circ \cdots \circ f$, composed n times.

This sequence has the property that $d(x_m, x_n) \geq r$ whenever $m \neq n$. In fact, for $m < n$ and using the assumption that f is an isometry,

$$d(x_m, x_n) = d(f^m(x_0), f^n(x_0)) = d(f(f^{m-1}(x_0)), f(f^{n-1}(x_0)))$$
$$= d(f^{m-1}(x_0), f^{n-1}(x_0)) = d(x_{m-1}, x_{n-1}).$$

Applying this m times, we find that $d(x_m, x_n) = d(x_0, x_{n-m})$ and the latter is at least r since x_{n-m} is in $f(X)$. It follows that the sequence $\{x_n\}_{n=1}^{\infty}$ has no convergent subsequence, which is impossible since X is compact.

▷ Suppose that f is a bijection. Then, for x and y in X,

$$d(x, y) = d(f(f^{-1}(x)), f(f^{-1}(y))) = d(f^{-1}(x), f^{-1}(y))$$

so that f^{-1} is an isometry. Hence, f^{-1} is continuous and f is a homeomorphism.

Example: Set $X = [0, \infty)$ with the Euclidean metric. Set $f(x) = x + 1$. Then f is an isometry ($|f(x) - f(y)| = |(x+1) - (y+1)| = |x - y|$) but f is not onto since $f(X) = [1, \infty)$.

3.20

▷ With $f(z) = g(z)(z - z_0)^{-n}$,

$$f'(z) = (z - z_0)^{-n-1}\big[(z - z_0)g'(z) - ng(z)\big].$$

Then,

$$f^\#(z) = \frac{2|(z - z_0)g'(z) - ng(z)|}{|z - z_0|^{n+1} + |g(z)|^2|z - z_0|^{1-n}}.$$

The numerator of this quotient approaches $2n|g(z_0)|$ as $z \to z_0$. If $n \geq 2$ the denominator tends to $+\infty$ and so $f^\#(z) \to 0$ as $z \to z_0$ in this case. If $n = 1$, the case of a simple pole, the denominator approaches $|g(z_0)|^2$ and so

$$f^\#(z) \to \frac{2}{|g(z_0)|} = \frac{2}{|\mathrm{Res}(f; z_0)|}.$$

▷ $f^\#$ is continuous wherever f doesn't have a pole, and is continuous at each pole by definition of continuity if defined in this way.

▷ This is a similar calculation to the first part. One finds that $f^\#(z_0) = 0$ if the order of the zero is 2 or higher. If the zero at z_0 is simple then $f^\#(z_0) = 2|g(z_0)|$. Since $1/g(z_0) = \mathrm{Res}(1/f; z_0)$, we see that $f^\#(z_0) = 2/|\mathrm{Res}(1/f; z_0)|$ in this case.

Solutions to the Exercises 311

▷ Suppose first that f has neither a pole nor a zero at z. Then,

$$(1/f)^{\#}(z) = \frac{2|(1/f)'(z)|}{1+|(1/f)(z)|^2} = \frac{2|f'(z)|/|f(z)|^2}{1+1/|f(z)|^2} = \frac{2|f'(z)|}{|f(z)|^2+1} = f^{\#}(z).$$

If f has a pole (zero) of order 2 or higher at z_0 then $1/f$ has a zero (pole) of order 2 or higher at z_0 and both spherical derivatives are zero in this case. If f has a simple pole (zero) at z_0 then $1/f$ has a simple zero (pole) at z_0 and the spherical derivatives have the common value $2/|\text{Res}(f; z_0)|$ ($2/|\text{Res}(1/f; z_0)|$).

3.21 A solution to this exercise is given in the text immediately after the exercise itself.

3.22

▷ Let

$$M(z) = \frac{az+b}{\bar{a}-\bar{b}z}, \quad \text{where } |a|^2 + |b|^2 = 1.$$

We need to check that $\rho(M(z))|M'(z)| = \rho(z)$. Now, $M'(z) = 1/(\bar{a}-\bar{b}z)^2$, so that

$$\rho(M(z))|M'(z)| = \frac{2}{1+\left|\frac{az+b}{\bar{a}-\bar{b}z}\right|^2} \cdot \frac{1}{|\bar{a}-\bar{b}z|^2}$$

$$= \frac{2}{|\bar{a}-\bar{b}z|^2 + |az+b|^2}$$

$$= \frac{2}{(|a|^2+|b|^2)(1+|z|^2)} = \frac{2}{1+|z|^2} = \rho(z).$$

▷ Yes, they form a subgroup of the full group of Möbius transformations since the composition of two isometries is again an isometry. In fact, writing $M_{a,b}$ for $M(z) = (az+b)/(\bar{a}-\bar{b}z)$, we find using Exercise 2.3 that

$$M_{a,b} \circ M_{c,d} = M_{ac-b\bar{d},ad+b\bar{c}}.$$

3.23 Referring back to Exercise 3.6, fix a great circle C given as the locus of $A|z|^2 + \bar{B}z + B\bar{z} - A = 0$ where A is real. Let M be an isometry of the spherical metric as described by (3.39). To show that $M(C)$ is also a great circle, we need to show that if $w \in M(C)$ then w also satisfies an equation of the form $A'|w|^2 + \bar{B'}w + B'\bar{w} - A' = 0$ where A' is real. Since $M^{-1}(w)$ lies on C, we have

$$A|M^{-1}(w)|^2 + \bar{B}M^{-1}(w) + B\overline{M^{-1}(w)} - A = 0.$$

If M is a rotation, $M(z) = e^{i\theta}z$, or if M is the inversion $M(z) = 1/z$, it is clear that w satisfies the required equation.

The main case is if $M(z) = (\alpha - z)/(1 + \overline{\alpha}z)$, $\alpha \in \mathbb{C}$, in which case $M^{-1}(w) = M(w)$. Substituting $M(w)$ for z in $A|z|^2 + \overline{B}z + B\overline{z} - A = 0$ we find, after a calculation, that w satisfies $A'|w|^2 + \overline{B'}w + B'\overline{w} - A' = 0$ with $A' = A(1 - |\alpha|^2) - \alpha\overline{B} - \overline{\alpha}B$ and $B' = -2\alpha A + \overline{B}\alpha^2 - B$ showing that, indeed, $M(C)$ is a great circle in this case also.

The condition that two points w_1 and w_2 are antipodal is that $w_1\overline{w_2} = -1$. That antipodal points are preserved under an isometry is again clear in the case of a rotation or an inversion. Now consider $M(z) = (\alpha - z)/(1 + \overline{\alpha}z)$, $\alpha \in \mathbb{C}$. Then,

$$M(-1/\overline{w}) = \frac{\overline{w}\alpha + 1}{\overline{w} - \overline{\alpha}} = -1/\overline{M(w)},$$

showing that the images of antipodal points under M are again antipodal.

3.24

▷ Here $\tilde{\gamma}(t) = e^{i\theta}\gamma(t)$, $t \in [a, b]$. Thus, $|\tilde{\gamma}'(t)| = |e^{i\theta}\gamma'(t)| = |\gamma'(t)|$ and so

$$L_\rho(\tilde{\gamma}) = \int_a^b \rho(\tilde{\gamma}(t)) |\tilde{\gamma}'(t)| \, dt$$

$$= \int_a^b \rho(e^{i\theta}\gamma(t)) |\gamma'(t)| \, dt = \int_a^b \rho(\gamma(t)) |\gamma'(t)| \, dt = L_\rho(\gamma).$$

Here we used the rotation invariance of ρ when equating $\rho(e^{i\theta}\gamma(t))$ with $\rho(\gamma(t))$.

▷ Suppose that Γ is a geodesic in the spherical metric and let θ be real. Let γ be a subcurve of Γ (as considered in Definition 3.7) that is the shortest curve between its endpoints. Then $e^{i\theta}\gamma$ is a subcurve of $e^{i\theta}\Gamma$ and is the shortest curve between its endpoints; if there were a shorter curve joining the endpoints of $e^{i\theta}\gamma$ then the rotation of this curve through $-\theta$ would result, by the first part of this exercise, in a curve shorter than γ and joining the endpoints of γ. This shows that $e^{i\theta}\Gamma$ is a geodesic if Γ is.

▷ Suppose that z and w are complex numbers with the same modulus r. With $z = re^{it_1}$ and $w = re^{it_2}$, we have $w = e^{i\theta}z$ with $\theta = t_2 - t_1$. Let γ be any curve joining 0 to z. Then $e^{i\theta}\gamma$ is a curve of the same ρ-length that joins 0 to w. Taking the infimum of the lengths of all such curves γ gives $d^\#(0, z)$. Thus,

$$d^\#(0, z) = \inf_\gamma L_\rho(\gamma) = \inf_\gamma L_\rho(e^{i\theta}\gamma) \geq \inf_{\tilde{\gamma}} L_\rho(\tilde{\gamma}) = d^\#(0, w),$$

where the last infimum is over *all* curves $\tilde{\gamma}$ joining 0 to w. Similarly, $d^\#(0, w) \geq d^\#(0, z)$ as $z = e^{-i\theta}w$. Thus, $d^\#(0, z) = d^\#(0, w)$, as required.

Solutions to the Exercises 313

▷ Again suppose that $|z| = |w|$ so that we can write $w = e^{i\theta}z$. Let γ be any curve joining z to ∞. Then $e^{i\theta}\gamma$ is a curve of the same ρ-length joining w to ∞. Let $\tilde{\gamma}$ be any curve joining w to ∞. Then $e^{-i\theta}\tilde{\gamma}$ is a curve of the same ρ-length joining z to ∞. Thus, for any curve joining z to ∞ we can find a curve of the same ρ-length joining w to ∞, and vice versa. It follows that the infimum of the ρ-lengths of all curves joining z to ∞ equals the infimum of the ρ-lengths of all curves joining w to ∞, that is $d^{\#}(z, \infty) = d^{\#}(w, \infty)$.

3.25 By Theorem 3.7, all geodesic arcs through the antipodal points 1 and -1 are images of a ray from the origin under the isometry $M(z) = (1-z)/(1+z)$ (which maps 1 to 0 and maps -1 to ∞—here $\alpha = 1$). Suppose that this ray makes an angle θ with the positive real axis, where $-\pi < \theta < \pi$. The image of the ray $R_\theta = \{re^{i\theta} : 0 \leq r \leq \infty\}$ under the Möbius map M is part of a circle that passes through the points $M(0) = 1$, $M(\infty) = -1$ and $M(e^{i\theta})$. Now

$$M(e^{i\theta}) = \frac{1 - e^{i\theta}}{1 + e^{i\theta}} = \frac{(1 - e^{i\theta})(1 - e^{-i\theta})}{(1 + e^{i\theta})(1 - e^{-i\theta})} = \frac{1 - \cos\theta}{i \sin\theta} = -i\tan(\theta/2).$$

For $-\pi < \theta < \pi$, this geodesic is the circular arc from 1 to $-i\tan(\theta/2)$ to -1 (when $\theta = 0$ this 'circular arc' reduces to the line segment from 1 to -1). For $\theta = \pi$, the image of the negative real axis under M is the curve from 1 along the positive real axis to the point at infinity followed by the negative real axis from the point at infinity to -1.

3.26 By (3.44),

$$d^{\#}(x, u) = 2\arctan\left|\frac{x-u}{1+xu}\right| \quad \text{and} \quad d^{\#}(u, y) = 2\arctan\left|\frac{u-y}{1+yu}\right|.$$

Since arctan is strictly increasing on $\mathbb{R}$, we require

$$\left|\frac{x-u}{1+xu}\right| = \left|\frac{u-y}{1+yu}\right|,$$

which leads to either

(a) $\dfrac{x-u}{1+xu} = \dfrac{u-y}{1+yu}$ or (b) $\dfrac{x-u}{1+xu} = -\dfrac{u-y}{1+yu}.$

The possibility (b) leads to $1 + u^2 = 0$, which has no solution. The possibility (a) leads to the quadratic

$$(x+y)u^2 + 2(1-xy)u - (x+y) = 0.$$

If $x+y=0$ this has as its only solution $u=0$, which lies midway between x and $y=-x$. Otherwise,

$$u = \frac{-1+xy \pm \sqrt{(1+x^2)(1+y^2)}}{x+y}.$$

There are now two cases depending on the sign of $x+y$. Note that, since $x<y$, we have $x^2 < y^2$ if $x+y>0$ and $x^2 > y^2$ if $x+y<0$.

If $x+y>0$ then $x<u<y$ becomes, after multiplying across by $x+y$,

$$1+x^2 < \pm\sqrt{(1+x^2)(1+y^2)} < 1+y^2.$$

This inequality doesn't hold if we take the minus sign, and it does if we take the plus sign since $x^2 < y^2$.

If $x+y<0$ then $x<u<y$ becomes, after multiplying across by $x+y$,

$$1+x^2 > \pm\sqrt{(1+x^2)(1+y^2)} > 1+y^2.$$

Again, this inequality doesn't hold if we take the minus sign, and it does if we take the plus sign since now $x^2 > y^2$. In either case,

$$u = \frac{-1+xy + \sqrt{(1+x^2)(1+y^2)}}{x+y},$$

lies equidistant between x and y.

3.27 Let $\hat{C}_1$ and $\hat{C}_2$ be great circles on the sphere. We show that their images under stereographic projection, say C_1 and C_2, intersect. Exercise 3.6 gives a complete description of the projection of a great circle on the sphere. If both C_1 and C_2 are distinct Euclidean lines that pass through 0, then they intersect at the antipodal points 0 and ∞.

If C_1, say, is a Euclidean line through the origin and C_2 is a Euclidean circle given, for some z_2 in $\mathbb{C}$, by

$$|z - z_2| = \sqrt{1+|z_2|^2},$$

then the point 0 lies inside C_2 (as $|z_2| < \sqrt{1+|z_2|^2}$) and so the line C_1 that passes through 0 must intersect C_2, and does so exactly twice.

Finally, suppose that C_1 is also a Euclidean circle of the form, for some z_1 in $\mathbb{C}$,

$$|z - z_1| = \sqrt{1+|z_1|^2}.$$

Solutions to the Exercises

To see that these circles intersect, notice that the distance between their centres is less than the sum of their radii, that is

$$|z_1 - z_2| \leq |z_1| + |z_2| < \sqrt{1 + |z_1|^2} + \sqrt{1 + |z_2|^2}.$$

In all cases, the points of intersection are antipodal points. Map one of the points of intersection to the origin by an isometry of the spherical metric. By Exercise 3.23, both C_1 and C_2 are mapped to great circles that pass through 0 and so are mapped to Euclidean straight lines whose second point of intersection is therefore the point at infinity, which is antipodal to 0. Since, again by Exercise 3.23, isometries preserve antipodal points, the original points of intersection of C_1 and C_2 are also antipodal.

Exercises from Chap. 4

4.1 Consider

$$h(z) = \frac{1}{r_2} f(r_1 z), \quad z \in \mathbb{D}.$$

Then $h : \mathbb{D} \to \mathbb{D}$ and $h(0) = 0$. By Schwarz's Lemma, $|h'(0)| \leq 1$, so $(r_1/r_2)|f'(0)| \leq 1$, that is $|f'(0)| \leq r_2/r_1$. Again by Schwarz's Lemma, $|h(z)| \leq |z|$, that is $|f(r_1 z)| \leq r_2|z|$ for $|z| < 1$. If $w \in D(0, r_1)$, apply this last with $z = w/r_1$ to deduce that $|f(w)| \leq r_2|z| = (r_2/r_1)|w|$.

4.2

▷ Suppose that $h(a_1) = h(a_2)$, with a_1 and a_2 in A. Then, $g(f(a_1)) = g(f(a_2))$ with $f(a_1)$ and $f(a_2)$ both in B. Since g is injective, we may deduce that $f(a_1) = f(a_2)$. Then, since f is injective, $a_1 = a_2$. Hence, h is injective.

To see that h is into, let $c \in C$ be arbitrary. Since g is an onto function, there is some b in B with $g(b) = c$. Then, since f is onto, there is some a in A with $f(a) = b$. Putting these together, $h(a) = g(f(a)) = g(b) = c$, and so h assumes every value in C.

▷ Aut(D) contains the identity map Id$(z) = z$, $z \in D$. If f and g are in Aut(D) then so is their composition. Also, if $f \in $ Aut(D) then f^{-1} is also analytic (as $f' \neq 0$ in D) and so $f^{-1} \in $ Aut(D). Hence, each f in Aut(D) has an inverse since $f^{-1} \circ f = $ Id on D.

4.3 Since $|\beta| = 1$,

$$|M(z)| < 1 \iff \left|\frac{\alpha - z}{1 - \overline{\alpha}z}\right| < 1$$

$$\iff |\alpha - z|^2 < |1 - \overline{\alpha}z|^2$$

$$\iff |\alpha|^2 + |z|^2 - 2\Re(\overline{\alpha}z) < 1 + |\alpha|^2|z|^2 - 2\Re(\overline{\alpha}z)$$

$$\iff 1 + |\alpha|^2|z|^2 - |\alpha|^2 - |z|^2 > 0$$

$$\iff (1 - |\alpha|^2)(1 - |z|^2) > 0.$$

Since $|\alpha| < 1$ by assumption, we conclude that $|M(z)| < 1$ if and only if $|z| < 1$.

Next we compute M^{-1}. We solve for z the equation $w = M(z) = \beta(\alpha - z)/(1 - \overline{\alpha}z)$ and find that

$$z = \frac{\beta\alpha - w}{\beta - \overline{\alpha}w}.$$

Noting that $\overline{\beta} = 1/\beta$, we set

$$M^{-1}(w) = \overline{\beta}\,\frac{\beta\alpha - w}{1 - (\overline{\beta}\alpha)w}.$$

This is of the form (4.5). So, given $w \in \mathbb{D}$, set $z = M^{-1}(w)$. Then, $z \in \mathbb{D}$ (by the previous part of this exercise) and (as one may check directly) $M(z) = w$. Thus, M maps $\mathbb{D}$ onto $\mathbb{D}$.

4.4 With $M(z) = \beta(\alpha - z)/(1 - \overline{\alpha}z)$, we have

$$M'(z) = \beta\,\frac{|\alpha|^2 - 1}{(1 - \overline{\alpha}z)^2}.$$

Putting $z = 0$ and $z = \alpha$ in turn leads to the given expressions for $M'(0)$ and $M'(\alpha)$.

4.5 Expanding $1/(1 - \overline{\alpha}z)$ in a geometric series leads to

$$M(z) = \frac{\alpha - z}{1 - \overline{\alpha}z}$$

$$= (\alpha - z)(1 + \overline{\alpha}z + \overline{\alpha}^2 z^2 + \overline{\alpha}^3 z^3 + \ldots)$$

$$= \alpha + (|\alpha|^2 - 1)(z + \overline{\alpha}z^2 + \overline{\alpha}^2 z^3 + \ldots).$$

Solutions to the Exercises   317

4.6 By Schwarz's Lemma, $|a_1| = |f'(0)| \le 1$ with equality if and only if $f(z) = \beta z$, that is when g is constant. Since this case has been put to one side, $|a_1| < 1$. Hence,

$$M_{a_1}(z) = \frac{a_1 - z}{1 - \overline{a_1}z}, \quad z \in \mathbb{D},$$

is an automorphism of the unit disk $\mathbb{D}$ by Proposition 4.1. Since g is a self-map of D so is $h = M_{a_1} \circ g$. Finally, $h(0) = M_{a_1}(g(0)) = M_{a_1}(a_1) = 0$, so h satisfies the assumptions of Schwarz's Lemma.

Thus, by Schwarz's Lemma, $|h'(0)| \le 1$. But, $h'(z) = M'_{a_1}(g(z))\, g'(z)$ and so

$$|h'(0)| = |M'_{a_1}(g(0))|\,|g'(0)| = \left|\frac{1}{|a_1|^2 - 1}\right||a_2| = \frac{|a_2|}{1 - |a_1|^2} \le 1.$$

That is, $|a_2| \le 1 - |a_1|^2$. By the case of equality in Schwarz's Lemma, equality holds if and only if $h(z) = \beta z$ for some β of modulus 1. That is, $(M_{a_1} \circ g)(z) = \beta z$ so that $g(z) = M_{a_1}^{-1}(\beta z) = M_{a_1}(\beta z)$. Since $g(z) = f(z)/z$ for $z \ne 0$, we find that

$$f(z) = z\frac{a_1 - \beta z}{1 - \overline{a_1}\beta z} = \beta z \frac{a_1\overline{\beta} - z}{1 - \overline{a_1}\beta z}, \quad z \in \mathbb{D},$$

for some β of modulus 1. If $g(z) = \beta$ with $|\beta| = 1$, then $|a_1| = 1$ and $|a_2| = 0$, so that equality holds in $|a_2| \le 1 - |a_1|^2$ in this case also.

Setting $\alpha = a_1\overline{\beta}$ we see that, in summary, if equality holds in (4.6) then there are α in $\mathbb{D}$ and β of modulus 1 such that

$$f(z) = \beta z \frac{\alpha - z}{1 - \overline{\alpha}z}, \quad z \in \mathbb{D}.$$

The converse follows from the power series expansion in Exercise 4.5.

4.7 The map $\phi(z) = Rz$ is a conformal map of the unit disk $\mathbb{D}$ onto the disk $D(0, R)$, with inverse $\phi^{-1}(z) = z/R$. Let $g \in \text{Aut}(\mathbb{D})$. Then $f = \phi \circ g \circ \phi^{-1}$ is in $\text{Aut}(D(0, R))$. Conversely, if $f \in \text{Aut}(D(0, R))$ then $g = \phi^{-1} \circ f \circ \phi$ is a map in $\text{Aut}(\mathbb{D})$ and $f = \phi \circ g \circ \phi^{-1}$. In summary, $f \in \text{Aut}(D(0, R))$ if and only if it has the form $\phi \circ g \circ \phi^{-1}$ where $g \in \text{Aut}(\mathbb{D})$.

By Theorem 3.2, if $g \in \text{Aut}(\mathbb{D})$ then g has the form $g(w) = \beta M_{\alpha'}(w)$, $w \in \mathbb{D}$, for some β of modulus 1 and some $\alpha' \in \mathbb{D}$. Then, for $z \in D(0, R)$,

$$f(z) = \left(\phi \circ g \circ \phi^{-1}\right)(z) = Rg(z/R)$$
$$= \beta R M_{\alpha'}(z/R)$$
$$= \beta R \frac{\alpha' - z/R}{1 - \overline{\alpha'}z/R} = \beta \frac{R\alpha' - z}{1 - \overline{\alpha'}z/R}$$

Thus, if we set $\alpha = R\alpha'$, we see that the general form of an automorphism of $D(0, R)$ has the form

$$f(z) = \beta \frac{\alpha - z}{1 - \bar{\alpha}z/R^2}, \quad z \in D(0, R),$$

where β has modulus 1 and $\alpha \in D(0, R)$.

4.8 For $\alpha \in \mathbb{D}$ and $|\beta| = 1$, we can write

$$\beta \frac{\alpha - z}{1 - \bar{\alpha}z} = (-\beta)\frac{\alpha - z}{\bar{\alpha}z - 1} = (-\beta)\frac{az + b}{\bar{b}z + \bar{a}}$$

where $a = -1/\sqrt{1 - |\alpha|^2}$ and $b = \alpha/\sqrt{1 - |\alpha|^2}$. Also, $|-\beta| = 1$ and $|a|^2 - |b|^2 = 1$.

Conversely, if $|\beta| = 1$ and $|a|^2 - |b|^2 = 1$ then

$$\beta \frac{az + b}{\bar{b}z + \bar{a}} = (-\beta)\frac{-az - b}{\bar{b}z + \bar{a}} = \left(-\beta \frac{a}{\bar{a}}\right)\frac{\alpha - z}{1 - \bar{\alpha}z}$$

with $\alpha = -b/a$. Of course, $-\beta a/\bar{a}$ has modulus 1 ($\bar{a}$ can't be zero since $|a|^2 = |b|^2 + 1 \geq 1$). Moreover, $|\alpha|^2 = |b|^2/|a|^2 = (|a|^2 - 1)/|a|^2 = 1 - 1/|a|^2 < 1$.

4.9 M is certainly analytic in $\mathbb{D}$ as the only point at which M is not analytic is 1. Also, since M is a Möbius transformation and Möbius transformations are univalent, so is M.

Now we show that M maps $\mathbb{D}$ into $\mathbb{H}$. Suppose that $z \in \mathbb{D}$. Then

$$\Im M(z) = \Re\left(\frac{1+z}{1-z}\right) = \frac{1}{2}\left(\frac{1+z}{1-z} + \frac{1+\bar{z}}{1-\bar{z}}\right) = \frac{1 - |z|^2}{|1 - z|^2},$$

which is positive. That is, $M(z)$ lies in the upper half-plane.

Finally, we show that $M: \mathbb{D} \to \mathbb{H}$ is onto. So let $w \in \mathbb{H}$ be arbitrary. Solving $M(z) = w$, we find that $z = (w - i)/(w + i)$. Then,

$$|z|^2 = \left|\frac{w - i}{w + i}\right|^2 = \frac{(w - i)(\bar{w} + i)}{(w + i)(\bar{w} - i)} = \frac{|w|^2 + 1 - 2\Im w}{|w|^2 + 1 + 2\Im w} < 1,$$

since, in general, $2\Im w = i\bar{w} - iw$ and here $\Im w > 0$. Thus, $z \in \mathbb{D}$ and $M(z) = w$, verifying that M is onto. Moreover,

$$M^{-1}(w) = \frac{w - i}{w + i}, \quad w \in \mathbb{H}.$$

Solutions to the Exercises

4.10 Suppose that

$$f(z) = \frac{az+b}{cz+d}, \quad z \in \mathbb{H},$$

where a, b, c and d are real and $ad - bc$ is positive. If $c = 0$ then f is linear and analytic everywhere. If $c \neq 0$, the only point where f is not analytic is $z = -d/c$ which lies on the real axis since c and d are real. Thus, f is analytic on $\mathbb{H}$.

Now we check that f maps $\mathbb{H}$ into $\mathbb{H}$. Let $z \in \mathbb{H}$. Then,

$$\Im f(z) = \frac{1}{2i}\left(\frac{az+b}{cz+d} - \frac{a\bar{z}+b}{c\bar{z}+d}\right) = \frac{(ad-bc)\Im z}{|cz+d|^2},$$

which is positive since $ad - bc$ and $\Im z$ are both positive.

Being a Möbius transformation, f is automatically univalent. Finally, we check that f maps $\mathbb{H}$ *onto* $\mathbb{H}$. Let $w \in \mathbb{H}$ and set $z = (dw-b)/(a-cw) = f^{-1}(w)$. Then, $\Im z = (ad-bc)\Im w/|a-cw|^2$, which is positive and shows that $z \in \mathbb{H}$. Since $f(z) = w$, we see that $f : \mathbb{H} \to \mathbb{H}$ is onto.

4.11 Let ϕ be an automorphism of the unit disk $\mathbb{D}$. By Theorem 4.2, there is a β of modulus 1 and an α in $\mathbb{D}$ for which ϕ has the form

$$\phi(z) = \beta\frac{\alpha - z}{1 - \bar{\alpha}z}.$$

Then,

$$\phi'(z) = \beta\frac{|\alpha|^2 - 1}{(1 - \bar{\alpha}z)^2},$$

and

$$\begin{aligned}\frac{|\phi'(z)|}{1 - |\phi(z)|^2} &= \frac{1 - |\alpha|^2}{|1 - \bar{\alpha}z|^2} \bigg/ \left(1 - \frac{|\alpha - z|^2}{|1 - \bar{\alpha}z|^2}\right) \\ &= (1 - |\alpha|^2) \bigg/ \left(|1 - \bar{\alpha}z|^2 - |\alpha - z|^2\right) \\ &= (1 - |\alpha|^2) \bigg/ \left[1 + |\alpha|^2|z|^2 - \bar{\alpha}z - \alpha\bar{z} - (|\alpha|^2 + |z|^2 - \bar{\alpha}z - \alpha\bar{z})\right] \\ &= (1 - |\alpha|^2) \bigg/ \left[(1 - |\alpha|^2)(1 - |z|^2)\right] \\ &= \frac{1}{1 - |z|^2}.\end{aligned}$$

Next, since $\beta\bar\beta = 1$,

$$\frac{\phi(z) - \phi(w)}{1 - \overline{\phi(z)}\phi(w)} = \beta \frac{\frac{\alpha-z}{1-\bar\alpha z} - \frac{\alpha-w}{1-\bar\alpha w}}{1 - \left(\frac{\bar\alpha-\bar z}{1-\alpha\bar z}\right)\left(\frac{\alpha-w}{1-\bar\alpha w}\right)}$$

$$= \beta \frac{(1-\alpha\bar z)\bigl[(\alpha-z)(1-\bar\alpha w) - (\alpha-w)(1-\bar\alpha z)\bigr]}{(1-\bar\alpha z)\bigl[(1-\alpha\bar z)(1-\bar\alpha w) - (\bar\alpha-\bar z)(\alpha-w)\bigr]}$$

$$= \beta \frac{(1-\alpha\bar z)\bigl[-z - |\alpha|^2 w + w + |\alpha|^2 z\bigr]}{(1-\bar\alpha z)\bigl[1 + |\alpha|^2 \bar z w - |\alpha|^2 - \bar z w\bigr]}$$

$$= \beta \frac{(1-\alpha\bar z)(1-|\alpha|^2)(w-z)}{(1-\bar\alpha z)(1-|\alpha|^2)(1-\bar z w)} = \beta \frac{(1-\alpha\bar z)(w-z)}{(1-\bar\alpha z)(1-\bar z w)}.$$

Since $|1 - \alpha\bar z| = |1 - \bar\alpha z|$,

$$\left|\frac{\phi(z) - \phi(w)}{1 - \overline{\phi(z)}\phi(w)}\right| = \left|\frac{z-w}{1-\bar z w}\right|,$$

which is (4.18).

4.12 Suppose that $\rho(\phi(z))\,|\phi'(z)| = \rho(z)$ for each $z \in \mathbb{D}$ and for each $\phi \in \operatorname{Aut}(\mathbb{D})$. If $|w| = |z|$, then $w = e^{i\theta}z$ for some θ. Choose $\phi(z) = e^{i\theta}z$ (a rotation) and conclude that $\rho(z) = \rho(\phi(z))|\phi'(z)| = \rho(w)$ since $|\phi'(z)| = 1$. This shows that $\rho(w) = \rho(z)$ if $|w| = |z|$ and so ρ is a radial function.

Now let $0 < r < 1$ and choose ϕ to be the automorphism $\phi(z) = (r-z)/(1-rz)$, for which $\phi(r) = 0$. Also, (see Exercise 4.4), $|\phi'(r)| = 1/(1-r^2)$. Thus,

$$\rho(r) = \rho(\phi(r))|\phi'(r)| = \rho(0)/(1-r^2).$$

Together these two facts show that, for any z in $\mathbb{D}$, $\rho(z) = \rho(|z|) = \rho(0)/(1-|z|^2)$.

4.13 The tangents to $\mathbb{T}$ and to $C(a, r)$ at $e^{i\theta}$ meet at right angles. Therefore, the radii of $\mathbb{T}$ and of $C(a, r)$ to $e^{i\theta}$ also meet at right angles. With the circles omitted, we have the following triangle:

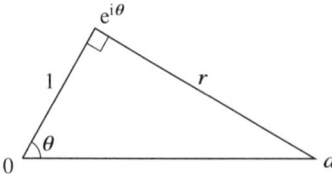

By the Sine Rule, $a = 1/\sin(\frac{\pi}{2}-\theta) = 1/\cos\theta$ and then, by Pythagorus, $1+r^2 = a^2 = 1/\cos^2\theta$, giving $r = \tan\theta$.

Solutions to the Exercises

In the general case, rotate by $e^{-i(\theta_1+\theta_2)/2}$. Then, $e^{i\theta_2}$ rotates to $e^{i\theta}$, and $e^{i\theta_1}$ rotates to $e^{-i\theta}$, where $\theta = (\theta_2 - \theta_1)/2$. Using the immediately preceding calculation and rotating back, the circle is $C(a, r)$ where $a = e^{i(\theta_1+\theta_2)/2}/\cos\theta$ and $r = \tan\theta$.

4.14 The inequality

$$|f'(\alpha)| \le \frac{1 - |f(\alpha)|^2}{1 - |\alpha|^2}.$$

can be rewritten as

$$\frac{|f'(\alpha)|}{1 - |f(\alpha)|^2} \le \frac{1}{1 - |\alpha|^2}.$$

In terms of the density ρ of the hyperbolic metric, this says that if f is a self-map of $\mathbb{D}$ then

$$\rho(f(z))|f'(z)| \le \rho(z), \quad z \in \mathbb{D},$$

with equality if and only if f is an automorphism of $\mathbb{D}$. This is (4.19) but for a general self-map of $\mathbb{D}$ rather than an automorphism alone.

4.15 For α and β in $\mathbb{H}$,

$$d(\alpha, \beta; \mathbb{H}) := d(g(\alpha), g(\beta); \mathbb{D}),$$

where g is a conformal map of $\mathbb{H}$ onto $\mathbb{D}$. Recall that $d(z, w; \mathbb{D})$ *is* a metric on $\mathbb{D}$.

(i) Since $d(z, w; \mathbb{D}) \ge 0$ whenever z and w are in $\mathbb{D}$, $d(\alpha, \beta; \mathbb{H}) \ge 0$ whenever α and β are in $\mathbb{H}$. Moreover, if $d(\alpha, \beta; \mathbb{H}) = 0$ then $d(g(\alpha), g(\beta); \mathbb{D}) = 0$, from which is follows that $g(\alpha) = g(\beta)$. Since g is injective, we deduce that $\alpha = \beta$.

(ii) For α and β in $\mathbb{H}$,

$$d(\alpha, \beta; \mathbb{H}) = d(g(\alpha), g(\beta); \mathbb{D}) = d(g(\beta), g(\alpha); \mathbb{D}) = d(\beta, \alpha; \mathbb{H}),$$

where we used $d(z, w; \mathbb{D}) = d(w, z; \mathbb{D})$ for z and w in $\mathbb{D}$.

(iii) Using the triangle inequality for the metric $d(\cdot, \cdot; \mathbb{D})$, we obtain the triangle inequality for $d(\cdot, \cdot; \mathbb{H})$ as follows. For α, β and γ in $\mathbb{H}$,

$$\begin{aligned} d(\alpha, \beta; \mathbb{H}) &= d(g(\alpha), g(\beta); \mathbb{D}) \\ &\le d(g(\alpha), g(\gamma); \mathbb{D}) + d(g(\gamma), g(\beta); \mathbb{D}) \\ &= d(\alpha, \gamma; \mathbb{H}) + d(\gamma, \beta; \mathbb{H}). \end{aligned}$$

Exercises from Chap. 5

5.1

▷ For f, g in $\mathscr{C}(X, Y)$, set $h(x) = d_Y(f(x), g(x))$, $x \in X$. By the triangle inequality, for x_1 and x_2 in X,

$$h(x_1) = d_Y(f(x_1), g(x_1))$$
$$\leq d_Y(f(x_1), f(x_2)) + d_Y(f(x_2), g(x_2)) + d_Y(g(x_2), g(x_1))$$
$$= d_Y(f(x_1), f(x_2)) + h(x_2) + d_Y(g(x_1), g(x_2)).$$

Interchanging the roles of x_1 and x_2, we find that

$$|h(x_1) - h(x_2)| \leq d_Y(f(x_1), f(x_2)) + d_Y(g(x_1), g(x_2)).$$

Since f and g are uniformly continuous on X, so is h.

▷ Clearly, $d(f, g) \geq 0$ and $d(f, g) = 0$ if $f = g$. If $d(f, g) = 0$ then $d_Y(f(x), g(x)) = 0$ for each x in X, hence $f(x) = g(x)$ for each x in X, and $f = g$.

The symmetry property, that $d(f, g) = d(g, f)$, is also clear.

If f, g and h are in $\mathscr{C}(X, Y)$ then, by the triangle inequality in Y,

$$d_Y(f(x), g(x)) \leq d_Y(f(x), h(x)) + d_Y(h(x), g(x)).$$

Taking the supremum over all x in X, and using the general property that $\sup(E + F) = \sup(E) + \sup(F)$, gives the triangle inequality for (5.1).

5.2 See, for example, [24, Sects. 24 & 25].

5.3

▷ Suppose that E is totally bounded and that ϵ is positive. Then there exist points $x_1, x_2, \ldots x_n$ in E such that $E \subseteq \bigcup_{i=1}^{n} B(x_i, \epsilon/2)$. If $x \in \mathrm{cl}(E)$ there is a point e in E with $d(x, e) \leq \epsilon/2$. In turn, there is some i such that $d(e, x_i) < \epsilon/2$. It follows that $d(x, x_i) < \epsilon$ so that $\mathrm{cl}(E) \subseteq \bigcup_{i=1}^{n} B(x_i, \epsilon)$.

Conversely, suppose that $\mathrm{cl}(E)$ is totally bounded and that ϵ is positive. There exist points $y_1, y_2, \ldots y_n$ in $\mathrm{cl}(E)$ such that $E \subset \mathrm{cl}(E) \subseteq \bigcup_{i=1}^{n} B(y_i, \epsilon/2)$. Since y_i is in the closure of E, there is some x_i in E with $d(y_i, x_i) \leq \epsilon/2$. Then, $B(y_i, \epsilon/2) \subseteq B(x_i, \epsilon)$ and so $E \subseteq \bigcup_{i=1}^{n} B(x_i, \epsilon)$.

▷ Suppose now that (M, d) is complete. Suppose that E is totally bounded. If $\{x_n\}$ is a Cauchy sequence in $\mathrm{cl}(E)$ then, since (M, d) is complete, it is convergent and the limit must belong to the closure of E since the limit of the sequence is a limit point of E. This shows that $\mathrm{cl}(E)$ is complete. Since E is totally bounded then, by the first part of this exercise, so is $\mathrm{cl}(E)$ and so $\mathrm{cl}(E)$ is compact by

Solutions to the Exercises

Proposition 5.2 (d). Conversely, if $cl(E)$ is compact then it is totally bounded and so, again by the first part of this exercise, so is E itself.

5.4 Suppose that F is relatively compact and that $E \subseteq F$. Then, $cl(E) \subseteq cl(F)$ hence $cl(E)$ is a closed subset of the compact set $cl(F)$ and so is itself compact. That is, E is relatively compact.

5.5 If $\mathscr{F}$ is equicontinuous on X then, trivially, $\mathscr{F}$ is equicontinuous at each point of X.

Conversely, suppose that $\mathscr{F}$ is equicontinuous at each point of X and let ϵ be positive. For each x_0 in X there is a positive $\delta = \delta(x_0)$ such that $d_Y(f(x), f(x_0)) \leq \epsilon/2$ whenever $d_X(x, x_0) \leq \delta(x_0)$ and $f \in \mathscr{F}$. The balls $B_X(x_0, \delta(x_0)/2)$, as x_0 ranges over X, form an open cover of X. Since X is compact, finitely many of these balls cover X, say $X \subseteq \bigcup_{i=1}^{n} B_X(y_i, \delta(y_i)/2)$ for points y_1 to y_n in X. Set $\delta = \min\{\delta(y_i)/2 : i = 1, \ldots n\}$, which is positive. Now suppose that x_1 and x_2 are in X with $d_X(x_1, x_2) \leq \delta$. There is some i for which $x_1 \in B_X(y_i, \delta(y_i)/2)$. By pointwise equicontinuity at y_i, $d_Y(f(x_1), f(y_i)) \leq \epsilon/2$ for all $f \in \mathscr{F}$. Next, $d_X(x_2, y_i) \leq d_X(x_2, x_1) + d_X(x_1, y_i) < \delta + \delta(y_i)/2 \leq \delta(y_i)$, so that $x_2 \in B_X(y_i, \delta(y_i))$. Again, by pointwise equicontinuity at y_i, $d_Y(f(x_2), f(y_i)) \leq \epsilon/2$, for $f \in \mathscr{F}$. By the triangle inequality, $d_Y(f(x_1), f(x_2)) \leq \epsilon$ for every $f \in \mathscr{F}$.

5.6 Suppose that E is a relatively compact subset of a metric space (X, d_X) and that $f : X \to Y$ is a continuous map from X to the metric space (Y, d_Y). Set $F = f(E)$. We want to show that F is relatively compact in Y. Since $cl(E)$ is compact and the continuous image of a compact set is compact, the set $f(cl(E))$ is compact in Y. We are therefore done if we show that $f(cl(E)) = cl(f(E))$.

Now $cl(E)$ comprises E together with all its limit points. We show that if x is a limit point of E then $f(x)$ is a limit point of $f(E)$. This is clear since, if $\{x_n\}_n$ is a sequence of points in E that converges to x then $\{f(x_n)\}_n$ is a sequence of points in $f(E)$ that converges to $f(x)$. Thus, $f(cl(E)) \subseteq cl(f(E))$. Conversely, if y is a limit point of $f(E)$ there is a sequence of points $\{f(x_n)\}_n$ in $f(E)$, so that each x_n is in E, that converges to y. Since $cl(E)$ is compact, the sequence $\{x_n\}_n$ has a subsequence that converges to a point x in $cl(E)$. The image of this subsequence under f converges to $f(x)$, since f is continuous, and also to y. Thus $y = f(x)$, that is $y \in f(cl(E))$ and we see that $cl(f(E)) \subseteq f(cl(E))$.

5.7 By the triangle inequality,

$$d_Y(f_n(x_n), f(x)) \leq d_Y(f_n(x_n), f(x_n)) + d_Y(f(x_n), f(x))$$
$$\leq d(f_n, f) + d_Y(f(x_n), f(x)).$$

As $n \to \infty$, $d(f_n, f) \to 0$ since $f_n \to f$ in $\mathscr{C}(X, Y)$, and $d_Y(f(x_n), f(x)) \to 0$ since $x_n \to x$ and f is continuous at x.

5.8 Since $\mathbb{R}^d$ is complete, Theorem 5.1 applies. Thus, $\mathscr{F} \subseteq \mathscr{C}(X)$ is relatively compact if and only if (a) and (b) of that result hold. With $Y = \mathbb{R}^d$, the condition (a) equates to the condition that, for each x in X, the set $\mathscr{F}(x) = \{f(x) : f \in \mathscr{F}\}$

is bounded. If $\mathscr{F}$ is bounded, then $\mathscr{F}(x)$ is certainly bounded for each individual x. Thus, if $\mathscr{F}$ is bounded and equicontinuous then (a) and (b) hold and so $\mathscr{F}$ is relatively compact.

In the other direction, suppose that $\mathscr{F}$ is relatively compact so that conditions (a) and (b) are satisfied with $Y = \mathbb{R}^d$. By equicontinuity of the family $\mathscr{F}$ with the choice $\epsilon = 1$, there is a positive δ such that $\|f(x) - f(y)\| \leq 1$ whenever $d_X(x, y) \leq \delta$ and $f \in \mathscr{F}$. Since X is compact it is totally bounded and so there are points $x_1, x_2, \ldots x_n$ such that $X = \bigcup_{i=1}^n B_X(x_i, \delta)$. Using (a), we see that for each i there is a number M_i such that $\|f(x_i)\| \leq M_i$ for all f in $\mathscr{F}$. With $M = \max\{M_1, M_2, \ldots, M_n\} + 1$, we then have $\|f(x)\| \leq M$ for all x in X and for all f in $\mathscr{F}$ (if $x \in B_X(x_i, \delta)$ and $f \in \mathscr{F}$ then $\|f(x)\| \leq \|f(x) - f(x_i)\| + \|f(x_i)\| \leq 1 + M_i \leq M$). Thus $\mathscr{F}$ is bounded and equicontinuous.

Finally, if $\mathscr{F} \subseteq \mathscr{C}(X)$ is compact then it is closed and, being relatively compact, it is bounded and equicontinuous. On the other hand, if $\mathscr{F}$ is closed, bounded and equicontinuous then $\mathrm{cl}(F)$ is compact and, since F is closed, $\mathrm{cl}(F) = F$.

5.9 When $D = \mathbb{D}$, the set K_n is simply the closed disk $\overline{D(0, (n-1)/n)}$. When $D = \mathbb{H}$, the set K_n is $\overline{D(0, n)} \cap \{z : \Im z \geq 1/n\}$.

5.10 The first part is a direct consequence of the fact that, by Proposition 5.1, $\mathscr{C}(K_n, Y)$ is a metric space.

As an example, take $D = \mathbb{D}$, K to be the compact set $\overline{D(0, 1/2)}$ and take f to be the function identically zero on $\mathbb{D}$. Now take $g(z) = h(|z|)$ where h is defined on the interval $[0, 1]$ by $h(x) = 0$ if $0 \leq x \leq 1/2$ and $h(x) = 2x - 1$ if $1/2 \leq x \leq 1$. Both f and g belong to $\mathscr{C}(\mathbb{D}, \mathbb{C})$. They are not the same function yet $\max\{|f(z) - g(z)| : z \in K\} = 0$.

5.11 It is clear that $\widetilde{d}(x, y) \geq 0$ for x and y in X and that $\widetilde{d}(x, y) = 0$ if and only if $d(x, y) = 0$ which occurs if and, if d is a metric, only if $x = y$. Symmetry of $\widetilde{d}$ is also clear. To verify the triangle inequality, for x, y and z in X,

$$\begin{aligned}
\widetilde{d}(x, z) + \widetilde{d}(z, y) &= \frac{d(x, z)}{1 + d(x, z)} + \frac{d(z, y)}{1 + d(z, y)} \\
&\geq \frac{d(x, z)}{1 + d(x, z) + d(z, y)} + \frac{d(z, y)}{1 + d(z, y) + d(x, z)} \\
&= \frac{d(x, z) + d(z, y)}{1 + d(x, z) + d(z, y)} \\
&\geq \frac{d(x, y)}{1 + d(x, y)} = \widetilde{d}(x, y).
\end{aligned}$$

At the last step we used the triangle inequality for d and the fact that $x/(1+x)$ is an increasing function on $[0, \infty)$.

5.12

▷ Suppose first that d_1 and d_2 are topologically equivalent metrics on X. For each x in X and positive r, the ball $E = B_{d_1}(x, r)$ is open in (X, d_1), therefore it is also open in (X, d_2). Since E is open in (X, d_2) and contains x, there is a ball $B_{d_2}(x, r_2)$ with r_2 positive such that $B_{d_2}(x, r_2) \subseteq E$. This is (i). The condition (ii) is deduced by reversing the roles of d_1 and d_2.

Conversely, suppose that (i) and (ii) hold and that E is open in (X, d_1). Each x in E is an interior point of E and so there is some ball $B_{d_1}(x, r)$ of positive radius r such that $B_{d_1}(x, r) \subseteq E$. Using (i), we find a ball $B_{d_2}(x, r_2)$ of positive radius r_2 such that $B_{d_2}(x, r_2) \subseteq B_{d_1}(x, r) \subseteq E$, so that x is an interior point of E from the point of view of the metric space (X, d_2). That is, E is open in (X, d_2) as well. Reversing the roles of d_1 and d_2 and using (ii) instead of (i) shows that E is open in (X, d_1) if it is open in (X, d_2).

▷ A sequence of points $\{x_n\}_{n=1}^\infty$ in a metric space (X, d) is convergent to x in (X, d) if and only if every neighbourhood of x contains all but finitely many terms in the sequence. If d_1 and d_2 are topologically equivalent metrics on X then every point x has the same neighbourhoods relative to either d_1 or d_2. It follows that a sequence $\{x_n\}_{n=1}^\infty$ converges to x in (X, d_1) if and only if it converges to x in (X, d_2).

▷ We show that (X, d_1) and (X, d_2) have the same closed sets. Suppose that x is a limit point of a set F relative to the metric d_1. That is, there is a sequence $\{x_n\}_{n=1}^\infty$ in F, $x_n \ne x$, for which $x_n \to x$ in the metric d_1, that is $d_1(x_n, x) \to 0$ as $n \to \infty$. Then, by assumption, $x_n \to x$ in the metric d_2 and so x is a limit point of F relative to the metric d_2. Reversing the roles of d_1 and d_2, we see that x is a limit point of F in the metric d_1 if and only if it is a limit point of F in the metric d_2. Thus, F is closed in (X, d_1) (that is, F contains all its limit points relative to d_1) if and only if F is closed in (X, d_2), as the set of limit points is the same in both cases.

5.13 Suppose that a sequence of functions $\{f_n\}_{n=1}^\infty$ converges locally uniformly to f on D, that is uniformly to f on each compact subset of D. For z in D, chose a closed disk of positive radius r such that $\overline{D(z, r)}$ is contained in D. Since this closed disk is compact, $f_n \to f$ uniformly on this disk.

Conversely, suppose that to each $z \in D$ there corresponds a positive $r = r_z$ such that $f_n \to f$ uniformly on the closed disk $\overline{D(z, r_z)}$. Let K be a compact subset of D. Then $K \subset \bigcup_{z \in K} D(z, r_z)$ and, being compact, can be covered by finitely many such disks, say $K \subseteq \bigcup_{i=1}^n D(z_i, r_{z_i})$. Since f_n converges to f uniformly on each disk $D(z_i, r_{z_i})$, we deduce that f_n converges to f uniformly on K. (Given ϵ positive, for each i there is N_i such that $d_Y(f_n(z), f(z)) \le \epsilon$ for each $z \in D(z_i, r_{z_i})$ and for each $n \ge N_i$. Then, $d_Y(f_n(z), f(z)) \le \epsilon$ for each $z \in K \subseteq \bigcup_{i=1}^n D(z_i, r_{z_i})$ and each $n \ge N$ where $N = \max\{N_1, N_2, \ldots, N_n\}$.)

5.14 Suppose that a family of functions $\mathscr{F}$ in $\mathscr{C}(D, \mathbb{C})$ is uniformly bounded on each compact subset of D. Then it is clearly locally bounded. In fact, take $K = \overline{D(z_0, r)}$ where r is positive yet small enough that $K \subset D$.

Conversely, suppose that the family of functions $\mathscr{F}$ in $\mathscr{C}(D, \mathbb{C})$ is locally bounded. Let K be a compact subset of D. Then $K \subset \bigcup_{z \in D} D(z, r_z)$ where r_z can be chosen positive for each z in K and such that $|f(w)| \leq M_z < \infty$ for $w \in D(z, r_z)$ and $f \in \mathscr{F}$. Being compact, a finite number of these disks cover K, say $K \subset \bigcup_{i=1}^n D(z_i, r_{z_i})$. Then f is bounded on K by $M = \max\{M_{z_1}, M_{z_2}, \ldots, M_{z_n}\}$, which is finite, this for every f in the family of functions $\mathscr{F}$.

5.15 The sequence of functions $\{f_n\}_{n=1}^\infty$ converges to f in $\mathscr{C}(D, \mathbb{C}_\infty)$ if and only if, to each compact set K and each positive ϵ, there corresponds a natural number N such that

$$d^\#(f_n(z), f(z)) \leq \epsilon \text{ for all } z \in K \text{ and all } n \geq N.$$

Now $d^\#(f_n(z), f(z)) = d^\#(1/f_n(z), 1/f(z))$, since $z \to 1/z$ is an isometry of the spherical metric. It follows that $f_n \to f$ in $\mathscr{C}(D, \mathbb{C}_\infty)$ if and only if $1/f_n \to 1/f$ in $\mathscr{C}(D, \mathbb{C}_\infty)$.

5.16 Suppose that $\{f_n\}_{n=1}^\infty$ converges to f in $\mathscr{H}(D)$ and that each function f_n is never zero in D. Suppose that the limit function f is not identically 0 but that f has a zero at z_0 in D. Since the zero of f at z_0 is isolated, it is possible to choose r positive so that the closed disk $\overline{D(z_0, r)}$ lies in D and f doesn't vanish on the circle $C(z_0, r)$. By Hurwitz's Theorem, for all large n the function f_n has the same number of zeros as f in the disk $D(z_0, r)$, that is at least one, which is impossible since f_n has no zeros. Thus, the limit function f is either identically zero or is zero-free. The example $f_n(z) = (1 - z^2)/n$, $z \in \mathbb{D}$, shows that the limit function may be identically zero.

5.17

▷ Suppose that $\mathscr{F}$ is a relatively compact family of functions in $\mathscr{M}^*(D)$. By Exercise 5.15, $1/\mathscr{F}$ is relatively compact in $\mathscr{C}(D, \mathbb{C}_\infty)$, in fact relatively compact in $\mathscr{M}^*(D)$ by Theorem 5.6.

▷ We need to show that if $f_n \to f$ in $\mathscr{M}^*(D)$ then $f_n + a \to f + a$ in $\mathscr{M}^*(D)$. Exercise 3.18 shows that there is a constant C depending only on $|a|$ such that

$$\frac{1}{C} d^\#(f_n(z), f(z)) \leq d^\#(f_n(z) + a, f(z) + a) \leq C d^\#(f_n(z), f(z)).$$

Then $f_n \to f$ spherically uniformly on a compact subset K of D if and only if $f_n + a \to f + a$ spherically uniformly on K.

▷ This is similar to the last part, but using the estimate

$$\frac{1}{C} d^\#(f_n(z), f(z)) \leq d^\#(af_n(z), af(z)) \leq C d^\#(f_n(z), f(z)).$$

▷ This follows from the previous parts as a general Möbius transformation can be written as a composition of translations, dilations and inversions.

5.18 By Montel's Theorem, it suffices to show that the family $\mathscr{F}$ is locally bounded, that is bounded on $K = \overline{D(0,r)}$ for each $r < 1$. To this end, fix $r < 1$ and set $\epsilon = (1/r - 1)/2$ so that $r(1+\epsilon) = (1+r)/2 = \rho < 1$. Now suppose that $f \in \mathscr{F}$ and that $z \in K$. By assumption, there is a natural number N such that $M_n \leq (1+\epsilon)^n$ for $n \geq N$. Then,

$$|f(z)| \leq \sum_{n=0}^{\infty} |a_n| r^n \leq \sum_{n=0}^{\infty} |M_n| r^n \leq \sum_{n=0}^{N-1} |M_n| r^n + \sum_{n=N}^{\infty} (1+\epsilon)^n r^n$$

$$\leq \sum_{n=0}^{N-1} |M_n| r^n + \sum_{n=N}^{\infty} \rho^n = C < \infty.$$

That is, $|f(z)|$ is bounded by C on K independently of f in $\mathscr{F}$.

Conversely, suppose that $\mathscr{F}$ is a relatively compact subset of $\mathscr{H}(\mathbb{D})$, that is $\mathscr{F}$ is locally bounded on $\mathbb{D}$. Thus, for each $r \in (0,1)$, there is an $M(r) < \infty$ such that $|f(z)| \leq M(r)$ for $|z| = r$, $f \in \mathscr{F}$. If $f \in \mathscr{F}$ with $f(z) = \sum_{n=0}^{\infty} a_n z^n$ then, by the Cauchy Integral Formula for derivatives, $|a_n| \leq M(r)/r^n$. Set $M_n = \sup\{|a_n|: f \in \mathscr{F}\}$. Then, $\sqrt[n]{M_n} \leq \sqrt[n]{M(r)}/r$ and so $\limsup_{n\to\infty} \sqrt[n]{M_n} \leq 1/r$. Since this holds for every r in $(0,1)$, $\limsup_{n\to\infty} \sqrt[n]{M_n} \leq 1$.

5.19 Suppose that $\mathscr{F}$ is not a normal family of analytic functions on the disk $D = D(z_0, r)$. There then exists a sequence of functions $\{f_n\}_{n=1}^{\infty}$ from $\mathscr{F}$ that has no convergent subsequence. Suppose to the contrary that $\bigcup_{f \in \mathscr{F}} f(D)$ is not dense in the complex plane. Then, $\bigcup_{n=1}^{\infty} f_n(D)$ is not dense either. It follows that there is some disk $D(a, \delta)$ of positive radius δ such that no function f_n takes values in $D(a, \delta)$. Each analytic function $1/(f_n - a)$ is bounded by $1/\delta$ on D. By Montel's Theorem, $\mathscr{G} = \{1/(f_n - a): n \geq 1\}$ is a normal family on D. By Exercise 5.17, the family of functions $\{f_n: n \geq 1\}$ is also a normal family on D. Hence, $\{f_n\}_{n=1}^{\infty}$ does have a convergent subsequence after all, which is a contradiction.

5.20 Suppose that $\{f_n\}_{n=1}^{\infty}$ is a sequence of meromorphic functions that converges to f spherically locally uniformly on a domain D, and that z_0 is in D. Suppose first that z_0 is not a pole of f. Then, by Proposition 5.5, f is analytic in some disk about z_0, f_n is analytic in this disk for sufficiently large n, and f_n converges to f uniformly on this disk. It follows from Theorem 5.5 that the sequence of derivatives f'_n also converges uniformly to f' in a disk about z_0. As a consequence, $f_n^{\#} = 2|f'_n|/(1+|f_n|^2)$ also converges uniformly to $f^{\#}$ on a disk about z_0.

If some point z_0 is a pole of f then f cannot be identically zero in D, hence neither can f_n (for sufficiently large n). Then $\{1/f_n\}_{n=1}^{\infty}$ is a sequence of meromorphic functions that converges spherically locally uniformly to $1/f$. Since $(1/f)(z_0) = 0$, the previous case applies and we find that $(1/f_n)^{\#}$ converges uniformly on some disk about z_0 to $(1/f)^{\#}$. Since, in general, $(1/f)^{\#} = f^{\#}$ (see Exercise 3.20), we see that $f_n^{\#}$ converges uniformly to $f^{\#}$ on a disk about z_0 in this case also.

Exercises from Chap. 6

6.1 Let γ_0, γ_1 and γ_2 be piecewise-smooth closed curves in D, each parameterised on $[0, 1]$.

Any closed curve γ is homotopic to itself via $\Gamma(t, s) = \gamma(t)$, $(t, s) \in I_2$, so that homotopy is reflexive.

Suppose that γ_0 is homotopic to γ_1 relative to the domain D via the homotopy $\Gamma_1(t, s)$, $(t, s) \in I_2$. Set $\tilde{\Gamma}_1(t, s) = \Gamma_1(t, 1 - s)$. Then, $\tilde{\Gamma}_1$ is a homotopy between γ_1 and γ_0 showing that homotopy is symmetric.

Finally, suppose that γ_1, in turn, is homotopic to γ_2 via the homotopy $\Gamma_2(t, s)$, $(t, s) \in I_2$. Set

$$\Gamma(t, s) = \begin{cases} \Gamma_1(t, 2s), & 0 \leq s \leq 1/2, \\ \Gamma_2(t, 2s - 1), & 1/2 \leq s \leq 1. \end{cases}$$

Then $\Gamma(t, s)$ is continuous on I_2 with $\Gamma(t, 1/2) = \gamma_1(t)$, $0 \leq t \leq 1$, and is seen to be a homotopy between γ_0 and γ_1. Thus, homotopy is transitive.

6.2 Let D be a domain that is star-shaped about a, so that the line segment $[a, z]$ lies in D whenever z lies in D. Note that, as s increases from 0 to 1, the point $sa + (1 - s)z$ moves from z back to a along this line segment.

Let γ, parameterised by $\gamma(t)$, $t \in [0, 1]$, be a closed piecewise-smooth curve in D. Set

$$\Gamma(t, s) = sa + (1 - s)\gamma(t), \quad (t, s) \in I_2.$$

Then, $\Gamma(t, s)$ is continuous on the square I_2 since γ is a continuous curve. This is Property (i) of homotopy. Since $\gamma(t)$ lies in D, and D is star-shaped about a, $\Gamma(t, s)$ lies in D which is Property (ii). Property (iv) holds since $\gamma(0) = \gamma(1)$ and so $\Gamma(0, s) = sa + (1 - s)\gamma(0)$ and $\Gamma(1, s) = sa + (1 - s)\gamma(1)$ are equal for each s. Finally, $\Gamma(t, 0) = \gamma(t)$ while $\Gamma(t, 1) = a$, $0 \leq t \leq 1$, which is Property (iii), showing that the curve γ is homotopic relative to D to the constant curve a. That is, γ is homotopic to zero.

6.3 Suppose that $\Gamma(t, s)$, $t \in I_2$, is a homotopy between γ_0 and γ_1 relative to D_1. It is straightforward that $\tilde{\Gamma}(t, s) = h(\Gamma(t, s))$, $t \in I_2$, is a homotopy between $h(\gamma_0)$ and $h(\gamma_1)$ relative to D_2.

Let $\tilde{\gamma}$ be a closed curve in D_2. Then $\gamma = h^{-1}(\tilde{\gamma})$ is a closed curve in D_1. If D_1 is simply connected, then there is a constant curve γ_1 in D_1 to which γ is homotopic. Say $\gamma_1(t) \equiv a$, with $a \in D_1$, is this constant curve. Then $h(\gamma) = \tilde{\gamma}$ is homotopic to the constant curve $h(\gamma_1) \equiv h(a)$ in D_2. Thus, every closed curve in D_2 is homotopic to zero, showing that D_2 is simply connected.

6.4 By the Riemann Mapping Theorem, there is a conformal map $f : \mathbb{D} \to D$ with $f(0) = a$. Set $g(z) = (f^{-1} \circ f_2^{-1} \circ f_1 \circ f)(z)$, $z \in \mathbb{D}$. Then g maps $\mathbb{D}$ into itself with $g(0) = 0$ since f_2^{-1} is defined on D_2 and $D_1 \subseteq D_2$. By Schwarz's Lemma, $|g'(0)| \leq 1$ with equality if and only if g is a rotation. Since

$$|g'(0)| = |(f^{-1})'(a)| \times |(f_2^{-1})'(f_2(a))| \times |f_1'(a)| \times |f'(0)|$$

$$= \frac{1}{|f'(0)|} \times \frac{1}{|f_2'(a)|} \times |f_1'(a)| \times |f'(0)| = \frac{|f_1'(a)|}{|f_2'(a)|},$$

we have $|f_1'(a)| \leq |f_2'(a)|$. The inequality is strict unless $g(z) = \beta z$, some $\beta \in \mathbb{T}$. In that case, g is an automorphism of $\mathbb{D}$ and then $h = f \circ g \circ f^{-1}$ is an automorphism of D. Since $h = f_2^{-1} \circ f_1$, we can write $f_1 = f_2 \circ h$ so that $f_1(D) = f_2(h(D)) = f_2(D)$. That is, $D_1 = D_2$. Conversely, if $D_1 = D_2$ then h is an automorphism of D and $g = f^{-1} \circ h \circ f$ is an automorphism of the unit disk that fixes 0. Then, $|g'(0)| = 1$ and so $|f_1'(a)| = |f_2'(a)|$. This shows that $|f_1'(a)| = |f_2'(a)|$ if and only if $D_1 = D_2$.

6.5 We should first check that D is actually a domain. It is certainly open, as it is the union of two open sets. It is also path connected as, if $z_1 \in D_1$ and $z_2 \in D_2$ we can choose a point z in the non-empty set $D_1 \cap D_2$ and join z_1 to z_2 in D by a path joining z_1 to z in D_1 followed by a path joining z to z_2 in D_2.

Now we verify that D satisfies Property (c) of Theorem 6.8. Suppose that f is analytic in D. Then f is analytic in both D_1 and D_2 so that, since these are simply connected domains, f has an antiderivative F_1 in D_1 and F_2 in D_2. On the non-empty open set $D_1 \cap D_2$, $F = F_2 - F_1$ has derivative zero and so is constant *on each component of* $D_1 \cap D_2$. Since $D_1 \cap D_2$ is assumed to be connected, $F_2 = F_1 + C$ on $D_1 \cap D_2$ some for constant C. We replace F_2 by $F_2 - C$, and set

$$F(z) = \begin{cases} F_1(z), & z \in D_1; \\ F_2(z) - C, & z \in D_2. \end{cases}$$

Then, F is well-defined in D, is analytic in D and its derivative is f.

An example in which D fails to be simply connected is $D_1 = A(2, 4) \setminus (2, 4)$, the annulus $A(2, 4) = \{z \colon 2 < |z| < 4\}$ slit along the line segment $(2, 4)$, and $D_2 = D(3, 1)$. Even though both D_1 and D_2 are simply connected, their union $D = D_1 \cup D_2$ is the annulus $A(2, 4)$ which is not simply connected. Here, $D_1 \cap D_2$ consists of two separate components, each a half-disk, and is not connected.

6.6 Let $f \colon D \to \mathbb{D}$ be analytic. Let $h \colon D \to \mathbb{D}$ be conformal so that, for $z \in D$, $\rho_D(z) = 2|h'(z)|/(1 - |h(z)|^2)$. Now $f \circ h^{-1}$ is a self-map of $\mathbb{D}$. By (4.16a) with $\alpha = h(z), z \in D$,

$$|(f \circ h^{-1})'(h(z))| \leq \frac{1 - |(f \circ h^{-1})(h(z))|^2}{1 - |h(z)|^2}.$$

That is

$$|f'(z)|\,|(h^{-1})'(h(z))| \le \frac{1-|f(z)|^2}{1-|h(z)|^2}$$

or, since $|(h^{-1})'(h(z))| = 1/|h'(z)|$,

$$\rho_{\mathbb{D}}(f(z))|f'(z)| = \frac{2|f'(z)|}{1-|f(z)|^2} \le \frac{2|h'(z)|}{1-|h(z)|^2} = \rho_D(z).$$

If $f(z) = 0$ and we arrange, as we may, that $h(z) = 0$ as well, this last inequality becomes $2|f'(z)| \le 2|h'(z)| = \rho_D(z)$. By the equality statement in (4.16a), equality holds if and only if $f \circ h^{-1}$ is an automorphism of $\mathbb{D}$ and this is the case if and only if $f = (f \circ h^{-1}) \circ h$ is a conformal map of D onto $\mathbb{D}$.

6.7 Let f be a conformal map of D onto $\mathbb{D}$ with $f(z_0) = 0$. Then, $\rho_D(z_0) = 2|f'(z_0)|$. Moreover, $g(z) = f(z/r)$ maps rD conformally onto the unit disk with $g(rz_0) = 0$. Thus,

$$\rho_{rD}(rz_0) = 2|g'(rz_0)| = 2|f'(rz_0/r)|\frac{1}{r} = \frac{2}{r}|f'(z_0)| = \frac{1}{r}\rho_D(z_0).$$

In particular, $\rho_{D(0,r)}(0) = \rho_{\mathbb{D}}(0)/r = 2/r$.

6.8 In general, $\Im(z^2) = \Im((x+iy)^2) = 2xy$. For (6.23), $\Re((1+z)/(1-z)) = \Re((1+z)(1-\bar{z}))/|1-z|^2$ and, similarly, $\Im((1+z)/(1-z)) = \Im((1+z)(1-\bar{z}))/|1-z|^2$. Since

$$(1+z)(1-\bar{z}) = 1 - |z|^2 + z - \bar{z} = 1 - |z|^2 + 2i\Im z,$$

$\Re((1+z)(1-\bar{z})) \times \Im((1+z)(1-\bar{z})) = 2\Im z(1-|z|^2)$, from which (6.23) follows.
For (6.24),

$$\rho_{\mathbb{D}_h}(z) = 4\frac{|1+z|}{|1-z|^3} \times \frac{|1-z|^4}{4\Im z\,(1-|z|^2)} = \frac{|1+z|\,|1-z|}{\Im z\,(1-|z|^2)} = \frac{|1-z^2|}{\Im z\,(1-|z|^2)}.$$

For (6.25),

$$|1 - r^2 e^{2i\theta}|^2 = |1 - r^2 \cos(2\theta) - ir^2 \sin(2\theta)|^2$$
$$= \left(1 - r^2 \cos(2\theta)\right)^2 + r^4 \sin^2(2\theta)$$
$$= 1 + r^4 - 2r^2 \cos(2\theta)$$
$$= 1 + r^4 - 2r^2(1 - 2\sin^2\theta)$$
$$= (1-r^2)^2 + 4r^2 \sin^2\theta.$$

Solutions to the Exercises

For (6.27), setting $z = re^{i\theta}$ in (6.26), we find that

$$\rho_{\mathbb{D}_s}(re^{i\theta}) = \frac{1}{2\sqrt{r}} \frac{|1+re^{i\theta}|}{\sqrt{r}\cos(\theta/2)(1-r)}$$

$$= \frac{\sqrt{(1+r\cos\theta)^2 + r^2\sin^2\theta}}{2r(1-r)\cos(\theta/2)}$$

$$= \frac{\sqrt{1+r^2+2r\cos\theta}}{2r(1-r)\cos(\theta/2)}.$$

Since $1 + r^2 + 2r\cos\theta = (1-r)^2 + 2r(1+\cos\theta) = (1-r)^2 + 4r\cos^2(\theta/2)$, (6.27) follows.

6.9 We scale each ray $[r_1 e^{i\theta}, R_1 e^{i\theta}]$ to the ray $[r_2 e^{i\theta}, R_2 e^{i\theta}]$. We want a homeomorphism $x(r)$ of the interval $[r_1, R_1]$ onto the interval $[r_2, R_2]$. If R_1 and R_2 are both finite, we could take a function $x(r) = ar + b$ for which $x(r_1) = r_2$ and $x(R_1) = R_2$. Solving the linear equations for a and b leads to $a = (R_2 - r_2)/(R_1 - r_1)$ and $b = (r_2 R_1 - r_1 R_2)/(R_1 - r_1)$. The homeomorphism is then

$$x(r) = \frac{R_2 - r_2}{R_1 - r_1} r + \frac{r_2 R_1 - r_1 R_2}{R_1 - r_1}.$$

If both R_1 and R_2 are infinite then $x(r) = r + (r_2 - r_1)$ works. If R_1 is infinite and R_2 is finite then

$$x(r) = \frac{R_2(r - r_1) + r_2(r_1 + 1)}{r + 1}$$

works. The inverse of this last function works if R_1 is finite and R_2 is infinite, specifically

$$x(r) = \frac{r + r_2(R_1 - r_1) - r_1}{R_1 - r}.$$

The explicit homeomorphism is then $h(z) = h(re^{i\theta}) = x(r)e^{i\theta}$ for the relevant homeomorphism of intervals $x(r)$.

6.10

▷ We need to find an automorphism $M_\alpha(z) = (\alpha - z)/(1 - \alpha z)$, $0 < \alpha < 1$, that maps the annulus $A(r_0, 1)$ onto D for suitable r_0. The map M_α will fix the unit circle and map the circle $C(0, r_0)$ to the circle $C(a, a)$. Now $C(0, r_0)$ is a hyperbolic circle. Since M_α preserves hyperbolic distances, it will map the hyperbolic centre of

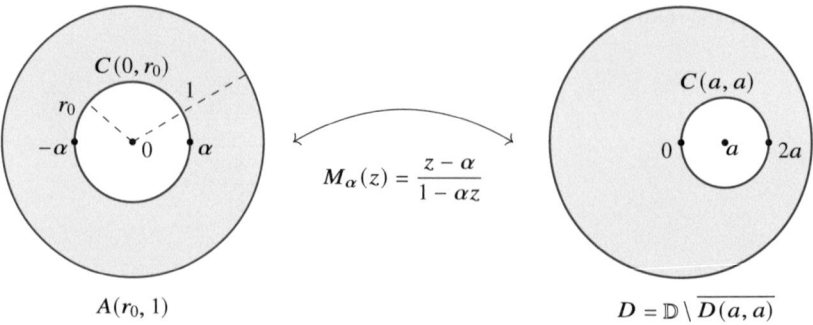

$C(0, r_0)$, that is 0, to the hyperbolic centre of $C(a, a)$. Since $M_\alpha(0) = \alpha$, we should choose α to be the hyperbolic centre of $C(a, a)$. This means that α should be the hyperbolic midpoint of $[0, 2a]$ so that

$$d(0, \alpha; \mathbb{D}) = \frac{1}{2} d(0, 2a; \mathbb{D}),$$

that is, by (4.22),

$$\log\left(\frac{1+\alpha}{1-\alpha}\right) = \frac{1}{2} \log\left(\frac{1+2a}{1-2a}\right).$$

We can solve this equation for α. First,

$$\frac{1+\alpha}{1-\alpha} = y \text{ where } y = \sqrt{\frac{1+2a}{1-2a}},$$

so that, after some computations,

$$\alpha = \frac{y-1}{y+1} = \frac{1}{2a}\left(1 - \sqrt{1-4a^2}\right).$$

Of course, M_α is its own inverse and $M_\alpha(0) = \alpha$ and, as one may check if necessary, $M_\alpha(2a) = -\alpha$, so that M_α maps the circle $C(a, a)$ to the circle $C(0, \alpha)$ and maps D onto $A(r_0, 1)$ with $r_0 = \alpha$. The modulus of the annulus D is then

$$\lambda = \lambda(a) = \frac{1}{2\pi} \log\left(\frac{1}{\alpha}\right) = \frac{1}{2\pi} \log\left[\frac{1}{2a}\left(1 + \sqrt{1-4a^2}\right)\right].$$

Note that $\alpha \to 1$ and $\lambda(a) \to 0$ as $a \to 1/2^-$.

▷ The map $w = g(z) = (1+z)/(1-z)$ maps the unit disk onto the right half-plane and maps the circle $C(1/2, 1/2)$ onto the line $\Re w = 1$. The domain $D = \mathbb{D} \setminus \overline{D(1/2, 1/2)}$ is then mapped to the strip $S = \{w : 0 < \Re w < 1\}$. In turn,

Solutions to the Exercises

S is mapped to the upper half-plane by $h(w) = \exp(\pi \mathrm{i} w)$. Composing these mappings, we see that D is mapped conformally onto the upper half-plane $\mathbb{H}$ by

$$f(z) = h(g(z)) = \exp\left(\pi \mathrm{i} \frac{1+z}{1-z}\right).$$

Exercises from Chap. 7

7.1 Let ϵ positive be given. With f given by (7.1), set $2\pi \mathrm{i} f_i(z) = \int_{\gamma_i} f(w)/(w-z) \, dw$ for $1 \le i \le N$. We find polynomials P_i, $1 \le i \le N$, such that $|f_i(z) - P_i(z)| \le \epsilon/N$ uniformly for z in K and for each i. The polynomial $P(z) = \sum_{i=1}^{N} P_i(z)$ then uniformly approximates f to within ϵ on K.

Fix i with $1 \le i \le N$. By Proposition 7.2, there is a rational function

$$R_i(z) = \sum_{k=1}^{M_i} \frac{\alpha_{i,k}}{\beta_{i,k} - z}$$

with poles $\beta_{i,k}$ lying only on γ_i such that $|f_i(z) - R_i(z)| \le \epsilon/(2N)$ for z in K. Now suppose that we can 'approximate, for z_0 in $D \setminus K$, the rational function $h(z) = 1/(z_0 - z)$ by polynomials, uniformly on K'. There are then polynomials $P_{i,k}$, $1 \le k \le M_i$, such that

$$\left| \frac{\alpha_{i,k}}{\beta_{i,k} - z} - P_{i,k}(z) \right| \le \frac{\epsilon}{2 M_i N}, \quad z \in K.$$

Set $P_i(z) = \sum_{k=1}^{M_i} P_{i,k}(z)$. Then, for all z in K,

$$|f_i(z) - P_i(z)| \le |f_i(z) - R_i(z)| + |R_i(z) - P_i(z)|$$

$$\le \frac{\epsilon}{2N} + \sum_{k=1}^{M_i} \left| \frac{\alpha_{i,k}}{\beta_{i,k} - z} - P_{i,k}(z) \right|$$

$$\le \frac{\epsilon}{2N} + \sum_{k=1}^{M_i} \frac{\epsilon}{2 M_i N} = \frac{\epsilon}{N}.$$

7.2 Since f and g are continuous on K, both are bounded on K, say by M. Let ϵ positive be given, with $\epsilon \le 1$. Since $P_n \to f$ uniformly on K, for all sufficiently large n we have

$$|P_n(z)| \le |P_n(z) - f(z)| + |f(z)| \le 1 + M, \quad z \in K.$$

Since $P_n \to f$ and $Q_n \to g$ uniformly on K, for all sufficiently large n we have

$$|P_n(z) - f(z)| \le \frac{\epsilon}{2M+1} \text{ and } |Q_n(z) - g(z)| \le \frac{\epsilon}{2M+1}, \quad z \in K.$$

Then, uniformly for z in K and for all sufficiently large n, we have

$$|P_n(z)Q_n(z) - f(z)g(z)| \le |P_n(z)Q_n(z) - P_n(z)g(z)| + |P_n(z)g(z) - f(z)g(z)|$$
$$\le |P_n(z)||Q_n(z) - g(z)| + |g(z)||P_n(z) - f(z)|$$
$$\le (M+1)|Q_n(z) - g(z)| + M|P_n(z) - f(z)|$$
$$\le (M+1)\frac{\epsilon}{2M+1} + M\frac{\epsilon}{2M+1} = \epsilon.$$

7.3 Choose α in K and set $g(z) = f(z)(z - \alpha)$. Since g is analytic in D, (7.5) applies and we see that

$$g(\alpha) = \sum_{n=1}^{M} \frac{1}{2\pi i} \int_{C_n} \frac{g(w)}{w - \alpha} \, dw.$$

Since $g(\alpha) = 0$ and $g(w)/(w - \alpha) = f(w)$, we find that

$$\sum_{n=1}^{M} \int_{C_n} f(w) \, dw = 0.$$

Exercises from Chap. 8

8.1

▷ For f in the class $\mathbf{S}$ with power series (8.1), and for $z \in \mathbb{U}$,

$$g(z) = \frac{1}{f(1/z)} = \frac{1}{1/z + a_2/z^2 + a_3/z^3 + \ldots} = \frac{z}{1 + a_2/z + a_3/z^2 + \ldots}.$$

By the geometric series for $1/(1+w)$ with $w = a_2/z + a_3/z^2 + a_4/z^3 + \ldots$ (noting that $w \to 0$ as $z \to \infty$),

$$g(z) = z\big[1 - (a_2/z + a_3/z^2 + a_4/z^3) + (a_2/z + a_3/z^2)^2 - (a_2/z)^3 + O(z^{-4})\big]$$
$$= z\big[1 - a_2/z + (-a_3 + a_2^2)/z^2 + (-a_4 + 2a_2 a_3 - a_2^3)/z^3 + O(z^{-4})\big]$$
$$= z - a_2 + (a_2^2 - a_3)/z + (2a_2 a_3 - a_2^3 - a_4)/z^2 + \ldots.$$

▷ For g in the class Σ with power series (8.2), and for $z \in \mathbb{D}$,

$$f(z) = \frac{1}{g(1/z)} = \frac{1}{1/z + b_0 + b_1 z + b_2 z^2 + \ldots}$$

$$= \frac{z}{1 + b_0 z + b_1 z^2 + b_2 z^3 + \ldots}.$$

By the geometric series $1/(1+w)$ with $w = b_0 z + b_1 z^2 + b_2 z^3 + \ldots$,

$$f(z) = z\big[1 - (b_0 z + b_1 z^2 + b_2 z^3) + (b_0 z + b_1 z^2)^2 - (b_0 z)^3 + O(z^4)\big]$$

$$= z\big[1 - b_0 z + (-b_1 + b_0^2)z^2 + (-b_2 + 2b_0 b_1 - b_0^3)z^3 + O(z^4)\big]$$

$$= z - b_0 z^2 + (b_0^2 - b_1)z^3 + (2b_0 b_1 - b_0^3 - b_2)z^4 + \ldots$$

8.2 With $f(z) = z/(1-z)$, we have $f(1/z) = 1/(z-1)$ and then $g(z) = z - 1$. Then $g(\mathbb{U}) = \mathbb{U} - 1$ or $g(\mathbb{U}) = \mathbb{C}_\infty \setminus \overline{D(-1, 1)}$.

With $k(z) = z/(1-z)^2$, we have $k(1/z) = z/(z-1)^2$ and then $g(z) = (z-1)^2/z = (z + 1/z) - 2$. The image of the ray $[-\infty, -1/4]$ on the real axis under inversion is the interval $[-4, 0]$ and so $g(\mathbb{U})$ is the complex plane slit along the real axis between -4 and 0, that is $g(\mathbb{U}) = \mathbb{C}_\infty \setminus [-4, 0]$. Another way to see this is to write z on the unit circle as $e^{i\theta}$ so that $z + 1/z = 2\cos\theta$. The image of the unit circle under $z \to z + 1/z$ is then the interval $[-2, 2]$.

8.3

▷ Let a_i, $1 \leq i \leq n$, be the zeros of Q counting multiplicity. Then,

$$Q(z) = b_n \prod_{i=1}^{n}(z - a_i) = (-1)^n b_n \prod_{i=1}^{n} a_i + \ldots + b_n z^n,$$

so that $1 = |b_n| \prod_{i=1}^{n} |a_i| > \prod_{i=1}^{n} |a_i|$. As a consequence, at least one of the zeros of Q must have modulus less than 1.

▷ With $P(z) = z + a_2 z^2 + a_3 z^3 + \ldots + a_n z^n$, we have $P'(z) = 1 + 2a_2 z + 3a_3 z^2 + \ldots + na_n z^n$. Applying the previous part with $Q(z) = P'(z)$, we see that if $n|a_n| > 1$ then P' has a zero in the unit disk and so P cannot be univalent there. That is, if P is univalent in $\mathbb{D}$ then $|a_n| \leq 1/n$.

The converse is false in general. For example, $P(z) = z + \frac{5}{3}z^2 + \frac{1}{3}z^3$ is not univalent in $\mathbb{D}$. In fact, $P'(z) = 1 + \frac{10}{3}z + z^2 = (z + \frac{1}{3})(z + 3)$, which has a zero at $-\frac{1}{3}$ in the unit disk.

▷ Fix $n \geq 2$ and set $P(z) = z + cz^n$. By the previous part, if P is univalent then $|c| \leq 1/n$. Alternatively, since $P'(z) = 1 + ncz^{n-1}$, if $n|c| > 1$ then P' vanishes at the roots of $z^{n-1} = -1/(nc)$, all of which lie inside the unit circle.

On the other hand, if $n|c| \leq 1$ and $f(z_1) = f(z_2)$ with both z_1 and z_2 in $\mathbb{D}$ then

$$0 = (z_1 - z_2) + c(z_1^n - z_2^n) = (z_1 - z_2)\left[1 + c\sum_{j=0}^{n-1} z_1^{n-1-j} z_2^j\right].$$

But

$$\left|\sum_{j=0}^{n-1} z_1^{n-1-j} z_2^j\right| \leq \sum_{j=0}^{n-1} |z_1|^{n-1-j} |z_2|^j < \sum_{j=0}^{n-1} 1 = n.$$

It follows that $1 + c\sum_{j=0}^{n-1} z_1^{n-1-j} z_2^j \neq 0$, hence $z_1 - z_2 = 0$ and P is univalent.

8.4 Since $\sum_{n=2}^\infty n|a_n| \leq 1$, we have that $n|a_n| \to 0$ as $n \to \infty$. In particular, $|a_n| \leq 1$ for all sufficiently large n so that the series for f converges everywhere in the unit disk.

Now to check for univalence. For distinct points z_1 and z_2 in $\mathbb{D}$, $f(z_1) - f(z_2) = (z_1 - z_2) + \sum_{n=2}^\infty a_n(z_1^n - z_2^n)$. Write

$$z_1^n - z_2^n = (z_1 - z_2)\sigma_n \text{ where } \sigma_n = \sum_{j=0}^{n-1} z_1^{n-1-j} z_2^j,$$

so that

$$f(z_1) - f(z_2) = (z_1 - z_2)\left[1 + \sum_{n=2}^\infty a_n \sigma_n\right].$$

Since $|\sigma_n| < n$, we have $\sum_{n=2}^\infty |a_n||\sigma_n| < \sum_{n=2}^\infty |a_n| n \leq 1$. Thus, the only way in which $f(z_1) - f(z_2) = 0$ is if $z_1 - z_2 = 0$.

8.5 We show that F' has a zero in the unit disk and so is not univalent. We have $f'(z) = 1/(1-z)^2$ and $g'(z) = 1/(1-iz)^2$. Then,

$$F'(z) = \frac{1}{2}\left[\frac{1}{(1-z)^2} + \frac{1}{(1-iz)^2}\right]$$

$$= \frac{1}{2(1-z)^2(1-iz)^2}\left[(1-iz)^2 + (1-z)^2\right]$$

$$= \frac{1}{(1-z)^2(1-iz)^2}\left[1 - (1+i)z\right].$$

Hence, F' vanishes at the point $1/(1+i)$, which lies in $\mathbb{D}$, and so F is not univalent in the unit disk.

8.6 Making use of (8.3),

$$k_\theta(z) = \frac{z}{(1-e^{-i\theta}z)^2} = \sum_{n=1}^{\infty} n e^{-i(n-1)\theta} z^n.$$

The image of the unit disk under k_θ is the image of the unit disk under the Koebe function but rotated anticlockwise by an angle θ, that is the complex plane slit along the ray away from the origin from $-\frac{1}{4}e^{i\theta}$ to infinity.

8.7 Near the origin $f(z)/w$ is small, less than 1 in modulus, and so we can expand $1/(1-f(z)/w)$ as a geometric series. Then,

$$f_4(z) = f(z) \times \frac{1}{1-f(z)/w}$$

$$= z\left[1 + a_2 z + a_3 z^2 + O(z^3)\right] \times \left[1 + \frac{1}{w}(z + a_2 z^2) + \frac{1}{w^2}z^2 + O(z^3)\right]$$

$$= z\left[1 + \left(a_2 + \frac{1}{w}\right)z + \left(a_3 + \frac{2a_2}{w} + \frac{1}{w^2}\right)z^2 + O(z^3)\right].$$

8.8 Since $k(z) = z/(1-z)^2$,

$$h(z) = z\sqrt{\frac{k(z^2)}{z^2}} = z\sqrt{\frac{1}{(1-z^2)^2}} = \frac{z}{1-z^2}.$$

Here we took the square root that has the value 1 at the origin. The image of the unit disk under h is the square root of the Koebe region, that is the complex plane slit along the positive imaginary axis from $i/2$ to infinity and along the negative imaginary axis from $-i/2$ to infinity.

8.9 Since h is an odd function in the class **S**, it has a power series expansion that begins $h(z) = z + b_3 z^3 + b_5 z^5 + \ldots$. By (8.7),

$$f(z^2) = z^2 + a_2 z^4 + a_3 z^6 + \ldots$$
$$= (z + b_3 z^3 + b_5 z^5 + \ldots)^2$$
$$= z^2 + 2b_3 z^4 + (b_3^2 + 2b_5)z^6 + \ldots.$$

Equating coefficients of like powers gives, in turn, $b_3 = a_2/2$ and $2b_5 = a_3 - b_3^2 = a_3 - a_2^2/4$. Then, $b_5 = a_3/2 - a_2^2/8$.

8.10 Suppose that h is an odd function in the class **S**. Then, h^2 is an even analytic function in the disk $\mathbb{D}$ and so has only even powers of z in its expansion, $h(z)^2 = z^2 + \sum_{n=2}^{\infty} c_{2n} z^{2n}$, say. Set $f(z) = z + \sum_{n=2}^{\infty} c_{2n} z^n$. Then f is analytic in $\mathbb{D}$, $f(0) = 0$ and $f'(0) = 1$. Of course, $f(z^2) = h(z)^2$.

It remains to show that f is univalent. If $f(z_1) = f(z_2)$ with z_1 and z_2 in $\mathbb{D}^*$ then $h(\sqrt{z_1})^2 = h(\sqrt{z_2})^2$. It follows that either $h(\sqrt{z_1}) = h(\sqrt{z_2})$ or $h(\sqrt{z_1}) = -h(\sqrt{z_2}) = h(-\sqrt{z_2})$. By the univalence of h, in the first case $\sqrt{z_1} = \sqrt{z_2}$ and in the second case $\sqrt{z_1} = -\sqrt{z_2}$. Squaring we see that in either case $z_1 = z_2$.

8.11

▷ We have

$$g(z) = z - (\alpha + \beta) + (\alpha\beta)/z = \frac{1}{z}(z - \alpha)(z - \beta),$$

so that

$$f(z) = 1/g(1/z) = \frac{z}{(1 - \alpha z)(1 - \beta z)}.$$

The choice $\alpha = \beta = 1$ gives the Koebe function $k(z) = z/(1 - z)^2$ while the choice $\alpha = 1, \beta = -1$ gives its square root transform $h(z) = z/(1 - z^2)$.

▷ Write $\alpha\beta = e^{i\theta}$ for $0 \leq \theta < 2\pi$ and consider $g_0(z) = z + e^{i\theta}/z$. As suggested, $g_0(e^{it}) = 2e^{i\theta/2}\cos(t - \theta/2)$. As t ranges over $[0, 2\pi]$, the image of the unit circle under g_0 is the line segment $[-2e^{i\theta/2}, 2e^{i\theta/2}]$ (described twice). Then, $g_0(\mathbb{U})$ is the complement of this line segment.

Writing $\alpha = e^{i\phi}$, we see that $\beta = e^{i\theta}/\alpha = e^{i(\theta-\phi)}$ and that

$$\alpha + \beta = e^{i\phi} + e^{i(\theta-\phi)} = e^{i\theta/2}\left(e^{i(\phi-\theta/2)} + e^{i(\theta/2-\phi)}\right) = 2e^{i\theta/2}\cos(\phi - \theta/2).$$

Then, $g(\mathbb{U}) = g_0(\mathbb{U}) - (\alpha+\beta)$, in other words the complement of the line segment $[-2e^{i\theta/2}(1 + \cos(\phi - \theta/2)), 2e^{i\theta/2}(1 - \cos(\phi - \theta/2))]$, which still contains the origin since the cosine term lies between -1 and 1.

8.12 By definition of $\delta_D(z)$, there is a sequence of points $\{w_n\}_{n=1}^\infty$ in $\mathbb{C} \setminus D$ with $|z - w_n| \to \delta_D(z)$ as $n \to \infty$. Since $\{w_n\}_{n=1}^\infty$ is a bounded sequence, it has a convergent subsequence $\{w_{n_k}\}_{k=1}^\infty$ with $w_{n_k} \to w$, say, as $k \to \infty$. The point w also lies in $\mathbb{C} \setminus D$ since the complement of D is closed. In addition, $|z - w| = \lim_{k \to \infty}|z - w_{n_k}| = \delta_D(z)$. Hence, $w \in \mathbb{C} \setminus D$ and $|z - w| = \delta_D(z)$.

8.13 With f convex univalent in the class $\mathbf{S}$, w an omitted value for f, and $h(z) = (f(z) - w)^2$, suppose that $h(z_1) = h(z_2)$, z_1 and z_2 being both in the unit disk. Then, $f(z_1)^2 + w^2 - 2wf(z_1) = f(z_2)^2 + w^2 - 2wf(z_2)$. The w^2 terms cancel and then $f(z_1) - f(z_2)$ is a factor so that

$$0 = 2[f(z_1) - f(z_2)]\left(\frac{f(z_1) + f(z_2)}{2} - w\right).$$

The second factor doesn't vanish as, if it did, the point w would then be the midpoint of two points $f(z_1)$ and $f(z_2)$ in the convex domain $D = f(\mathbb{D})$ and would then have to be in D itself as well. Hence, $f(z_1) - f(z_2) = 0$ and so $z_1 = z_2$ since f is univalent.

Since $h(0) = w^2$ and $h'(0) = -2w$, the normalised form of h is

$$g(z) = \frac{h(z) - h(0)}{h'(0)} = \frac{h(z) - w^2}{-2w}.$$

The function f never takes the value w, so h never takes the value 0, and so $w/2$ is an omitted value for the function g in **S**. It follows from the Koebe 1/4-Theorem that $|w/2| \geq 1/4$ and so $|w| \geq 1/2$.

8.14 For each univalent function $f \in \mathscr{F}$, set $g(z) = (f(z) - f(0))/f'(0)$, $z \in \mathbb{D}$, so that g is in the class **S**. By the Growth Theorem, $|g(z)| \leq r/(1-r)^2$ for $z = re^{i\theta}$ in $\mathbb{D}$. Then,

$$|f(z)| \leq |f(z) - f(0)| + |f(0)| \leq |f'(0)||g(z)| + C_1 \leq C_2 \frac{r}{(1-r)^2} + C_1.$$

As in the proof of Theorem 8.10, the family $\mathscr{F}$ is uniformly bounded on compact subsets of $\mathbb{D}$ and is therefore a normal family by Montel's Theorem.

Exercises from Chap. 9

9.1 The given expressions for u_r and u_θ follow directly from the Chain Rule. For example,

$$u_r = u_x \frac{\partial x}{\partial r} + u_y \frac{\partial y}{\partial r} = u_x \cos\theta + u_y \sin\theta.$$

Multiplying the expression for u_r by $r\sin\theta$ and the expression for u_θ by $\cos\theta$ and adding gives

$$ru_r \sin\theta + u_\theta \cos\theta = ru_y \sin^2\theta + ru_y \cos^2\theta = ru_y.$$

Similarly, multiplying the expression for u_r by $r\cos\theta$ and the expression for u_θ by $-\sin\theta$ and adding gives

$$ru_r \cos\theta - u_\theta \sin\theta = ru_x \cos^2\theta + ru_x \sin^2\theta = ru_x.$$

Next,
$$u_{xx} = (u_x)_x = (u_x)_r \cos\theta - \frac{1}{r}(u_x)_\theta \sin\theta$$
$$= (u_r \cos\theta - \frac{1}{r}u_\theta \sin\theta)_r \cos\theta - \frac{1}{r}(u_r \cos\theta - \frac{1}{r}u_\theta \sin\theta)_\theta \sin\theta$$
$$= \left(u_{rr} \cos\theta - \frac{1}{r}u_{r\theta} \sin\theta + \frac{1}{r^2}u_\theta \sin\theta\right) \cos\theta$$
$$- \frac{1}{r}\left(u_{r\theta} \cos\theta - u_r \sin\theta - \frac{1}{r}u_{\theta\theta} \sin\theta - \frac{1}{r}u_\theta \cos\theta\right) \sin\theta$$
$$= u_{rr} \cos^2\theta + \frac{1}{r^2}u_{\theta\theta} \sin^2\theta + \frac{1}{r}u_r \sin^2\theta + \left(\frac{1}{r^2}u_\theta - \frac{1}{r}u_{r\theta}\right) \sin(2\theta).$$

In a similar manner, one finds that
$$u_{yy} = u_{rr} \sin^2\theta + \frac{1}{r^2}u_{\theta\theta} \cos^2\theta + \frac{1}{r}u_r \cos^2\theta + \left(\frac{1}{r}u_{r\theta} - \frac{1}{r^2}u_\theta\right) \sin(2\theta).$$

Adding these expressions gives
$$\Delta u = u_{xx} + u_{yy} = u_{rr} + \frac{1}{r^2}u_{\theta\theta} + \frac{1}{r}u_r,$$

as required.

9.2 A function $u(r,\theta) = f(r)$ is harmonic if and only if, by (9.3), $\frac{d}{dr}(rf'(r)) = 0$, that is if and only if $rf'(r) = a$, a constant. Then, $f(r) = a\log r + b$. Note that this solution has a logarithmic singularity at the origin if $a \neq 0$.

9.3 Here, $u(r,\theta) = \log\rho(r,\theta) = \log 2 - \log(1+r^2)$, so that $u_r = -2r/(1+r^2)$. Then,
$$\Delta u = \frac{1}{r}\frac{\partial}{\partial r}(ru_r) = \frac{1}{r}\frac{\partial}{\partial r}\left(\frac{-2r^2}{1+r^2}\right) = -\frac{4}{(1+r^2)^2} = -\rho^2 = -e^{2u}.$$

9.4 We may assume that D contains the lower half-plane $\overline{\mathbb{H}}$. Choose a point x_0 on the real axis that lies in D (if no such choice is possible, then $D = \overline{\mathbb{H}}$ and there is nothing to prove). Set $w = M(z) = 1/(x_0 - z)$, $z \in D$, and set $\Omega = M(D \setminus \{x_0\})$. The complement of Ω is bounded since D contains a disk about x_0. Also, M fixes the real axis and maps the lower half-plane to itself.

As in the proof of Theorem 9.3, set $u_D(z) = \log\rho_D(z)$, set $f(w) = M^{-1}(w) = x_0 - 1/w$, and set $v(w) = u_D(f(w)) + \log|f'(w)| = u_D(x_0 - 1/w) + \log(1/|w|^2)$, $w \in \Omega$. As before, v satisfies the conditions of Proposition 9.3 so that $v(w) \geq v(\overline{w})$ for $w \in \Omega \cap \mathbb{H}$. For z in $D \cap \mathbb{H}$, the point $w = M(z)$ lies in $\Omega \cap \mathbb{H}$ and so
$$u_D(z) + \log|x_0 - z|^2 \geq u_D(\overline{z}) + \log|x_0 - \overline{z}|^2,$$

so that $u_D(z) \geq u_D(\overline{z})$ or $\rho_D(z) \geq \rho_D(\overline{z})$.

The case of equality is already dealt with at the end of the proof of Theorem 9.3.

Solutions to the Exercises

9.5

▷ By rotation and translation, we may assume that D is symmetric in the real axis and that $[\alpha, \beta]$ is the line segment $[0, x]$ ($x > 0$), contained in D. Let $f(z) = \sum_{n=1}^{\infty} a_n z^n$ be the Riemann map of the unit disk $\mathbb{D}$ onto D, so that $f(0) = 0$ and $f'(0) > 0$. Consider

$$g(z) = \overline{f(\bar{z})} = \sum_{n=1}^{\infty} \overline{a_n} z^n, \quad z \in \mathbb{D}.$$

Then, g is also univalent and maps $\mathbb{D}$ onto $\overline{D}$. But $\overline{D} = D$ since D is symmetric in the real axis. Also, $g(0) = 0$ and $g'(0) = \overline{a_1} = \overline{f'(0)} = f'(0)$ since $f'(0) > 0$. By the uniqueness of the Riemann map, $g = f$, and then by the uniqueness of the coefficients in power series, $a_n = \overline{a_n}$ for each n. It follows that these coefficients are real and so f maps the interval $(-1, 1)$ in $\mathbb{D}$ into the real axis. With $f(r) = x$, it follows that $f([0, r]) = [0, x]$. Since $[0, r]$ is a geodesic arc in the unit disk it follows that $[0, x]$ is a geodesic arc in D (see Theorem 6.10).

▷ By rotation and translation, we may assume that L is the real axis and that the Euclidean line segment $[0, x]$ is the geodesic arc joining 0 to x in D. Let f be a conformal map of the unit disk $\mathbb{D}$ onto D with $f(0) = 0$. By precomposing f with a rotation, we may arrange it so that $f^{-1}(x)$ lies between 0 and 1. Writing r for $f^{-1}(x)$, the image of the geodesic arc joining 0 to r in $\mathbb{D}$, that is the Euclidean line segment $[0, r]$, is mapped to the geodesic arc joining 0 to x in D, that is $[0, x]$.

For $0 < t < r$, $f'(t) = \lim_{h \to 0}(f(t + h) - f(t))/h$. Considering only real h, and since f takes real values on $[0, r]$, we see that $f'(t)$ is real on $(0, r)$. Then, $f''(t) = \lim_{h \to 0}(f'(t + h) - f'(t))/h$ is similarly also real on $(0, r)$. By induction, all derivatives of f on $(0, r)$ are real. Hence, $f^{(n)}(0)$ is real for each n. With the power series for f being $f(z) = \sum_{n=1}^{\infty} a_n z^n$, and since $a_n = f^{(n)}(0)/n!$, all coefficients of f are real. Hence, f satisfies $f(\bar{z}) = \overline{f(z)}$. This, in turn, implies that $D = f(\mathbb{D})$ is symmetric in the real axis, for if $w \in D$, say $w = f(z)$, $z \in \mathbb{D}$, then $\overline{w} = \overline{f(z)} = f(\bar{z})$ also lies in D.

Exercises from Chap. 10

10.1 Any covering space (Ω, f) of D is isomorphic to itself via the identity map on Ω.

If g is an isomorphism of (Ω_1, f_1) to (Ω_2, f_2) then g^{-1} (which makes sense since g is a homeomorphism) is an isomorphism of (Ω_2, f_2) to (Ω_1, f_1). In fact, since $f_1 = f_2 \circ g$ on Ω_1, precomposing with g^{-1} gives $f_1 \circ g^{-1} = f_2$ on Ω_2.

As effectively noted just prior to this exercise, if (Ω_1, f_1), (Ω_2, f_2) and (Ω_3, f_3) are all covering spaces of D, if g is a isomorphism from (Ω_1, f_1) to (Ω_2, f_2) and

h is a isomorphism from (Ω_2, f_2) to (Ω_3, f_3) then $h \circ g$ is a isomorphism from (Ω_1, f_1) to (Ω_3, f_3).

10.2 If g was a homomorphism of $(\mathbb{C}_0, z^n)$ to $(\mathbb{C}, \exp)$ then we would have g analytic in $\mathbb{C}_0$ (this is a consequence of Proposition 10.1) with $e^{g(z)} = z^n$. The function g would have an isolated singularity at 0 which would have to be removable. If not, either $g(z) \to \infty$ as $z \to 0$ if the singularity was a pole or $g(D(0, \epsilon))$ is dense in $\mathbb{C}$ for each ϵ positive if the singularity was essential. In either case, this would be incompatible with $e^{g(z)} = z^n \to 0$ as $z \to 0$. Thus, g is entire and

$$0 = \lim_{z \to 0} z^n = \lim_{z \to 0} e^{g(z)} = e^{g(0)},$$

which is impossible as the exponential function never takes the value zero.

10.3 If the covering space $(\mathbb{C}_0, f_n)$ is homomorphic to $(\mathbb{C}_0, f_m)$, where $n \in \mathbb{N}$ and $m \in \mathbb{N}$, there is an analytic map $g: \mathbb{C}_0 \to \mathbb{C}_0$ such that $f_n(z) = f_m(g(z))$, that is $z^n = (g(z))^m$, for every $z \in \mathbb{C}_0$. Since g is analytic in $\mathbb{C}_0$, it has an isolated singularity at 0. This must be a removable singularity since $(g(z))^m \to 0$ as $z \to 0$ (if not and g has a pole at 0 then $|g(z)| \to \infty$ as $z \to 0$ and, if g has an essential singularity at 0, then the Casorati-Weierstrass Theorem implies that g cannot have a limit at 0). Moreover, g takes the value 0 at the removable singularity at 0. Let $g(z) = \sum_{i=k}^{\infty} a_i z^i$ be the power series expansion of g about 0 where $k \geq 1$ and $a_k \neq 0$. Then, $(g(z))^m$ has an expansion that begins with the non-vanishing term $a_k^m z^{km}$. We must have $a_k^m z^{km} = z^n$ so that $km = n$ (and $a_k^m = 1$). Thus, m divides n and $k = n/m$. Not only but it must be that $g(z) = a_k z^k$. If the expansion of g had a further non-zero term, the next one being $a_j z^j$, then $(g(z))^m$ would have the additional non-zero term $m a_k^{m-1} a_j z^{(m-1)k+j}$, which is not possible since $(g(z))^m$ is simply z^n.

In summary, the covering space $(\mathbb{C}_0, f_n)$ is homomorphic to $(\mathbb{C}_0, f_m)$ only if m divides n. The homomorphism g is then $g(z) = \alpha z^{n/m}$ where α is an mth-root of unity. The converse is clear.

10.4 The 'c' side of $T_{2,l}$ is the semicircle $C^+(1/6, 1/6)$. By the formula (2.15) for reflection,

$$R_c\left(\tfrac{1}{2}\right) = \frac{1}{6} + \frac{(1/6)^2}{1/2 - 1/6} = \frac{1}{4}.$$

The 'b' side of $T_{2,l}$ is the semicircle $C^+(5/12, 1/12)$. By the formula (2.15) for reflection,

$$R_b(0) = \frac{5}{12} + \frac{(1/12)^2}{0 - 5/12} = \frac{2}{5}.$$

Solutions to the Exercises

The reflected regions $T_{2,l,l}$ and $T_{2,l,r}$ are shown in the figure below. A similar

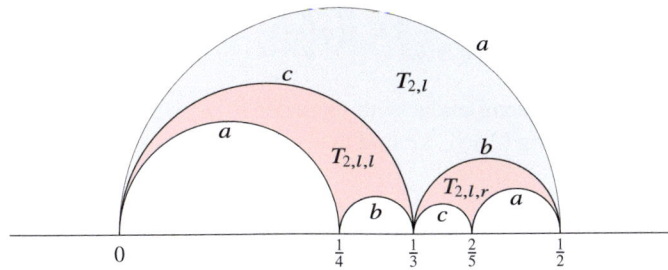

calculation starting with the hyperbolic triangle $T_{2,r}$ gives rise to the hyperbolic triangle $T_{2,r,l}$ with vertices $\{\frac{1}{2}, \frac{3}{5}, \frac{2}{3}\}$ and the hyperbolic triangle $T_{2,r,r}$ with vertices $\{\frac{2}{3}, \frac{3}{4}, 1\}$.

10.5

▷ Suppose that $x|(p+r)$ and that $x|(q+s)$. Then x divides $q(p+r) - p(q+s) = qr - ps = 1$, hence $x = 1$.

▷ Since $qr > ps$, we have $q(p+r) > p(q+s)$ and so $\frac{p}{q} < \frac{p}{q} \oplus \frac{r}{s}$. Similarly, $r(q+s) > s(p+r)$ and so $\frac{p}{q} \oplus \frac{r}{s} < \frac{r}{s}$.

By definition, $\frac{p}{q}$ and $\frac{p}{q} \oplus \frac{r}{s}$ are Farey neighbours if and only if $q(p+r) - p(q+s) = 1$, which is the case since $qr - ps = 1$. Similarly, $\frac{p}{q} \oplus \frac{r}{s}$ and $\frac{r}{s}$ are Farey neighbours.

▷ Suppose that $\frac{m}{n}$ lies between $\frac{p}{q}$ and $\frac{r}{s}$. Then,

$$\frac{1}{qs} = \frac{r}{s} - \frac{p}{q} = \left(\frac{r}{s} - \frac{m}{n}\right) + \left(\frac{m}{n} - \frac{p}{q}\right)$$
$$= \frac{rn - ms}{sn} + \frac{mq - pn}{nq} \geq \frac{1}{sn} + \frac{1}{nq} = \frac{q+s}{nsq}.$$

Multiplying across by nsq, we deduce that $n \geq q + s$. It is simple to check that $\frac{p}{q} \oplus \frac{r}{s}$ is the only fraction with denominator $q + s$ that lies between $\frac{p}{q}$ and $\frac{r}{s}$.

10.6

▷ The automorphism

$$M(z) = \frac{z - p/q}{r/s - p/q} = sq\left(z - \frac{p}{q}\right) = s(qz - p)$$

of the upper half-plane sends p/q to 0 and sends r/s to 1. Here we used $qr - ps = 1$.

▷ Again using $qr - ps = 1$, we find that

$$x = M\left(\frac{p+r}{q+s}\right) = s\left(q\frac{p+r}{q+s} - p\right) = \frac{s}{q+s}.$$

▷ Writing e' for the (semi)circle with diameter $[0, x]$ which has centre $x/2$ and radius $x/2$, and using (2.15), we find that

$$R_{e'}(1) = \frac{x}{2} + \frac{x^2/4}{1 - x/2} = \frac{x}{2-x} = \frac{s}{2q+s}.$$

Writing f' for the (semi)circle with diameter $[x, 1]$, which has centre $(1+x)/2$ and radius $(1-x)/2$, and using (2.15), we find that

$$R_{f'}(0) = \frac{1+x}{2} + \frac{(1-x)^2/4}{0-(1+x)/2} = \frac{2x}{1+x} = \frac{2s}{q+2s}.$$

▷ Bearing in mind that points symmetric in a generalised circle are preserved under Möbius transformations we see that, with the notation of Lemma 10.4, $R_e(r/s) = M^{-1}(R_{e'}(1))$ and that $R_f(p/q) = M^{-1}(R_{f'}(0))$. Since $M^{-1}(z) = p/q + w/(qs)$, we find that

$$R_f\left(\frac{p}{q}\right) = M^{-1}(R_{f'}(0)) = \frac{p}{q} + \frac{R_{f'}(0)}{qs} = \frac{p}{q} + \frac{2s}{qs(q+2s)}$$
$$= \frac{2ps + pq + 2}{q(q+2s)} = \frac{2ps + pq + 2(qr - ps)}{q(q+2s)} = \frac{p+2r}{q+2s},$$

which is (10.7). Similarly,

$$R_e\left(\frac{r}{s}\right) = M^{-1}(R_{e'}(1)) = \frac{p}{q} + \frac{R_{e'}(1)}{qs} = \frac{p(2q+s)+1}{q(2q+s)}.$$

Replacing '1' by $qr - ps$ leads directly to (10.4).

10.7 Reflection in the 'c' side of T_0 is $R_c(z) = \bar{z}/(2\bar{z}-1)$. Reflection in the 'b' side of T_1, which is the semicircle $C^+(\frac{3}{4}, \frac{1}{4})$, is

$$R_b(z) = \frac{3}{4} + \frac{(1/4)^2}{\bar{z} - 3/4} = \frac{3\bar{z} - 2}{4\bar{z} - 3}.$$

Then,

$$\tau(z) = R_b(R_a(z)) = \frac{3z - 2(2z-1)}{4z - 3(2z-1)} = \frac{z-2}{2z-3}.$$

Solutions to the Exercises

Now we write τ in terms of τ_1 and τ_2 via

$$\tau(z) = \frac{z-2}{2(z-2)+1} = (\tau_1 \circ \tau_2^{-1})(z).$$

10.8 The 'a' side of T_1 is the semicircle $C^+(\frac{1}{4}, \frac{1}{4})$ and so

$$R_c(z) = \frac{(1/4)\bar{z}}{\bar{z}-1/4} = \frac{\bar{z}}{4\bar{z}-1}.$$

The 'b' side of $T_{2,l}$ is the semicircle $C^+(\frac{5}{12}, \frac{1}{12})$ and so

$$R_b(z) = \frac{5}{12} + \frac{(1/12)^2}{\bar{z}-5/12} = \frac{5\bar{z}-2}{12\bar{z}-5}.$$

The composition of these reflections is

$$\tau(z) = R_b(R_c(z)) = \frac{5z - 2(4z-1)}{12z - 5(4z-1)} = \frac{3z-2}{8z-5}.$$

After some experimentation, one finds that $\tau = \tau_1 \circ \tau_2 \circ \tau_1^{-1}$.

10.9 The 'a' side of $T_{2,r}$ (Fig. 10.7) is the semicircle $C^+(\frac{7}{12}, \frac{1}{12})$ and so

$$R_a(z) = \frac{7}{12} + \frac{(1/12)^2}{\bar{z}-7/12} = \frac{7\bar{z}-4}{12\bar{z}-7}.$$

The 'c' side of $T_{2,r,l}$ is the semicircle $C^+(\frac{19}{30}, \frac{1}{30})$ and so

$$R_c(z) = \frac{19}{30} + \frac{(1/30)^2}{\bar{z}-19/30} = \frac{19\bar{z}-12}{30\bar{z}-19}.$$

The composition of these reflections is

$$\tau(z) = R_c(R_a(z)) = \frac{19(7z-4) - 12(12z-7)}{30(7z-4) - 19(12z-7)} = \frac{11z-8}{18z-13}.$$

Applying this τ to the automorphism that sends T_0 to $T_{2,r}$, we obtain the automorphism $\tilde{\tau}$ that sends T_0 to its grand-grandchild $T_{2,r,l,r}$. The automorphism that sends T_0 to $T_{2,r}$ is, by Exercise 10.7, $(\tau_1 \circ \tau_2^{-1})(z) = (z-2)/(2z-3)$. Then,

$$\tilde{\tau}(z) = \tau\left(\frac{z-2}{2z-3}\right) = \frac{11(z-2) - 8(2z-3)}{18(z-2) - 13(2z-3)} = \frac{5z-2}{8z-3}.$$

Again, with some experimentation, $\tilde{\tau}(z) = (\tau_1 \circ \tau_2^{-1} \circ \tau_1^{-1})(z)$. Then, $\tau = \tau_1 \circ \tau_2^{-1} \circ \tau_1^{-1} \circ \tau_2 \circ \tau_1^{-1}$ is the automorphism that sends $T_{2,r}$ to its grandchild $T_{2,r,l,r}$.

10.10 We again wish to send p/q to ∞, and so we set

$$M(z) = \frac{p/q - r/s}{p/q - z} = \frac{1}{s(p - qz)},$$

where we used $ps - qr = 1$. Then, $M^{-1}(w) = p/q - 1/(qsw)$. The 'd' side of T in Fig. 10.10 is again sent to the line $\Re z = 1$. Since

$$M\left(\frac{p}{q} \oplus \frac{r}{s}\right) = \frac{q + s}{s},$$

the 'f' side of T is sent to the line $\Re z = (q+s)/s$. Exactly as in the proof of Lemma 10.7, $\tilde{R}_l(w) = w + 2q/s$. Then, as was the case for $R_r(z)$,

$$R_l(z) = (M^{-1} \circ \tilde{R}_l \circ M)(z) = M^{-1}\left(M(z) + \frac{2q}{s}\right) = M^{-1}\left(\frac{1}{s(p - qz)} + \frac{2q}{s}\right)$$

$$= M^{-1}\left(\frac{(1 + 2pq) - 2q^2 z}{s(p - qz)}\right) = \frac{p}{q} - \frac{s(p - qz)}{qs[(1 + 2pq) - 2q^2 z]}$$

which simplifies to (10.11).

10.11 For the (2, 1) entry,

$$2q^2(1 + 2rs) + 2s^2(1 - 2pq) = 2q^2 + 2s^2 + 4q^2 rs - 4s^2 pq$$
$$= 2q^2 + 2s^2 + 4qs(qr - ps)$$
$$= 2(q^2 + s^2 + 2qs) = 2(q + s)^2.$$

For the (2, 2) entry,

$$(2q^2)(-2r^2) + (1 - 2pq)(1 - 2rs) = 1 + 4pqrs - 2pq - 2rs - 4q^2 r^2$$
$$= 1 + 4qr(ps - qr) - 2pq - 2rs$$
$$= 1 - 4qr - 2pq - 2rs$$
$$= 1 - 2qr - 2(1 + ps) - 2pq - 2rs$$
$$= -1 - 2(p + r)(q + s).$$

10.12

▷ The double reflection $R_{r,0}$ sends T_0 to the hyperbolic triangle with vertices 0, $\frac{1}{3}$ and $\frac{1}{2}$. A second double reflection $R_{r,0}$ sends this hyperbolic triangle to the

Solutions to the Exercises

hyperbolic triangle T with vertices 0, $\frac{1}{5}$ and $\frac{1}{4}$. Combined, this gives $\tau = R_{r,0}^2 = \tau_1^2$ as $R_{r,0} = \tau_1$.

▷ If we follow $R_{r,0}$ with the double reflection $R_{r,1/3}$, we find that T_0 is mapped to the hyperbolic triangle T with vertices $\frac{1}{3}, \frac{3}{8}$ and $\frac{2}{5}$. Now $\frac{1}{3} = \frac{0}{1} \oplus \frac{1}{2}$ as Farey neighbours. Then, by (10.13) and (10.14),

$$R_{r,1/3} = R_{l,1/2} \circ R_{l,0} = (R_{r,0} \circ R_{r,1}) \circ R_{l,0}.$$

The required automorphism τ is then

$$\tau = R_{r,1/3} \circ R_{r,0} = (R_{r,0} \circ R_{r,1} \circ R_{l,0}) \circ R_{r,0} = R_{r,0} \circ R_{r,1}$$

since $R_{l,0} = R_{r,0}^{-1}$. Then, $\tau = \tau_1 \circ \tau_2 \circ \tau_1^{-1}$.

▷ To get from T_0 to the triangle T with vertices $\frac{3}{5}, \frac{5}{8}, \frac{2}{3}$, we first apply the double reflection $R_{l,1}$ then the double reflection $R_{l,2/3}$. Since $\frac{2}{3}$ is the Farey sum of the Farey neighbours $\frac{1}{2}$ and $\frac{1}{1}$, we can compute $R_{l,2/3}$ as

$$R_{l,2/3} = R_{r,1/2} \circ R_{r,1} = (R_{l,1} \circ R_{l,0}) \circ R_{r,1}.$$

Then,

$$\tau = R_{l,2/3} \circ R_{l,1} = R_{l,1} \circ R_{l,0} = \tau_1 \circ \tau_2^{-1} \circ \tau_1^{-1}.$$

10.13

▷ The double reflection in Exercise 10.7 is $R_{l,1} = \tau_1 \circ \tau_2^{-1}$.
▷ The double reflection in Exercise 10.8 is

$$R_{l,1/2} = R_{r,0} \circ R_{r,1} = \tau_1 \circ (\tau_2 \circ \tau_1^{-1}).$$

▷ The double reflection in Exercise 10.9 is

$$R_{l,2/3} = R_{r,1/2} \circ R_{r,1} = (R_{l,1} \circ R_{l,0}) \circ R_{r,1} = (\tau_1 \circ \tau_2^{-1}) \circ \tau_1^{-1} \circ (\tau_2 \circ \tau_1^{-1}).$$

References

1. Ahlfors, L.: Conformal Invariants: Topics in Geometric Function Theory. Reprinted by AMS Chelsea Publishing. American Mathematical Society Providence, Rhode Island (2010)
2. Ahlfors, L., Sario, L.: Riemann Surfaces. Princeton Mathematical Series 26, Princeton University Press, Princeton (1960)
3. Beliaev, D.: Conformal Maps and Geometry. Advanced Textbooks in Mathematics. World Scientific Publishing Co. Pte. Ltd., Hackensack (2020)
4. Brown Flinn, B.: Hyperbolic convexity and level sets of analytic functions. Indiana Univ. Math. J. **32**(6), 831–841 (1983)
5. Burckel, R.B.: An Introduction to Classical Complex Analysis, vol. 1. Pure Appl. Math., 82. Academic Press, Inc. [Harcourt Brace Jovanovich, Publishers], New York-London (1979)
6. Collingwood, E.F., Lohwater, A.J.: The Theory of Cluster Sets. Cambridge Tracts in Mathematics and Mathematical Physics, vol. 56. Cambridge University Press, Cambridge (1966)
7. Conway, J.B.: Functions of One Complex Variable. Graduate Texts in Mathematics, vol. 11, 2nd edn. Springer, New York-Berlin (1978)
8. Conway, J.B.: Functions of One Complex Variable II. Graduate Texts in Mathematics, vol. 159. Springer, New York (1995)
9. de Branges, L.: A proof of the Bieberbach conjecture. Acta Math. **154**(1–2), 137–152 (1985)
10. Duren, P.L.: Univalent Functions. Grundlehren der mathematischen Wissenschaften, vol. 259. Springer, New York (1983)
11. Fisher, Y., Hubbard, J.H., Wittner, B.S.: A proof of the uniformisation theorem for arbitrary plane domains. Proc. Am. Math. Soc. **104**(2), 413–418 (1988)
12. Greenberg, M.J.: Euclidean and non-Euclidean Geometries, Development and History, 2nd edn. W. H. Freeman and Company, New York (1980)
13. Hayman, W.K.: The asymptotic behaviour of p-valent functions. Proc. Lond. Math. Soc. (3) **5**, 257–284 (1955)
14. Hayman, W.K.: Meromorphic Functions. Oxford Mathematical Monographs. Clarendon Press, Oxford (1964)
15. Hayman, W.K.: Multivalent Functions, 2nd edn. Cambridge Tracts in Mathematics, vol. 110. Cambridge University Press, Cambridge (1994)
16. Howie, J.M.: Complex Analysis. Springer Undergraduate Mathematics Series. Springer-Verlag London, Ltd., London (2003)
17. Jørgensen, V.: On an inequality for the hyperbolic measure and its applications in the theory of functions. Math. Scand. **4**, 113–124 (1956)
18. Kodaira, K.: Complex Analysis. Cambridge Studies in Advanced Mathematics, vol. 107. Cambridge University Press, Cambridge (2007)

19. Krantz, S.: Complex Analysis: The Geometric Viewpoint, 2nd edn. Carus Mathematical Monographs, vol. 23. Mathematical Association of America, Washington (2004)
20. Nehari, Z.: Conformal Mapping. Reprinting of the 1952 edition. Dover Publications, Inc., New York (1975)
21. Pommerenke, Ch.: Univalent Functions: With a Chapter on Quadratic Differentials by Gerd Jensen. Studia Mathematica/Mathematische Lehrbücher, Band XXV. Vandenhoeck & Ruprecht, Göttingen (1975)
22. Pommerenke, Ch.: Boundary Behaviour of Conformal Maps. Grundlehren der mathematischen Wissenschaften, vol. 299. Springer, Berlin (1992)
23. Remmert, R.: Classical Topics in Complex Function Theory. Graduate Texts in Mathematics, vol. 172. Springer, New York (1997)
24. Simmons, G.F.: Introduction to Topology and Modern Analysis. Reprint of the 1963 original. Robert E. Krieger Publishing Co., Inc., Melbourne (1983)
25. Stein, E.M., Shakarchi, R.: Complex Analysis. Princeton Lectures in Analysis 2, Princeton University Press, Princeton (2003)
26. Suffridge, T.J.: On univalent polynomials. J. Lond. Math. Soc. **44**, 496–504 (1969)
27. Thomas, D.K., Tuneski, N., Vasudevarao, A.: Univalent Functions. A Primer. De Gruyter Studies in Mathematics, vol. 69. De Gruyter, Berlin (2018)
28. Zalcman, L.: A heuristic principle in complex function theory. Am. Math. Mon. **82**, 813–817 (1975)
29. Zalcman, L.: Normal families: new perspectives. Bull. Am. Math. Soc. (N. S.) **35**, 215–230 (1998)
30. Zakeri, S.: A Course in Complex Analysis. Princeton University Press, Princeton (2021)

Index

A
Alexandroff one-point compactification, 28
Area Theorem, 211
Arzelà-Ascoli Theorem, 104, 111, 119
Automorphism
    of the disk $\mathbb{D}$, 84
    of a domain, 81
    of half-plane $\mathbb{H}$, 89
    of a simply connected domain, 160

B
Bieberbach Conjecture, 214–216
Bieberbach's Coefficient Estimate, 214

C
Carathéodory kernel convergence, 234
Carathéodory kernel theorem, 234
Casorati-Weierstrass Theorem, 25, 103, 138, 158
Cauchy's Integral Formula, 143, 195
Cauchy's Theorem
    for a disk, 142
    homotopic form, 143, 145–148
Conformal mapping, 6, 81
    conformally equivalent annuli, 182
    conformally equivalent domains, 158, 182
    prime end, 175
Conformal radius, 163
Convex set, 237
Covering space, 258
    analytic covering map, 258
    analytic covering space, 258
    automorphism, 269

covering map, 258
evenly covered, 258
fibre, 258
homomorphism, 264
isomorphism, 264
lift of a curve, 262
universal covering space, 269
Curve, 6
    closed, 6
    piecewise-smooth, 6, 47
    smooth, 6
    winding number, 197

D
Domain, 6
    convex, 218
    geometrically simple, 175
    hyperbolic, 257, 288
    proper simply connected, 141
    simply connected, 145, 199
    star-shaped, 142

E
Equicontinuity, 109
    equicontinuity at a point, 112
Euclid's Postulates, 74, 77
    Parallel Postulate, 74, 77, 95
Even function, 209
Exhaustion by compact subsets, 114
Extended complex plane, 13, 33
    extended real line, 13
    point at infinity, 13

F
Farey addition, 274
  Farey neighbours, 274

G
Generalised circle, 8, 13
Geodesic, 65
  in disk $\mathbb{D}$, 95
  spherical geodesic, 72
  spherical metric, 68
Geodesic arc, 65
  in disk $\mathbb{D}$, 93
Great Picard Theorem, 25, 103, 137–139, 257

H
Harmonic function
  Laplace Operator, 237
  maximum principle, 252
  mean-value property, 253
Homeomorphism, 81
Homotopy, 144
  of closed curves, 144
  fixed endpoint, 159
  homotopic to zero, 145
Hurwitz's Theorem, 127, 228, 230
Hyperbolically convex set, 237
Hyperbolic distance
  in the disk $\mathbb{D}$, 91, 93
  in half-plane $\mathbb{H}$, 98, 101
Hyperbolic metric, 91, 236
  conformal invariance of, 167
  on the disk $\mathbb{D}$, 91
  on a half-disk, 172
  on a half-plane, 169
  monotonicity of, 168
  on a sector, 171
  on a simply connected domain, 161
  on a slit-disk, 174
  on a slit-plane, 172
  on a strip, 170

I
Induced topology, 37
Initial topology, 37

K
Kernel of a sequence of domains w.r.t. a point, 231
Koebe 1/4-Theorem, 216

L
Latitude, 30
Little Picard Theorem, 103, 137, 257, 287
Locally bounded, 121
Locally uniform convergence, 118, 121–123
Locally uniform spherical convergence, 121, 123–126
Longitude, 30

M
Möbius transformation
  dilation, 11
  inversion, 11
  rotation, 11
  translation, 11
Marty's Theorem, 104, 131–132
Meromorphic function, 43
  on Riemann sphere, 41, 45
  spherical derivative, 63
Metric space isometry, 59
Möbius transformation, 11
Modular function, 257, 271–280
  automorphism group, 280–287
Monodromy Theorem, 262
Montel Theorem
  Montel's First Theorem, 104, 129, 135
  Montel's Second Theorem, 104, 136–137, 287

N
Neighbourhood, 6
Non-Euclidean geometry, 75, 95
Normal family, 103, 128, 136

O
Odd function, 209
Open base for a topology, 36

P
Poincaré metric, 91
Precompact, 106
Pre-kernel of a sequence of domains, 231
Pseudo-hyperbolic distance, 94
Pseudo-spherical distance, 69
Pseudometric, 115

R
Relatively compact, 106
Riemann map, 152
Riemann Mapping Theorem, 104, 152–157

Index 353

Riemann sphere, 27, 28, 33, 38
    antipodal points, 33
    great circle, 36, 73
    north pole, 29
    point at infinity, 33
        analytic at infinity, 40
        essential singularity at infinity, 42
        pole at infinity, 42
        zero at infinity, 40
    spherical cap, 36
    spherical metric, 46, 55
        completeness of, 58
        great circle, 59
        isometry of, 64, 65
        spherical distance, 69
    stereographic projection, 28, 30, 33
        conformality of, 51, 53
Riemann surface, 38
    chart, 39
Riemannian metric, 46
    ρ-length of a curve, 48
    conformal metric, 48
    distance between points, 49
    isometry of, 59
Runge's Theorem, 189–195

S
Schwarz Lemma, 78–80
Schwarz-Pick Lemma
    First Version, 89
    Second Version, 98
    Third Version, 166
Schwarz Reflection Principle, 15, 183, 186
    for a circle, 23
    for a line, 16

Self-map, 80
Sequentially compact, 106
Simply connected domain, 148
Singularity, 24
    essential, 25
    Laurent series, 24
    pole, 25
    removable, 24
Symmetry with respect to a circle, 21

T
Topologically equivalent metrics, 117
Totally bounded, 106

U
Uniformisation theorem, 29
Univalent function, 6, 127, 203
    conjugation of, 209
    convex univalent function, 218
    Koebe function, 206
    local univalence, 203
    omitted value transformation, 209
    rotation of, 208
    square root transformation, 209
    univalence criterion, 207

W
Weierstrass Approximation Theorem, 189

Z
Zalcman's Lemma, 132–135

MIX
Papier aus verantwortungsvollen Quellen
Paper from responsible sources
FSC® C105338

If you have any concerns about our products,
you can contact us on
ProductSafety@springernature.com

In case Publisher is established outside the EU,
the EU authorized representative is:
**Springer Nature Customer Service Center GmbH
Europaplatz 3, 69115 Heidelberg, Germany**

Printed by Libri Plureos GmbH
in Hamburg, Germany